FUNDAMENTALS OF ELECTRIC CIRCUITS

FOURTH EDITION

FUNDAMENTALS OF ELECTRIC CIRCUITS

David A. Bell
*Lambton College
of Applied Arts and Technology
Sarnia, Ontario, Canada*

Prentice Hall
Englewood Cliffs, New Jersey 07632

Library of Congress Cataloging-in-Publication Data

Bell, David A.
  Fundamentals of electric circuits.

  Includes index.
  1. Electric circuits. I. Title.
TK454.B3987   1988   621.319'2   87-14401
ISBN 0-13-336645-6

Editorial/production supervision and
    interior design: *Ann L. Mohan*
Cover design: *Bruce Kenselaar*
Cover photo: © *Michael Simpson, FPG*
Manufacturing buyer: *Peter Havens*

Printed in the United States of America

10 9 8 7 6 5 4 3 2 1

ISBN 0-13-336645-6

Prentice-Hall International (UK) Limited, *London*
Prentice-Hall of Australia Pty. Limited, *Sydney*
Prentice-Hall Canada Inc., *Toronto*
Prentice-Hall Hispanoamericana, S.A., *Mexico*
Prentice-Hall of India Private Limited, *New Delhi*
Prentice-Hall of Japan, Inc., *Tokyo*
Simon & Schuster Asia Pte. Ltd., *Singapore*
Editora Prentice-Hall do Brasil, Ltda., *Rio de Janeiro*

CONTENTS

6
PARALLEL RESISTIVE CIRCUITS 122

7
SERIES-PARALLEL CIRCUITS 145

8
NETWORK ANALYSIS TECHNIQUES 164

9
NETWORK THEOREMS   194

10
VOLTAGE CELLS, BATTERIES AND DC POWER SUPPLIES   216

11
MAGNETISM   251

12
MAGNETIC CIRCUITS   279

13
DC MEASURING INSTRUMENTS   313

14
INDUCTANCE   339

15
CAPACITANCE   377

16
INDUCTANCE AND CAPACITANCE IN DC CIRCUITS   406

20
SERIES AND PARALLEL AC CIRCUITS 543

21
POWER IN AC CIRCUITS 569

22
AC NETWORK ANALYSIS 595

PREFACE

Since my earliest times as a student, I have often been shocked to find that many technical concepts which at first seemed impossibly complex were really quite simple when I finally figured them out. Occasionally, I was almost ready to believe that there was a conspiracy to deliberately make technical material difficult to understand! As an electronics writer, I have strived to present the subject matter in the simplest possible way. My approach has led some instructors to, at first, classify previous editions of this book as "low level," and then to be surprised to find that the material is covered to at least the same depth as in many so-called "higher level" books.

Basic electricity and magnetism courses present a mountain of technical material that students have to "dig their way through" before arriving at more interesting topics. So, every effort has been made to make the subject matter in this book interesting, as well as understandable. As many appropriate illustrations as possible are included, to help students understand new concepts and circuits. Each illustration is carefully discussed, and explanatory captions are provided, rather than just a figure number and title.

Technical material is most readily absorbed through the process of answering review questions and solving problems. Many practical worked-out examples are provided (which are neither too simple nor too complex). Numbered step-by-step analysis procedures are offered to assist students in learning to analyze complex networks. Practice problems are provided throughout the text, with answers given at the chapter ends. A summary of formulas and extensive sets of review questions and problems are also included at the end of each chapter, with answers to odd-numbered problems at the back of the book.

I am very grateful to those individuals who made suggestions for improvements to previous editions of this book. Comments concerning the fourth edition would be welcome.

ORGANIZATION OF THE BOOK

Assuming that the reader has had no previous electrical instruction, the first chapter offers explanations for electrical phenomena, and introduces the basic electrical units without getting into precise definitions.

Because the first laboratory experiments performed by students normally involve the use of ammeters, voltmeters, and ohmmeters, it is important that they acquire an early knowledge of how to use these instruments. Methods of correctly connecting and reading such instruments, both deflection-type and digital, are discussed in Chapter 2. Explanation of how the instruments operate is left until later in the book.

Chapter 3 explains Ohm's law, and offers many example calculations of current, voltage, resistance, and power dissipation. Conductors, insulators, and resistors are the subject of Chapter 4. Differences between the materials, insulator breakdown, conductor resistivity, thermal effects, etc. are discussed.

Three chapters (5, 6, and 7) are employed to study resistive circuits as: series circuits, parallel circuits, and series-parallel circuits. This leads into network analysis and network theorems, treated in the next two chapters. It is important to acquire a thorough understanding of dc circuits and circuit analysis methods before studying alternating current circuits. Therefore, the circuit analysis topics are first covered as purely resistive circuits with dc voltage inputs. Later, they are again treated for ac impedance circuits.

Voltage cells, batteries, and dc power supplies are the topic of Chapter 10. Coverage includes cell construction, battery construction and performance, and the use of laboratory-type dc power supplies.

In covering magnetism and magnetic circuits (Chapters 11 and 12), every effort is again made to explain the basic phenomena in a simple, understandable way. As always, appropriate illustrations and step-by-step worked examples are employed at each stage.

Once an understanding of magnetism is achieved, the operation of dc measuring instruments is explained Chapter 13. Then, inductance and capacitance are introduced in Chapters 14 and 15 respectively, and the performance of inductive and capacitive components in dc circuits is treated in the succeeding chapter.

The generation of an alternating voltage is explained in Chapter 17 in terms of a simple two-conductor generator. The sine wave instantaneous value, frequency, phase angle, rms value, etc. are all explained, once again using appropriate calculation examples. When the study of alternating voltage and current is commenced, students must become familiar with the cathode ray oscilloscope. To facilitate this, the front panel of a typical dual-trace oscilloscope is presented as part of Chapter 17. The oscilloscope controls are discussed, and its basic application to the study of waveforms is covered. No attempt is made to explain how an oscilloscope works.

Complex numbers and rectangular/polar conversions are gently introduced in Chapter 18. Then, in Chapter 19, the behavior of inductors and capacitors in ac circuits is investigated. This includes series-connected and parallel-connected RLC circuits, but resonance is left to a later chapter. With complex numbers and ac RLC circuits understood, ac circuit analysis and power dissipation in ac circuits are studied in Chapters 20, 21, and 22.

The study of resonance (Chapter 23) commences with an explanation of the basic phenomenon, and works through series and parallel resonance, resistance damping, tuned coupled coils, and simple resonance filters. Transformers are investigated next (Chatper 24), and here the treatment includes open-circuit and short-circuit tests, voltage regulation, efficiency, and referred (or reflected) quantities. As always, practical worked examples are again presented.

At this stage, ac measuring instruments can be sensibly discussed (Chapter 25). Rectifier voltmeters, rectifier ammeters (using current transformers), and ac applications of electrodynamic instruments are all explained.

The final two chapters cover three-phase ac systems and nonsinusoidal waveforms. Chapter 26 progresses from a basic three-phase generator through three-phase circuit analysis, phase sequence measurement, and power factor correction. Chapter 27 presents harmonic analysis methods of investigating nonsinusoidal waveforms.

NEW TO THE FOURTH EDITION

The entire text and illustrations have been completely revised and updated where appropriate. In particular, the first three chapters have been rewritten, in order to introduce students quickly to the use of basic instruments and practical calculations. Instead of precisely defining SI units, the basic electrical units are explained and typical measured values are discussed. Unit definitions are presented in Appendix 1.

Objectives have been included at the beginning of each chapter. Practice problems have been added at section ends, with the answers provided at end of the chapter. Many more problems have been included at the chapter ends, and all problems are now identified according to text section numbers. Computer programming problems are included at the end of problem sets, so that they can be assigned or ignored at the discretion of the instructor. Answers to only the odd-numbered problems are given in the appendixes. A memory aid for the color code and typical standard component values are also included in the appendixes.

SUPPLEMENTARY MATERIALS

Laboratory Manual containing practical experiments, suggested laboratory rules, and recommendations for report writing. Each experiment has:

- A *title*.
- A one-paragraph *introduction*.
- A numbered step-by-step *procedure*.
- *Circuits* and *connection diagrams*.
- An *analysis* section.
- A unique *laboratory record sheet* which teaches students good data recording methods (instead of boxes throughout the procedure).

Instructor's Resource Manual containing:

- Solutions for all practice problems and end-of-chapter problems in the textbook.
- A list of the laboratory equipment and components required to perform all experiments in the laboratory manual.
- Model results for all experiments in the laboratory manual.

Test Manual containing:

- Thirteen sets of double tests (Test A and Test B) covering the material in the textbook.
- Model answers for all tests.

Set of Transparency Masters containing:

- One hundred of the most important illustrations from the book.

Computer Program Manual containing:

- 13 *BASIC* programs, and
- 13 *Pascal* programs

for solving many of the example problems in the textbook. Commencing with very simple but practical programming examples, progressively more difficult programs are offered. Each new program is a slight modification of the previous one, so that students find it easy to absorb programming techniques.

1

THE NATURE OF ELECTRICITY

Objectives   You will be able to:

Explain electrification by friction, and the concept of positive and negative electricity.

State the fundamental law of electrification by friction.

Describe the planetary atom, and sketch a plane diagram to represent an atom and its component parts.

Describe what occurs when an electric current flows.

Define a conductor, an insulator, and resistance.

Explain potential difference, and its relationship to current and resistance.

Sketch the basic construction of a voltage cell and discuss its operation.

Sketch the graphic symbols used to represent basic circuit components, and draw simple circuit diagrams.

Explain conventional current direction and electron flow.

Define the difference between direct current (*dc*) and alternating current (*ac*).

Explain how electric shock occurs, and discuss some of the situations in which electricity can be hazardous to humans.

Write numbers and perform calculations using scientific notation, metric prefixes, and engineering notation.

Introduction

Human knowledge of electricity began with the study of the phenomenon known to experimenters as electrification by friction. The early concept of an electric fluid being transferred from one body to another is now understood as a motion of electrons that have become detached from their atoms. A flow of electrons constitutes an electric current, and the pressure that produces the electron flow is a potential difference between two bodies, or between two terminals. A potential difference between two bodies is produced by an excess of electrons on one body, and/or a

1

deficiency of electrons on the other. In the simplest source of electricity, the excess and deficiency of electrons are the result of chemical action. When an electric current flows, heat is generated by the electron motion. In certain circumstances, light may also be produced—hence the electric lamp.

The various components and interconnections of an electrical system comprise what is described as an electric circuit, and a circuit diagram is a graphic representation of an electric circuit.

Because electricity can be dangerous, it is important to use it carefully and to take immediate action when someone suffers an electric shock.

1-1
ELECTRIFICA-TION BY FRICTION

The study of electricity began with investigations of what became known as *electrification by friction*. The ancient Greeks knew that when amber was rubbed with wool, it could attract lightweight particles of other material, such as feathers or lint. The Greek word *elektron*, which means amber, is the origin of the word *electricity*.

Early in the seventeenth century it was discovered that amber was not the only material that had this property. Glass rubbed with silk, and ebonite rubbed with fur, were both found capable of attracting small particles of other materials. It was also discovered that when a silk-rubbed glass rod was suspended on thread and a fur-rubbed ebonite rod was brought near it, the glass was attracted to the ebonite. When another silk-rubbed glass rod was brought near the suspended glass rod, it was found that the two were repelled from each other (see Figure 1-1). Similarly, a suspended ebonite rod that had first been rubbed with fur was found to be repelled from another fur-rubbed ebonite rod and attracted to silk-rubbed glass. The effect is most easily demonstrated by *charging* lightweight pith or styrofoam balls by touching them with the silk-rubbed and fur-rubbed rods. As illustrated in Figure 1-2, repulsion and attraction occur.

From these results it was concluded that there were two types of electricity, and that those materials *charged* with the same type of electricity repelled each other, while those charged with different types of electricity attracted each other. This conclusion produced a fundamental law:

Like charges repel; unlike charges attract.

Around the middle of the eighteenth century, Benjamin Franklin* suggested that when glass was rubbed with silk, some kind of *electric fluid* passed from the silk to the glass, and that this gave the glass an

---

* American statesman and scientist (1706–1790).

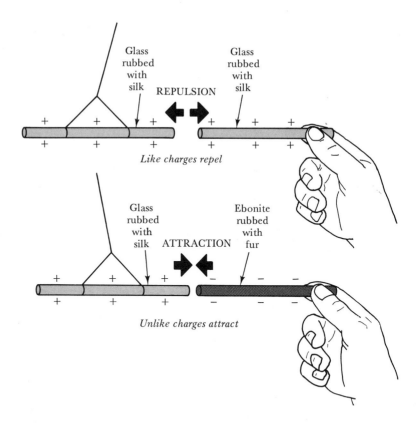

FIGURE 1-1. Glass and ebonite rods can be used to demonstrate electrification by friction. When rubbed with silk, the glass rod becomes positively charged. A fur-rubbed ebonite rod becomes negatively charged.

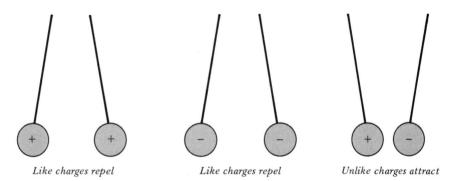

FIGURE 1-2. Thread-suspended lightweight pith or styrofoam balls demonstrating that *like charges repel* and *unlike charges attract*.

Sec. 1-1   Electrification by Friction

increased amount of electric fluid, or *positive charge*. Conversely, when the ebonite was rubbed with fur, the electric fluid passed from the ebonite to the fur, he argued. Thus, the ebonite acquired a reduced amount of electricity, or *negative charge*.

1-2
PLANETARY ATOM

The modern explanation of electrification by friction utilizes our present understanding that the atom consists of a central *nucleus* surrounded by orbiting *electrons*. The diagram in Figure 1-3(a) illustrates the concept of the *planetary atom*. The electrons have a negative charge, and relative to the nucleus they are extremely small particles. The nucleus [Figure 1-3(b)] is largely a cluster of two types of particles, *protons* and *neutrons*, each of which has a mass approximately *1800 times the mass of the*

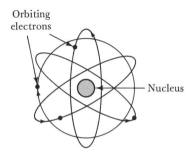

(a) Nucleus with orbiting electrons

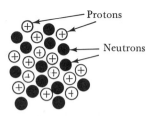

(b) The nucleus is largely a cluster of protons and neutrons

FIGURE 1-3. An atom consists of a central nucleus surrounded by orbiting electrons. Electrons have a negative charge, and the protons in the nucleus have a positive charge.

electron. Neutrons have no charge at all, and protons have a positive charge equal in magnitude to the negative charge on an electron. Thus, the nucleus is positively charged.

The three basic particles—*proton, neutron,* and *electron*—are similar from one atom to another. Differences between atoms are due to dissimilar numbers and arrangements of the three particles. Different materials are made up of different types of atoms, or combinations of several types of atoms.

It has been found that electrons can occupy only certain orbital rings or *shells* at fixed distances from the nucleus, and that each shell can contain only a certain number of electrons. The outer shell is termed the *valence shell,* and the electrons that occupy it are referred to as *valence electrons.* These outer-shell electrons largely determine the electrical (and chemical) characteristics of an atom.

Since the protons and orbital electrons of an atom are equal in number and equal and opposite in charge, they neutralize each other electrically. Consequently, each atom is normally electrically neutral; that is, it exhibits neither a positive nor a negative charge. If an atom loses an electron, thus losing some negative charge, it exhibits a positive charge. In this case the atom is referred to as a *positive ion.* Similarly, an atom that gains an additional electron becomes negatively charged and is then termed a *negative ion.*

Benjamin Franklin's *electric fluid* (Section 1-1) can now be looked upon as being composed of electrons passing from one material to another. That material which loses electrons becomes positively charged as the result of the loss of negative charges. The material that gains electrons becomes negatively charged. Thus, glass rubbed with silk acquires its positive charge by losing some electrons to the silk, and the negative charge on the ebonite is acquired by its gaining electrons from the fur with which it was rubbed.

1-3

POTENTIAL DIFFERENCE, CURRENT, AND RESISTANCE

When two oppositely charged bodies make contact, electrons flow from the negatively charged body to the positively charged body. Thus, electrons flow from a location with an excess of electrons to one with a deficiency of electrons. Electrons will also flow from a negatively charged body to an uncharged body, or to one with a lower negative charge.

The movement of electrons constitutes a flow of electric current.

The flow of charge carriers (i.e., electrons) also occurs if the two charged bodies are connected by a piece of metallic material (Figure 1-4). The flow does not occur when nonmetallic material is employed.

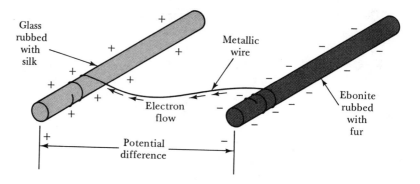

FIGURE 1-4. A potential difference exists between a negatively charged body and a positively charged body. This produces a flow of electrons from negative to positive when the bodies are connected by a metallic wire.

> *Those materials which readily allow electrons to pass through them are termed **conductors**.*

> *Those which do not permit electron flow are known as **insulators**.*

Since some materials are better conductors than others, it can be said that

> *they offer lower **resistance** to the flow of electric current.*

Conductors obviously have much lower resistance than insulators.

Electric current is measured in *amperes,* or *amps* (symbol *A*), and resistance is measured in *ohms* (symbol Ω—Greek letter omega).

The ability of two oppositely charged bodies to produce a flow of electricity between them may be thought of as a *potential* for the production of electric current. Thus, a positively charged body is said to have a *positive potential,* and a plus sign (+) is used to identify this potential on all diagrams. Similarly, a negatively charged body is described as having a *negative potential,* and a minus sign (−) is used for its identification. Two oppositely charged bodies are said to have a *potential difference* between them (see Figure 1-4). It is also possible for a potential difference to exist between two similarly charged bodies, if one is charged to a higher potential than the other.

Potential difference is measured in *volts,* (symbol *V*), and the term *voltage* is commonly applied for potential difference. Another, more descriptive, term for potential difference is *electromotive force (emf)*.

> *Potential difference is an electrical pressure that tends to cause current to flow.*

When a suitable conducting path is provided, current will flow as a result of the electrical pressure.

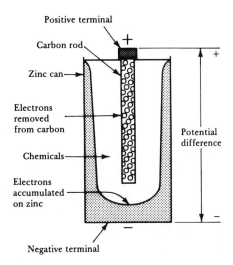

Positive terminal

Carbon rod

Zinc can

Electrons removed from carbon

Chemicals

Electrons accumulated on zinc

Potential difference

Negative terminal

FIGURE 1-5. Cross-section of a zinc-carbon voltage cell. Chemical action removes electrons from the carbon and accumulates them on the zinc. The carbon is the positive terminal of the cell, and the zinc is the negative terminal.

1-4

BASIC SOURCE OF ELECTRICITY

The simplest modern source of electricity is a *zinc-carbon voltage cell.* As illustrated in Figure 1-5, the cell consists basically of a zinc can containing certain chemicals, and having a centrally located carbon rod (see Chapter 10). Chemical action removes electrons from the carbon, and causes them to accumulate on the zinc. Because the carbon has lost (negatively charged) electrons, it is positively charged. Conversely, the accumulation of electrons makes the zinc negatively charged. Therefore, the carbon is the *positive terminal* of the cell, and the zinc is the *negative terminal.* The side of the zinc can is usually insulated, while the bottom is left uncovered to act as the negative terminal. A potential difference, or voltage, exists between the positive and negative terminals of the cell. If a conducting path is provided, electrons will flow from the negative terminal to the positive.

1-5

ELECTRIC LAMP

Figure 1-6(a) illustrates the movement of an electron in a conductor when a potential difference is applied to the ends of the conductor. The electron is repelled from the negative terminal and attracted toward the positive terminal. However, the atoms within the conductor impede the motion of the electron, so that it cannot simply accelerate from the negative end to the positive end. Instead, the electron bounces about from one atom to another and merely drifts toward the positive end (see

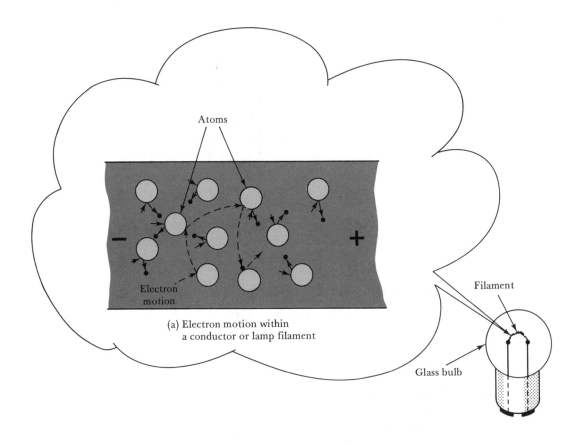

Atoms

(a) Electron motion within
a conductor or lamp filament

Electron
motion

Filament

Glass bulb

FIGURE 1-6. When a current of electrons flows in the filament of an electric lamp, a great many electron-atom collisions occur. These collisions produce sufficient heat to cause the filament to glow.

the broken line in the illustration). Each time the electron strikes an atom, it dissipates energy in the form of heat. When a great many electrons are involved in a flow of current, the heat dissipation can be considerable.

An electric lamp consists of a thin tungsten wire, or *filament*, contained in a glass bulb from which the air has been evacuated [see Figure 1-6(b)]. When a current of electrons is passed through the filament, many electron-atom collisions occur within the filament. The resulting heat makes the filament glow white hot, thus emitting light.

1-6
ELECTRIC
CIRCUIT

The simple electric circuit shown in Figure 1-7 is made up of a voltage cell, a switch, an electric lamp, and connecting conductors. The voltage cell and lamp have already been discussed. The conductors are copper

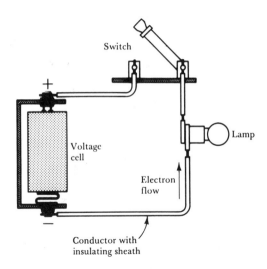

FIGURE 1-7. Simple electric circuit. The potential difference at the voltage cell terminals is an electric pressure that causes a current of electrons to flow when a suitable conducting path is provided.

wire with an insulating plastic or rubber sheath. The switch illustrated is a bar of copper hinged at one terminal, with a spring clip to grip the bar tightly at the other terminal.

Recall from Section 1-3 that *a potential difference is an electric pressure that causes current to flow when a suitable conducting path is provided.* When the switch is open, the conducting path between the cell terminals is interrupted, so that no current flows. When the switch is closed, the potential difference between the terminals of the voltage cell causes a current of electrons to flow from the negative terminal through the conductors, the lamp filament, and the switch, to the positive terminal of the cell.

1-7
CIRCUIT
DIAGRAMS

The circuit in Figure 1-7 can be represented by a *circuit diagram* consisting of graphic symbols. Figure 1-8(a) shows that the symbol for a conductor is a straight line, while that for a voltage cell is a long bar and a short bar side by side. The long bar is the positive terminal and the short bar is the negative terminal. The lamp and switch symbols are also shown, and a *resistor* circuit symbol is illustrated. As will be explained later, a resistor is a component which is constructed to have a particular resistance value. Other circuit symbols are given in Appendix 1.

Figure 1-8(b) shows the complete diagram for the circuit in Figure 1-7. In Figure 1-8(c), a resistor is substituted in place of the lamp, and the switch is shown closed. As illustrated on the diagram, the letter

Conductor   Voltage cell   Lamp   Switch   Resistor

(a) Graphic symbols for circuit components

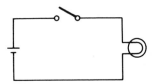

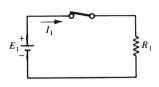

(b) Circuit diagram for the circuit in Figure 1-7

(c) Circuit diagram with a resistor

FIGURE 1-8. An electric circuit can be represented by a circuit diagram in which graphic symbols are employed for each component. The letter symbols E, I, and R are used to represent voltage, current, and resistance.

symbol E is used to represent voltage (sometimes V is used instead of E). The symbol I represents current, and R is used for resistance.

1-8
CURRENT
DIRECTION

In the early days of electrical research it was believed that a positive charge represented an increased quantity of electricity, while a negative charge was seen as a reduced quantity. Consequently, it was assumed that current flowed *from positive to negative.* This is a convention that remains in use today, even though current flow has long been accepted as a movement of electrons from negative to positive.

*Current flow from positive to negative is known as **conventional current direction.***

*Electron flow from negative to positive is termed **the direction of electron flow.***

It is important to understand and to be able to think both in terms of conventional current direction and in terms of electron flow. Electron flow is used to explain how electronic devices operate. But *the graphic symbol for every electronic device employs an arrowhead that points in the direction of conventional current.*

Figure 1-9 shows the symbol for the simplest electronic device—a *diode.* A diode is a *one-way device.* Current flows through a diode only

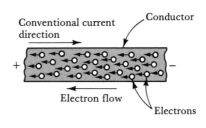

Conventional current direction

Conductor

Electron flow

Electrons

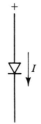

+

I

−

The circuit symbol used for a diode is an arrowhead and bar. The direction of the arrowhead indicates the conventional direction of current flow through the device.

FIGURE 1-9. Electric current flow is a movement of electrons from negative to positive. Conventional current direction is taken as a current flow from positive to negative. All circuit symbols for electronic devices employ arrowheads which indicate conventional current direction.

when the polarity of the applied voltage is such that the arrowhead points from positive to negative. When the voltage is reversed (arrowhead pointing from negative to positive), current cannot flow. Thus, the arrowhead indicates the direction of conventional current flow.

Electronic component and equipment manufacturers normally use conventional current direction when they indicate current on a circuit diagram. Conventional current direction is employed throughout this book.

1-9
DIRECT CURRENT AND ALTERNATING CURRENT

Current that flows continuously in one direction is termed *direct current* (abbreviated *dc*). This is the kind of current supplied by a voltage cell. Since the voltage at the terminals of a voltage cell does not reverse polarity but remains substantially constant, a voltage cell can be termed a *direct voltage source*. Figure 1-10(a) shows that the circuit symbol for a direct voltage source is the voltage cell symbol already discussed. The graph of direct current plotted versus time is also illustrated.

Alternating current (abbreviated *ac*) is current that flows first in one direction for a brief time, and then reverses to flow in the opposite direction for a similar time. This is illustrated in Figure 1-10(b), where the graph of alternating current is plotted versus time. The circuit symbol for an *alternating voltage source* is also shown. The study of alternating voltage and current commences in Chapter 17.

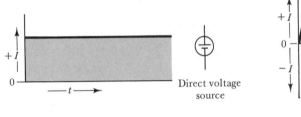

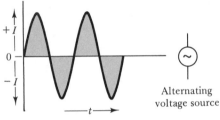

(a) Direct current or dc

(b) Alternating current or ac

FIGURE 1-10. A direct current (*dc*) plotted versus time shows that the current level remains constant. An alternating current (*ac*) continuously reverses, alternately flowing in one direction and then in the other.

1-10
ELECTRIC SHOCK

Electricity can kill, and for that reason it is very important to treat electrical supplies and appliances with great care. Perhaps the greatest domestic danger exists in bathrooms and kitchens, because of the presence of water and water taps. Water pipes are connected to other pipes which are buried in the ground; they provide an electrical connection to ground. Normally, there is a high potential difference between at least one of the electric supply conductors and ground. Consequently, a person holding a faulty appliance in one hand and a grounded water tap in the other provides a path for current to flow from the conductor to ground [see Figure 1-11(a)]. In this case the person can experience a severe electric shock, and may find it impossible to open either hand!

A similar danger exists when someone is using a portable appliance, such as an electric drill, if, for example, the supply cable is worn or has temporary splices. While holding the appliance in one hand and pulling the cable, the other hand might close around a bare wire [Figure 1-11(b)]. Here again it may be impossible to open either hand as the electricity flows through the body.

An even more lethal situation exists when someone taking a bath reaches for an electrical appliance [Figure 1-11(c)]. In this case the person's body is grounded via the water. If the appliance is faulty, *death is virtually inevitable.*

In a case of electric shock, it is most important to immediately get the victim away from the source of the shock. To avoid endangering a rescuer, the electricity supply must be switched *off* or disconnected before the victim is touched. Medical help should be summoned, and the victim should be made comfortable and kept calm. If the victim has stopped breathing, artificial respiration should be commenced and continued until medical help arrives. If the victim's heart has stopped beating, external heart compression should be applied by a trained individual.

Chap. 1   The Nature of Electricity

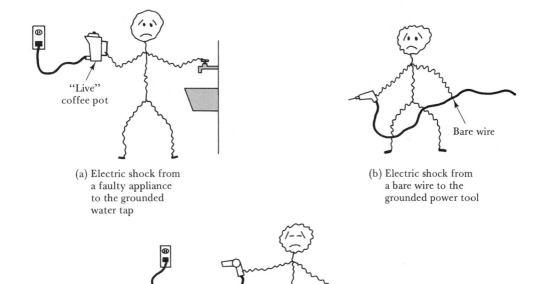

(a) Electric shock from a faulty appliance to the grounded water tap

(b) Electric shock from a bare wire to the grounded power tool

(c) Very severe shock from a faulty appliance when in a bath tub

FIGURE 1-11. Dangerous electrical situations can occur when equipment or appliances are in need of repair. *Electric shocks can be severe enough to kill.*

1-11
SCIENTIFIC NOTATION, METRIC PREFIXES, AND ENGINEERING NOTATION

SCIENTIFIC NOTATION. Most electrical and electronic calculations involve very large or very small numbers. These are most conveniently expressed using *scientific notation*, which means that they are written as numbers raised to the power of 10. Consider a resistance of 1000 Ω (1000 ohms) and a current of 0.001 A (0.001 amps), which are both quite common.

Since, $\qquad 1000 \ \Omega = 1 \times 10 \times 10 \times 10 \ \Omega$

it can be written as, $\qquad 1 \times 10^3 \ \Omega$

Similarly, $\qquad 0.001 \ A = \dfrac{1}{10 \times 10 \times 10} \ A$

or, $\qquad 1 \times 10^{-3} \ A$

Also, $\qquad 100 \ V = 1 \times 10 \times 10 \ V$
$\qquad\qquad\qquad = 1 \times 10^2 \ V$

and, $\qquad 0.0001 \ A = \dfrac{1}{10 \times 10 \times 10 \times 10} \ A$
$\qquad\qquad\qquad\quad = 1 \times 10^{-4} \ A$

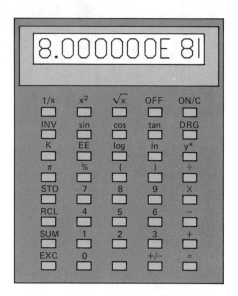

FIGURE 1-12. *Scientific notation* on a pocket calculator. The calculator display does not have sufficient numerals to show this number without using scientific notation. Electrical and electronics students should possess calculators that can display a result in *engineering notation*.

Figure 1-12 shows a pocket calculator displaying the number 8×10^{81}. Clearly, the calculator display could not show the number without using scientific notation.

Note that in the SI system (or international system) of units listed in Appendix 2, spaces are used instead of commas when writing large numbers. Four-numeral numbers are an exception. One thousand is written as *1000*, while ten thousand is *10 000*.

METRIC PREFIXES. Metric prefixes and the letter symbols for the various multiples and submultiples of 10 are listed in Table 1-1, with those most commonly used with electrical units shown bold. The prefixes are employed to simplify the writing of very large and very small quantities. Thus, *1000 Ω* can be expressed as *1 kilohm*, or *1 kΩ*. Here *kilo* is the prefix that represents *1000*, and *k* is the symbol for kilo. Similarly, 1×10^{-3} A can be written as *1 milliamp*, or *1 mA*.

ENGINEERING NOTATION. As already discussed, *1 kΩ* is 1×10^3 Ω, and *1 mA* is 1×10^{-3} A. Note also from Table 1-1 that 1×10^6 Ω is expressed as *1 MΩ*, and 1×10^{-6} A can be written as *1 μA*. These quantities, and most of the metric prefixes in Table 1-1, involve multiples of 10^3 or 10^{-3}. Quantities that use 10^3 or 10^{-3} are said to be written in *engineering notation*. A quantity such as 1×10^4 Ω is more conveniently expressed as 10×10^3 Ω, or *10 kΩ*. Also, 47×10^{-4} A is best written as

TABLE 1-1

VALUE	SCIENTIFIC NOTATION	PREFIX	SYMBOL
1 000 000 000 000	10^{12}	tera	T
1 000 000 000	10^{9}	**giga**	**G**
1 000 000	10^{6}	**mega**	**M**
1 000	10^{3}	**kilo**	**k**
100	10^{2}	hecto	h
10	10	deka	da
0.1	10^{-1}	deci	d
0.01	10^{-2}	centi	c
0.001	10^{-3}	**milli**	**m**
0.000 001	10^{-6}	**micro**	**μ**
0.000 000 001	10^{-9}	**nano**	**n**
0.000 000 000 001	10^{-12}	**pico**	**p**

4.7 × 10⁻³ A, or *4.7 mA.* For electrical calculations, engineering notation is more convenient than ordinary scientific notation.

When purchasing a *pocket calculator,* electrical and electronics technology students should *select one that can display a result in engineering notation.*

SIGNIFICANT FIGURES. Consider the number 816, which is said to be accurate to three significant figures. If the number is rewritten as accurate to two significant figures, it becomes 820. This is because 816 is clearly closer to 820 than it is to 810. If the number 814 were rewritten accurate to two significant figures, it would be 810. In the case of 820 and 810, the zero is not a significant figure.

The number 2.03, written accurate to two significant figures, becomes 2.0. In this case the zero is a significant figure, and it obviously implies that the number is greater than 2.0 but less than 2.1. It also implies that the exact value is closer to 2.0 than to 2.1.

Now consider the result of using an electronic calculator to resolve 5.4/2.3:

$$\frac{5.4}{2.3} = 2.347826087$$

Clearly, it does not make sense to have an answer containing ten significant figures when each of the original numbers has only two significant figures. Suppose that the two original numbers are assumed to be accurate to ±1%, which would not be unusual in electrical calculations. Then the accuracy of the answer can be shown to be not better than ±2%. Therefore,

$$\frac{5.4 \pm 1\%}{2.3 \pm 1\%} = 2.347826087 \pm 2\%$$
$$= 2.347826087 \pm 0.0469565$$

The only reasonable way to write this answer is

$$2.35 \pm 0.05$$

Consequently, the problem and answer become

$$\frac{5.4}{2.3} = 2.35$$

The answer now has three significant figures, although the numerator and denominator each have only two significant figures but were assumed accurate to $\pm 1\%$. Where the accuracy is known to be better than $\pm 1\%$, the answer may have a larger number of significant figures.

A reasonable approach to dealing with the problem of significant figures and decimal places, is to express all quantities in the form used in digital instruments (see Section 2-5). When a digital meter can display a number up to 999, it is said to have a *three-digit display*. The largest number that most digital meters can display is 1999, with a decimal point placed anywhere after the 1. The first numeral can only be 1 or a blank, so this is termed a *three-and-a-half digit display*. Therefore, since few electrical measurements are going to be more accurate than shown on a $3\frac{1}{2}$ digit display, quantities should normally be expressed in this form. When the number begins with 1, use four significant figures. When it begins with a numeral other than 1, use three significant figures.

PRACTICE PROBLEMS

1-11.1   Express the following quantities using scientific notation:
(a) 0.0003   (b) 300×257 (c) $(156 \times 10^4)/2000$

1-11.2   Solve the following, and give the answers using engineering notation:
(a) $(6 \times 10^{-6}) \times (3.3 \times 10^3)$
(b) $0.0047 \times 5 \times 10^5$
(c) $0.05/(2 \times 10^3)$

REVIEW QUESTIONS

1-1   Explain what is meant by electrification by friction. Also define positive and negative electricity as related to electrification by friction.

1-2   State the fundamental law that originated from the study of electrification by friction. Briefly explain the law.

1-3   Draw a plane diagram of the planetary atom. Identify the three basic particles that constitute an atom and define their relative quantities of charge and mass.

Chap. 1   The Nature of Electricity

1-4 Describe a valence electron and discuss the effect on an atom when it loses an electron and the effect when it gains an electron.

1-5 Describe what occurs when an electric current flows between two charged bodies. Briefly define a conductor, an insulator, and resistance.

1-6 Explain what constitutes a potential difference and how it is related to electric current.

1-7 Sketch the basic construction of a voltage cell. Explain how it produces positive and negative terminals.

1-8 Explain how electric current can produce light and briefly describe an electric lamp.

1-9 Sketch a diagram for an electric circuit consisting of a voltage cell, a lamp, and a switch. Briefly explain the operation of the circuit.

1-10 Refer to the circuit symbols in Appendix 1. Sketch an electric circuit consisting of a *generator,* a *single-pole switch,* a *resistor,* and a *lamp.*

1-11 Explain the difference between conventional current direction and direction of electron flow. Sketch the circuit diagram of a lamp supplied from a voltage cell and identify the directions of conventional current flow and electron flow.

1-12 Sketch the circuit symbol for a diode showing the polarity of applied voltage for current flow. Show the direction of conventional current flow and the direction of electron movement.

1-13 Define direct current, alternating current, emf, potential difference, and voltage.

1-14 Draw graphs to show the difference between direct current and alternating current. Briefly explain.

1-15 Describe some of the most dangerous situations in which a person can suffer an electric shock. List the emergency actions that should be taken to aid a victim of electric shock.

1-16 List the names of the various metric prefixes and the corresponding symbols. Also, list the value represented by each prefix, both in the long form and by scientific notation.

1-17 When an electronic calculator is used to determine 7.2/3.1, the answer displayed is 2.322580645. How would you give the answer if 7.2 and 3.1 were both known to be accurate to ±1%? Explain.

PROBLEMS

1-1 Express each of the following quantities in scientific notation: (a) 0.015, (b) 16 000, (c) 6260, (d) 0.0007, (e) 989 000.

1-2 Perform the following calculations, and give the answers in scientific notation: (a) $0.15 \times 1600/0.004$, (b) $87\,500/6.25$, (c) $0.5^2 \times 400$, (d) $2^{12}/\sqrt{256}$, (e) $0.003^3/1500$.

1-3 Express each of the following quantities in engineering notation: (a) 7800 Ω, (b) 0.000 06 A, (c) 0.019 A, (d) 91 500 Ω, (e) 0.05 A.

1-4 Perform the following calculations, and give the answers in engineering notation: (a) 1290×620, (b) 0.000 06/36, (c) 27.9/0.009, (d) $6.25/(125 \times 10^6)$

ANSWERS TO PRACTICE PROBLEMS

1-11.1 (a) 3×10^{-4} (b) 7.71×10^4 (c) 7.8×10^2

1-11.2 (a) 19.8×10^{-3} (b) 2.35×10^3 (c) 25×10^{-6}

2

MEASURING CURRENT, VOLTAGE, AND RESISTANCE

Objectives   You will be able to:

Correctly connect an ammeter into a circuit, and select the most suitable range on a multirange ammeter.

Correctly connect a voltmeter into a circuit, and select the most suitable range on a multirange voltmeter.

Correctly read the indicated current or voltage on deflection instruments.

Correctly connect an ohmmeter to measure resistance, select the most suitable range, and read the indicated resistance.

Correctly set the range and function controls on deflection and digital multimeters to measure current, voltage, or resistance. Also, correctly connect and read the instrument.

Estimate the accuracy of current, voltage, and resistance measurements.

Sketch circuit diagrams involving ammeters and voltmeters.

Introduction

The first experiments performed by students of electricity and electronics involve measurement of current, voltage, and resistance. This requires knowledge of the correct method of connecting each of the basic electrical instruments, and selection of the most suitable range. It is also important to be able to read the meter scale correctly and to be aware of the possible error in the measured quantity.

The use of multifunction instruments involves setting the function switch and the range switch, as well as making correct connections and reading the scale. Digital multimeters are easier to read than deflection instruments. The accuracy of measurement should be carefully assessed.

2-1
CURRENT MEASUREMENT

UNIT OF CURRENT

*The **ampere** (symbol A), or **amp,** is the unit of electric current.*

The definition of the amp is given in Appendix 2. At this stage, the definition is less important than an understanding of typical current levels. Most electronic devices pass currents ranging from 1 mA to 50 mA. Some devices take currents as high as several amps, while many have input currents less than 1 μA. A typical electric lamp used in a home might have a current of approximately 1 A.

AMMETER. The instrument used to measure electric current is called an *ammeter* (shortened form of *amp-meter*). Different types of ammeters are usually required for measuring *direct current* (dc) and *alternating current* (ac). However, many *multifunction instruments* can be employed to measure either type of current.

Ignoring its internal components, a *deflection type* ammeter consists basically of a *calibrated scale* over which a pointer is deflected to indicate the measured current, two *terminals* identified as + and −, and a *range switch* to select the current range for each particular measurement. Such an instrument is shown in Figure 2-1.

TERMINAL CONNECTIONS. An ammeter must be connected so that the current to be measured flows through the instrument. Consequently,

ammeters are always connected in series with the component in which current is to be measured.

At this stage of study, it is helpful for most students to simply think of an ammeter as equivalent to a piece of connecting wire.

In Figure 2-1, a dc ammeter is shown measuring the current *I* passing through an electric lamp supplied from a battery. Note that the circuit is arranged so that the current flow (in the conventional direction, see Section 1-8) is from the battery + terminal, into the + terminal of the ammeter, and out of the ammeter − terminal. From the ammeter, the current continues through the lamp back to the battery − terminal.

When a *dc* ammeter is connected so that the current flow is into the instrument − terminal, and out of the ammeter + terminal, the pointer deflects to the left of the 0 mark on the scale, instead of moving to the right over the scale. In this case, the ammeter is said to be connected with the wrong *polarity*. For an *ac* ammeter (see Chapter 25), the terminal polarity is not important; the pointer normally deflects over the scale regardless of the polarity of the terminals.

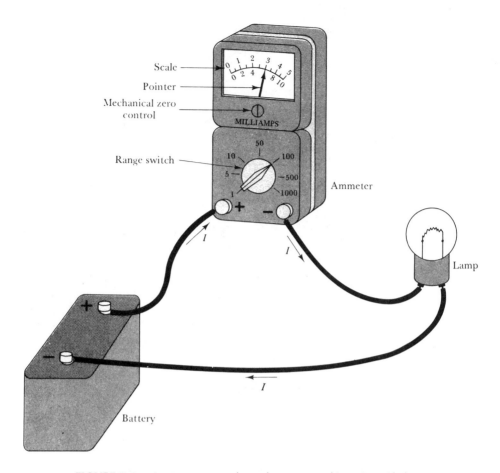

Scale

Pointer

Mechanical zero
control

Range switch

Ammeter

Lamp

Battery

FIGURE 2-1. Ammeters must always be connected *in series* with the component in which the current is to be measured, so that the current flows through the ammeter. Dc deflection instruments must be connected with the correct terminal polarity.

ZEROING A METER. Before using any deflection instrument, the pointer position should always be checked to see that it indicates exactly zero when the instrument is still unconnected. If the pointer is not at zero, it should be *zeroed* by adjusting the *mechanical zero control* (see Figure 2-1). When zeroing, the glass cover on the instrument scale should be lightly tapped with a finger to relieve friction within the meter.

AMMETER SCALE AND RANGE. The ammeter illustrated in Figure 2-1 has its scale marked in *milliamps (mA)*, and its range switch set to the *100* position. This means that when the pointer is at the right-hand side of the scale (full-scale deflection), the current measured is 100 mA. Using the scale calibrations of *0 to 10* (bottom scale), 10 represents 100 mA, and the pointer indication of 6 gives the measured current as 60 mA. When the range switch is set to 50 mA, the *0 to 5* calibration (upper scale) is used, and *5* represents full-scale deflection of 50 mA. The *0 to 5* scale is always used on the 5 mA, 50 mA, and 500 mA ranges, while the *0 to 10* scale is used with the 1 mA, 10 mA, 100 mA, and 1000 mA ranges.

When the approximate level of current to be measured is unknown,

the instrument should be set to its highest range before being connected into the circuit.

(The ammeter illustrated in Figure 2-1 should be set to the 1000 mA range.) The range can then be switched down until the greatest *on-scale* deflection is obtained.

Figure 2-2 shows the pointer positions that might result when a current of approximately 0.9 mA is measured using five different ranges of the instrument. In Figure 2-2(a) the range is 100 mA, and there is almost no readable movement of the pointer away from the 0 position. On the 50 mA range, Figure 2-2(b), the current can be read only as something less than 2.5 mA. On the 10 mA range illustrated in Figure 2-2(c), the pointer indicates 0.8 mA to 0.9 mA; while on the 5 mA range, Figure 2-2(d), the meter can be read with a little more certainty as

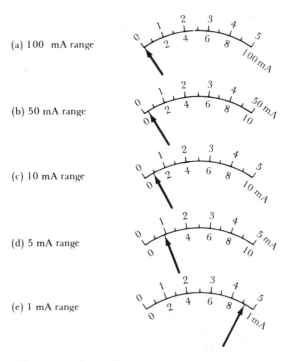

FIGURE 2-2. Illustration of the effect of meter range selection upon the pointer position of a deflection meter. Except in the case of an ohmmeter, the most accurate measurement is made with the pointer position that gives the largest on-scale deflection.

approximately 0.9 mA. The 1 mA range indication illustrated in Figure 2-2(e) clearly shows that the current is greater than 0.9 mA. The reading could be reliably stated as 0.93 mA. (The process of estimating the exact position of the pointer between two scale markings is referred to as *interpolation*.) Obviously,

for the greatest measurement accuracy the best range to use is the one that gives the largest deflection not exceeding full scale.

EXAMPLE 2-1   Determine the measured current for each of the pointer positions shown in Figure 2-3.

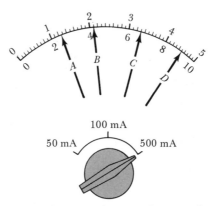

FIGURE 2-3.   Meter indications for Example 2-1.

SOLUTION

The range switch is at 500 mA; *therefore use the upper scale and take* 500 mA *as full-scale deflection. This means that the 5 (full-scale) reading must be multiplied by* 100. *Thus, all readings must be multiplied by* 100.

$$measured\ current\ at\ A = 1.25\ mA \times 100 = \textbf{125 mA}$$

$$measured\ current\ at\ B = 2.1\ mA\ \times 100 = \textbf{210 mA}$$

$$measured\ current\ at\ C = 3.3\ mA\ \times 100 = \textbf{330 mA}$$

$$measured\ current\ at\ D = 4.5\ mA\ \times 100 = \textbf{450 mA}$$

AMMETERS IN CIRCUITS.   The circuit diagram for the battery, ammeter, and lamp shown in Figure 2-1 is drawn in Figure 2-4(a). Note that the circuit symbol for the ammeter is an A with a circle around it. In Figure 2-4(b), a circuit diagram is shown for two lamps supplied from a single battery. Two ammeters are included; A_1 measures the current I_1 through L_1, while A_2 indicates the current I_2 through L_2.

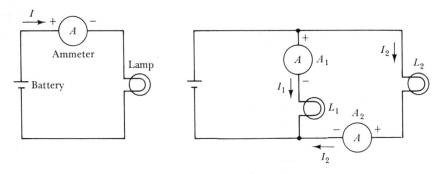

(a) Circuit diagram for **Fig.** 2-1

(b) Circuit diagram for a battery, two lamps, and two ammeters

FIGURE 2-4. The graphic symbol for an ammeter is a circle with an *A*. The ammeter terminal polarity should be shown on the circuit diagram.

PRACTICE PROBLEMS

2-1.1 Determine the measured current for each of the pointer positions shown in Figure 2-3, if the range switch is set to 100 mA.

2-1.2 If the range switch in Figure 2-3 is set to 25 mA, what would the shown pointer positions indicate?

2-2
VOLTAGE MEASUREMENT

UNIT OF VOLTAGE

The volt (symbol V) is the unit of potential difference, or voltage.

Another, more descriptive, name for the electrical pressure from a voltage cell or other source is *electromotive force (emf)*. The unit of emf, potential difference, or voltage, is defined in Appendix 2. Here again, as in the case of current, a knowledge of typical voltage levels is more important than the definition. Electronic circuits typically use supply voltages ranging from 5 V to 50 V, and within a circuit, potential differences as low as 10 mV might be measured. In North America, the normal residential supply voltages are 115 V for lighting and 220 V for stoves and heating.

VOLTMETER. The instrument used to measure the potential difference, or electrical pressure, between two points in a circuit is called a *voltmeter*. As in the case of ammeters, different types of instruments are usually employed for measuring *direct voltage* (dc volts) and *alternating voltage* (ac volts). The basic deflection-type dc voltmeter (see Figure 2-5) is very similar in appearance to a dc ammeter. The instrument has two terminals

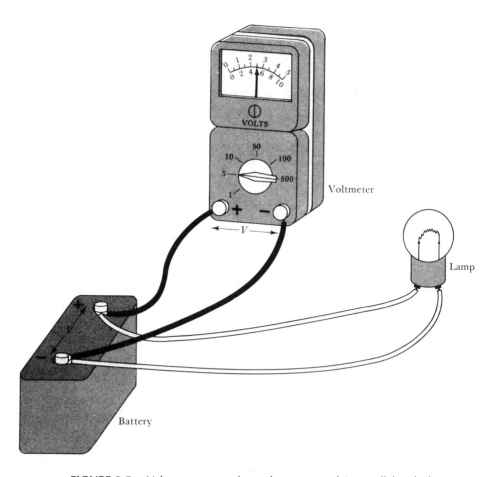

Voltmeter

Lamp

Battery

FIGURE 2-5. Voltmeters must always be connected *in parallel* with the component which is to have its voltage measured, so that the voltage is also applied to the voltmeter. Dc deflection instruments must be connected with the correct terminal polarity.

identified as + and −, a range switch, and a pointer which moves over a scale calibrated in volts.

TERMINAL CONNECTIONS. Figure 2-5 shows a dc voltmeter connected to measure the terminal voltage V of a battery that is supplying current to an electric lamp. Note that the voltage being measured appears directly *across,* or in parallel with, the voltmeter terminals.

Voltmeters are always connected in parallel with the voltage to be measured.

Note also that the + terminal of the battery is connected directly to the + terminal of the voltmeter, and that the battery − terminal is connected to the voltmeter − terminal. Thus, as in the case of the dc ammeter, a dc voltmeter must be connected with the correct terminal polarity, in order to obtain a positive (on-scale) deflection. With an ac voltmeter, a positive deflection is obtained regardless of the terminal polarity.

VOLTMETER SCALE AND RANGE. Like all deflection instruments, the voltmeter pointer should be checked before use to ensure that it indicates exactly zero. Where necessary, the mechanical zero control should be adjusted. Also, when the approximate value of the voltage to be measured is not known,

the instrument should always be set to its highest range before connecting it into a circuit.

The range can then be switched down until the greatest on-scale deflection is obtained.

The instrument shown in Figure 2-5 has six voltage ranges: 1 V, 10 V, and 100 V, all for use with the *0 to 10* scale (lower scale); and 5 V, 50 V, and 500 V, all for use with the *0 to 5* scale (upper scale). With the meter on its 5 V range, as illustrated, the pointer position shows that the battery voltage is 2.5 V. As explained for the ammeter,

the greatest voltmeter accuracy is obtained when the range is selected to give a deflection approaching full scale.

EXAMPLE 2-2 Figure 2-6 shows two pointer positions on a voltmeter scale marked *0 to 3* and *0 to 10*. Determine each indicated voltage:
a. When the range switch is set to 0.1 V.
b. When the range is 30 V.

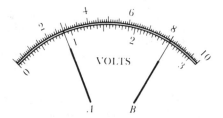

FIGURE 2-6. Meter indications for Example 2-2.

SOLUTION

a. 0.1 V range

   pointer position A indication = 2.8 V/100 = **0.028 V**
   pointer position B indication = 8.2 V/100 = **0.082 V**

b. 30 V range

   pointer position A indication = 0.9 V × 10 = **9 V**
   pointer position B indication = 2.6 V × 10 = **26 V**

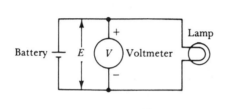

(a) Circuit diagram for Fig. 2-5

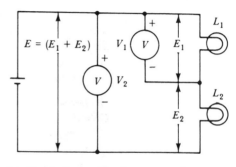

(b) Circuit diagram for a battery, two lamps, and two voltmeters

FIGURE 2-7. The graphic symbol for a voltmeter is a circle with a V. The voltmeter terminal polarity should be shown on the circuit diagram.

VOLTMETERS IN CIRCUITS. The circuit diagram for the circuit in Figure 2-5 is shown in Figure 2-7(a). The circuit symbol for the voltmeter is seen to be simply a V with a circle around it. In Figure 2-7(b) a two-lamp circuit is shown, with a battery and two voltmeters. Voltmeter V_1 is seen to be measuring voltage E_1 across lamp L_1, while voltmeter V_2 indicates the total battery voltage $E = E_1 + E_2$.

PRACTICE PROBLEMS

2-2.1 Suppose the range switch for the voltmeter in Figure 2-6 was set to 50 V, what voltage would be indicated by the pointer positions as shown.

2-2.2 If the voltage range is 0.3 V for the meter in Figure 2-6, what are the readings for pointer positions A and B?

2-3

RESISTANCE MEASUREMENT

UNIT OF RESISTANCE

*The **ohm** (symbol Ω) is the unit of resistance.*

As in the case of the other units, the ohm is defined in Appendix 2, and once again, a knowledge of typical resistance values is more important than the definition. The resistance of a conductor is normally less than 0.01 Ω, while the resistance of an electric lamp filament might be approximately 1 Ω. In electronic circuits, resistances of 2.7 Ω to 22 MΩ are found, with values in the range of 100 Ω to 100 kΩ being the most common.

OHMMETER. Resistance is measured by means of an *ohmmeter* (ohm-meter). The ohmmeter has its own battery or internal power supply for the purpose of passing a current through the resistance to be measured. Any externally applied voltage may damage the instrument, or at least give an incorrect measurement. Thus,

an ohmmeter must never be connected to a circuit in which a supply is switched ON.

The use of an ohmmeter to measure the resistance of a component is illustrated in Figure 2-8. The ohmmeter terminals are connected in parallel with the resistance to be measured [Figure 2-8(a)]. The polarity of the connection is unimportant. In the case of a component that forms part of a circuit, the circuit supply must be OFF, and one terminal of the component must be disconnected from the circuit before the ohmmeter is connected [see Figure 2-8(b)].

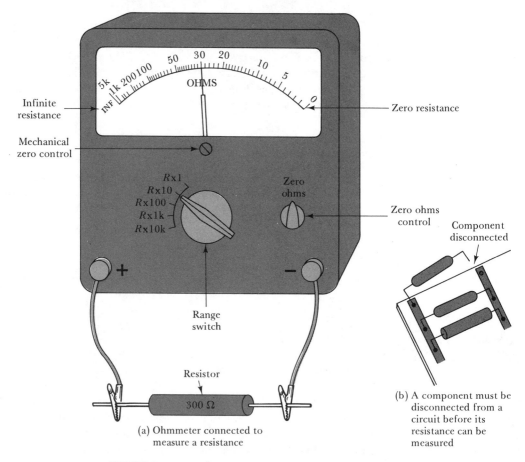

(a) Ohmmeter connected to measure a resistance

(b) A component must be disconnected from a circuit before its resistance can be measured

FIGURE 2-8. An ohmmeter is connected in parallel with the component which is to have its resistance measured. At least one end of the component must be disconnected from the circuit when measuring its resistance.

OHMMETER SCALE AND RANGE. Like ammeters and voltmeters, ohm-meters usually have a range switch and a pointer that moves over a calibrated scale. However, as Figure 2-8 shows, the ohmmeter scale is different from that on other instruments. The *0 ohms* position is seen to be on the right-hand side of the scale instead of the left. The left-hand end of the scale is marked *INF,* which means *infinity*—that is, an extremely large value. Note that the length of the scale representing a 10 Ω change from 0 Ω to 10 Ω is greater than the scale length from 10 Ω to 20 Ω, which also represents a 10 Ω change in resistance. Moving from right to left, the ohmmeter scale becomes more and more cramped, and the scale is said to be *nonlinear*. The ammeter and voltmeter scales already discussed are termed *linear scales* because equal lengths on each scale represent equal current or voltage changes. However, some ammeters and voltmeters have nonlinear scales.

ZEROING AN OHMMETER. The ohmmeter has a *ZERO OHMS* control, as well as a mechanical zero control (see Figure 2-8). As with other deflection instruments, the mechanical zero control should be used to set the pointer exactly at the left-hand end of the scale when the instrument is unconnected. Before measuring a resistance, the ohmmeter terminals should be connected together (that is, *short circuited*), and the *ZERO OHMS* control should be adjusted until the pointer indicates exactly 0 Ω on the right-hand side of the scale. This procedure should also be repeated whenever the range of the ohmmeter is altered. The ohmmeter terminals should never be left in a short-circuited condition because this would discharge the internal battery.

READING THE SCALE. When the ohmmeter range switch is at $R \times 1$, the pointer indicates the measured resistance value directly in ohms. With the range switch at $R \times 10$ and the pointer indicating 30, as illustrated in Figure 2-8(a), the measured resistance is (30×10) Ω $= 300$ Ω. On this range, the small length of scale from 200 to INF represents all resistance values above 2 kΩ. Obviously, resistances greater than 2 kΩ cannot be measured accurately on this range of the ohmmeter. (For example, where would 5 MΩ be indicated on the scale?) Similarly, the *0 to 5* section of the scale represents all resistance values below 50 Ω when the range is $R \times 10$. Again, these resistance values cannot be measured with any degree of accuracy. (Where would the 0.05 Ω position be found?) From this discussion it is seen that

the ohmmeter measures resistances most accurately when its pointer position is around center scale.

The ohmmeter described above is normally not available as a single insrument. Instead, it is usually found as part of a *multimeter,* which is described in the following section. Ohmmeters are not shown as part of a circuit diagram because they are never left connected permanently in a circuit.

2-4
DEFLECTION MULTIMETERS

FUNCTIONS AND SCALES. The *multimeter* is a multifunction instrument capable of measuring voltage (dc or ac), current (dc or ac), and resistance. Some multimeters can also measure additional quantities, such as inductance and capacitance. Another name applied to the instruments is *VOM*, which is short for *volt-ohm-milliammeter*.

A typical deflection-type multimeter is shown in Figure 2-9. The instrument illustrated has three scales, one of which (the upper scale) is used only for resistance measurements. The other two scales may be used for either voltage or current measurements. A *VOM* that is capable of measuring other quantities (that is, in addition to voltage, current, and resistance) may have several more scales.

Two terminals identified as + and − are used for all measurements. The terminal polarity is important only when measuring dc quantities.

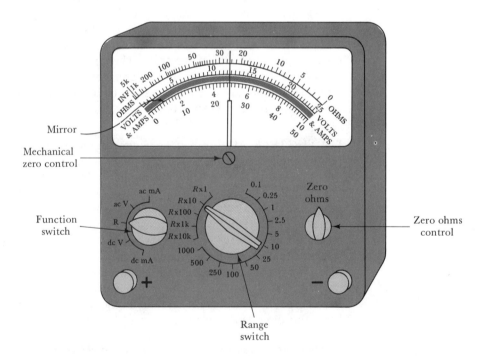

FIGURE 2-9. A deflection multimeter usually has a function switch to select current measurement (ac or dc), voltage measurement (ac or dc), or resistance measurement. A range switch is also provided to select the maximum current, voltage, or resistance that the meter is to indicate.

Some multimeters have additional terminals for use when unusually high voltages or currents are to be measured.

VOLTAGE AND CURRENT RANGES. The main controls on the multimeter shown in Figure 2-9 are a *function switch* and a *range switch*. The function switch sets the instrument to the type of quantity which is to be measured: *dc mA* (direct current in milliamps), *dc V* (direct voltage), *R* (resistance), *ac V* (alternating voltage), or *ac mA* (alternating current in milliamps). The range switch selects the instrument range to be used. Moving clockwise around the range switch from the top, the *0.1 through 1000* positions are current and voltage ranges, both ac and dc. If the function switch were at *dc V* and the range switch at *250*, the meter could measure up to 250 V dc at full-scale deflection. When the function switch is set to *ac mA* and the range switch is *1000*, alternating currents up to a level of 1 A (1000 mA) can be measured.

RESISTANCE RANGES. Resistance measurements are made using the *R × 10k* to *R × 1* range positions. With the function switch set at *R* and the range switch at *R × 1*, the resistance scale reads directly in ohms. If the range switch is at *R × 1k*, all the resistance values indicated on the scale must be multiplied by 1 kΩ. For example, a pointer indication of *30* would be read as $(30 \times 1\ k\Omega) = 30\ k\Omega$. The *ZERO OHMS* control is used as described in Section 2-3.

OVERLOAD AND REVERSE. Some multimeters are equipped with an overload protection circuit. When the current passing through the instrument exceeds a maximum safe-level, internal connections are automatically open-circuited, and a *cut-out* button pops up. The cut-out device is reset by pushing the button down again.

A *REVERSE* button is also sometimes provided on a multimeter. This can be used when a dc current or voltage is not connected with the correct polarity. Pushing (and holding) the *REVERSE* button causes the meter polarity to be reversed internally, so that the pointer deflects to the right of *0*.

PARALLAX ERROR. The multimeter illustrated in Figure 2-9 has a fine *knife-edge* pointer for accurate scale readings. A mirror is also provided alongside the scale; the purpose of this is to eliminate the scale-reading error (parallax error) that occurs when the viewer's eye is not directly in front of the pointer. To take a reading, the operator closes one eye and then moves his/her head until the pointer reflection in the mirror disappears behind the pointer. The operator's eye is now directly in front of the pointer, and the meter can be read with great accuracy.

EXAMPLE 2-3 Determine the readings of the multimeter for the pointer positions and the switch settings illustrated in Figures 2-9 and 2-10.

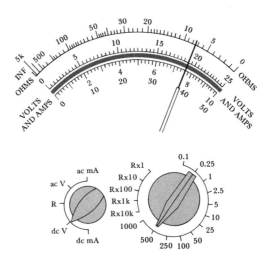

FIGURE 2-10. Meter indication for Example 2-3.

SOLUTION

In Figure 2-9 the function switch is at R and the range switch is at R × 10. Therefore, read the pointer position on the resistance scale and multiply by 10.

$$meter\ reading = 25\ \Omega \times 10 = \textbf{250}\ \boldsymbol{\Omega}$$

In Figure 2-10,

$$full\text{-}scale\ deflection = 500\ \text{V dc}$$

Using the bottom scale,

$$the\ meter\ reading = 38\ \text{V} \times 10 = \textbf{380 V dc}$$

PRACTICE PROBLEM

2-4.1 Determine the reading illustrated in Figure 2-10: (a) if the function switch is set to *dc mA*, and the range switch to *25*, (b) if the function switch is set to *R*, and the range switch to *R × 1 kΩ*.

2-5
DIGITAL MULTIMETERS

The multimeters illustrated in Figure 2-11 are known as *digital instruments*, because, instead of a pointer and scale, electronic *digital read-out devices* are used to display the measured quantity in numerical form.

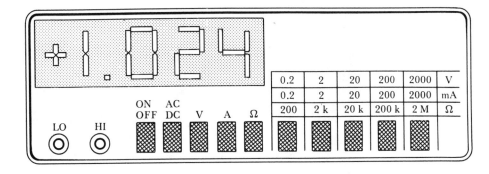

(a) Front panel of a digital instrument

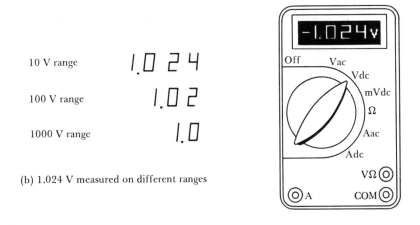

10 V range

100 V range

1000 V range

(b) 1.024 V measured on different ranges

(c) Pocket-size digital multimeter

FIGURE 2-11. Digital multimeters use digital read-out devices to display the measured quantity in numerical form. Plus or minus signs are displayed to indicate the terminal polarity of the quantity.

The meter shown in Figure 2-11(a) uses push-button switches for range and function selection. Two connecting terminals are provided, identified as *LO* (low potential) and *HI* (high potential). An *ON/OFF* switch is included because this instrument either has an external ac supply or is battery operated.

The measured voltage, current, or resistance is shown in the form of a three- or four-digit number (sometimes more than four) with a properly placed decimal point. When dc quantities are measured, the polarity is identified by means of a + or − sign to the left of the number. The read-out devices which display the three right-hand numerals on the meters in Figure 2-11 each consist of seven segments which are selectively energized. If only those three read-out units were available, the meter could display a number up to *999,* and it would be said to have a *three-digit display.* Each additional seven-segment unit requires a considerable amount of electronic circuitry. However, an additional *1* unit only needs simple circuitry to switch it *on* or *off.* Including *1* at the left side of the display allows the meter to indicate up to *1999.* Since the first (left) numeral can only be *1* or blank, it is termed a *half-digit,* and the complete display is called a *three-and-a-half digit display.*

Figure 2-11(b) shows the displays that would be obtained when measuring 1.204 V on each of several ranges. As in the case of the deflection instrument, *the most accurate measurements are achieved when the lowest possible range is used.* Many digital instruments employ *automatic ranging circuits* that automatically change the instrument range to give the most accurate measurement of a given input.

The pocket-size digital multimeter illustrated in Figure 2-11(c) can measure *alternating voltage (Vac), direct voltage (Vdc), resistance (Ω), alternating current (Aac),* and *direct current (Adc).* The instrument is auto-ranging, except for a low dc voltage range *(mVdc).* Note that a single *common (COM)* terminal is used with separate *current (A)* and *voltage/resistance (V/Ω)* terminals.

A digital multimeter could easily be substituted in place of the deflection ammeter and voltmeter in Figures 2-1 and 2-5. In both cases, the terminal polarity of the digital instrument would be unimportant, because a minus sign would be displayed if the polarity was incorrect.

Digital instruments are obviously much easier to read than deflection-type meters. Also, they are usually more accurate than deflection instruments, and less likely to be damaged by overload conditions.

2-6
INSTRUMENT ACCURACY

DEFLECTION INSTRUMENT ACCURACY. The accuracy of all deflection instruments is specified by the manufacturer as a percentage of full-scale deflection (FSD). For example, a voltmeter with an FSD of 100 V might have its accuracy stated as ±2%. In this case, the maximum possible error at all points on the scale is ±2% of 100 V, or ±2 V. Thus, when the pointer is indicating exactly 100 V, as in Figure 2-12(a), the measured voltage is correctly stated as 100 V ± 2 V, or 98 V to 102 V. The actual level of the measured voltage might be anywhere from 98 V to 102 V.

When the pointer of this voltmeter indicates 50 V, as illustrated in Figure 2-12(b), the actual measured voltage must be taken as 50 V ± (2% of full scale), or 50 V ± 2 V. This means that the measured voltage is somewhere beween 48 V and 52 V. The ±2 V error is ±4% of 50 V. Consequently, the measurement percentage error increases as the pointer moves further away from full-scale deflection.

(a) Voltmeter or ammeter at FSD
0
100V
for ± 2% accuracy:
V = 100 V ± (2% of 100 V)
= 98 V to 102 V
(b) Voltmeter or ammeter at ½ FSD
50V
0
100V
for ± 2% accuracy:
V = 50 V ± (2% of 100 V)
= 48 V to 52 V
(c) Ohmmeter at ½ FSD
25Ω
INF
0Ω
for ± 2% multimeter accuracy:
R = 25 Ω ± (4 x 2%)
= 25 Ω ± 8%
= 23 Ω to 27 Ω

FIGURE 2-12. Errors in a deflection instrument are related to the full-scale deflection. A meter that has a possible ±1% error at full-scale will have a possible error of ±2% at half-scale.

EXAMPLE 2-4 If the instrument illustrated in Figure 2-10 has an accuracy of ±1%, determine the upper and lower limits of the measured voltage.

SOLUTION

From Example 2-3:

$$\textit{meter reading} = 380\ \text{V}$$

and,

$$\textit{full-scale deflection} = 500\ \text{V}$$
$$\textit{error} = \pm 1\%\ \text{of}\ 500\ \text{V}$$
$$= \pm 5\ \text{V}$$

measured voltage becomes

$$380\ \text{V} \pm 5\ \text{V} = \mathbf{375\ V\ to\ 385\ V}$$

Deflection-type multimeters typically have an accuracy of ±2% for voltage and current measurements, although more accurate instruments are available. As illustrated in Figure 2-12(c), the error becomes considerably greater when the instrument is used for resistance measurements. It has already been explained in Section 2-3 that ohmmeter measurements are most accurate when the pointer is near mid-scale. However,

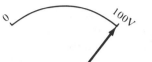

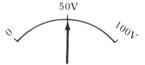

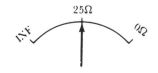

even at mid-scale, the resistance measurement error is typically $\pm(4x\%$ of mid-scale resistance), where x is the multimeter specified accuracy. Thus, for a multimeter with an accuracy of $\pm2\%$, the mid-scale accuracy of the ohmmeter function is $\pm(8\%$ of mid-scale resistance). When measurements are made away from mid-scale, the error is increased.

DIGITAL INSTRUMENT ACCURACY. The accuracy of a digital multi-meter might be stated by the manufacturer as *±0.5% Rdg ± 1d*. This means $\pm(0.5\%$ *of the reading*$) \pm 1$ *digit,* where the *1 digit* refers to the right hand numeral of the display. For example, a display of 100.0 V would be accurate to:

$$\pm(0.5\% \text{ of } 100.0 \text{ V}) \pm 000.1 \text{ V}$$

$$= \pm(0.5 \text{ V}) \pm 000.1 \text{ V}$$

$$= \pm0.6 \text{ V}$$

Therefore, the measured voltage could be stated as:

$$100 \text{ V} \pm 0.6 \text{ V} = 99.4 \text{ V to } 100.6 \text{ V}$$

EXAMPLE 2-5   A digital multimeter indicates 5.432 V. If the instrument accuracy is specified as $\pm0.1\%$ Rdg $\pm$ 1d, calculate the upper and lower limits of the measured voltage.

SOLUTION

$$error = \pm(0.1\% \text{ of } 5.432 \text{ V}) \pm 0.001 \text{ V}$$

$$\cong \pm0.005 \text{ V} \pm 0.001 \text{ V}$$

$$\cong \pm0.006 \text{ V}$$

The measured voltage becomes:

$$5.432 \text{ V} \pm 0.006 \text{ V} = \textbf{5.426 V to 5.438 V}$$

PRACTICE PROBLEMS

2-6.1   For the ohmmeter indication in Figure 2-8, determine the upper and lower limits of the measured resistance if the multimeter is accurate to $\pm3\%$ at full scale.

2-6.2   What are the upper and lower limits of the voltage measurements found in Example 2-2(b) if the meter is accurate to $\pm3\%$ at full scale?

2-6.3   If the accuracy of the digital meter in Figure 2-11(a) is specified as *±0.3% rdg ± 1d,* determine the upper and lower limits of the indicated quantity.

Summary of Basic Electrical Instrument Characteristics

Instrument	Circuit Symbol	Quantity Measured	Instrument Resistance	Instrument Connection
Voltmeter	(V)	Electrical pressure in volts	Very high	In parallel with the points at which the voltage is to be measured
Ammeter	(A)	Electric current in amps	Very low	In series with the circuit in which the current is to be measured
Ohmmeter	(Ω)	Resistance in ohms	Depends upon range setting	In parallel with the component which is to have its resistance measured

REVIEW QUESTIONS

2-1 How should an ammeter be connected in a circuit? Explain briefly. Discuss the terminal polarity for dc and ac ammeters, and the effect of incorrect connection in each case.

2-2 Explain the purpose and use of the mechanical zero control on a deflection instrument.

2-3 An ammeter has the following ranges: 1 mA, 2.5 mA, 50 mA, and 100 mA. Which range should be selected to give the most accurate measurement of:
(a) 55 mA,
(b) 3.9 mA,
(c) 0.9 mA,
(d) 19 mA.
What would be the effects of selecting a range which is too high and a range which is too low?

2-4 Sketch a circuit diagram showing an ammeter measuring the current supplied from a battery to three series-connected electric lamps. Identify the polarity of the instrument, and show the (conventional) direction of current flow.

2-5 Briefly explain how a voltmeter should be connected into a circuit. Discuss the importance of terminal polarity for ac and dc voltmeters, and state the effect of incorrect connection in each case.

2-6 A voltmeter has the following ranges: 0.3 V, 1 V, 3 V, 10 V, and 30 V. The meter is to be used to measure the terminal voltage of a battery, which is known to be around 1.5 V. What range should be selected? What would be the effect of selecting each of the other ranges?

2-7 Sketch a circuit diagram showing a voltmeter measuring the terminal voltage of a battery that is supplying current to two electric lamps in parallel. Show the current direction in the circuit and the terminal polarity of the instrument.

2-8 Discuss the precautions necessary when using an ohmmeter to measure the resistance of a component connected in a circuit. How are the mechanical zero control and the *ZERO OHMS* control used?

2-9 What is meant by linear and nonlinear meter scales? At what part of a scale are the most accurate measurements made:
(a) On a voltmeter?
(b) On an ohmmeter?
(c) On an ammeter?
Explain briefly.

2-10 Sketch the scale of an ohmmeter suitable for measuring resistance values around 100 Ω. Show the zero, infinity, and 100 Ω positions.

2-11 List the quantities that can be measured on a typical multimeter. Also list the controls usually available and the purpose of each, and explain the use of a mirror scale.

2-12 Explain how a digital multimeter differs from a deflection-type instrument.

2-13 Explain the accuracy specification of a measuring instrument:
(a) For a deflection instrument specified as accurate to ±1%.
(b) For a digital instrument with an accuracy of ±0.3% Rdg ± 1d.

PROBLEMS

SECTION 2-1

2-1 Determine the measured current for each of the pointer positions shown in Figure 2-13 when the range switch is at 1000 mA.

2-2 If the instrument range is 50 mA for the pointer positions shown in Figure 2-13, determine the indicated current levels.

FIGURE 2-13.

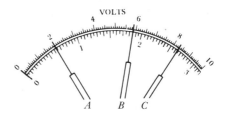

FIGURE 2-14.

2-3   If the range of the instrument illustrated in Figure 2-13 were 250
        mA, what would be the current indicated by each pointer position
        shown?

2-4   Determine the current levels indicated by the pointer positions in
        Figure 2-3 if the instrument range is set to 1000 mA.

SECTION 2-2

2-5   If the range and function switches for Figure 2-13 are set to
        50 V, determine the meter readings for each of the three pointer
        positions.

2-6   Determine the measured voltage levels indicated by the pointer
        positions in Figure 2-6 if the instrument is set to a 150 V range.

2-7   For each pointer position shown in Figure 2-14, write the mea-
        sured voltage:
        (a) When the voltmeter range is 100 V.
        (b) When the range is 300 mV.

2-8   If the instrument in Figure 2-14 is set to a 200 V range, determine
        the indicated voltage levels.

SECTION 2-3

2-9   Figure 2-15 shows an ohmmeter scale and range switch. Which
        ranges should be used when measuring resistances that have the
        following approximate values:
        (a) 560 Ω.
        (b) 12 kΩ.
        (c) 2 MΩ.
        (d) 10 Ω.

2-10   Determine the measured resistance values for the ohmmeter scale
        and pointer positions illustrated in Figure 2-15 if the instrument
        range is set to $R \times 100$.

2-11   What are the resistance values indicated by each of the pointer
        positions shown in Figure 2-15 when the instrument range is
        $R \times 10$?

Problems

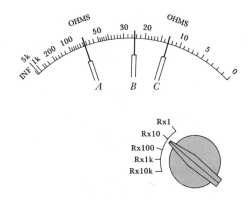

FIGURE 2-15.

2-12 Determine the indicated resistance values in Figure 2-15 if the meter range is $R \times 10$ k.

SECTION 2-4

2-13 Determine the readings of the multimeter for the pointer positions and switch settings in Figure 2-16.

2-14 Determine the indicated quantities in Figure 2-16 if the range switch is set to 0.1.

2-15 If the range switch in Figure 2-16 is at 50, what do the three pointer positions indicate?

2-16 Show the pointer positions on Figure 2-16 when the instrument is measuring 4.5 mA on the following ranges: (a) 100 mA, (b) 25 mA, (c) 5 mA.

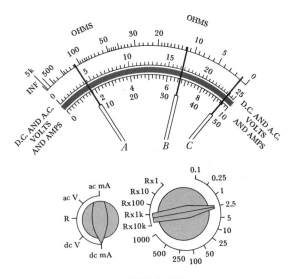

FIGURE 2-16.

Chap. 2   Measuring Current, Voltage, and Resistance

2-17 With the range switch at $R \times 1$ and the function switch at R, state the meter readings for the three pointer positions in Figure 2-16.

2-18 On the meter scale in Figure 2-13, show the pointer position for 8.7 V when the meter range is set for:
(a) 10 V.
(b) 25 V.
(c) 50 V.

2-19 If the instrument in Figure 2-16 is set to measure resistance on a range of $R \times 10$ k, determine the measured resistances for the pointer positions illustrated.

2-20 On the ohmmeter scale in Figure 2-15, show the pointer position for a resistance of 7.5 kΩ when the meter range is set for:
(a) $R \times 1$.
(b) $R \times 10$.
(c) $R \times 100$.
(d) $R \times 1$ kΩ.
(e) $R \times 10$ kΩ.

2-21 Determine the quantities indicated in Figure 2-16 if the instrument is set to a 250 V range.

SECTION 2-6

2-22 Assuming a 10 V range and an instrument accuracy of ±1%, calculate the maximum and minimum levels of the voltages indicated in Figure 2-14.

2-23 Calculate the maximum and minimum values of the quantities indicated in Figure 2-16 if the multimeter accuracy is specified as ±2%.

2-24 Determine the maximum and minimum currents measured for Problem 2-1, if the instrument accuracy is ±3%.

2-25 If the resistance values determined in Problems 2-11 and 2-12 were made on a multimeter with an accuracy of ±3%, calculate the maximum and minimum resistances for pointer position B.

2-26 Determine the maximum and minimum measured quantities for Problem 2-15 if the instrument accuracy is ±0.5%.

2-27 A digital instrument has a specified accuracy of ±0.2% Rdg ± 1d. Calculate the maximum and minimum values of the voltage when the instrument indicates:
(a) 9.418 V.
(b) 1.290 V.
(c) 3.067 V.

2-28 A digital instrument with an accuracy of ±0.3% Rdg ± 1d indicates the following current levels: (a) 7.240 mA, (b) 1.893 mA,

(c) 4.235 mA. Calculate the maximum and minimum quantities in each case.

2-29 The accuracy of the resistance ranges on a digital multimeter are specified as $\pm0.6\%$ Rdg $\pm$ 1d. Determine the upper and lower limits of the measured resistance when the instrument indication is:
(a) 4.663 kΩ.
(b) 560.1 Ω.
(c) 27.77 Ω.

2-30 Determine the maximum and minimum readings in Problem 2-28 if the instrument accuracy is $\pm0.5\%$ Rdg $\pm$ 2d.

ANSWERS TO PRACTICE PROBLEMS

2-1.1 25 mA, 42 mA, 66 mA, 90 mA

2-1.2 6.25 mA, 10.5 mA, 16.5 mA, 22.5 mA

2-2.1 14 V, 41 V

2-2.2 90 mV, 260 mV

2-4.1 19 mA, 7.5 kΩ

2-6.1 264 Ω to 336 Ω

2-6.2 8.1 V to 9.9 V, 25.1 V to 26.9 V

2-6.3 1.020 to 1.028

3

OHM'S LAW AND ELECTRICAL CALCULATIONS

Objectives   You will:

Gain an understanding of resistance and conductance, and be able to convert between the two.

Be able to state Ohm's law and write the Ohm's law equation.

Given any two of resistance, current, or potential difference, be able to calculate the third quantity.

Understand electrical power dissipation, and be able to determine power dissipation in a circuit or component.

Understand and be able to calculate efficiency where it applies to electronic circuits.

Understand electrical energy, and be able to calculate electrical energy consumption.

Introduction

Current flow in an electric circuit is the result of an applied voltage. The opposition to current flow which exists in every circuit is termed *resistance,* and its converse is *conductance.* The relationship between voltage, current, and resistance is defined by the simple but extremely important *Ohm's law.*

Electrical power is dissipated when current flows in a circuit or component. In many cases, this power is wasted as unwanted heat. Electrical energy is used when power is dissipated over a period of time.

**RESISTANCE
AND
CONDUCTANCE**

CONDUCTOR RESISTANCE

*The resistance of a conductor is a measure of its opposition to current
flow.*

When a potential difference is applied to the ends of a conductor, the
quantity of current that can flow depends upon the applied voltage, the
type of metal used, the length of the conductor, and the cross-sectional
area of the conductor. The type of metal is important, because different
metals offer different resistances to current flow. (This is investigated
further in Chapter 4.) The cross-sectional area is involved, because, just
as a thick water pipe can carry a larger flow of water than a thin one, so
a thick conductor passes more electric current than a thin one of the
same material [See Figure 3-1(a) and (b)].

How the conductor length affects its resistance is best understood
by taking another look at the current flow process discussed in Section
1-5. Figure 1-6 shows that the resistance of a conductor has a lot to do
with the atoms with which the electrons are continually colliding. Figure
3-1(c) illustrates the fact that an electron moving through a 2 m conductor has to get past twice as many atoms as in the case of a 1 m con-

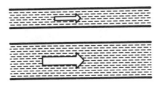

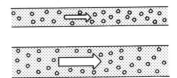

(a) A thick water pipe passes a larger
    current of water than a thin pipe.

(b) A thick conductor passes a larger electric
    current than a thin one of the same material.
    The thick conductor has a lower resistance
    than the thin one.

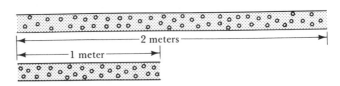

(c) Electrons flowing along a 2 m conductor have to pass twice as many atoms
    as those travelling through a similar 1 m conductor. The resistance of a
    2 m conductor is twice that of the a similar 1 m conductor.

FIGURE 3-1. Conductor resistance. A thick conductor offers a lower resistance
to current flow than a thin one of similar length and material. A short conductor
has a lower resistance than a similar long one.

Chap. 3   Ohm's Law and Electrical Calculations

ductor (assuming the same cross-sectional area and material for each conductor). There are likely to be twice as many electron-atom collisions in the longer conductor. Consequently, it can be stated that a 2 m conductor has twice the resistance of a similar 1 m conductor.

As already mentioned in Section 2-3, the *ohm* (symbol Ω) is the unit of resistance. The symbol used to identify resistances is *R*. Several resistances in a circuit might be numbered R_1, R_2, R_3, etc.

CONDUCTANCE. In some circumstances it is convenient to use the reciprocal of resistance instead of the actual resistance value. The term *conductance* (symbol *G*) is applied to the reciprocal of resistance. While resistance is a measure of opposition to current flow, *conductance is a measure of the ease with which a conductor or resistor will pass a current*. The same graphic symbol is employed in circuit diagrams for conductance and resistance.

The siemens (symbol S) is the unit of conductance.

$$conductance = \frac{1}{resistance}$$

or,

$$\boxed{G = \frac{1}{R}}$$

(3-1)

where G is in siemens and R is in ohms.

Another (outdated) term for the unit of conductance is *mho* ("ohm" spelled backwards). The symbol for *mhos* is an upsidedown omega ($\mho$) .

EXAMPLE 3-1   Determine the conductance of a 100 Ω resistor.

SOLUTION

$$G = \frac{1}{R} = \frac{1}{100 \ \Omega}$$

$$= 10 \ mS$$

PRACTICE PROBLEMS

3-1.1   Convert 4.7 kΩ and 5.6 Ω into conductances.

3-1.2   Express 500 S and 33 μS as resistances.

3-2

OHM'S LAW

Consider Figure 3-2(a), in which an emf E is shown producing a current I through a resistance R. If $E = 1$ V and $R = 1$ Ω, as illustrated, the current that flows is 1 A. This comes directly from the definition of the ohm in Appendix 2:

One ohm is that resistance which permits a current of 1 A to flow when a potential difference of 1 V is applied to the resistance.

Now consider the effect of doubling the applied voltage [Figure 3-2(b)]. If an electrical pressure of 1 V causes 1 A to flow through 1 Ω, then doubling the pressure doubles the current to 2 A. Similarly, when E remains 1 V, and the resistance is doubled, as in Figure 3-2(c), the current flow is halved (i.e., $I = 0.5$ A).

The above discussion shows that the current flow through a resistance is directly proportional to the applied voltage, and inversely proportional to the value of the resistance. When the voltage is doubled, the current is doubled. When the resistance is doubled, the current is halved. Recall that *voltage is an electrical pressure that tends to cause current flow when a suitable conducting path is provided*, and that *resistance is opposition to current flow:* The relationship between current, voltage, and resistance is stated in the important fundamental law of electricity known as *Ohm's law:*

Ohm's Law

$$current = \frac{voltage}{resistance}$$

or,

$$\boxed{I = \frac{E}{R}}$$

(3-2)

where I is in amperes, E is in volts, and R is in ohms.

Equation (3-2) can be rewritten as

$$E = I \times R \quad \text{and} \quad R = \frac{E}{I}$$

EXAMPLE 3-2 Calculate the new current levels in the circuits shown in Figure 3-2 if the resistances are changed to 2.5 kΩ.

SOLUTION

(a)

$$I = \frac{E}{R} = \frac{1 \text{ V}}{2.5 \text{ k}\Omega} = \textbf{400 } \boldsymbol{\mu}\textbf{A}$$

(b)

$$I = \frac{E}{R} = \frac{2 \text{ V}}{2.5 \text{ k}\Omega} = \textbf{800 } \boldsymbol{\mu}\textbf{A}$$

(c)

$$I = \frac{E}{R} = \frac{1 \text{ V}}{5 \text{ k}\Omega} = \textbf{200 } \boldsymbol{\mu}\textbf{A}$$

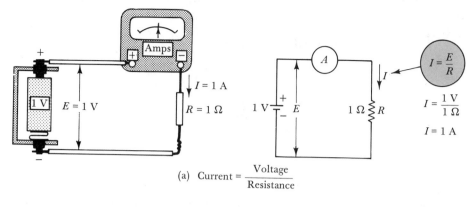

(a) $\text{Current} = \dfrac{\text{Voltage}}{\text{Resistance}}$

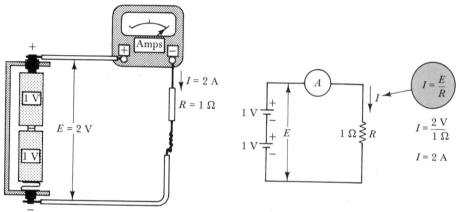

(b) Doubling the voltage causes the current to be doubled.

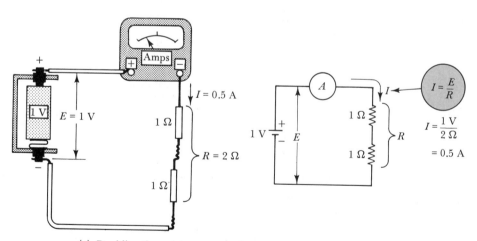

(c) Doubling the resistance causes the current to be halved.

FIGURE 3-2. Illustration of Ohm's law. Current is directly proportional to volt-age, and inversely proportional to resistance. *Current = voltage/resistance.* When the voltage across a resistance is doubled, the current is doubled. When the resistance is doubled, the current level is halved.

3-2.1 Calculate the current that flows in a 2.5 kΩ resistance when it is connected to a 12 V supply.

3-2.2 The voltage measured across a 22 Ω resistance is 3.3 V. Determine the current level.

3-3
APPLICATION OF OHM'S LAW

When Ohm's law is employed to calculate the current through a resistance, the voltage used must be the voltage that appears across that particular resistance, and the current must be the current that flows through that resistance. These quantities must not be confused with the circuit supply voltage and current, although in Figure 3-2 they are the same quantities. Consider the circuit in Figure 3-3. The voltage that appears across resistor R_3 is V_3, and the current through R_3 is I_3. Thus,

$$I_3 = \frac{V_3}{R_3}$$

EXAMPLE 3-3 The measured voltages in the circuit in Figure 3-3 are: $V_1 = 9$ V, $V_2 = 10$ V, $V_3 = 10$ V. The resistance values are: $R_1 = 600$ Ω, $R_2 = 4$ kΩ, $R_3 = 800$ Ω. Calculate the individual currents.

SOLUTION

$$I_1 = \frac{V_1}{R_1} = \frac{9 \text{ V}}{600 \text{ Ω}} = \textbf{15 mA}$$

$$I_2 = \frac{V_2}{R_2} = \frac{10 \text{ V}}{4 \text{ kΩ}} = \textbf{2.5 mA}$$

$$I_3 = \frac{V_3}{R_3} = \frac{10 \text{ V}}{800 \text{ Ω}} = \textbf{12.5 mA}$$

EXAMPLE 3-4 A resistance of 100 Ω has a potential difference of 10 V across its terminals.
(a) Calculate the current.
(b) If the resistance is changed to 250 Ω, determine the new voltage that must be applied to give the same level of current.
(c) If the emf is altered to 15 V, calculate the new value of resistance that will set the current at 300 mA.

SOLUTION

(a) $$I = \frac{E}{R} = \frac{10 \text{ V}}{100 \text{ }\Omega} = \textbf{0.1 A}$$

(b) $$E = IR = 0.1 \text{ A} \times 250 \text{ }\Omega = \textbf{25 V}$$

(c) $$R = \frac{E}{I} = \frac{15 \text{ V}}{300 \text{ mA}} = \textbf{50 }\boldsymbol{\Omega}$$

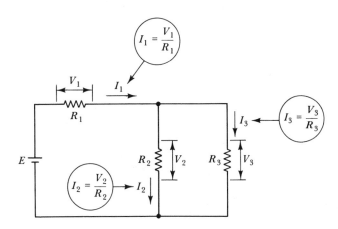

FIGURE 3-3. When Ohm's law is used to calculate the current through a resistance, the voltage across the particular resistance and the resistance value must be used in the calculation.

Memory Aid for Ohm's Law

Ohm's Law Circle

To recall the equation for any quantity, just cover that quantity on the circle. Then read the remaining quantities as the equation.

For current, cover I. Then, $I = \dfrac{E}{R}$

For voltage, cover E: $\quad E = IR$

For resistance, cover R: $\quad R = \dfrac{E}{I}$

EXAMPLE 3-5 Calculate the value of conductance that passes 250 mA when an emf of 50 V is applied across its terminals, as illustrated in Figure 3-4.

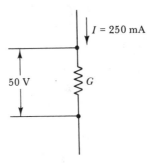

FIGURE 3-4. Circuit for Example 3-5. Conductance is determined as the inverse of resistance.

SOLUTION

$$R = \frac{E}{I} = \frac{50 \text{ V}}{250 \text{ mA}} = 200 \ \Omega$$

$$G = \frac{1}{R} = \frac{1}{200 \ \Omega} = 5 \text{ mS}$$

EXAMPLE 3-6 An emf of 120 V is applied to one end of two electrical conductors, as illustrated in Figure 3-5. The voltage measured at the other ends is 110 V when the current level is 25 A. Calculate the total resistance of the conductors.

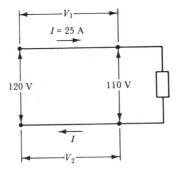

FIGURE 3-5. Circuit for Example 3-6. Voltage drop occurs along conductors when a current flows through them. Ohm's law can be applied to determine the resistance of the conductors from a knowledge of the current and voltage drop.

SOLUTION

The total voltage drop is,

$$V_1 + V_2 = 120 \text{ V} - 110 \text{ V}$$

$$= 10 \text{ V}$$

$$R = \frac{V_1 + V_2}{I} = \frac{10 \text{ V}}{25 \text{ A}} = 0.4 \ \Omega$$

PRACTICE PROBLEMS

3-3.1   (a) Calculate the current through a 3.9 kΩ resistor when the voltage across it is measured as 22 V.

       (b) If the current through the 3.9 kΩ resistor is to be exactly 4.5 mA, determine the required potential difference.

       (c) If a 4.5 mA current is to flow through a resistance when 22 V is applied across it, determine the required resistance.

3-3.2   The voltages in the circuit of Figure 3-3 are measured as $V_1 = 5$ V, $V_2 = 12$ V, and $V_3 = 12$ V. The resistance values are: $R_1 = 2.5$ kΩ, $R_2 = 10$ kΩ, $R_3 = 15$ kΩ. Calculate the currents levels.

3-3.3   An electric lamp supplied from a battery has 11.5 V at its terminals. The battery terminal voltage is 12 V. If the lamp current is 2.5 A, calculate the resistance of the lamp filament, and the resistance of the cables connecting the lamp to the battery.

3-4

ELECTRICAL POWER AND ENERGY

POWER DISSIPATION.   When an electric current flows in a conductor, heat is dissipated as a result of the work done in moving electrons past atoms (see Section 1-5). Work is also performed in a lamp filament when the electric current is converted into light. Similarly, work is done when a speaker converts electric current into sound, and when an electric motor produces motion.

Power (P) is the time rate of doing work.

If a certain amount of work W is done in a time t, then $P = W/t$. It can be demonstrated that the power supplied to any electrical component is,

Power = voltage $\times$ current,

or,

$$\boxed{P = EI} \tag{3-3}$$

The unit of electric power is the Watt (W).

Thus, when an applied potential difference of 1 V causes a current of 1 A to flow in a circuit or device, the power supplied is,

$$P = EI = 1 \text{ V} \times 1 \text{ A}$$
$$= 1 \text{ W}$$

From Ohm's law,

$$I = \frac{E}{R} \quad \text{and} \quad E = I \times R$$

Substituting for I in Equation (3-3) gives,

$$P = E \times \frac{E}{R}$$

$$\boxed{P = \frac{E^2}{R}} \tag{3-4}$$

And substituting for E in Equation (3-3) gives,

$$P = IR \times I$$

or, $$\boxed{P = I^2 R} \tag{3-5}$$

From Equations (3-3), (3-4), and (3-5), it is seen that power can be calculated from a knowledge of any two of *E, I,* and *R*.

Electric heating elements are designed to convert electricity into heat for boiling water, space heating, etc. As illustrated in Figure 3-6, power supplied to an electric lamp is converted into light, while a speaker converts input power into sound.

Power dissipated in a conductor is wasted power [Figure 3-6(c)]. Excessive power dissipation in a conductor can produce overheating to the point at which the conductor insulation might be damaged or destroyed.

Power dissipated in electronic devices is also wasted power [Figure 3-6(d)]. Here again, too much power dissipation can destroy an electronic device. Consequently, every device has a mximum power rating specified by the manufacturer.

To protect equipment and circuits from excessive power dissipation, a *fuse* may be connected into the circuit. A fuse is simply a short piece of thin copper wire or strip. Power is dissipated in the fuse as the circuit current flows through it ($P = I^2 R$). When the current flow ex-

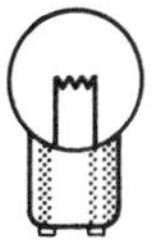

(a) Power dissipated in a lamp is converted into light.

(b) Power supplied to a speaker is converted into sound.

(c) Power dissipated in a conductor is wasted power.

(d) Power dissipated in an electronic device is wasted power.

FIGURE 3-6. When an electric current flows, power is dissipated. Power can be converted into light, heat, sound, and mechanical motion. Power dissipation in conductors and electronic devices is unwanted and wasteful.

ceeds the design limit, the power dissipated in the fuse causes it to melt, and thus interrupt the current flow to the circuit.

EXAMPLE 3-7   Determine the power dissipated in each of the resistors in Example 3-3.

SOLUTION

From Example 3-3: $V_1 = 9$ V, $V_2 = 10$ V, $V_3 = 10$ V, $I_1 = 15$ mA, $I_2 = 2.5$ mA, $I_3 = 12.5$ mA.

Equation (3-3), $P_1 = V_1 I_1 = 9$ V $\times$ 15 mA

           $= \mathbf{135\ mW}$

           $P_2 = V_2 I_2 = 10$ V $\times 2.5$ mA

           $= \mathbf{25\ mW}$

           $P_3 = V_3 I_3 = 10$ V $\times 12.5$ mA

           $= \mathbf{125\ mW}$

EXAMPLE 3-8   A 120 Ω resistor has a specified maximum power dissipation of 1 W. Calculate the maximum current level.

SOLUTION

Equation (3-5), $P = I^2 R$

Therefore, $I^2 = \dfrac{P}{R}$

giving, $I = \sqrt{\dfrac{P}{R}} = \sqrt{\dfrac{1\ W}{120\ \Omega}}$

$= \mathbf{91.3\ mA}$

EXAMPLE 3-9   An electric heater with a resistance of 5.5 Ω is supplied from a 115 V source via two conductors that each have a resistance of 0.125 Ω (see Figure 3-7). Calculate the power dissipated in the heater and the power wasted in the conductors.

SOLUTION

Total resistance,

$$R = R_c + R_h + R_c$$

$$= 0.125\ \Omega + 5.5\ \Omega + 0.125\ \Omega$$

$$= 5.75\ \Omega$$

$$I = \frac{E}{R} = \frac{115\ V}{5.75\ \Omega}$$

$$= \mathbf{20\ A}$$

For the heater,

$$P_h = I^2 R_h = (20\ A)^2 \times 5.5\ \Omega$$

$$= \mathbf{2.2\ kW}$$

For the conductors,

$$P_c = I^2 (R_c + R_c) = (20\ A)^2 \times 0.25\ \Omega$$

$$= \mathbf{100\ W}$$

EFFICIENCY.   Not all of the power supplied to electrical equipment is converted into useful activity. As illustrated by Example 3-9, some power is wasted in the conductors. However, usually there is a further

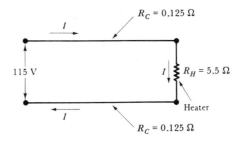

FIGURE 3-7. Power is dissipated in a heater and in conductors when a current flows through them. The heater power dissipation is $P_h = I^2R_h$, and the total conductor power dissipation is $P_c = I^2(R_c + R_c)$.

wastage of some of the power delivered to the terminals of a circuit or appliance. The ability of an electric motor, electronic amplifier, or other component to convert power into a useful output is defined in terms of its *efficiency*. Thus, if 80% of the power supplied to an audio amplifier is delivered to a speaker, and 20% is wasted as heat, then the amplifier has an efficiency of 80%.

The symbol for efficiency is η (Greek letter eta).

$$efficiency = \frac{power\ output}{power\ input} \times 100\%$$

or,

$$\boxed{\eta = \frac{P_o}{P_i} \times 100\%} \tag{3-6}$$

EXAMPLE 3-10 An electronic amplifier receives a supply of 1.2 W, and delivers 700 mW of power to a speaker. Calculate its efficiency.

SOLUTION

Equation (3-6),

$$\eta = \frac{P_o}{P_i} \times 100\%$$

$$= \frac{700\ mW}{1.2\ W} \times 100\%$$

$$= 58\%$$

ELECTRICAL ENERGY

When 1 W of power is delivered to an electrical appliance for a period of 1 hour, the energy consumed is 1 watt-hour (Wh).

The watt-hour is too small for most practical purposes, so the *kilowatt-hour (kWh)* is normally used as the unit of electrical energy. The energy consumed, or work done is,

$$\boxed{W = Pt} \qquad \qquad (3\text{-}7)$$

EXAMPLE 3-11   For the heater and cables in Example 3-9, calculate the energy consumed in a 24 hour period (a) by the heater, (b) by the cables.

SOLUTION

(a) From Equation (3-7),

$$W_h = P_h t$$
$$= 2.2 \text{ kW} \times 24 \text{ h}$$
$$\mathbf{= 52.3 \text{ kWh}}$$

(b) $$W_c = P_c t$$
$$= 100 \text{ W} \times 24 \text{ h}$$
$$\mathbf{= 2.4 \text{ kWh}}$$

PRACTICE PROBLEMS

3-4.1   An electric heater has a terminal voltage of 115 V, and a current of 21.7 A. Calculate the power dissipated. Also determine the required current and voltage if the heater is to dissipate 1 kW.

3-4.2   An electric lamp supplied from a battery takes a current of 20 A, and its terminal voltage is 11 V. If the battery terminal voltage is 12.5 V, determine the power dissipated in the lamp, and the power wasted in the cables.

3-4.3   An electronic power supply has an output current of 500 mA, and an output voltage of 12 V. The input is 115 V and 78 mA. Calculate the efficiency of the power supply.

3-4.4   A house has three 100 W lamps, four 60 W lamps, and two 25 W lamps. If all of the lights are *on* for an average of three hours each day, determine the energy consumed in one week.

BASIC PROGRAM

Computer programs can tremendously simplify the solution of problems. Once a program is written, it can be used again many times. New component values and circuit quantities can be substituted into the program each time to almost instantaneously obtain results.

The following BASIC program was written to solve the type of problem presented in Example 3-9. The quantities from Example 3-9 were entered when the program was run.

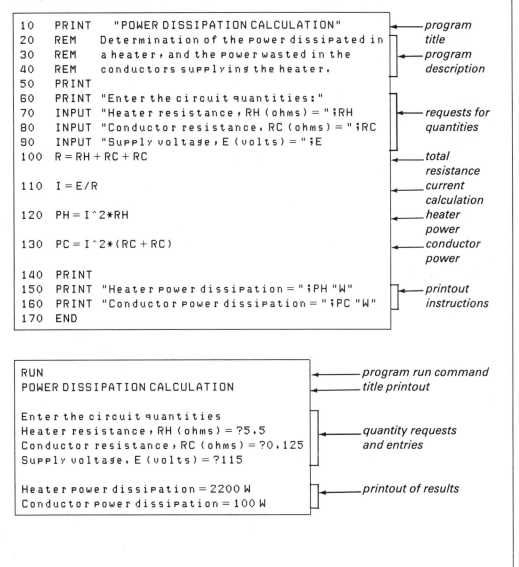

```
10   PRINT   "POWER DISSIPATION CALCULATION"          program
20   REM     Determination of the power dissipated in   title
30   REM     a heater, and the power wasted in the     program
40   REM     conductors supplying the heater.          description
50   PRINT
60   PRINT "Enter the circuit quantities:"
70   INPUT "Heater resistance, RH (ohms) = ";RH        requests for
80   INPUT "Conductor resistance, RC (ohms) = ";RC     quantities
90   INPUT "Supply voltage, E (volts) = ";E
100  R = RH + RC + RC                                   total
                                                        resistance
110  I = E/R                                            current
                                                        calculation
120  PH = I^2*RH                                        heater
                                                        power
130  PC = I^2*(RC + RC)                                 conductor
                                                        power
140  PRINT
150  PRINT "Heater power dissipation = ";PH "W"         printout
160  PRINT "Conductor power dissipation = ";PC "W"      instructions
170  END
```

```
RUN                                                    program run command
POWER DISSIPATION CALCULATION                          title printout

Enter the circuit quantities
Heater resistance, RH (ohms) = ?5.5                    quantity requests
Conductor resistance, RC (ohms) = ?0.125               and entries
Supply voltage, E (volts) = ?115

Heater power dissipation = 2200 W                       printout of results
Conductor power dissipation = 100 W
```

PASCAL PROGRAM

Pascal is a popular alternative computer program to BASIC. As in the case of a BASIC program, once a Pascal program is written, it can be used again many times, substituting new component values and circuit quantities into the program each time.

The following Pascal program was written to solve the type of problem presented in Example 3-9. The quantities from Example 3-9 were entered when the program was run.

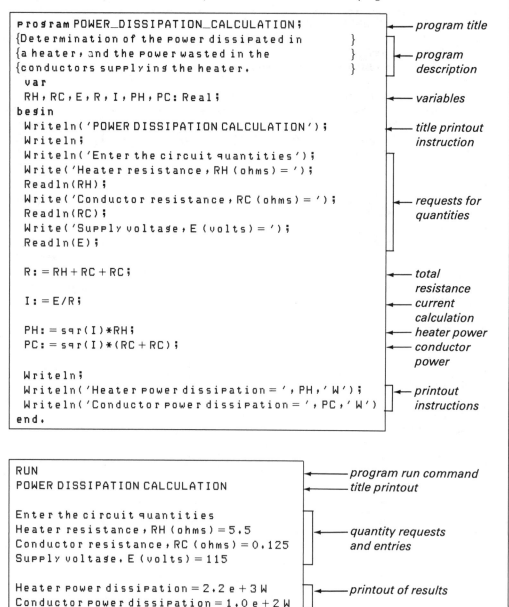

```
program POWER_DISSIPATION_CALCULATION;          ← program title
{Determination of the power dissipated in    }
{a heater, and the power wasted in the       }  ← program
{conductors supplying the heater.            }     description
  var
  RH,RC,E,R,I,PH,PC: Real;                        ← variables
begin
  Writeln('POWER DISSIPATION CALCULATION');       ← title printout
  Writeln;                                           instruction
  Writeln('Enter the circuit quantities');
  Write('Heater resistance, RH (ohms) = ');
  Readln(RH);
  Write('Conductor resistance, RC (ohms) = ');    ← requests for
  Readln(RC);                                        quantities
  Write('Supply voltage, E (volts) = ');
  Readln(E);

  R: = RH + RC + RC;                              ← total
                                                     resistance
  I: = E/R;                                       ← current
                                                     calculation
  PH: = sqr(I)*RH;                                ← heater power
  PC: = sqr(I)*(RC + RC);                         ← conductor
                                                     power
  Writeln;
  Writeln('Heater power dissipation = ',PH,' W');  ← printout
  Writeln('Conductor power dissipation = ',PC,' W')   instructions
end.
```

```
RUN                                             ← program run command
POWER DISSIPATION CALCULATION                   ← title printout

Enter the circuit quantities
Heater resistance, RH (ohms) = 5.5              ← quantity requests
Conductor resistance, RC (ohms) = 0.125            and entries
Supply voltage, E (volts) = 115

Heater power dissipation = 2.2 e + 3 W          ← printout of results
Conductor power dissipation = 1.0 e + 2 W
```

SUMMARY OF FORMULAS

Conductance:

$$conductance = \frac{1}{resistance} \qquad G = \frac{1}{R}$$

Ohm's Law:

$$current = \frac{emf}{resistance} \qquad I = \frac{E}{R}$$

Power:

$$power = emf \times amperes \qquad P = EI$$

Power:

$$power = \frac{(emf)^2}{resistance} \qquad P = \frac{E^2}{R}$$

Power:

$$power = (current)^2 \times resistance \qquad P = I^2R$$

Efficiency:

$$efficiency = \frac{power\ output}{power\ input} \times 100\% \qquad \eta = \frac{P_o}{P_i} \times 100\%$$

Energy consumed:

$$energy\ consumed = power \times time \qquad W = Pt$$

REVIEW QUESTIONS

3-1 Define *resistance*, and discuss the relative resistances of thin and thick, long and short, conductors.

3-2 Define *conductance* and the unit of conductance. State the relationship between resistance and conductance.

3-3 Define the unit of resistance. Explain the relationship between current, voltage, and resistance.

3-4 State *Ohm's law*, and write it in the forms for calculating each of the three quantities.

3-5 Define *power*, and discuss the effects of power dissipation in conductors and electronic devices.

3-6 Write three forms of the equation for electrical power.

3-7 Explain what is meant by *efficiency*, as applied to electrical equipment. Write the equation for efficiency.

3-8 Define electrical *energy*, and discuss units used in the measurement of energy.

PROBLEMS

SECTION 3-1

3-1 Convert the following resistances into conductances:
(a) 1 MΩ, (b) 0.1 Ω, (c) 1.5 kΩ, (d) 270 Ω, (e) 33 kΩ.

3-2 Express the following conductances as resistances:
(a) 100 S, (b) 0.1 S, (c) 5 μS, (d) 750 S, (e) 2 mS.

3-3 Determine the current that flows in each of the resistances in Problem 3-1 when a potential difference of 15 V is applied to the terminals of each resistance.

3-4 An emf of 100 V is applied across a resistance of 5 kΩ.
(a) Calculate the current that flows in the resistance.
(b) If the emf is changed to 110 V, determine the new value of resistance required to maintain the original current level.

3-5 Calculate the emf required to pass a current of 10 mA through a resistance of 3.3 kΩ.

3-6 A current of 3 A is to be passed through a 100 Ω resistor. Calculate the required voltage.

3-7 An electric lamp takes a current of 15 A when connected to a 12 V supply. Calculate the resistance of the lamp filament.

3-8 Calculate the conductance of a heating element that takes a current of 20 A when connected to a 120 V supply.

3-9 Determine the voltage required to pass a current of 100 mA through a resistance of 1 kΩ.

3-10 Determine the level of current that flows in a 2.2 kΩ resistor when it is connected across a 12 V battery.

3-11 Calculate the conductance of a circuit which takes 45 mA when its terminal voltage is 25 V. Also, determine the current that flows when the conductance is doubled.

SECTION 3-3

3-12 A circuit arranged as in Figure 3-3 has the following voltage and current levels: $V_1 = 12$ V, $V_2 = V_3 = 8$ V, $I_1 = 10$ mA, $I_2 = 3$ mA, $I_3 = 7$ mA. Calculate the resistor values.

3-13 The voltage measured at the terminals of an electrical appliance is 113 V, the supply voltage is 115 V, and the current is 18 A. Calculate the resistance of the supply cables.

3-14 A 9 V battery supplies 100 mA of current to an electronic circuit. The circuit terminal voltage is 8.85 V. Calculate the resistance of the conductors connecting the battery to the circuit.

SECTION 3-4

3-15 Determine the power dissipated in the lamp in Problem 3-7.

3-16 Calculate the power dissipated in the heater in Problem 3-8.

3-17 Calculate the power dissipated in the resistance in Problem 3-9.

3-18 Calculate the power dissipated in each resistance in Problem 3-12.

3-19 Calculate the current taken by a 100 W lamp from a 115 V supply. Also determine the resistance of the lamp filament.

3-20 For Problem 3-13, determine the power supplied to the appliance and the power dissipated in the cables.

3-21 A certain resistance dissipates 10 W when connected to a 50 V supply. Calculate the current that flows and the value of the resistance.

3-22 Determine the conductance of a lamp that dissipates 40 W when connected to a 120 V supply. Also, calculate the lamp current.

3-23 An electric lamp consumes 20 W when passing a current of 100 mA. Calculate the required supply voltage and the resistance of the lamp.

3-24 An electronic amplifier delivers an output of 10 W to a speaker. The amplifier supply voltage and current are 30 V and 470 mA. Determine the efficiency of the amplifier.

3-25 An electrical appliance dissipates 1250 W of power when connected to a 115 V supply. Determine the current taken by the appliance, and calculate the electric energy consumed over a period of 5 hours.

3-26 The appliance in Problem 3-13 is used 8 hours each day for 365 days. Calculate the energy consumed (a) by the appliance, (b) by the cables.

3-27 A 230 V electrical appliance absorbs 2 kW and has an efficiency of 75%. Determine the supply current, and the resistance offered by the appliance. Also, determine the energy employed usefully when the appliance is operated for 30 minutes.

3-28 A house with a 115 V supply has ten 100 W lamps, a 3.4 kW oven, a 300 W electric motor on the furnace, and a 200 W motor on the refrigerator. Calculate the supply current that flows when all the electrical equipment is in use at the same time. Also, determine the energy used daily if everything is switched *on* for an average of 4 hours each day.

COMPUTER PROBLEMS

3-29 Write a computer program to solve the type of problem presented in Example 3-3. Run the program using the quantities in the example.

3-30 Write a computer program to solve the kind of problem presented in Example 3-7. Use the quantities from the example when you run the program.

ANSWERS TO PRACTICE PROBLEMS

3-1.1 213 μS, 179 mS

3-1.2 0.002 Ω, 30.3 kΩ

3-2.1 4.8 mA

3-2.2   150 mA

3-3.1   5.64 mA, 17.55 V, 4.9 kΩ

3-3.2   2 mA, 1.2 mA, 800 μA

3-3.3   4.6 Ω, 0.2 Ω

3-4.1   2.5 kW, 13.7 A, 73 V

3-4.2   220 W, 30 W

3-4.3   66.9%

3-4.4   12.39 kWh

Chap. 3   Ohm's Law and Electrical Calculations

4

CONDUCTORS, INSULATORS, AND RESISTORS

Objectives You will be able to:

Describe the kinds of atomic bonding found in conductors, insulators, and semiconductors.

Explain the relationship between insulator breakdown, insulator thickness, applied potential difference, and electric field strength. Solve problems involving these quantities.

Discuss the relationship between conductor resistance, conductor length, cross-sectional area, and specific resistance. Solve problems involving these quantities.

Solve problems involving conductor resistance and temperature.

Describe the various types of electronics resistors, and use the color code to identify the resistance of each component.

Solve problems involving resistor power dissipation.

Solve problems involving resistor variation with temperature change.

Introduction

Conductors carry electric current. Insulators protect conductors, and protect people from conductors. Whether a material is a conductor or an insulator depends upon its atoms, and upon the relationship of each atom to its surrounding atoms.

Insulators may break down if subjected to excessive voltages. Similarly, conductors may be destroyed if too much current is passed through them. The best conductor materials have the lowest *resistivity*, that is, resistance per cubic meter. The best insulator materials have the highest breakdown voltage for a given thickness.

The resistance of any conductor can be calculated from a knowledge of its cross-sectional area, its length, and the resistivity of the material. Once the resistance is known, the voltage drop along a conductor and the power dissipated in it can be determined for any given current.

A device constructed to have a certain value of resistance is known as a *resistor*. Large quantities of resistors are used in electronic circuits for providing voltage drops, current limiting, and other functions. These are manufactured in standard resistance values and various ranges of accuracy and power dissipation capability. Small resistors use a *color code* to identify the resistance value and accuracy.

The *power rating* of a resistor is important because it limits the maximum voltage that should be applied and the maximum current that should flow through the component. How a resistor value varies with temperature is defined by its *temperature coefficient*.

4-1
ATOMIC BONDING

VALENCE SHELL. Consider the planetary atom again as described in Section 1-2, and recall that all atoms consist of a central nucleus surrounded by orbiting electrons. It has been found that electrons can occupy only certain orbital rings or *shells* at fixed distances from the nucleus, and that each shell can contain a particular number of electrons. The force that holds each electron in orbit around its nucleus depends upon the distance of the electron from the nucleus. Those electrons which are closest to the nucleus are securely held in orbit. Those that are in the outermost orbit are less tightly bound to the atom.

The atoms of two important electrical materials are illustrated by the two-dimensional diagrams in Figure 4-1. Since it is the electrons in the outer shell (or *valence shell*) that determine the electrical characteristics of an atom, only the outer shells need be examined. Referring to Figure 4-1(a), it is seen that the outer shell of the copper atom contains only one electron. This electron is so weakly attached to the nucleus that it easily escapes and drifts off through the spaces between the atoms that make up a piece of copper.

Now look at the silicon atom illustrated in Figure 4-1(b). Observe that the outer shell of the silicon atom contains four electrons but has the facility to contain eight electrons. Thus, the valence shell of silicon is said to have four electrons and four holes. A *hole* is defined simply as the absence of an electron in a shell where one could exist. The outer-shell electrons are more strongly attached to the silicon atom than in the case of copper. However, these electrons can be separated from the atom by the application of small amounts of energy. Thus, the silicon atom can either accept electrons into the holes in the valence shell, or give up electrons to drift about within the material.

METALLIC BONDING. Whether a material is a conductor, a semiconductor, or an insulator depends largely upon what happens to the outer-shell electrons when the atoms bond themselves together to form a solid. In the case of copper, the easily detached valence electrons are given up by the atoms. This creates a great mass of free electrons (or *electron gas*) drifting about through the spaces between the copper atoms. Since each atom has lost a (negative) electron, it becomes a *positive ion*. The electron gas is, of course, negatively charged; consequently, an electrostatic

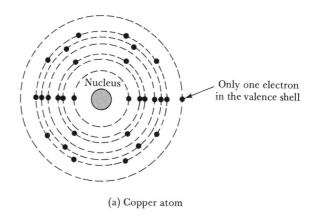

(a) Copper atom

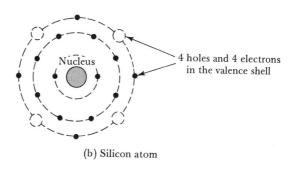

(b) Silicon atom

FIGURE 4-1. Two-dimensional diagrams of copper and silicon atoms. Copper has one electron in its outer, or valence, shell. Silicon has four valence shell electrons and four holes.

force of attraction exists between the positive ions and the electron gas. This is the *bonding force* that holds the material together. In the case of copper and other metals, the bonding force is termed *metallic bonding*, or sometimes *electron gas bonding*. This type of bonding is illustrated in Figure 4-2(a).

Since the free electrons in the electron gas can be given motion by the application of an electric field, current flow is easily achieved. Thus, it is obvious that the electron gas makes copper an excellent conductor of electricity. In general, all metals are good conductors, but some are better than others. Silver is the very best conductor (i.e., lowest resistance) followed closely by copper. Gold is also a relatively good conductor, as is aluminum.

COVALENT BONDING. In the case of silicon, which has four outer-shell electrons and four holes, the bonding arrangement is a little more com-

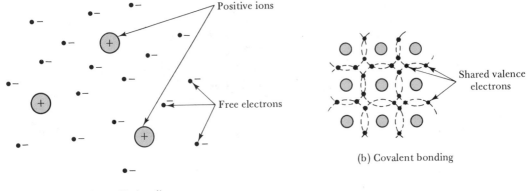

(a) Metallic bonding

(b) Covalent bonding

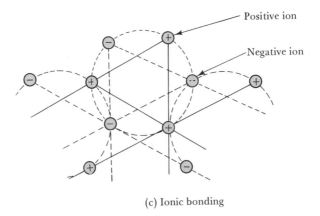

(c) Ionic bonding

FIGURE 4-2. Atomic bonding in conductors, semiconductors, and insulators. Metallic bonding is the force that holds the atoms together in conductors. In semiconductors, the bonding process is covalent. Ionic bonding occurs in some insulators, such as glass.

plicated than for copper. Atoms in a solid piece of silicon are so close to each other that the outer-shell electrons behave as if they were orbiting in the valence shells of two atoms. In this way each valence-shell electron fills one of the holes in the valence shell of a neighboring atom. This arrangement, illustrated in Figure 4-2(b), forms a bonding force known as *covalent bonding*. In covalent bonding every valence shell of every atom appears to be filled, and consequently there are no holes and no free electrons drifting about within the material.

Because there are no free electrons to permit conduction of an electric current, silicon should be a very poor conductor of electricity. At very low temperatures, silicon does, in fact, behave like an insulator. However, even at normal room temperature, some of the electrons have been given enough thermal energy to escape from the silicon atoms.

Consequently, small electric currents can be made to flow through the silicon, and so it is said to behave as a *semiconductor*.

IONIC BONDING. In some insulating materials, notably rubber and plastics, the bonding process is also covalent. The valence electrons in these bonds are very strongly attached to their atoms, so the possibility of current flow is virtually zero. In other types of insulating materials, some atoms have parted with outer-shell electrons, but these have been accepted into the orbit of other atoms. Thus, the atoms are *ionized*; those which gave up electrons have become *positive ions,* and those which accepted the electrons become *negative ions*. This creates an electrostatic bonding force between the atoms, termed *ionic bonding*. The situation is illustrated in Figure 4-2(c), which shows how the negative and positive ions might be arranged together in groups. Ionic bonding is found in such materials as glass and porcelain. Because there are virtually no free electrons, no current can flow, and the material is an insulator.

4-2
INSULATORS

Figure 4-3 shows some typical arrangements of conductors and *insulators*. Electric cable usually consists of conducting copper wire surrounded by an insulating sheath of rubber or plastic. Sometimes there is more than one conductor, and these are, of course, individually insulated.

Even in the best of insulating materials there are some free electrons drifting about between the atoms. Therefore, a very small electric current can flow through an insulator. In normal circumstances this current is so small that it is absolutely negligible. A current can also flow along the surface of an insulator, especially if the insulator is dirty or wet.

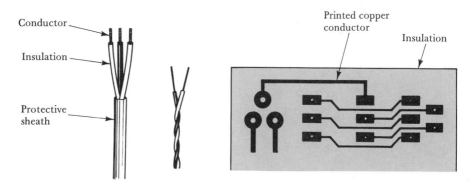

FIGURE 4-3. Conductors employed for industrial and domestic purposes normally have stranded copper wires with rubber or plastic insulation. In electronics equipment, flat cables of fine wires and thin printed circuit conductors are widely used.

This is particularly evident with the porcelain insulators on high-voltage overhead cables during a rain storm.

If a sufficiently high potential difference is applied across an insulator, electrons can be *pulled out* of the atoms, and substantial current flow may occur. This is known as *insulator breakdown,* and it normally results in the destruction of the insulator. To help avoid breakdown, all insulating materials (e.g., sheaths of electric cables, etc.) are rated according to the maximum voltage that may be safely applied. Table 4-1 lists typical *breakdown voltages* (or *dielectric strengths*) for various insulating materials. It is seen that air has the lowest breakdown voltage, at 30 kV/cm, while mica has the highest, at 2000 kV/cm. The figures given vary to some extent with the actual thickness of the insulating material.

The *electric field strength* is a measure of the electric stress set up in an insulator. The field strength should always be substantially lower than that which might cause insulator breakdown. The field strength is simply the applied voltage divided by the insulator thickness.

$$\boxed{\text{electric field strength} = \frac{E}{d}} \qquad \text{(4-1)}$$

When E is in volts and d is in meters, the units of field strength are volts/meter (V/m).

EXAMPLE 4-1   Two electric conductors encased in rubber are exactly 0.25 cm apart.
  a. Calculate the electric field strength in the insulation when the potential difference between the conductors is 25 V.
  b. What is the minimum potential difference between conductors that may cause insulator breakdown?

SOLUTION

a. From Equation (4-1),

$$\text{electric field strength} = \frac{E}{d} = \frac{25 \text{ V}}{0.25 \text{ cm}} = \textbf{100 V/cm}$$

b. From Table 4-1, insulation breakdown for rubber occurs at 270 kV/cm.

$$E = d \times (\text{breakdown voltage})$$
$$= 0.25 \text{ cm} \times 270 \text{ kV/cm}$$
$$= \textbf{67.5 kV}$$

TABLE 4-1 Typical Breakdown Voltages of Some Insulating Materials

MATERIAL	BREAKDOWN VOLTAGE (kV/cm)
Air	30
Porcelain	70
Rubber	270
Glass	1200
Mica	2000

PRACTICE PROBLEMS

4-2.1 Determine the electric field strength (in V/cm) in a printed circuit board 2 mm thick if 45 V is applied between conductors on opposite sides.

4-2.2 Calculate the voltage that may cause breakdown of the air between two terminals spaced 4 mm apart.

4-2.3 Calculate the electric field strength in 0.15 mm of mica which separates two conductors with a potential difference of 330 V. Is it likely that insulation breakdown may occur?

4-3
CONDUCTORS

The function of a *conductor* is to conduct current from one point to another in an electric circuit. As already discussed, electric cables usually consist of copper conductors sheathed with rubber or plastic insulating material. Cables that have to carry large currents must have relatively thick conductors. Where very small currents are involved, the conductor may be a thin strip of copper or even an aluminum film [see Figure 4-3]. Between these two extremes a wide range of conductors exist for various applications.

Because each conductor has a finite resistance, a current passing through it causes a voltage drop from one end of the conductor to the other. When conductors are long and/or carry large currents, the conductor voltage drop may cause unsatisfactory performance of the equipment supplied. Power (I^2R) is also dissipated in every current-carrying conductor, and this is, of course, wasted power. In extreme cases the power dissipated in the conductors may generate sufficient heat to destroy the insulation or even melt the conductor. Where the resistance per unit length of a conductor is known, the conductor voltage drop and power dissipation are easily calculated.

The table in Appendix 4 gives standard *American Wire Gauge* (*AWG*) sizes, together with the metric equivalents. Diameters are given in *mm* and in *mils* (inches × 10⁻³), and resistances (for copper) are given in Ω/km and in $\Omega/1000\ ft$.

EXAMPLE 4-2 Two conductors are to carry a current of 500 mA to an electronic circuit. The supply voltage is 30 V, and the voltage at the terminals of the circuit is to be not less than 29.5 V. The conductors are each to be 25 cm long. Select a suitable wire gauge from the table in Appendix 4.

SOLUTION

The total voltage drop along the conductors is:

$$E = 30\ V - 29.5\ V$$
$$= 0.5\ V$$

Total conductor (maximum) resistance:

$$R = \frac{E}{I} = \frac{0.5\ V}{500\ mA}$$
$$= 1\ \Omega$$

Total conductor length:

$$\ell = 2 \times 25\ cm = 50\ cm$$
$$Resistance\ per\ meter = \frac{1\ \Omega}{50\ cm} = 2\ \Omega/m$$
$$= 2000\ \Omega/km$$

Referring to Appendix 4:

for #38, $R = 2147\ \Omega/km$.

This is larger than the required 2 kΩ/km, so the voltage drop along it would be too large.

#37 *has* $R = 1715\ \Omega/km$

When I = 500 mA, the voltage along this conductor will be less than the 0.5 V maximum, so #37 (or lower) is suitable.

EXAMPLE 4-3   An electric motor is supplied with 1 A of current from a 115 V source (E_1), as shown in Figure 4-4. Each of the cables used is 40 m long, and the resistance of the cables is 0.025 Ω/m. Determine the potential difference (E_2) at the motor terminals. Also, calculate the power dissipated in the cables.

SOLUTION

For 40 m of cable:

$$R = 40 \text{ m} \times 0.025 \text{ } \Omega/\text{m}$$

$$= 1 \text{ } \Omega$$

Voltage drop along each cable:

$$V_1 = V_2 = I \times R$$

$$= 1 \text{ A} \times 1 \text{ } \Omega$$

$$= 1 \text{ V}$$

Motor terminal voltage:

$$E_2 = E_1 - V_1 - V_2$$

$$= 115 \text{ V} - 1 \text{ V} - 1 \text{ V}$$

$$\mathbf{= 113 \text{ V}}$$

Power dissipated in the cables:

$$P = (V_1 + V_2) \times I$$

$$= 2 \text{ V} \times 1 \text{ A}$$

$$\mathbf{= 2 \text{ W}}$$

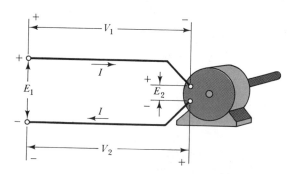

FIGURE 4-4. Voltage drops occur along current-carrying conductors, depending upon the conductor resistance and the current level.

4-3.1 A 2 kW electric heater is supplied from a 115 V source using 15 m of cable with two #14 copper wire conductors. Calculate the terminal voltage at the heater.

4-3.2 If the electric motor in Example 4-3 can operate satisfactorily with a terminal voltage of 110 V, select a suitable wire gauge for the supply cables.

4-3.3 Determine the power dissipated in the cables in: (a) Problem 4-3.1, (b) Problem 4-3.2.

4-4
CONDUCTOR RESISTIVITY

The resistance per unit length of a given conductor is not always available. Therefore, to allow for all possible combinations of cable length and cross-sectional area, the *specific resistance* of the conducting material is employed.

*The **specific resistance** (or **resistivity**) of a material is the resistance of 1 meter cube of the material.*

Obviously, since some metals are better conductors than others, each has its own specific resistance. Table 4-2 lists the specific resistances of various metals.

*The symbol used for **specific resistance** is ρ (Greek lowercase letter rho), and the units of ρ are ohm meters ($\Omega \cdot$ **m**).*

The origin of the units will become apparent shortly.

Consider the meter cube of copper illustrated in Figure 4-5(a). From Table 4-2, the specific resistance of copper is $\rho = 1.72 \times 10^{-8}\ \Omega \cdot$ m.

If the length of copper is doubled or tripled by placing blocks side-

TABLE 4-2 Specific Resistance for Electrical Conducting Metals

METAL	SPECIFIC RESISTANCE AT 20°C ($\Omega \cdot$ m)
Silver	1.64×10^{-8}
Copper (annealed)	1.72×10^{-8}
Gold	2.45×10^{-8}
Aluminum	2.83×10^{-8}
Tungsten	5.5×10^{-8}
Nickel	7.8×10^{-8}
Constantan	49×10^{-8}

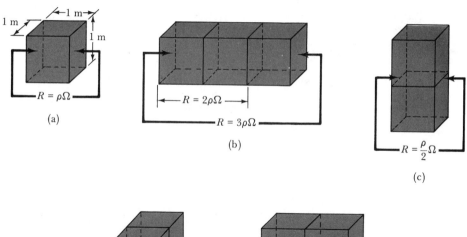

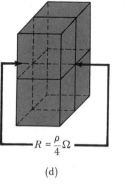

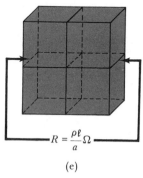

FIGURE 4-5. If a one meter cube of copper has a resistance of ρ Ω, three cubes in series will have a resistance of 3ρ Ω, and the resistance of two in parallel will be ρ/2 Ω. This gives: Resistance = (ρ × length)/(cross-sectional area).

by-side as shown in Figure 4-5(b), the resistance is increased. (Blocks, or other components, connected together in this way are said to be *connected in series*. Series connected circuits are discussed in Chapter 5.) The resistance of two series-connected blocks is 2ρ Ω, and that of three is 3ρ Ω, and so on. From this it is derived that the total resistance of any length of copper with a cross-sectional area of 1 m² is,

$$R = \rho \times (\text{length in meters})$$

Now consider the effect of doubling the cross-sectional area of the copper by placing two blocks as illustrated in Figure 4-5(c). (Blocks, or other components, connected together in this way are said to be *connected in parallel*. Parallel connected circuits are discussed in Chapter 6.) Because the cross-sectional area is doubled, the resistance is halved. If four blocks are parallel-connected [Figure 4-5(d)], the resistance is quartered. In the case of blocks connected in parallel, the resistance can be written,

$$R = \frac{\rho}{\text{cross-sectional area in m}^2}$$

When blocks are put in series and parallel [Figure 4-5(e)], the two equations above are combined to give

$$\boxed{R = \frac{\rho\ell}{a}} \qquad\qquad \textbf{(4-2)}$$

where *R* is the resistance in ohms, ℓ is the length in meters, and *a* is the cross-sectional area in m^2.

The resistance of any length of conductor of any cross-sectional area can be calculated from Equation (4-2). Where *R*, ℓ, and *a* are known, the equation can be rewritten to determine the specific resistance:

$$\rho = \frac{Ra}{\ell}$$

The units of ρ are

$$\rho = \frac{\Omega \times m^2}{m} = \Omega \cdot m = \text{ohm meters}$$

EXAMPLE 4-4   Determine the resistance of 20 m of annealed copper wire 2 mm in diameter.

SOLUTION

$$\text{cross-sectional area, } a = \frac{\pi D^2}{4}$$

Therefore,
$$a = \frac{\pi \times (2 \times 10^{-3})^2}{4}$$

$$= \pi \times 10^{-6} \ m^2$$

From Equation (4-2):
$$R = \frac{\rho\ell}{a}$$

From Table 4-2, for copper
$$\rho = 1.72 \times 10^{-8}$$

Thus,
$$R = \frac{1.72 \times 10^{-8} \times 20 \ m}{\pi \times 10^{-6} \ m^2}$$

$$= \textbf{0.109 } \boldsymbol{\Omega}$$

EXAMPLE 4-5 Calculate the resistance per meter length of aluminum wire 5 mm in diameter.

SOLUTION

$$cross\text{-}sectional\ area,\ a = \frac{\pi D^2}{4}$$

$$= \frac{\pi \times (5 \times 10^{-3})^2}{4}$$

$$= 6.25\pi \times 10^{-6}\ m^2$$

From Table 4-2, for aluminum $\rho = 2.83 \times 10^{-8}$

$$R = \frac{\rho\ell}{a}$$

$$= \frac{2.83 \times 10^{-8} \times 1\ m}{6.25\pi \times 10^{-6}\ m^2}$$

$$\mathbf{= 1.44 \times 10^{-3}\ \Omega}$$

EXAMPLE 4-6 An electric heater takes a current of 15 A from a 115 V source. The cables connecting the heater to the supply are each 43 m long. If the total voltage drop along the cables is not to exceed 12 V, determine the diameter of suitable copper wire, and select a suitable wire size from Appendix 4.

SOLUTION

The maximum allowable resistance of the cable is

$$R = \frac{E}{I} = \frac{\text{voltage drop along cable}}{I}$$

$$= \frac{12\ V}{15\ A} = 0.8\ \Omega$$

$$total\ length\ of\ wire = 2 \times 43\ m$$

$$= 86\ m$$

From Equation (4-2),

$$a = \frac{\rho\ell}{R} = \frac{1.72 \times 10^{-8} \times 86\ m}{0.8\ \Omega}$$

$$= 1.849 \times 10^{-6}\ m^2$$

and

$$a = \frac{\pi}{4} D^2$$

or,

$$D = \sqrt{\frac{4a}{\pi}} = \sqrt{\frac{4 \times 1.849 \times 10^{-6}}{\pi}}$$

$$= 1.53 \text{ mm}$$

The minimum wire diameter must be 1.53 mm. *From Appendix 4, #14 has* $D = 1.63$ mm, *so* **#14 *is suitable.***

PRACTICE PROBLEMS

4-4.1 Using the specific resistance of copper and the wire diameter, calculate the resistance per kilometer of #14 copper wire.

4-4.2 Determine the resistance of 1.25 m of nickel wire 1 mm in diameter.

4-4.3 A printed circuit board has copper 0.25 mm thick. When all unwanted copper is removed, the conducting strips are to carry 100 mA of current with a maximum voltage drop of 5 mV per meter. Determine the minimum width for the strips.

4-5
TEMPERATURE EFFECTS ON CONDUCTORS

The values of specific resistance stated in Table 4-2 refer only to conducting materials at a temperature of 20°C. Thus, in all resistance calculations using the specific resistance, the conductor temperature is assumed to remain constant at 20°C. Since the resistance of metals changes with temperature, it is necessary to be able to calculate the resistance values at higher or lower temperature levels.

The resistance of all pure metals tends to increase as the temperature of the metal rises. This may be explained in terms of the atoms actually vibrating at elevated temperatures, and thus becoming greater obstructions in the path of moving electrons. Because the resistance increases with increasing temperatures, metals are said to have a *positive temperature coefficient* (PTC). Some materials, notably semiconductors, exhibit a decrease in resistance as their temperature rises. These have a *negative temperature coefficient* (NTC). Over the normal range of operating temperatures, all metals exhibit a nearly linear relationship between resistance and temperature.

In Figure 4-6 the variation in resistance of copper is plotted versus temperature. Note that the resistance of copper appears to go to zero at a temperature of −234.5°C. The actual graph is not completely linear,

Chap. 4   Conductors, Insulators, and Resistors

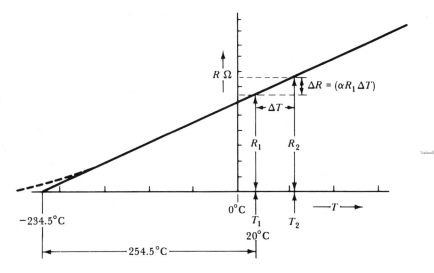

FIGURE 4-6. Resistance of copper plotted versus temperature. Because the resistance appears to go to zero at $-234.5°C$, the resistance change from R_1 at $20°C$ is $R_1/(234.5 + 20)$ $\Omega/°C$.

as shown by the broken line. The straight line passing through $-234.5°C$ is a projection of the linear portion of the graph. If the resistance R_1 at a temperature of $20°C$ is known, the new value of resistance at any other temperature can be calculated.

Let $R_1 = 1$ Ω. If the temperature is reduced to $-234.5°C$, the temperature change is

$$\Delta T = 20°C - (-234.5°C)$$

$$= 254.5°C$$

and the resistance change per degree of temperature change is

$$\frac{\Delta R_1}{\Delta T} = \frac{1\ \Omega}{254.5°C}$$

$$= 0.003\ 93\ \Omega/°C$$

If R_1 were 10 Ω at $20°C$, the resistance change with temperature would be

$$\frac{\Delta R_1}{\Delta T} = \frac{10\ \Omega}{254.5°C}$$

$$= 10\ \Omega \times 0.003\ 93\ \Omega/°C$$

$$= 0.0393\ \Omega/°C$$

Therefore, for any value of R_1,

$$\frac{\Delta R_1}{\Delta T} = R_1 \times 0.003\ 93\ \Omega/°C$$

The quantity $0.003\ 93\ \Omega/°C$ is the *temperature coefficient for copper at 20°C*. The symbol α (Greek lowercase letter *alpha*) is used for temperature coefficient. Table 4-3 lists the temperature coefficients for various metals at 20°C. Note that the alloy *constantan* has an extremely small temperature coefficient. This characteristic makes constantan useful in applications where resistance is required to remain as constant as possible when the temperature changes. Temperature coefficients can also be determined at temperatures other than 20°C. Column 3 in Table 4-3 lists temperature coefficients at 0°C.

As explained above, for any conducting material having a resistance R_1,

$$\frac{\Delta R_1}{\Delta T} = \alpha R_1$$

or $$\Delta R_1 = \alpha R_1 \Delta T \qquad \text{(see Figure 4-6)}$$

and resistance R_2 at a new temperature T_2 is

$$R_2 = R_1 + \Delta R_1$$
$$= R_1 + \alpha R_1 \Delta T \qquad \text{(see Figure 4-6)}$$

Therefore

$$\boxed{R_2 = R_1 (1 + \alpha \Delta T)} \qquad (4\text{-}3)$$

where R_1 is the resistance at 20°C, α is the temperature coefficient of the material, and ΔT is the temperature change from 20°C.

If R_1 and R_2 are known, Equation (4-3) can be rearranged to determine ΔT or α:

$$\Delta T = \frac{1}{\alpha} \left(\frac{R_2}{R_1} - 1 \right)$$

TABLE 4-3 Temperature Coefficients for Metals

MATERIAL	α AT 20°C	α AT 0°C
Silver	0.003 8	0.004 12
Copper	0.003 93	0.004 26
Gold	0.003 4	0.003 65
Aluminum	0.003 9	0.004 24
Tungsten	0.004 5	0.004 95
Nickel	0.006	
Constantan	0.000 008	

$$\alpha = \frac{1}{\Delta T}\left(\frac{R_2}{R_1} - 1\right)$$

The temperature coefficient at 0°C may also be used in Equation (4-3) if the resistance R_1 is measured at 0°C and ΔT is the temperature increase from 0°C.

EXAMPLE 4-7 The resistance of a length of copper cable is 5.8 Ω at 20°C. Determine the new resistance of the cable at 125°C.

SOLUTION

From Equation (4-3):

$$R_2 = R_1(1 + \alpha\Delta T)$$

$$\Delta T = T_2 - T_1$$

$$= 125°C - 20°C$$

$$= 105°C$$

For copper at 20°C, $\alpha = 0.003\,93$ (*see* Table 4-3). *Therefore,*

$$R_2 = 5.8\ \Omega[1 + (0.003\,93 \times 105°C)]$$

$$= \mathbf{8.2\ \Omega}$$

EXAMPLE 4-8 The resistance of a coil of nickel wire is 1 kΩ at 20°C. After being submerged in a liquid for some time, the resistance falls to 880 Ω. Calculate the temperature of the liquid.

SOLUTION

From Equation (4-3):

$$\Delta T = \frac{1}{\alpha}\left(\frac{R_2}{R_1} - 1\right)$$

For nickel at 20°C, $\alpha = 0.006$ (*see* Table 4-3). *Therefore,*

$$\Delta T = \frac{1}{0.006}\left(\frac{880\ \Omega}{1\ k\Omega} - 1\right)$$

$$= -20°C$$

and $\qquad \Delta T = T_2 - T_1$

Therefore, $\qquad T_2 = \Delta T + T_1$

$$= -20°C + 20°C$$

$$= \mathbf{0°C}$$

EXAMPLE 4-9   A certain length of wire has its resistance measured as 330 Ω at 20°C and 448.8 Ω at 100°C. Calculate the temperature coefficient and identify the material.

SOLUTION

From Equation (4-3):

$$\alpha = \frac{1}{\Delta T}\left(\frac{R_2}{R_1} - 1\right)$$

$$= \frac{1}{(100-20)°C}\left(\frac{448.8\ \Omega}{330\ \Omega} - 1\right)$$

$$= \mathbf{0.0045}$$

From Table 4-3, $\alpha = 0.0045$ *identifies the material as* **tungsten.**

PRACTICE PROBLEMS

4-5.1   A tungsten filament lamp has a resistance of 11.8 Ω at 20°C. Determine its resistance at 2500°C.

4-5.2   A resistor is constructed of nickel wire to have a resistance of 1.25 kΩ at 20°C. What will its resistance be at 55°C?

4-5.3   A fine wire has a resistance of 0.11 Ω at 20°C, and 0.116 Ω at 36°C. Determine the temperature coefficient and identify the material.

4-6
RESISTOR CONSTRUCTION

FIXED VALUE RESISTORS.   The function of a resistor is to offer a particular resistance to current flow. For a given current and known resistance, the voltage drop across the resistance can be predicted using Ohm's law. Similarly, for a given applied voltage, the current that flows may be predetermined by selection of the resistor value. The required power dissipation largely dictates the construction and physical size of a resistor.

Two common types of electronics resistors are *wire-wound* and *carbon composition* construction. A typical wire-wound resistor consists of a length of nickel wire wound on a ceramic tube and covered with porcelain. Low-resistance connecting wires are provided, and the resistance value is usually printed on the side of the component. Figure 4-7(a) illustrates the construction of a typical wire-wound resistor. Carbon composition resistors are constructed by molding mixtures of powdered carbon and insulating materials into a cylindrical shape [see Figure 4-7(b)]. An outer sheath of insulating material affords mechanical

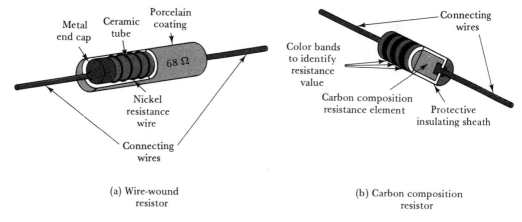

FIGURE 4-7. Individual resistors are typically wire-wound or carbon composition construction. Wire-wound resistors are used where high power dissipation is required. Carbon composition type are the least expensive.

and electrical protection, and copper connecting wires are provided at each end. Carbon composition resistors are smaller and less expensive than the wire-wound type. However, the wire-wound type is the more rugged of the two, and is able to survive much larger power dissipations than the carbon composition type.

VARIABLE RESISTORS. Most resistors have standard fixed values, so they can be termed *fixed resistors. Variable resistors,* or *adjustable resistors,* are also used a great deal in electronics. The construction of a typical variable resistor is illustrated in Figure 4-8(a), together with two frequently employed graphic symbols for a variable resistor in Figure 4-8(b).

The illustration in Figure 4-8(a) shows a coil of closely wound insulated resistance wire formed into a partial circle. The coil has a low-resistance terminal at each end, and a third terminal is connected to a movable contact with a shaft adjustment facility. The movable contact can be set to any point on a connecting track which extends over one (uninsulated) edge of the coil. Using the adjustable contact, the resistance from either end terminal to the center terminal may be adjusted from zero to the maximum coil resistance.

Another type of variable resistor, known as a *decade resistance box,* is shown in Figure 4-8(c). This is a laboratory component which contains precise values of switched series-connected resistors. As illustrated, the first switch (from the right) controls resistance values in 1 Ω steps from 0 Ω to 9 Ω, while the second switches values of 10 Ω, 20 Ω, 30 Ω, and so on. The decade box shown can be set within ± 1 Ω of any value from 0 Ω to 9 999 Ω. Other decade boxes are available with different resistance ranges.

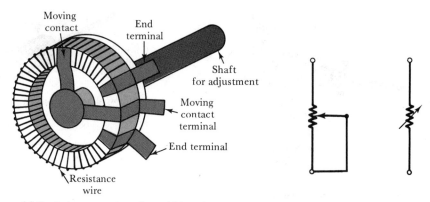

(a) Typical construction of a variable resistor (and potentiometer)

(b) Circuit symbols for a variable resistor

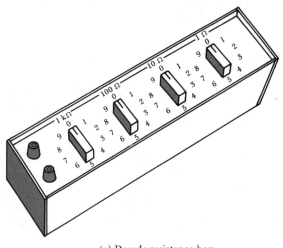

(c) Decade resistance box

FIGURE 4-8. Small variable resistors are used in electronic circuit construction. Large decade resistance boxes are employed in electronics laboratories.

4-7

COLOR CODE

Standard (fixed-value) resistors normally range from 2.7 Ω to 22 MΩ. (See Appendix 6.) The resistance *tolerances* on these standard values are typically ±20%, ±10%, ±5%, or ±1%. A tolerance of ±10% on a 100 Ω resistor means that the actual resistance may be as high as 100 Ω + 10% (i.e., 110 Ω), or as low as 100 Ω − 10% (i.e., 90 Ω). Obviously, the resistors with the smallest tolerance are the most accurate and the most expensive.

Because carbon composition resistors are physically small (some are less than 1 cm in length), it is not convenient to print the resistance

14 Pin dual-in-
line package

RESISTOR NETWORKS

Resistor networks are available in *integrated circuit* type *dual-in-line* packages. One construction method uses a *thick film* technique in which conducting solutions are deposited in the required form.

Internal
resistor
arrangement

PHOTOCONDUCTIVE CELL

This is simply a resistor constructed of photoconductive material (cadmium selenide or cadmium sulfide). When dark, the cell resistance is very high. When illuminated, the resistance decreases in proportion to the level of illumination.

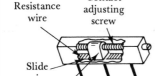

Resistance
wire

Contact
adjusting
screw

Slide
wire

Sliding
contact

Wire terminals

TRIMMING POTENTIOMETER

A small potentiometer suitable for mounting directly on a circuit board. A threaded shaft, which is adjustable by a screwdriver, sets the position of the moving contact on a resistance wire.

HIGH POWER RESISTORS

High power resistors are usually wire-wound on the surface of a ceramic tube. Air flow through the tube helps to keep the resistor from overheating.

value on the side. Instead, a *color code* in the form of colored bands is employed to identify the resistance value and tolerance. The color code (which uses the colors of the visible spectrum) is illustrated in Figure 4-9. Starting from one end of the resistor, the first two bands identify the first and second digits of the resistance value, and the third band indicates the number of zeros. An exception to this is when the third band is either silver or gold, which indicates a 0.01 or 0.1 multiplier, respectively. The fourth band is always either silver or gold, and in this position silver indicates a ±10% tolerance and gold indicates ±5% tolerance. Where no fourth band is present, the resistor tolerance is ±20%.

EXAMPLE 4-10 Identify the values of resistors with the following color codes: brown–black–red–silver, red–red–black–gold, and red–violet–gold–silver.

SOLUTION

$$\left.\begin{array}{cccc}\text{brown--} & \text{black--} & \text{red--} & \text{silver} \\ \downarrow & \downarrow & \downarrow & \downarrow \\ 1 & 0 & 00 & \pm 10\% \end{array}\right\} = 1000\ \Omega \pm 10\%$$

$$\left.\begin{array}{cccc}\text{red--} & \text{red--} & \text{black} & \text{-- gold} \\ \downarrow & \downarrow & \underline{\downarrow} & \downarrow \\ 2 & 2 & \text{no zeros} & \pm 5\% \end{array}\right\} = 22\ \Omega \pm 5\%$$

$$\left.\begin{array}{cccc}\text{red--} & \text{violet--} & \text{gold} & \text{-- silver} \\ \downarrow & \downarrow & \underline{\downarrow} & \downarrow \\ 2 & 7 & \times 0.1 & \pm 10\% \end{array}\right\} \begin{array}{l} = (27\ \Omega \times 0.1) \pm 10\% \\ = 2.7\ \Omega \pm 10\% \end{array}$$

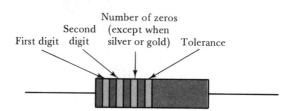

Number of zeros
Second   (except when
First digit   digit   silver or gold)   Tolerance

First Three Bands				Fourth Band	
Black	– 0	Blue	– 6	Gold	± 5%
Brown	– 1	Violet	– 7	Silver	± 10%
Red	– 2	Grey	– 8	none	± 20%
Orange	– 3	White	– 9		
Yellow	– 4	Silver	0.01		
Green	– 5	Gold	0.1		

FIGURE 4-9. The resistance value and accuracy of small resistors are identified by means of coded colored bands. The colors are numbered according to the visible spectrum. A memory aid for the color code is given in Appendix 5.

4-8

RESISTOR POWER RATINGS

Typical power ratings for wire-wound resistors start at 1 W and range to 10 W and much larger. Large surface areas are required to dissipate the heat generated; consequently, high power ratings result in physically large resistors. Carbon composition resistors are essentially low-power components; excessive power dissipation destroys them rapidly. The usual range of power ratings for carbon composition resistors is $\frac{1}{8}$ W, $\frac{1}{4}$ W, $\frac{1}{2}$ W, 1 W, and 2 W.

The maximum current that may be permitted to flow through a resistor, and the maximum voltage that may be applied across it, are limited by the specified maximum power dissipation. Using Equation (3-4), $P = E^2/R$, and Equation (3-5), $P = I^2R$, the maximum levels of current and voltage are easily calculated.

EXAMPLE 4-11 A 2.2 kΩ resistor has a specific maximum power dissipation of 1 W. Determine the maximum current that may be passed through the resistor and the maximum voltage that may be applied to its terminals.

SOLUTION

Equation (3-5):

$$P = I^2R$$

Therefore,

$$I = \sqrt{\frac{P}{R}}$$

and

$$I_{max} = \sqrt{\frac{1 \text{ W}}{2.2 \text{ k}\Omega}}$$

$$I = 21.3 \text{ mA}$$

Equation (3-4):

$$P = \frac{E^2}{R}$$

Therefore,

$$E = \sqrt{PR}$$

and

$$E_{max} = \sqrt{1 \text{ W} \times 2.2 \text{ k}\Omega}$$

$$E = 46.9 \text{ V}$$

PRACTICE PROBLEMS

4-8.1 Determine the maximum current that may be passed by 100 Ω and 100 kΩ, 1/8 W resistors.

4-8.2 A 6.8 Ω resistor is used in a circuit in which it may have a maximum of 3.3 V across its terminals. Determine the minimum power rating for the resistor.

4-9
TEMPERATURE COEFFICIENT OF RESISTORS

Wire-wound resistors behave much the same as conductors when their temperature increases or decreases (see Section 4-5). Thus, they have the kind of resistance/temperature characteristic illustrated in Figure 4-6. Carbon composition resistors, on the other hand, usually have the characteristic illustrated in Figure 4-10. In this case the resistance tends to increase when the temperature varies *above* or *below* room temperature.

The *temperature coefficient* (T.C.) of resistance is usually expressed in *parts per million per degree Celsius* (*ppm/°C*). To understand what this means, consider a resistor with a resistance of R Ω and a T.C. of ±100 ppm/°C. For every 1°C change in temperature, the maximum resistance change (ΔR) is

$$\Delta R = \pm \frac{R}{1\,000\,000} \times 100$$

i.e., *a change of 100 parts in one million parts of R.*

When the temperature change is ΔT°C, the resistance change is

$$\boxed{\Delta R = \pm \frac{R}{1\,000\,000} \times T.C. \times \Delta T} \qquad (4\text{-}4)$$

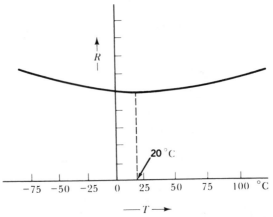

FIGURE 4-10. Typical variation of resistance with temperature change for carbon composition resistors. The temperature coefficient of resistors is usually expressed in *parts per million per degree Celsius* (*ppm/°C*).

Chap. 4   Conductors, Insulators, and Resistors

EXAMPLE 4-12   Calculate the maximum change in resistance for a 470 Ω resistor with a T.C. of ±500 ppm/°C, when its temperature increases from 20°C to 75°C.

SOLUTION

$$\Delta T = 75°C - 20°C$$

$$= 55°C$$

Using Equation (4-4):

$$\Delta R = \pm \frac{470 \ \Omega}{1\,000\,000} \times 500 \times 55°C$$

$$\cong \pm 12.9 \ \Omega$$

The resistance change calculated from the temperature coefficient is the maximum change in resistance that should occur over a given temperature range. With some resistors, the actual resistance change may be only a small fraction of the maximum, while other resistors of the same type may exhibit the maximum possible change. There are circuit applications in which it is important that two (or more) resistors change by approximately the same proportion as temperature increases or decreases. In this case it is said that the resistors are required to *track* each other. When resistors are constructed to fulfill this requirement, the manufacturer specifies the *matching* of the temperature coefficients. Sometimes the term *tracking temperature coefficient* is used. Thus, the T.C. for a certain type of resistor may be specified as:

temperature coefficient = ±20 ppm/°C matched to ±2 ppm/°C

or

tracking temperature coefficient = ±2 ppm/°C

PRACTICE PROBLEMS

4-9.1   Two 2.2 kΩ resistors each have a temperature coefficient of ±200 ppm/°C. Determine the maximum resistance difference that may develop between the two due to a 35°C temperature change.

4-9.2   If the resistors in Problem 4-9.1 have a tracking temperature coefficient of ±10 ppm/°C, determine the maximum resistance difference that may develop due to the 35°C temperature change.

SUMMARY OF FORMULAS

Insulator:

$$\textit{Electric field strength} = \frac{\text{applied voltage}}{\text{insulation thickness}} = \frac{E}{d}$$

Resistance:

$$\textit{resistance} = \frac{\text{resistivity} \times \text{length}}{\text{cross-sectional area}}$$

$$R = \frac{\rho\ell}{a}$$

Temperature effect on resistance:

$$R_2 = R_1(1 + \alpha\Delta T)$$

Resistance change:

$$\Delta R = \pm \frac{R}{1\,000\,000} \times \text{T.C.} \times \Delta T$$

REVIEW QUESTIONS

4-1 Sketch the plane diagram of a copper atom and of a silicon atom. Identify the valence shell and the electrons and holes.

4-2 Using illustrations, explain the various kinds of atomic bonding, and identify the kinds of bonding found in conductors, insulators, and semiconductors.

4-3 Explain the function of various types of insulators and discuss insulator breakdown.

4-4 Explain the function of a conductor and discuss the reasons for the various sizes of available conductors.

4-5 Define *specific resistance,* and write the equation relating the resistance of a conductor to its length and cross-sectional area. Identify the units for each quantity in the equation.

4-6 Using illustrations, explain how temperature affects the resistance of a metallic conductor. Derive an equation relating conductor resistance at an elevated temperature to resistance at 20°C.

4-7 Describe the two commonest types of electronic resistor construction and discuss their relative advantages and disadvantages.

4-8 Sketch the construction of a variable resistor, identify each part of the device and explain its operation.

4-9 Explain what is meant by T.C. = 100 ppm/°C. Sketch the approximate shape of the resistance/temperature characteristic of a carbon composition resistor. Define tracking temperature coefficient.

4-10 Draw a sketch to show how resistors are color coded. List the colors and identify the significance of each.

PROBLEMS

4-1   Two conductors 0.5 cm apart in an electric cable have a potential difference of 500 V. Calculate the electric field strength in the insulation between the conductors.

4-2   Determine the voltage that may cause breakdown in 4 mm of rubber insulation that separates two conductors.

4-3   The electric field strength in the insulation between the outer metal plate and the heating element in an electric smoothing iron is not to exceed 1000 kV/cm. Calculate the minimum insulation thickness if a 115 V supply is used.

4-4   Two conductors in a single cable have 0.9 cm of rubber insulation between them and a potential difference of 450 V. Calculate the electric field strength in the insulation, and determine the voltage that may cause insulation breakdown.

4-5   Two terminals are placed 1.5 cm apart in air. Referring to Table 4-1, calculate the potential difference between them that may cause breakdown.

4-6   Copper bus-bars that have a potential of 120 V with respect to ground are supported by porcelain insulators 2 cm thick. Determine the electric field strength in the porcelain, and calculate the voltage that may cause breakdown (a) in the insulators, (b) in the surrounding air.

4-7   Two conductors are separated by 0.2 cm of rubber insulation. Determine the minimum potential difference between them that could cause insulation breakdown.

4-8   A porcelain insulator supports overhead electric cables 15 cm underneath horizontal bars on a grounded metal pylon. If the surrounding air sometimes breaks down, allowing current to leak along the surface of the insulator, estimate the cable voltage with respect to ground.

4-9   The glass insulation between a heating element and a metal plate broke down when a 500 V potential difference was applied. Calculate the maximum thickness of the glass.

SECTION 4-3

4-10   An electronic circuit which takes a current of 470 mA is connected to a power supply via two conductors which are each 10 m long. If the total voltage drop along the conductors is not to exceed 100 mV, determine the gauge of suitable copper wire conductors.

4-11   An electric heater is supplied with 5 A of current from a 120 V source. The cables used to connect the heater are each 50 m long and have a resistance of 0.01 Ω/m. Calculate the potential difference at the heater terminals and the power dissipated in the cables.

4-12 The cables in Problem 4-11 are replaced with #14 copper wire. Calculate the new terminal voltage at the heater, and the power dissipated in the cables.

4-13 An automobile starter motor is connected to a 12 V battery via the chassis and a 1 m long insulated copper conductor. When *on*, the motor takes a current of 80 A, and its terminal voltage is to be not less than 11.5 V. Assuming that the chassis resistance is negligible, select a suitable wire gauge for the copper conductor.

4-14 An electric lamp has a terminal voltage of 113.6 V and a current of 1.13 A when connected to a 115 V supply via a 75 m twin conductor cable. Determine the wire gauge of the conductor.

SECTION 4-4

4-15 Calculate the resistance per meter of copper conductors with the following diameters: 1 mm, 2 mm, 3 mm, 4 mm, and 5 mm.

4-16 The resistance of a 65 cm length of wire 1 mm in diameter is measured as 0.0136 Ω. Determine the material of the wire.

4-17 Determine the resistance of copper and aluminum conductors 6 mm in diameter and 100 m long.

4-18 A strip of copper on a printed circuit board is 10 cm long and 0.3 cm across. If the resistance of the copper is measured as 0.0035 Ω, determine its thickness.

4-19 A piece of electronic equipment takes 2.5 A from a 120 V supply. The cables connecting the supply to the equipment are 10 m long. If the terminal voltage at the equipment is to be not less than 115 V, determine the smallest diameter of suitable copper conductors.

4-20 Copper wire with a #13 gauge is to be replaced by aluminum wire having an equal or lower resistance. Select a suitable gauge for the aluminum wire.

4-21 The copper conductors on a printed circuit board are 0.05 mm thick and 2 mm across. Determine the volts drop per centimeter along the conductors when a current of 300 mA flows.

4-22 A 0.001 Ω ammeter shunt is to be constructed of copper which is 1 cm wide and 2 mm thick. Calculate the required length of copper.

4-23 A 750 W electric motor is to be connected to a 130 V supply via two cables which are each 64 m in length. If the voltage at the motor terminals is to be not less than 125 V, determine the minimum thickness of the copper conductors in the cables.

4-24 A 330 Ω resistor is to be constructed of #40 gauge constantan wire. Determine the required length of wire.

SECTION 4-5

4-25 An aluminum conductor has a resistance of 10 Ω at 20°C. Determine the resistance of the conductor at 100°C.

4-26 The printed circuit board conductors described in Problem 4-21 are raised to a temperature of 65°C from the normal temperature level of 20°C. Determine the new value of voltage drop per centimeter when the current is 300 mA.

4-27 A coil of nickel wire is used to monitor the temperature of a liquid in which it is submerged. The coil resistance is 500 Ω at 20°C. Prepare a table of coil resistance values for liquid temperatures at 10° intervals ranging from 0°C to 100°C.

4-28 Two lengths of wire each have resistances of 1 kΩ at 20°C. At 75°C the resistance values are measured as $R_a = 1.209$ kΩ and $R_b = 1.187$ kΩ. Calculate the temperature coefficient of each wire and identify the materials.

4-29 The two conductors supplying current to a 250 V, 10 kW electric kiln normally operate at an average temperature of 150°C. If the conductors are each 9 m long, and the total voltage drop along them is not to exceed 4 V, select a suitable wire gauge (from Appendix 4) for copper conductors.

4-30 50 A of current is applied to the starter motor of an automobile. The terminal voltage of the motor is to be not less than 11 V. Calculate the minimum thickness and select a suitable wire gauge from Appendix 4, for two 3 m copper conductors that connect the starter to the 12 V battery. Maximum temperature of the conductors is 50°C.

SECTION 4-7

4-31 Identify the values of resistors with the following color codes: green–blue–orange, gray–red–brown–silver, and orange–white–green–gold.

4-32 State the identifying colors for the following 10% tolerance resistors: 47 kΩ, 22 Ω, 1 MΩ, 820 Ω, 330 Ω, 5.6 Ω.

4-33 Calculate the maximum current and voltage limits for:
(a) A $\frac{1}{4}$ W, 820 Ω resistor.
(b) A $\frac{1}{2}$ W, 1 MΩ resistor.

4-34 Calculate the minimum power rating for each of the following resistors, when they are used in a circuit in which they may have 30 V applied across their terminals: 33 kΩ, 2.2 kΩ, 560 Ω, 1.2 kΩ, 180 Ω.

4-35 Calculate the minimum power rating for each of the following resistors, when they are used in a circuit in which they may carry a maximum current of 33 mA: 270 Ω, 4.7 kΩ, 820 Ω, 5.6 kΩ, 22 Ω.

SECTION 4-9

4-36 A 560 Ω resistor with a T.C. of ±200 ppm/°C has its temperature raised from 20°C to 55°C. Determine the resistance change that occurs.

4-37 A 3.3 kΩ precision resistor with a T.C. of 10 ppm/°C is to operate over a temperature range of 20°C to 100°C. Calculate the maximum change in resistance that can occur over the temperature range.

4-38 Two 3.9 kΩ resistors are to track each other within 0.01% over a temperature change of 50°C. Calculate the required tracking temperature coefficient.

4-39 Calculate the new value of resistance for each of the resistors referred to in Problem 4-31 when its temperature changes from 20°C to 75°C. Take the T.C. of each resistor as +150 ppm/°C.

COMPUTER PROBLEMS

4-40 Write a computer program to determine the diameter of a suitable copper wire for the type of problem presented in Example 4-6.

4-41 Write a computer program to calculate the temperature coefficient of a given material when the measured resistance at two temperature levels is known, as in Example 4-9.

ANSWERS TO PRACTICE PROBLEMS

4-2.1	225 V/cm
4-2.2	12 kV
4-2.3	22 kV/cm, no
4-3.1	110.8 V
4-3.2	#22
4-3.3	70.2 W, 4.2 W
4-4.1	8.24 Ω/km
4-4.2	0.124 Ω
4-4.3	1.4 mm
4-5.1	143.5 Ω
4-5.2	1.51 kΩ
4-5.3	0.0034 Ω/°C, gold

4-7.1 yellow–violet–red, orange–orange–brown, brown–green–green, green–blue–gold

4-8.1 35.3 mA, 1.1 mA

4-8.2 1.6 W

4-9.1 30.8 Ω

4-9.2 0.77 Ω

5

SERIES
RESISTIVE
CIRCUITS

Objectives   You will be able to:

Sketch circuit diagrams for series-connected resistors and series-connected lamps.

Indicate current directions and voltage polarities throughout series circuits.

Write equations for total equivalent resistance, current levels, and voltage drops in series resistive circuits.

State Kirchhoff's voltage law.

Solve problems involving equivalent resistance, current levels, voltage drops, and power dissipations in series resistive circuits.

Explain voltage dividers and potentiometers, and solve problems involving voltage dividers and potentiometers.

Explain voltage-dropping resistors and current-limiting resistors, and solve problems involving such components.

Discuss the effects of open-circuits and short-circuits in series resistive circuits, and solve problems involving open-circuits and short-circuits.

Introduction

Resistors are said to be in *series* when they are connected in such a way that there is only one path through which current can flow. This means that the *current in a series circuit is the same in all parts* of the circuit.

    The voltage drop across each component in a series circuit is dependent upon the current level and the component resistance. Two or more series-connected resistors can be used as a *voltage divider*. The *potentiometer* is an adjustable resistor used as a variable voltage divider.

    The total power supplied to a *series circuit* is the sum of the powers dissipated in the individual components.

    Resistors may be connected in series with electrical equipment for the purpose of *voltage dropping or current limiting*.

5-1
CURRENT IN A SERIES CIRCUIT

Three *series-connected* resistors are shown in Figure 5-1(a) together with a battery to supply emf and current. It is seen that the resistors are connected end to end in such a way that there is only one path through which current can flow. The current path is from the positive terminal of the battery (conventional current direction), through resistor R_1 from top to bottom, and similarly through R_2 and R_3, and then into the negative terminal of the battery. Clearly, all of the current that flows into one end of R_1 must flow out at the other end. This same current flows into one end of R_2 and out the other end, and then through R_3. Thus, it is seen that

The current is the same in all parts of a series circuit.

Figure 5-1(b) shows the circuit diagram for the three resistors and battery. The total resistance connected across the battery terminals is obviously

$$R = R_1 + R_2 + R_3$$

For any series circuit with n resistors:

$$\boxed{R = R_1 + R_2 + R_3 + \cdots + R_n} \tag{5-1}$$

Using Ohm's law, the current through the series circuit is calculated as

$$\boxed{I = \frac{E}{R_1 + R_2 + R_3 + \cdots + R_n}} \tag{5-2}$$

An *equivalent circuit* can be drawn for the series resistance circuit as shown in Figure 5-1(c). The equivalent circuit consists simply of a voltage source and a resistance R equal to the sum of all resistors in series.

EXAMPLE 5-1   Calculate the current that flows through three resistors connected in series, as illustrated in Figure 5-1. The supply voltage is $E = 9$ V, and the resistors are $R_1 = 15\ \Omega$, $R_2 = 25\ \Omega$, and $R_3 = 5\ \Omega$.

SOLUTION

From Equation (5-2):

$$I = \frac{E}{R_1 + R_2 + R_3}$$

$$= \frac{9\text{ V}}{15\ \Omega + 25\ \Omega + 5\ \Omega} = \frac{9\text{ V}}{45\ \Omega}$$

$$= \mathbf{0.2\ A}$$

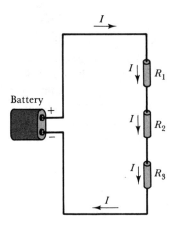

(a) Three series-connected
resistors supplied
by a battery

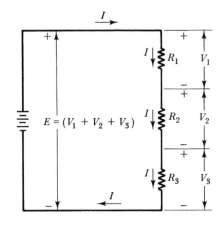

(b) Circuit diagram for three
series-connected resistors
and battery

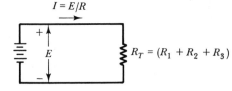

(c) Equivalent circuit

FIGURE 5-1. Resistors in series are connected end-to-end, so that there is only a single current path which passes through each resistor in turn. The total resistance is $R_T = R_1 + R_2 + \cdots$, and the current is $I = E/R_T$.

___

PRACTICE PROBLEMS

5-1.1 Five series-connected resistors have a supply of 12 V, and resistor values of $R_1 = 7$ kΩ, $R_2 = 5$ kΩ, $R_3 = 8$ kΩ, and $R_4 = 4$ kΩ. Determine the current through each resistor.

5-1.2 A series-connected group of three 3.3 kΩ resistors is to have a current level of 2.7 mA. Calculate the required supply voltage.

___

5-2
VOLTAGE DROPS IN A SERIES CIRCUIT

Referring again to Figure 5-1(b), it is seen that the current flow causes a *voltage drop,* or potential difference, across each resistor. If there was no potential difference between the terminals of each resistor, there would be no current flow.

Across R_1:

$$V_1 = IR_1$$

Across R_2:

$$V_2 = IR_2$$

Across R_3:

$$V_3 = IR_3$$

Note that the polarity of the resistor voltage drops is always such that the (conventional) current direction is from positive to negative. Thus, for the circuit as shown, the polarity is + at the top of each resistor, − at the bottom. Also note that the most positive end of each resistor is the end nearest the positive terminal of the battery (i.e., via the current path). The most negative end of each resistor is nearest the battery negative terminal. The sum of the resistor voltage drops is $V_1 + V_2 + V_3$, and, as shown in Figure 5-1(b), these must be equal to the applied emf E. For any series circuit,

$$E = V_1 + V_2 + V_3 + \cdots + V_n$$

or

$$\boxed{E = IR_1 + IR_2 + IR_3 + \cdots + IR_n} \qquad (5\text{-}3)$$

Therefore,

$$\boxed{E = I(R_1 + R_2 + R_3 + \cdots + R_n)} \qquad (5\text{-}4)$$

The relationship between the applied emf and the resistor voltage drops in a series circuit is defined by *Kirchhoff's voltage law* [*]:

Kirchhoff's Voltage Law **In any closed electric circuit, the algebraic sum of the voltage drops must equal the algebraic sum of the applied emf's.**

---

[*] Formulated by the German physicist Gustav Kirchhoff (1824–1887).

EXAMPLE 5-2   Using the component values given in Example 5-1, determine the voltage drops across each resistor in the circuit in Figure 5-1.

SOLUTION

From Example 5-1, I = 0.2 A

$$V_1 = IR_1 = 0.2 \text{ A} \times 15 \ \Omega$$

$$= 3 \text{ V}$$

$$V_2 = IR_2 = 0.2 \text{ A} \times 25 \ \Omega$$

$$= 5 \text{ V}$$

$$V_3 = IR_3 = 0.2 \text{ A} \times 5 \ \Omega$$

$$= 1 \text{ V}$$

$$V_1 + V_2 + V_3 = 9 \text{ V} = E$$

Where more than one battery or other source of emf is involved, Kirchhoff's voltage law still applies. Consequently, for the circuit shown in Figure 5-2(a),

$$E_1 + E_2 = IR_1 + IR_2 + IR_3 + IR_4$$

The two batteries in Figure 5-2(a) are connected *series-aiding*. That is, they are connected in such a way that they tend to produce currents in the same direction; they aid each other. When sources of emf are so connected that they tend to produce currents in opposite directions, as in Figure 5-2(b), they are termed *series-opposing*; they oppose each other. (See Section 10-4.)

The voltage equation for the circuit in Figure 5-2(b) is,

$$E_1 - E_2 = IR_1 + IR_2 + IR_3 + IR_4$$

ANALYSIS PROCEDURE FOR A SERIES CIRCUIT
1. *Determine the total applied voltage, $E = E_1 + E_2 + \cdots$*
2. *Calculate the total series resistance Equation (5-1).*
3. *Calculate the circuit current, Equation (5-2).*
4. *Determine the voltage drop across each component:*

$$V_1 = IR_1, \qquad V_2 = IR_2, \text{ etc.}$$

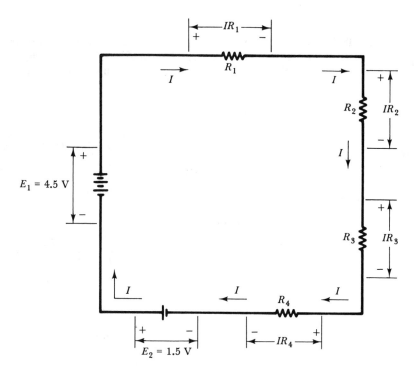

(a) Series circuit with emf sources
connected series-aiding

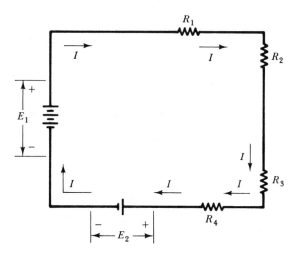

(b) Series circuit with emf sources
connected series-opposing

FIGURE 5-2. Series circuit with two sources of emf. The sources may assist each other (series-aiding), as in (a), or oppose each other (series-opposing), as in (b). For (a), $I = (E_1 + E_2)/R_T$. For (b), $I = (E_1 - E_2)/R_T$.

EXAMPLE 5-3 The four resistors in Figure 5-2(a) have the following values: $R_1 = 5\ \Omega$, $R_2 = 13\ \Omega$, $R_3 = 25\ \Omega$, and $R_4 = 17\ \Omega$. The emf are $E_1 = 4.5$ V and $E_2 = 1.5$ V. Determine the circuit current and the resistor voltage drops:

SOLUTION

$$E_1 + E_2 = 4.5\text{ V} + 1.5\text{ V} = 6\text{ V}$$

$$R_1 + R_2 + R_3 + R_4 = 5\ \Omega + 13\ \Omega + 25\ \Omega + 17\ \Omega$$

$$= 60\ \Omega$$

From Equation (5,2),

$$I = \frac{E_1 + E_2}{R_1 + R_2 + R_3 + R_4} = \frac{6\text{ V}}{60\ \Omega}$$

$$= \mathbf{0.1\ A}$$

$$V_1 = IR_1 = 0.1\text{ A} \times 5\ \Omega$$

$$= \mathbf{0.5\ V}$$

$$V_2 = IR_2 = 0.1\text{ A} \times 13\ \Omega$$

$$= \mathbf{1.3\ V}$$

$$V_3 = IR_3 = 0.1\text{ A} \times 25\ \Omega$$

$$= \mathbf{2.5\ V}$$

$$V_4 = IR_4 = 0.1\text{ A} \times 17\ \Omega$$

$$= \mathbf{1.7\ V}$$

THE VOLTMETER IS A SERIES CIRCUIT

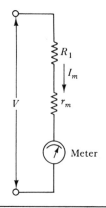

The voltmeter (discussed in Section 13-3) is a series circuit. The meter current (I_m), meter resistance (r_m), and the applied voltage (V), determine the value of the series resistor (R_1),

$$R_1 = \frac{V}{I_m} - r_m$$

EXAMPLE 5-4 Using the component values in Example 5-3, determine the current and voltage levels when E_2 is reversed, as in Figure 5-2(b).

SOLUTION

$$E_1 + E_2 = 4.5 \text{ V} + (-1.5 \text{ V}) = 3 \text{ V}$$

$$R_1 + R_2 + R_3 + R_4 = 60 \text{ }\Omega$$

From Equation (5-2), $I = \dfrac{E_1 + E_2}{R_1 + R_2 + R_3 + R_4} = \dfrac{3 \text{ V}}{60 \text{ }\Omega}$

$$I = 50 \text{ mA}$$

$$V_1 = IR_1 = 50 \text{ mA} \times 5 \text{ }\Omega$$
$$= 250 \text{ mV}$$

$$V_2 = IR_2 = 50 \text{ mA} \times 13 \text{ }\Omega$$
$$= 650 \text{ mV}$$

$$V_3 = IR_3 = 50 \text{ mA} \times 25 \text{ }\Omega$$
$$= 1.25 \text{ V}$$

$$V_4 = IR_4 = 50 \text{ mA} \times 17 \text{ }\Omega$$
$$= 850 \text{ mV}$$

PRACTICE PROBLEMS

5-2.1 Calculate the voltage drops across each resistor in Problem 5-1.1.

5-2.2 If a 4.5 V battery is connected series-opposing with the 12 V supply in Problem 5-1.1, determine the new current level and the voltage drop across each resistor.

5-3
VOLTAGE DIVIDER

CIRCUIT AND EQUATIONS. It has been shown that the voltage drops across a *string* of series resistors add up to the value of the supply emf E. Another way of looking at this is that the applied emf is divided up between the series resistors.

Figure 5-3 shows two series-connected resistors used as a *voltage divider* or potential divider. From previous studies,

$$I = \frac{E}{R_1 + R_2}$$

Also, $$V_1 = IR_1$$

Therefore,
$$V_1 = \frac{E}{R_1 + R_2} \times R_1$$

or

$$\boxed{V_1 = E \times \frac{R_1}{R_1 + R_2}} \qquad (5\text{-}5)$$

Where more than two series resistors are involved, the voltage drop across sany one resistor R_n is:

$$\boxed{V_n = E \times \frac{R_n}{R_1 + R_2 + R_3 + \cdots}} \qquad (5\text{-}6)$$

Voltage Divider Theorem **In a series circuit, the portion of applied emf developed across each resistor is the ratio of that resistor's value to the total series resistance.**

The voltage divider theorem (illustrated by Equations 5-5 and 5-6) is important because it is applied over and over again in electronic circuits. A surprisingly large amount of electronic circuit designs is merely selection of appropriate resistor values for voltage divider networks.

EXAMPLE 5-5 For the voltage divider circuit in Figure 5-3, the applied emf is $E = 100$ V, and the resistor values are $R_1 = 22\ \Omega$ and $R_2 = 28\ \Omega$. Calculate the values of V_1 and V_2.

SOLUTION

Equation (5-5), $V_1 = E \times \dfrac{R_1}{R_1 + R_2}$

$$= 100\ \text{V} \times \frac{22\ \Omega}{28\ \Omega + 22\ \Omega}$$

$V_1 = \textbf{44 V}$

$$V_2 = E \times \frac{R_2}{R_1 + R_2}$$

$$= 100\ \text{V} \times \frac{28\ \Omega}{28\ \Omega + 22\ \Omega}$$

$= \textbf{56 V}$

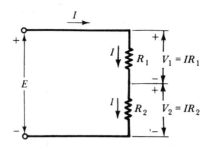

FIGURE 5-3. Two series-connected resistors can be used as a voltage divider. The circuit current is $I = E/(R_1 + R_2)$, and the resistor voltages are IR_1 and IR_2, or $V_1 = ER_1/(R_1 + R_2)$, and $V_2 = ER_2/(R_1 + R_2)$.

EXAMPLE 5-6   Apply the voltage divider theorem to determine the voltages across each resistor in Figure 5-1, as reproduced in Figure 5-4.

SOLUTION

From Equation (5-6), $V_1 = E \times \dfrac{R_1}{R_1 + R_2 + R_3}$

$$= 9\text{ V} \times \frac{15\ \Omega}{15\ \Omega + 25\ \Omega + 5\ \Omega}$$

$$= \mathbf{3\ V}$$

$$V_2 = E \times \frac{R_2}{R_1 + R_2 + R_3}$$

$$= 9\text{ V} \times \frac{25\ \Omega}{15\ \Omega + 25\ \Omega + 5\ \Omega}$$

$$= \mathbf{5\ V}$$

$$V_3 = E \times \frac{R_3}{R_1 + R_2 + R_3}$$

$$= 9\text{ V} \times \frac{5\ \Omega}{15\ \Omega + 25\ \Omega + 5\ \Omega}$$

$$= \mathbf{1\ V}$$

Compare these results to those obtained in Example 5-2.

VOLTAGE DIVIDER DESIGN. Equations 5-5 and 5-6 are correct only if no output (load) current is drawn from the junction of the resistors. This is because, as illustrated in Figure 5-5, output current I_0 flows through resistor R_1, but not through R_2. Thus,

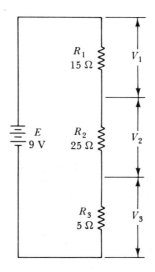

FIGURE 5-4. Circuit for Example 5-6.

$$V_{R1} = I_1 R_1 = R_1 (I_2 + I_0)$$

and

$$V_{R2} = I_2 R_2$$

If the resistor current is very much larger than the output current $(I_2 \gg I_0)$, then $(I_2 + I_0) \approx I_2$, and Equations 5-5 and 5-6 are approximately correct.

Design of a potential divider circuit normally commences with selection of I_2 very much larger than I_0. A ratio of 100 : 1 is frequently used, unless a resistor voltage precision greater than 1% is required. The resistor values are then calculated using I_2 and the desired voltage drops. Finally, suitable standard value components are selected (see Appendix 6).

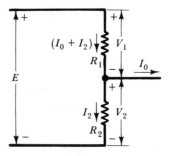

FIGURE 5-5. The resistor voltages of a potential divider are affected by the presence of an output current I_0. The effect is minimized by making $I_2 \gg I_0$.

Chap. 5 Series Resistive Circuits

EXAMPLE 5-7 A two-resistor potential divider, as in Figure 5-5, has $E = 16$ V. The output voltage is to be $V_2 = 6$ V, and the maximum output current is $I_0 = 100$ μA. Determine suitable resistor values.

SOLUTION

$$I_2 \gg I_0$$

let $I_2 = 100 \; I_2 = 100 \times 100 \; \mu A$

$$= 10 \text{ mA}$$

$$R_2 = \frac{V_2}{I_2} = \frac{6 \text{ V}}{10 \text{ mA}}$$

$$= \mathbf{600 \; \Omega} \text{ (use 560 } \Omega \text{ standard value, see Appendix 6)}$$

$$R_1 \simeq \frac{E - V_2}{I_2} = \frac{16 \text{ V} - 6 \text{ V}}{10 \text{ mA}}$$

$$\simeq \mathbf{1 \; k\Omega} \text{ (use 1 k}\Omega \text{ standard value)}$$

PRACTICE PROBLEMS

5-3.1 A voltage divider as in Figure 5-3 is to be designed to produce 7.5 V from a 12 V supply. Using a 10 mA resistor current, calculate suitable resistor values.

5-3.2 A three-resistor voltage divider as in Figure 5-4 has: $R_1 = 1.8$ kΩ, $R_2 = 2.7$ kΩ, $R_3 = 3.9$ kΩ, and $E = 15$ V. Calculate the values of V_1, V_2, and V_3.

5-3.3 A two-resistor voltage divider uses a 12 V battery supply, and has $R_1 = 820$ Ω, and $R_2 = 330$ Ω. If the junction of R_1 and R_2 is grounded, determine the voltage at each battery terminal with respect to ground.

5-4
POTENTI-OMETER

The circuit diagram of a variable resistor used as a *potentiometer* is illustrated in Figure 5-6. The construction of such as resistor is shown in Figure 4-8(a), and discussed in Section 4-6.

The potentiometer is essentially a single resistor with terminals at each end, and a moving contact that can be set to any point on the resistor. Thus, when the moving contact is exactly halfway between the two end terminals, the resistance from the moving contact to each end terminal is exactly half the total resistance of the potentiometer (see Figure 5-6). In this situation the *output voltage between points C and B is*

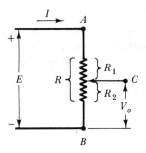

FIGURE 5-6. A potentiometer is a resistor with two fixed end terminals and one movable contact. When connected across a supply of E volts, the output is adjustable from $V_{0(min)} = 0$ to $V_{0(max)} = E$.

$$V_0 = E \times \frac{R_2}{R_1 + R_2}$$

$$= E \times \frac{R/2}{R/2 + R/2}$$

$$= E \times \tfrac{1}{2}$$

If the moving contact is adjusted until $R_2 = \tfrac{1}{4}R$, then $R_1 = \tfrac{3}{4}R$, and the output voltage becomes

$$V_0 = E \times \frac{\tfrac{1}{4}R}{\tfrac{3}{4}R + \tfrac{1}{4}R}$$

$$= E \times \tfrac{1}{4}$$

When $R_2 = \tfrac{3}{4}R$, the output voltage is

$$V_0 = E \times \tfrac{3}{4}$$

and when $R_2 = R$,

$$V_0 = E$$

Thus, it is seen that the potentiometer can be adjusted to give an output voltage ranging from 0 V to E.

Ideally, the resistance between the moving contact on a potentiometer and either one of the end terminals should increase or decrease smoothly as the device is adjusted. Because of the construction of most potentiometers, however, this resistance change normally occurs in small jumps. How smoothly any given potentiometer changes its resistance when adjusted is defined in terms of the potentiometer *resolution*. The resolution is usually stated as a number of steps. Thus, the step changes in resistance of a 1 kΩ potentiometer with a resolution of 400 steps is

$$\text{resistance steps} = \frac{1 \text{ k}\Omega}{400} = 2.5 \ \Omega$$

A resolution of 400 steps (or 1 in 400) also means that the step changes in the *output voltage* are

$$\Delta V = \frac{E}{400}$$

EXAMPLE 5-8 Calculate the minimum and maximum values of V_0 that can be obtained from the cirucit of Figure 5-7.

SOLUTION

With the moving contact at the lower end of R_2:

$$V_0 = E \times \frac{R_3}{R_1 + R_2 + R_3}$$

$$= 30 \text{ V} \times \frac{1 \text{ k}\Omega}{1 \text{ k}\Omega + 500 \ \Omega + 3.5 \text{ k}\Omega}$$

$$= \mathbf{6 \ V}$$

With the moving contact at the upper end of R_2:

$$V_0 = E \times \frac{R_2 + R_3}{R_1 + R_2 + R_3}$$

$$= 30 \text{ V} \times \frac{1 \text{ k}\Omega + 500 \ \Omega}{1 \text{ k}\Omega + 500 \ \Omega + 3.5 \text{ k}\Omega}$$

$$= \mathbf{9 \ V}$$

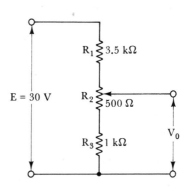

FIGURE 5-7. A potentiometer R_2 connected in series with two resistors R_1 and R_3 gives an output voltage which ranges from V_{R3} to $V_{R2} + V_{R3}$.

The output current effects discussed in Section 5-3 apply also in the case of the potentiometer. Like the design procedure for a potential divider, calculation of the value of a potentiometer for a particular application commences with selection of a potentiometer current very much larger than any output (load) current.

PRACTICE PROBLEMS

5-4.1 If the potentiometer circuit in Figure 5-7 has: $E = 18$ V, $R_1 = 4.7$ kΩ, $R_2 = 2$ kΩ, and $R_3 = 5.6$ kΩ, calculate the maximum and minimum levels of V_0.

5-4.2 The circuit of Figure 5-7 is to be altered to make V_0 adjustable from 2.5 V to 3.75 V. Determine the new resistance value for R_1 that will effect this change.

5-5

POWER IN A SERIES CIRCUIT

Using Equation (3-3), (3-4), or (3-5), it is possible to calculate the power dissipated in a resistor from a knowledge of any two of the three quantities: current, voltage, and resistance. Thus, in Figure 5-8 power dissipated in R_1 is

$$P_1 = V_1 I$$

or

$$P_1 = \frac{V_1^2}{R_1}$$

or

$$P_1 = I^2 R_1$$

The power dissipated in R_2 is calculated in exactly the same way, and the total power dissipated in the circuit is the sum of the individual resistor power dissipations. For any series resistance circuit, the total power dissipated is

$$\boxed{P = P_1 + P_2 + P_3 + \cdots + P_n} \qquad (5\text{-}7)$$

and

$$P = V_1 I + V_2 I + V_3 I + \cdots + V_n I$$

$$= I(V_1 + V_2 + V_3 + \cdots + V_n)$$

$$P = IE = \text{battery output power}$$

The total power can also be calculated as

$$\boxed{P = \frac{E^2}{R_1 + R_2 + R_3 + \cdots + R_n}} \qquad (5\text{-}8)$$

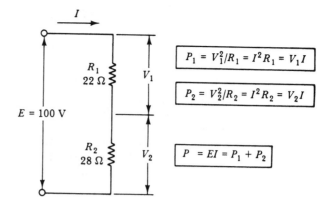

FIGURE 5-8. The power dissipated in series-connected resistors is I^2R_1, I^2R_2, I^2R_3, or V_1^2/R_1, V_2^2/R_2, etc. Each resistor must be able to dissipate this power as a minimum.

where E is the supply voltage
or

$$P = I^2(R_1 + R_2 + R_3 + \cdots + R_n)$$ (5-9)

EXAMPLE 5-9 Determine the total power dissipation and the power dissipated in each resistor in Figure 5-8 when the component values are as specified in Example 5-5: $E = 100$ V, $R_1 = 22$ Ω, $R_2 = 28$ Ω, $V_1 = 44$ V, $V_2 = 56$ V.

SOLUTION

$$P_1 = \frac{V_1^2}{R_1} = \frac{(44 \text{ V})^2}{22 \text{ Ω}}$$

$$= \textbf{88 W}$$

$$P_2 = \frac{V_2^2}{R_2} = \frac{(56 \text{ V})^2}{28 \text{ Ω}}$$

$$= \textbf{112 W}$$

$$P = P_1 + P_2 = 88 \text{ W} + 112 \text{ W}$$

$$= \textbf{200 W}$$

or

$$P = \frac{E^2}{R_1 + R_2} = \frac{(100 \text{ V})^2}{22 \text{ Ω} + 28 \text{ Ω}}$$

$$= \textbf{200 W}$$

The physical size and type of construction of a resistor (see Section 4-6) determines the maximum power that it may dissipate. The maximum power that may be safely dissipated in any component is specified by the manufacturer, and is referred to as its *power rating*. A wide range of resistors is available with various power ratings. Typical ratings for resistors usually employed in electronic circuits are: $\frac{1}{8}$ W, $\frac{1}{4}$ W, $\frac{1}{2}$ W and 1 W. The power ratings for small potentiometers and variable resistors typically range from $\frac{1}{2}$ W to 5 W. Every time the value of a resistor is calculated for a particular application, its power dissipation should also be determined. Where a component power dissipation exceeds its rating, the component is likely to *burn out*.

EXAMPLE 5-10   A 2 kΩ potentiometer is connected across a supply voltage of $E = 100$ V. Determine the minimum power rating for the potentiometer.

SOLUTION

$$minimum\ power\ rating,\ P = \frac{E^2}{R}$$

$$= \frac{(100\ V)^2}{2\ k\Omega}$$

$$= 5\ W$$

PRACTICE PROBLEMS

5-5.1   Calculate the power dissipation in each resistor in the circuit of Problem 5-4.1.

5-5.2   Calculate the minimum power rating for each resistor in the circuit of Figure 5-7.

5-6

VOLTAGE DROPPING AND CURRENT LIMITING

Sometimes a resistor is included in series with a circuit or electronic device to drop the supply voltage down to a required level. In other circumstances this same kind of series resistor arrangement can be thought of as a current-limiting resistor.

In the circuit shown in Figure 5-9(a), series resistor R_s provides a voltage drop to operate an electronic circuit which requires a voltage level lower than the source voltage. In Figure 5-9(b) R_s limits the current that flows through the three series-connected lamps L_1, L_2, and L_3, It can also be shown that in Figure 5-9(b), R_s drops the supply voltage down to the level required by the three lamps.

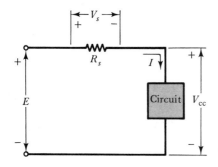

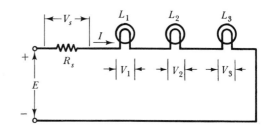

(a) Use of a series voltage-dropping resistor

(b) Use of a series current-limiting resistor

FIGURE 5-9. Resistors can be used for voltage-dropping or current-limiting. In each case the resistor value is calculated as $R_s = V_s / I$, and its power dissipation is $I^2 R_s$.

EXAMPLE 5-11 A certain electronic circuit takes a current of $I = 20$ mA when supplied with $V_{CC} = 6$ V. If the circuit is to be supplied from a source of $E = 24$ V, as illustrated in Figure 5-9(a), determine the required value of series resistance. Also calculate the power rating of the series resistor.

SOLUTION

$$V_s = E - V_{CC} = 24 \text{ V} - 6 \text{ V}$$

$$= \mathbf{18 \text{ V}}$$

$$R_s = \frac{V_s}{I_s} = \frac{18 \text{ V}}{20 \text{ mA}}$$

$$= \mathbf{900 \ \Omega}$$

$$P_s = I^2 R_s = (20 \text{ mA})^2 \times 900 \ \Omega$$

$$= \mathbf{0.36 \text{ W}}$$

Either an 820 Ω or a 1 kΩ standard value might be selected for R_s (see Appendix 6). An 820 Ω resistor will pass a larger current than 20 mA, while a 1 kΩ resistor will limit the current to a lower level.

EXAMPLE 5-12 Three 6 V, 3 W lamps connected in series as shown in Figure 5-9(b) are to be supplied from a 50 V source. Calculate the resistance value and power rating for the series current-limiting resistor.

SOLUTION

For each lamp,

$$P = VI$$

Therefore,

$$I = \frac{P}{V} = \frac{3 \text{ W}}{6 \text{ V}}$$

$$= 0.5 \text{ A}$$

Total voltage across the three lamps is

$$V = V_1 + V_2 + V_3$$

$$= 6 \text{ V} + 6 \text{ V} + 6 \text{ V}$$

$$= 18 \text{ V}$$

$$V_s = E - (V_1 + V_2 + V_3)$$

$$= 50 \text{ V} - 18 \text{ V}$$

$$= 32 \text{ V}$$

$$R_s = \frac{V_s}{I} = \frac{32 \text{ W}}{0.5 \text{ A}}$$

$$= \mathbf{64 \ \Omega} \text{ (use 68 } \Omega \text{ standard value)}$$

$$P_s = V_s I = 32 \text{ V} \times 0.5 \text{ A}$$

$$= \mathbf{16 \text{ W}}$$

PRACTICE PROBLEMS

5-6.1 An electronic device has a voltage of 2.4 V across it when its current level is 15 mA. If the device is to be connected to a 12 V supply, determine the resistance and power rating of a suitable current-limiting resistor.

5-6.2 The circuit in Figure 5-9(a) takes a constant current of 25 mA. If $E = 18$ V, and $R_s = 270 \ \Omega$, calculate V_{CC}.

5-7

OPEN-CIRCUITS AND SHORT-CIRCUITS IN A SERIES CIRCUIT

An *open-circuit* occurs in a series resistance circuit when one of the resistors becomes disconnected from an adjacent resistor [see Figure 5-10(a)]. An open circuit can also occur when one of the resistors has been destroyed by excessive power dissipation.

CURRENT LIMITING RESISTOR

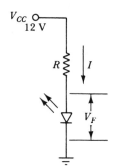

A current limiting resistor is used with a *light emitting diode* (LED) (an electronic device which glows when energized). Typically a LED requires a current of $I = 20$ mA, and has a voltage drop of $V_F = 1.2$ V. Using the available supply voltage (V_{CC}), the current limiting resistor is determined as,

$$R = \frac{V_{CC} - V_F}{I} = \frac{12 \text{ V} - 1.2 \text{ V}}{20 \text{ mA}}$$

$$= 540 \ \Omega \quad \text{(use 560 } \Omega \text{ standard value)}$$

VOLTAGE DIVIDING RESISTOR

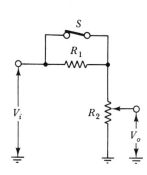

The output voltage from electronic signal generators is usually made adjustable by means of a potentiometer (R_2), as illustrated. With an additional resistor (R_1) switched into the circuit by opening switch S, the output voltage is further divided. When $R_1 = 9\,R_2$, the output is divided by a factor of 10.

with S closed

$$V_o = 0 \text{ to } V_i$$

with S open and $R_1 = 9\,R_2$

$$V_o = 0 \text{ to } V_i/10$$

In the circuit of Figure 5-10(a), the open circuit can be thought of as another resistance in series with R_1, R_2, and R_3. Thus, instead of the current being $I = E/(R_1 + R_2 + R_3)$, it becomes

$$I = \frac{E}{R_1 + R_2 + R_3 + R_{oc}}$$

Suppose that $R_{oc} = 100\ 000$ MΩ and $E = 100$ V; then, with $R_{oc} \gg (R_1 + R_2 + R_3)$

$$I \cong \frac{100 \text{ V}}{100\ 000 \text{ M}\Omega}$$

$$= 1 \text{ nA}$$

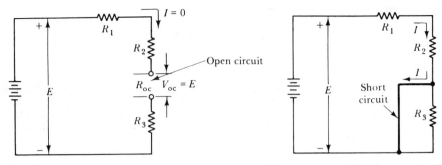

(a) Open-circuit condition in a series circuit   (b) Short-circuit condition in a series circuit

FIGURE 5-10. An open-circuit in any part of a series circuit interrupts current flow throughout the circuit. A short-circuit increases the current level and the power dissipation in those components not shorted.

This small current causes an insignificant voltage drop along R_1, R_2, and R_3. With virtually zero voltage drop across the resistors, the voltage across the open circuit is

$$V_{oc} \cong E$$

Figure 5-10(b) shows a series resistance circuit with resistor R_3 *short-circuited*. In this case the resistance between the terminals of R_3 is effectively zero. Consequently, instead of the current being $I = E/(R_1 + R_2 + R_3)$, it becomes

$$I = \frac{E}{R_1 + R_2}$$

It is obvious that open-circuits and short-circuits seriously affect the current flow through a series resistance circuit.

EXAMPLE 5-13 The circuit of Figure 5-9(b) has three 6 V, 3 W lamps, $E = 50$ V, and $R_s = 64 \ \Omega$ as calculated in Example 5-12. (a) Determine the voltage at the terminals of lamp L_2 if the lamp becomes open-circuited. (b) If L_2 becomes short-circuited, calculate the current flow through the circuit and the power dissipated in each of the other two lamps.

SOLUTION

a. With L_2 *open-circuited, no current flows and no voltage drop occurs across R_s, L_1, and L_3.*

$$L_2 \text{ terminal voltage} = E = 50 \text{ V}$$

b. *When the lamps are operating correctly:*

$$\text{lamp power} = \frac{(\text{lamp voltage})^2}{\text{lamp resistance}}$$

$$P = \frac{V^2}{R_L}$$

$$\text{or lamp resistance, } R_L = \frac{V^2}{P} = \frac{(6 \text{ V})^2}{3 \text{ W}}$$

$$= 12 \text{ } \Omega$$

When L_2 is short-circuited,

$$\text{total resistance, } R = R_s + R_{L1} + R_{L3}$$

$$= 64 \text{ } \Omega + 12 \text{ } \Omega + 12 \text{ } \Omega$$

$$= 88 \text{ } \Omega$$

and

$$I = \frac{E}{R} = \frac{50 \text{ V}}{88 \text{ } \Omega}$$

$$= 0.568 \text{ A}$$

$$P_{L1} = P_{L2} = I^2 R_L$$

$$= (0.568 \text{ A})^2 \times 12 \text{ } \Omega$$

$$= 3.87 \text{ W}$$

(Recall from Section 4-5 that the resistance of a tungsten lamp filament changes with temperature. Since the heating effect of the current is not taken into account, the above results are approximations.)

PRACTICE PROBLEMS

5-7.1 Calculate the output voltage in the circuit in Figure 5-7 when the junction of R_2 and R_3 becomes open-circuited.

5-7.2 If R_3 in Figure 5-7 becomes short-circuited, determine the new levels of $V_{0(\text{min})}$ and $V_{0(\text{max})}$. Also, calculate the new power dissipation in each resistor.

SUMMARY OF FORMULAS

Total series resistance:

$$R = R_1 + R_2 + R_3 + \cdots + R_n$$

Current in series circuit:

$$I = \frac{E}{R_1 + R_2 + R_3 + \cdots + R_n}$$

Total voltage:

$$E = IR_1 + IR_2 + IR_3 + \cdots + IR_n$$

$$E = I(R_1 + R_2 + R_3 + \cdots + R_n)$$

For a voltage divider:

$$V_1 = E \times \frac{R_1}{R_1 + R_2}$$

$$V_n = E \times \frac{R_n}{R_1 + R_2 + R_3 + \cdots}$$

Total power:

$$P = P_1 + P_2 + P_3 + \cdots + P_n$$

$$P = \frac{E^2}{R_1 + R_2 + R_3 + \cdots + R_n}$$

$$P = I^2(R_1 + R_2 + R_3 + \cdots + R_n)$$

REVIEW QUESTIONS

5-1 Sketch the circuit diagram for four series-connected resistors supplied from a battery. Show the current direction and indicate the polarity of all voltage drops. Define a series circuit in terms of the current.

5-2 Write the equations for total resistance and current in a series resistance circuit.

5-3 Write the equations for the voltage drops across the resistors in a series circuit and for the total voltage. State Kirchhoff's voltage law.

5-4 Sketch the circuit of three series resistors supplied by two batteries which are connected series-opposing. Explain the effect that the series-opposing connection has on the circuit compared to the situation when the batteries are connected series-aiding.

5-5 Sketch the circuit of two resistors employed as a voltage divider. Derive equations for the voltage drop across each resistor in terms of the supply voltage. State the voltage divider theorem.

5-6 Sketch the circuit diagram of a potentiometer supplied with voltage from a battery. Explain the principle of the potentiometer and discuss what is meant by the resolution of a potentiometer.

5-7 Write equations for the total power supplied to a series circuit in terms of:
a. The power dissipated in each component.
b. The supply voltage and the resistor values.
c. The current and the resistor values.

5-8 Explain what is meant by a voltage-dropping resistor. Sketch the diagram of a circuit using a voltage-dropping resistor. Explain briefly.

5-9   Five 12 V lamps are to be supplied from a 120 V source. Sketch a circuit diagram to show how the lamps should be connected and explain briefly.

5-10   Using circuit diagrams, explain open-cirucits and short-circuits, and discuss the effect that each has on a series resistance circuit.

PROBLEMS

SECTION 5-1

5-1   Calculate current that flows through four series-connected resistors when $R_1 = 150\ \Omega$, $R_2 = 250\ \Omega$, $R_3 = 125\ \Omega$, $R_4 = 75\ \Omega$, and the battery voltage is $E = 12$ V.

5-2   When a new resistor is substituted in place of R_1 in Problem 5-1 the current level is measured as 25 mA. Calculate the new value of R_1.

5-3   A battery with an open-circuit voltage of 9 V and an internal resistance of 1 Ω has two resistors ($R_1 = 4.7\ \Omega$ and $R_2 = 3.3\ \Omega$) connected in series across its terminals. Determine the circuit current.

5-4   A 4.7 kΩ resistor and a 2.2 kΩ resistor are connected in series to a 6 V supply. The supply is made up of four series-connected 1.5 V cells. Calculate the circuit current (a) when all of the cells are connected series-aiding, (b) when one cell is connected series-opposing.

SECTION 5-2

5-5   Calculate the voltage drops across each resistor in Problem 5-1 and show that they add up to equal the supply voltage.

5-6   Five resistors are connected in series across a 30 V supply. $R_1 = 1.8$ kΩ, $R_2 = 680\ \Omega$, $R_3 = 1.2$ kΩ, $R_4 = 1.5$ kΩ, and $R_5 = 470\ \Omega$. Determine the circuit current and the voltage drop across each resistor.

5-7   Refer to the circuit of Figure 5-1. Determine E, when: $R_1 = 6.8$ kΩ, $R_2 = 12$ kΩ, $V_3 = 2$ V, and $I = 370\ \mu$A.

5-8   Calculate E_2 in the circuit of Figure 5-11.

5-9   In Figure 5-12 calculate the voltage at terminals A, B, and C with respect to ground.

SECTION 5-3

5-10   Referring to Problem 5-4, calculate the voltage across each resistor for case (a) and case (b).

5-11   A voltage divider uses two resistors connected in series across a 75 V supply. If the resistor values are $R_1 = 37\ \Omega$ and $R_2 = 88\ \Omega$, calculate the output voltage that may be taken across each resistor.

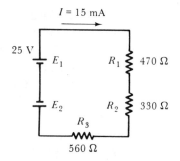

FIGURE 5-11.

5-12 A voltage divider is made up of two resistors R_1 and R_2 connected across a 15 V supply. R_2 has a resistance of 2.7 kΩ, and the voltage across it is to be 6.5 V. Determine the required resistance for R_1.

SECTION 5-4

5-13 A 100 kΩ potentiometer is connected in series with a resistance R_x. The two are supplied from a 15 V source. Determine the value of R_x if the maximum output voltage from the potentiometer is to be 12 V.

5-14 A 5 kΩ potentiometer (R_2) is to have an output voltage of 3.3 V when its moving contact is exactly halfway between its two fixed terminals. If the potentiometer is connected to an 18 V supply via a resistor R_1, calculate the required resistance for R_1.

5-15 A 3 kΩ potentiometer is connected as the center resistor in series with two 5.6 kΩ resistors to a ±6 V supply. Calculate the range of output voltage that may be obtained from the potentiometer.

SECTION 5-5

5-16 Three series-connected resistors are supplied from a 25 V source. $R_1 = 5$ kΩ, $R_2 = 13$ kΩ, and $R_3 = 11$ kΩ. Calculate the voltage drop across each resistor and the power dissipated in each.

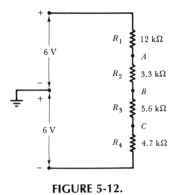

FIGURE 5-12.

Chap. 5   Series Resistive Circuits

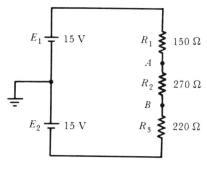

FIGURE 5-13.

5-17 For the circuit described in Problem 5-16, V_{R1} is to become 7 V. Calculate the new value of supply voltage required.

5-18 For the circuit of Figure 5-13; calculate the voltage levels at terminals A and B with respect to ground. Also determine the power dissipated in each resistor.

5-19 For the circuit of Figure 5-13, recalculate the voltage levels and power dissipated when E_2 becomes short-circuited.

5-20 Determine the voltage drop across L_1 in the circuit of Figure 5-14, and calculate the resistance of L_1 and the total power supplied to the circuit, if $I = 117$ mA.

5-21 The terminal voltage of a 12 Ω electric heater is measured as 110 V. The heater is supplied via conductors from a 115 V source. Calculate the resistance of the conductors and the power dissipated in them.

5-22 Determine the maximum and minimum values of V_0 in Figure 5-15. Also calculate the power dissipated in each resistor.

5-23 Calculate the minimum and maximum power dissipated in L_1 in the circuit of Figure 5-16.

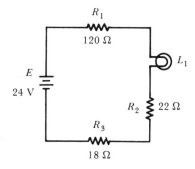

FIGURE 5-14.

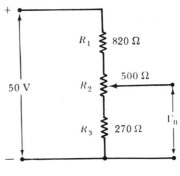

FIGURE 5-15.

SECTION 5-6

5-24 For the circuit described in Review Question 5-4, take $R_1 = 12\ \Omega$, $R_2 = 22\ \Omega$, and $R_3 = 16\ \Omega$. Also, let the battery voltages be $E_1 = 9\ V$ and $E_2 = 6\ V$. Determine the circuit current and resistor voltage drops:
(a) When the batteries are connected series-aiding.
(b) When they are connected series-opposing.

5-25 Calculate the power dissipated in each resistor in the circuit of Problem 5-24 for both case (a) and case (b) connections of the batteries as described. Also, calculate the total power dissipation in each case.

5-26 If the lamps described in Review Question 5-9 are each rated at 25 W, calculate the size and power rating of the required current-limiting resistor.

5-27 A small transistor radio normally takes 15 mA from a 9 V battery. Calculate the value of the resistor that must be connected in series with the radio if it is to be supplied from a 12 V battery. Also, calculate the power rating for the series resistor.

5-28 Twelve 6 V, 5 W series-connected lamps are to be supplied from a 230 V source. Calculate the resistance and power rating of the necessary voltage-dropping resistance.

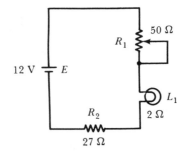

FIGURE 5-16.

Chap. 5   Series Resistive Circuits

SECTION 5-7

5-29 For Problem 5-6 recalculate the current and the voltage drops when R_3 becomes short-circuited.

5-30 Determine the levels of voltage at terminals A and B in the circuit of Figure 5-13, (a) when R_3 is open-circuited, (b) when R_3 is short-circuited.

5-31 In the circuit described in Review Question 5-9 and Problem 5-26, one of the lamps becomes short-circuited. Calculate the new level of current and the power dissipated in each lamp.

5-32 If R_2 in Figure 5-16 becomes short-circuited, determine the new minimum and maximum power dissipation in L_1.

5-33 For Problem 5-9 recalculate the voltage levels:
(a) When R_2 becomes short-circuited.
(b) When R_2 is open-circuited.

5-34 For Figure 5-15, calculate the maximum and minimum levels of V_0 when R_3 becomes short-circuited.

COMPUTER PROBLEMS

5-35 Write a program to solve the problem presented in Example 5-7. Resistor power dissipations should also be determined.

5-36 Write a computer program to solve Problem 5-15 for any plus–minus supply voltage and any combination of resistor values.

ANSWERS TO PRACTICE PROBLEMS

5-1.1 0.5 mA

5-1.2 26.73 V

5-2.1 3.5 V, 2.5 V, 4 V, 2 V

5-2.2 312.5 μA, 2.2 V, 1.6 V, 2.5 V, 1.25 V

5-3.1 750 Ω, 450 Ω

5-3.2 3.2 V, 4.8 V, 7 V

5-3.3 +8.6 V, −3.4 V

5-4.1 8.2 V, 11.1 V

5-4.2 10.5 kΩ

5-5.1 10 mW, 4.3 mW, 12 mW

5-5.2 126 mW , 18 mW, 36 mW

5-6.1 640 Ω, 144 mW

5-6.2 11.25 V

5-7.1 30 V

5-7.2 0 V, 3.75 V, 197 mW, 28.1 mW

6

PARALLEL RESISTIVE CIRCUITS

Objectives You will be able to:

Sketch circuit diagrams for parallel-connected resistors and parallel-connected lamps.

Indicate current directions and voltage polarities throughout parallel circuits.

Write equations for total equivalent resistance, current levels, and voltage drops in parallel resistive circuits.

State Kirchhoff's current law.

Solve Problems involving equivalent resistance, current levels, voltage drops, and power dissipations in parallel resistive circuits.

Explain the current divider, and solve problems involving current divisions.

Convert parallel resistive circuits into parallel conductive circuits, and solve problems involving conductances in parallel.

Discuss the effects of open-circuits and short-circuits in parallel resistive circuits.

Introduction

Resistors are said to be connected in parallel when the same voltage appears across every component. With different resistance values, different currents flow through each resistor. The total current taken from the supply is the sum of all the individual resistor currents. The equivalent resistance of a parallel resistor circuit is most easily calculated by using the reciprocal of each individual resistance value (i.e., using conductance values).

Two resistors connected in parallel may be used as a *current divider*. In a parallel circuit, as in a series circuit, the total power supplied is the sum of the powers dissipated in the individual components. Open-circuit and short-circuit conditions in a parallel circuit have an effect on the total supply current.

6-1

VOLTAGE AND CURRENT IN A PARALLEL CIRCUIT

Resistors are connected in parallel when the circuit has two terminals which are common to every resistor. Figure 6-1(a) shows two resistors, R_1 and R_2, connected in parallel and supplied by a battery. It is seen that the battery voltage E appears across R_1 and across R_2. Thus, it can be stated:

Resistors are connected in parallel when the same voltage is applied across each resistor.

This should be contrasted to the situation in a series circuit where the same current flows through each resistor.

Figure 6-1(b) shows the circuit diagram for the two parallel-connected resistors and battery. Note that in a parallel circuit, *different* currents flow through each parallel component. From consideration of Figure 6-1, the current through each of the resistors is

$$I_1 = \frac{E}{R_1}$$

and

$$I_2 = \frac{E}{R_2}$$

Now look at the current directions with respect to junction A. I_1 flowing through R_1 is flowing *away* from junction A, and I_2 flowing through R_2 is also flowing away from A. The battery current I is flowing toward A. Also, I, I_1, and I_2 are the only currents entering or leaving the junction. Consequently,

$$I = I_1 + I_2$$

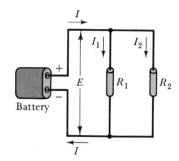

(a) Two parallel-connected resistors supplied by a battery

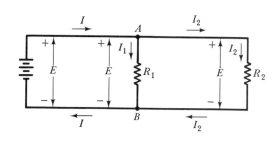

(b) Circuit diagram for two parallel-connected resistors and battery

FIGURE 6-1. Resistors are connected in parallel when the same voltage appears across each resistor. The current through each component is $I_1 = E/R_1$, $I_2 = E/R_2$, etc.

The same reasoning at junction B, where I_1 and I_2 are entering B and I is leaving B, also gives

$$I = I_1 + I_2$$

In the case where there are n resistors in parallel, the battery current is

$$\boxed{I = I_1 + I_2 + I_3 + \cdots + I_n} \tag{6-1}$$

The rule about currents entering and leaving a junction is defined in *Kirchhoff's current law:*

Kirchhoff's   **The algebraic sum of the currents entering a point in an electic circuit must equal**
Current Law   **the algebraic sum of the currents leaving that point.**

EXAMPLE 6-1   The parallel resistors shown in Figure 6-1 have values of $R_1 = 12\ \Omega$ and $R_2 = 15\ \Omega$. The battery voltage is $E = 9$ V. Calculate the current that flows through each resistor and the total current drawn from the battery.

SOLUTION

$$I_1 = \frac{E}{R_1} = \frac{9\text{ V}}{12\ \Omega}$$

$$= \mathbf{0.75\ A}$$

$$I_2 = \frac{E}{R_2} = \frac{9\text{ V}}{15\ \Omega}$$

$$= \mathbf{0.6\ A}$$

$$I = I_1 + I_2 = 0.75\text{ A} + 0.6\text{ A}$$

$$= \mathbf{1.35\ A}$$

EXAMPLE 6-2   Calculate the individual resistor currents in the circuit shown in Figure 6-2. Also, calculate the total current that must be supplied by the battery.

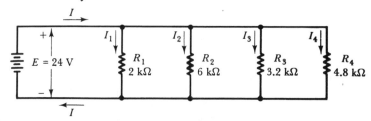

FIGURE 6-2. For four parallel-connected resistors $I_1 = E/R_1$, $I_2 = E/R_2$, $I_3 = E/R_3$, and $I_4 = E/R_4$. The total supply current is $I = I_1 + I_2 + I_3 + I_4$.

SOLUTION

$$I_1 = \frac{E}{R_1} = \frac{24\ V}{2\ k\Omega}$$

$$= 12\ mA$$

$$I_2 = \frac{E}{R_2} = \frac{24\ V}{6\ k\Omega}$$

$$= 4\ mA$$

$$I_3 = \frac{E}{R_3} = \frac{24\ V}{3.2\ k\Omega}$$

$$= 7.5\ mA$$

$$I_4 = \frac{E}{R_4} = \frac{24\ V}{4.8\ k\Omega}$$

$$= 5\ mA$$

$$I = I_1 + I_2 + I_3 + I_4$$

$$= 12\ mA + 4\ mA + 7.5\ mA + 5\ mA$$

$$= 28.5\ mA$$

PRACTICE PROBLEMS

6-1.1 Three parallel-connected resistors with a 5 V supply have values of $R_1 = 7.5\ k\Omega$, $R_2 = 5\ k\Omega$, and $R_3 = 8\ k\Omega$. Determine the branch currents and the total supply current.

6-1.2 If the three parallel-connected resistors in Problem 6-1.1 are to have a total supply current of 5 mA, determine the required supply voltage.

6-2
PARALLEL EQUIVALENT CIRCUIT

Consider the case of four resistors in parallel, as reproduced in Figure 6-3(a). From Equation (6-1), the battery current is

$$I = I_1 + I_2 + I_3 + I_4$$

which can be rewritten

$$I = \frac{E}{R_1} + \frac{E}{R_2} + \frac{E}{R_3} + \frac{E}{R_4}$$

or

$$I = E\left(\frac{1}{R_1} + \frac{1}{R_2} + \frac{1}{R_3} + \frac{1}{R_4}\right)$$

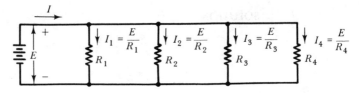

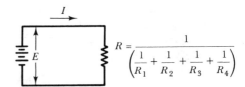

(a) Currents in parallel resistor circuit

$$R = \cfrac{1}{\left(\cfrac{1}{R_1} + \cfrac{1}{R_2} + \cfrac{1}{R_3} + \cfrac{1}{R_4}\right)}$$

(b) Equivalent circuit for four resistors in parallel

FIGURE 6-3. In the equivalent circuit for the four parallel-connected resistors supplied by a battery, the resistors are replaced by a single equivalent resistance $R = 1/(1/R_1 + 1/R_2 + 1/R_3 + 1/R_4)$.

For n resistors in parallel, this becomes

$$I = E\left(\frac{1}{R_1} + \frac{1}{R_2} + \frac{1}{R_3} + \cdots + \frac{1}{R_n}\right) \tag{6-2}$$

If all the resistors in parallel could be replaced by just one resistance that could draw the same current from the battery, the equation for current would be written

$$I = \frac{E}{R}$$

or

$$I = E\left(\frac{1}{R}\right)$$

where

$$\frac{1}{R} = \frac{1}{R_1} + \frac{1}{R_2} + \frac{1}{R_3} + \cdots + \frac{1}{R_n} \tag{6-3}$$

Thus, it is seen that

The reciprocal of the equivalent resistance of several resistors in parallel is equal to the sum of the reciprocals of the individual resistances.

Chap. 6 Parallel Resistive Circuits

Equation (6-3) can be rearranged to give

$$R = \frac{1}{\dfrac{1}{R_1} + \dfrac{1}{R_2} + \dfrac{1}{R_3} + \cdots + \dfrac{1}{R_n}}$$

(6-4)

The equivalent circuit of the parallel resistors and battery can now be drawn as illustrated in Figure 6-3(b).

THE AMMETER IS A PARALLEL CIRCUIT

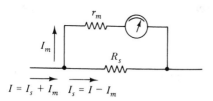

$I = \overrightarrow{I_s + I_m} \quad \overrightarrow{I_s = I - I_m}$

The ammeter (discussed in Section 13-2) is a parallel circuit. The meter current (I_m), meter resistance (r_m), and the current to be measured, determine the value of the parallel resistance (R_s),

$$R_s = \frac{I_m r_m}{I - I_m}$$

EXAMPLE 6-3 Determine the equivalent resistance of the four parallel resistors in Figure 6-2 and use it to calculate the total current drawn from the battery.

SOLUTION

From Equation (6-4):

$$R = \frac{1}{\dfrac{1}{R_1} + \dfrac{1}{R_2} + \dfrac{1}{R_3} + \dfrac{1}{R_4}}$$

$$= \frac{1}{\dfrac{1}{2\ k\Omega} + \dfrac{1}{6\ k\Omega} + \dfrac{1}{3.2\ k\Omega} + \dfrac{1}{4.8\ k\Omega}}$$

$$\cong \mathbf{842\ \Omega}$$

$$I = \frac{E}{R} = \frac{24\ V}{842\ \Omega}$$

$$= \mathbf{28.5\ mA}$$

ANALYSIS PROCEDURE FOR PARALLEL CIRCUITS:

1. Calculate the current through each resistor in turn:

$$I_1 = \frac{E}{R_1}, \qquad I_2 = \frac{E}{R_2}, \text{ etc.}$$

2. Calculate the total battery current:

$$I = I_1 + I_2 + \cdots$$

Alternatively:

1. Use Equation (6-4) to determine the equivalent resistance (R) of all the resistors in parallel.

2. Calculate the total battery current:

$$I = \frac{E}{R}$$

PRACTICE PROBLEMS

6-2.1 Three resistors in parallel have values of $R_1 = 68$ kΩ, $R_2 = 56$ kΩ, and $R_3 = 82$ kΩ. Calculate the equivalent resistance of the three, and determine the current they would take from a 14 V supply.

6-2.2 Determine the resistance of a fourth resistor in parallel with the three resistors in Problem 6-2.1 if the supply current is increased to 1 mA.

6-3
CONDUCT-ANCES IN PARALLEL

As discussed in Section 3-1, conductance is the reciprocal of resistance, and its unit is the *siemens* (S). In the case of parallel circuits, it is sometimes more convenient to use the conductance values of the resistors involved instead of the resistance values (see Example 6-4).

When Equation (6-3) is rewritten interms of conductances, it becomes

$$\boxed{G = G_1 + G_2 + G_3 + \cdots + G_n} \qquad (6\text{-}5)$$

It will be recalled that Ohm's law as applied to conductances is changed from

$$I = \frac{E}{R} \quad \text{to} \quad I = EG$$

Consequently, the current in each branch of the parallel circuit in Figure 6-2 can be calculated as

$$I_1 = EG_1, \qquad I_2 = EG_2, \text{ etc.}$$

and the current drawn from the battery becomes

$$I = EG$$

EXAMPLE 6-4   For the circuit shown in Figure 6-4 (reproduced from Figure 6-2), express each resistance as a conductance, and, using the conductance values, calculate the current through each resistor and the total current taken from the battery.

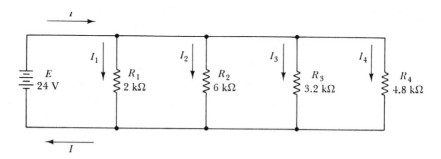

FIGURE 6-4.   In a parallel resistor circuit, resistors R_1, R_2 etc., can be expressed as conductances G_1, G_2, etc. The total conductance is $G = G_1 + G_2 + G_3 + G_4$ and the supply current is $I = EG$.

SOLUTION

$$G_1 = \frac{1}{R_1} = \frac{1}{2 \text{ k}\Omega} = 500 \ \mu\text{S}$$

$$G_2 = \frac{1}{R_2} = \frac{1}{6 \text{ k}\Omega} = 167 \ \mu\text{S}$$

$$G_3 = \frac{1}{R_3} = \frac{1}{3.2 \text{ k}\Omega} = 313 \ \mu\text{S}$$

$$G_4 = \frac{1}{R_4} = \frac{1}{4.8 \text{ k}\Omega} = 208 \ \mu\text{S}$$

$$I_1 = E \times G_1 = 24 \text{ V} \times 500 \ \mu\text{S}$$
$$= 12 \text{ mA}$$

$$I_2 = E \times G_2 = 24 \text{ V} \times 167 \ \mu\text{S}$$
$$= 4 \text{ mA}$$

$$I_3 = E \times G_3 = 24 \text{ V} \times 313 \text{ μS}$$
$$= 7.5 \text{ mA}$$

$$I_4 = E \times G_4 = 24 \text{ V} \times 208 \text{ μS}$$
$$= 5 \text{ mA}$$

$$G = G_1 + G_2 + G_3 + G_4$$
$$= (500 + 167 + 313 + 208) \text{ μS}$$
$$= 1.188 \text{ mS}$$

$$I = E \times G = 24 \text{ V} \times 1.188 \text{ mS}$$
$$= 28.5 \text{ mA} = I_1 + I_2 + I_3 + I_4$$

PRACTICE PROBLEMS

6-3.1 Convert each of the resistors in Problem 6-2.1 into conductances, and determine the total conductance of the circuit.

6-3.2 Determine the conductance of each of the three resistors in Problem 6-1.1 and the total conductance of the three in parallel.

6-4
CURRENT DIVIDER

Refer again to a two-resistor parallel circuit as illustrated in Figure 6-5. Such a parallel combination of two resistors is sometimes termed a *current divider,* because the supply current is divided between the two *branches* (or parallel sections) of the circuit. For this circuit

$$I_1 = \frac{E}{R_1} \quad \text{and} \quad I_2 = \frac{E}{R_2}$$

Also,
$$I = I_1 + I_2$$
$$= \frac{E}{R_1} + \frac{E}{R_2}$$

or
$$I = E \left(\frac{1}{R_1} + \frac{1}{R_2} \right)$$

Using $(R_1 \times R_2)$ as the common denominator for $1/R_1$ and $1/R_2$, the equation becomes

$$\boxed{I = E \left(\frac{R_1 + R_2}{R_1 \times R_2} \right)}$$

(6-6)

or
$$I = E \bigg/ \left(\frac{R_1 \times R_2}{R_1 + R_2} \right)$$

Also,
$$I = \frac{E}{R}$$

where R is the equivalent resistance of R_1 and R_2 in parallel $(R_1 \| R_2)$. Therefore, for two resistors in parallel the equivalent resistance is

$$\boxed{R = \frac{R_1 \times R_2}{R_1 + R_2}} \tag{6-7}$$

From Equation (6-6),

$$E = I \left(\frac{R_1 \times R_2}{R_1 + R_2} \right)$$

and substituting for E in

$$I_1 = \frac{E}{R_1}$$

gives
$$I_1 = \frac{I \left(\dfrac{R_1 \times R_2}{R_1 + R_2} \right)}{R_1}$$

or
$$I_1 = \frac{I}{R_1} \left(\frac{R_1 \times R_2}{R_1 + R_2} \right)$$

Therefore,

$$\boxed{I_1 = I \left(\frac{R_2}{R_1 + R_2} \right)} \tag{6-8}$$

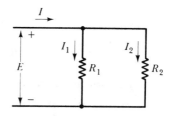

FIGURE 6-5. Two resistors can be connected in parallel to function as a current divider, giving $I_1 = IR_2/(R_1 + R_2)$, and $I_2 = IR_1/(R_1 + R_2)$.

and similarly,

$$I_2 = I\left(\frac{R_1}{R_1 + R_2}\right) \qquad (6\text{-}9)$$

Note that the expression for I_1 has R_2 on its top line, and that for I_2 has R_1 on its top line.

Equations (6-8) and (6-9) can be used to determine how a known supply current divides into two individual resistor currents.

EXAMPLE 6-5 Calculate the equivalent resistance and the branch currents for the circuit in Figure 6-5 when $R_1 = 12\ \Omega$, $R_2 = 15\ \Omega$, and $E = 9$ V.

SOLUTION

From Equation (6-7):

$$R = \frac{R_1 \times R_2}{R_1 + R_2} = \frac{12\ \Omega \times 15\ \Omega}{12\ \Omega + 15\ \Omega}$$

$$\cong \mathbf{6.67\ \Omega}$$

$$I = \frac{E}{R} = \frac{9\ \text{V}}{6.67\ \Omega}$$

$$= \mathbf{1.35\ A}$$

From Equation (6-8):

$$I_1 = I\left(\frac{R_2}{R_1 + R_2}\right)$$

$$= 1.35\ \text{A} \times \frac{15\ \Omega}{12\ \Omega + 15\ \Omega}$$

$$= \mathbf{0.75\ A}$$

From Equation (6-9):

$$I_2 = I\left(\frac{R_1}{R_1 + R_2}\right)$$

$$= 1.35\ \text{A} \times \frac{12\ \Omega}{12\ \Omega + 15\ \Omega}$$

$$= \mathbf{0.6\ A}$$

Compare this to the results of Example 6-1.

Chap. 6   Parallel Resistive Circuits

It is important to note that Equations (6-8) and (6-9) refer only to a circuit with two parallel branches. They are **not applicable to circuits with more than two parallel branches.** However, similar equations can be derived for the current division in a multibranch parallel circuit.

Consider the circuit in Figure 6-6, which has four resistors connected in parallel. The total current (I) splits into four components, as illustrated. From Equation (6-4), the parallel resistance for the whole circuit is:

$$R = 1 \Big/ \left[\frac{1}{R_1} + \frac{1}{R_2} + \frac{1}{R_3} + \frac{1}{R_4} \right]$$

The voltage drop across the parallel combination (and across each individual resistor) is:

$$E_R = IR$$

and,

$$I_1 = \frac{E_R}{R_1}, \qquad I_2 = \frac{E_R}{R_2}, \text{ etc.}$$

Therefore,

$$I_1 = \frac{IR}{R_1}, \qquad I_2 = \frac{IR}{R_2}, \text{ etc.}$$

For any multibranch parallel resistor circuit, the current in branch n is:

$$\boxed{I_n = I \, \frac{R}{R_n}} \qquad \text{(6-10)}$$

Because $R = 1/G$ and $R_n = 1/G_n$, Equation (6-10) can be rewritten in the form of conductances:

$$I_n = I \, \frac{G_n}{G}$$

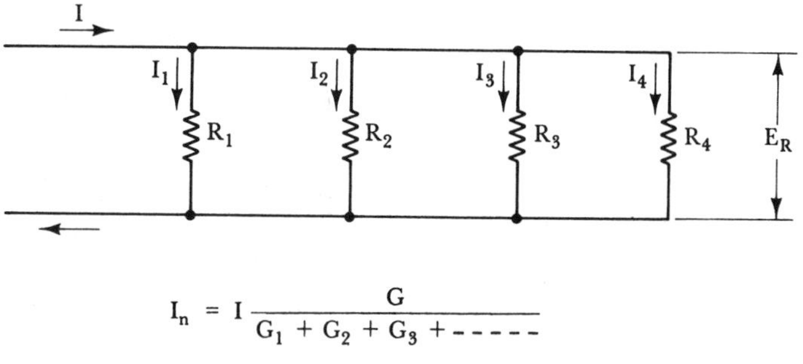

$$I_n = I \, \frac{G}{G_1 + G_2 + G_3 + - - - - -}$$

FIGURE 6-6. For a multiresistor current divider circuit, the current through any resistor R_n can be calculated from $I_n = IR/R_n$, where R is the equivalent of all the resistors in parallel. Alternatively, $I_n = IG_n/G$.

$$\text{or} \qquad \boxed{I_n = I\,\frac{G_n}{G_1 + G_2 + G_3 + \cdots}} \qquad (6\text{-}11)$$

Current divider Equation (6-11) is similar in form to voltage divider Equation (5-6), except that conductances are involved instead of resistances.

EXAMPLE 6-6 Use the current divider equation to determine the branch currents in the circuit of Figure 6-6. The component values are $R_1 = 2$ kΩ, $R_2 = 6$ kΩ, $R_3 = 3.2$ kΩ, and $R_4 = 4.8$ kΩ. The supply current is $I = 28.5$ mA.

SOLUTION

From Equation (6-4),

$$R = 1 \Big/ \left[\frac{1}{R_1} + \frac{1}{R_2} + \frac{1}{R_3} + \frac{1}{R_4}\right]$$

$$= 1 \Big/ \left[\frac{1}{2\text{ k}\Omega} + \frac{1}{6\text{ k}\Omega} + \frac{1}{3.2\text{ k}\Omega} + \frac{1}{4.8\text{ k}\Omega}\right]$$

$$= 842\ \Omega$$

From Equation (6-10),

$$I_1 = I\,\frac{R}{R_1}$$

$$= 28.5\text{ mA} \times \frac{842\ \Omega}{2\text{ k}\Omega}$$

$$\simeq 12\text{ mA}$$

$$I_2 = 28.5\text{ mA} \times \frac{842\ \Omega}{6\text{ k}\Omega}$$

$$\cong 4\text{ mA}$$

$$I_3 = 28.5\text{ mA} \times \frac{842\ \Omega}{3.2\text{ k}\Omega}$$

$$= 7.5\text{ mA}$$

$$I_4 = 28.5\text{ mA} \times \frac{842\ \Omega}{4.8\text{ k}\Omega}$$

$$= 5\text{ mA}$$

These results are comparable to the current levels calculated in Example 6-2, which involves the same circuit as Figure 6-6.

6-4.1 A total current of 55 mA flows through two resistors in parallel. If the resistor values are $R_1 = 680 \ \Omega$ and $R_2 = 560 \ \Omega$, determine the individual resistor currents.

6-4.2 Use the current divider equation to determine the individual resistor currents for a three-resistor parallel circuit with $R_1 = 33$ kΩ, $R_2 = 47$ kΩ, and $R_3 = 27$ kΩ. The supply current is 2.13 mA.

6-5

POWER IN PARALLEL CIRCUITS

Whether a resistor is connected in a series circuit or a parallel circuit, the power dissipated in the resistor is (resistor current) × (resistor voltage). For the circuit in Figure 6-7,

$$P_1 = EI_1$$

or

$$P_1 = \frac{E^2}{R_1}$$

or

$$P_1 = I_1^2 R_1$$

The power dissipated in resistor R_2 is calculated in a similar way. The total power output from the battery is, of course,

$$P = EI = E(I_1 + I_2)$$

$$= EI_1 + EI_2$$

or,

$$P = P_1 + P_2$$

For any parallel (or series) combination of n resistors, Equation (5-7) applies:

$$P = P_1 + P_2 + P_3 + \cdots + P_n$$

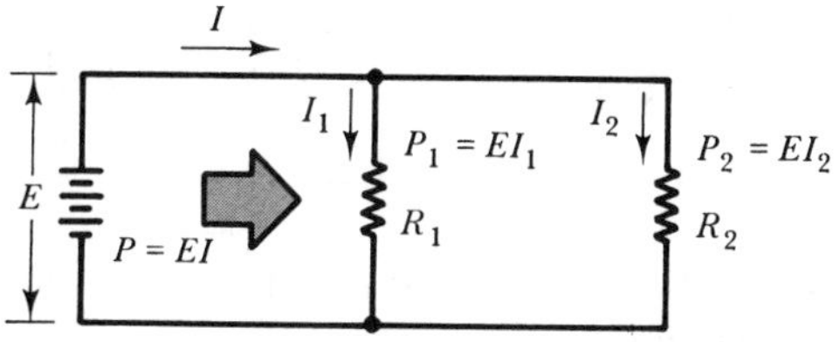

FIGURE 6-7. The power dissipated in a parallel resistor circuit is $P = EI$, or $P = P_1 + P_2 + \cdots$, where $P_1 = EI_1$, $P_2 = EI_2$, etc.

EXAMPLE 6-7 For the circuit described in Example 6-5, calculate the power dissipations in R_1 and R_2 and the total power supplied from the battery.

SOLUTION

$$P_1 = \frac{E^2}{R_1} = \frac{(9\ \text{V})^2}{12\ \Omega}$$

$$= 6.75\ \text{W}$$

$$P_2 = \frac{E^2}{R_2} = \frac{(9\ \text{V})^2}{15\ \Omega}$$

$$= 5.4\ \text{W}$$

Alternatively, $P_1 = I_1^2 R_1 = 0.75\ \text{A}^2 \times 12\ \Omega$

$$= 6.75\ \text{W}$$

$$P_2 = I_2^2 R_2 = 0.6\ \text{A}^2 \times 15\ \Omega$$

$$= 5.4\ \text{W}$$

$$P = EI = 9\ \text{V} \times 1.35\ \text{A}$$

$$= 12.15\ \text{W}$$

Also, $P = P_1 + P_2 = 6.75\ \text{W} + 5.4\ \text{W}$

$$= 12.15\ \text{W}$$

EXAMPLE 6-8 Three 12 V lamps are connected in parallel to a 12 V supply, as illustrated in Figure 6-8. The lamp ratings are $P_1 = 3$ W, $P_2 = 5$ W, and $P_3 = 10$ W. Determine the current that flows through each lamp and the total power delivered by the battery.

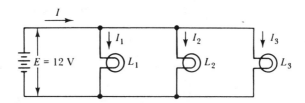

FIGURE 6-8. Parallel-connected lamps can be treated exactly like parallel-connected resistors.

SOLUTION

$$P_1 = EI_1$$

Therefore,
$$I_1 = \frac{P_1}{E} = \frac{3 \text{ W}}{12 \text{ V}}$$

$$= \mathbf{0.25 \text{ A}}$$

$$I_2 = \frac{P_2}{E} = \frac{5 \text{ W}}{12 \text{ V}}$$

$$\cong \mathbf{0.42 \text{ A}}$$

$$I_3 = \frac{P_3}{E} = \frac{10 \text{ W}}{12 \text{ V}}$$

$$\cong \mathbf{0.83 \text{ A}}$$

Total battery power output:
From Equation (5-7):

$$P = P_1 + P_2 + P_3$$

$$= 3 \text{ W} + 5 \text{ W} + 10 \text{ W}$$

$$= \mathbf{18 \text{ W}}$$

or,
$$P = E(I_1 + I_2 + I_3)$$

$$= 12 \text{ V} (0.25 \text{ A} + 0.42 \text{ A} + 0.83 \text{ A})$$

$$= \mathbf{18 \text{ W}}$$

PRACTICE PROBLEMS

6-5.1 Calculate the power dissipated in each resistor in Problem 6-2.1 and the total power taken from the supply.

6-5.2 A third resistor connected in parallel with the two in Problem 6-4.1 causes the power supplied to the circuit to increase to 1.5 W. Determine the power dissipated in each resistor.

6-6

OPEN-CIRCUITS AND SHORT-CIRCUITS IN A PARALLEL CIRCUIT

When one of the components in a parallel resistance circuit is *open-circuited,* as illustrated in Figure 6-9(a), no current flows through that branch of the circuit. The other branch currents are not affected by such an open circuit, because each of the other resistors still has the full supply voltage applied across its terminals. When I_1 goes to zero, the total current drawn from the battery is reduced from

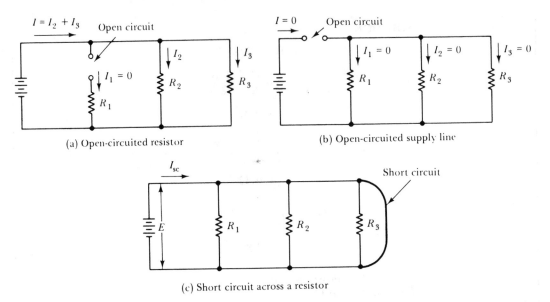

FIGURE 6-9. An open-circuit in one branch of a parallel circuit interrupts the current flow in that branch. An open-circuit in the supply line interrupts the supply current. A short-circuit in any branch shorts the supply.

$$I = I_1 + I_2 + I_3$$

to

$$I = I_2 + I_3$$

An open-circuit can also occur in the supply line to the parallel resistor combination, as shown in Figure 6-9(b). In this case the open circuit is effectively an infinite resistance in series with the battery and resistors. The result is that no supply current can flow, and consequently I_1, I_2, and I_3 are all zero.

Figure 6-9(c) shows a short-circuit across resistor R_3. This has the same effect whether it is across R_1, R_2, or R_3, or the battery terminals. In this case, the current that flows through each resistor is effectively zero. However, the battery now has a short circuit across its terminals. Consequently, the battery short-circuit current flows:

$$I_{sc} = \frac{E}{r_i}$$

where r_i is the battery internal resistance (see Section 10-2). In this situation an abnormally large current flows, and the battery could be seriously damaged.

Chap. 6   Parallel Resistive Circuits

PRACTICE PROBLEM

6-6.1 For the circuit described in Problem 6-4.2, determine the supply voltage, the total current taken from the supply when R_1 is open-circuited, and the voltage measured at the open-circuit.

SUMMARY OF FORMULAS

Total supply current to parallel circuit:

$$I = I_1 + I_2 + I_3 + \cdots + I_n$$

$$I = E \left(\frac{1}{R_1} + \frac{1}{R_2} + \frac{1}{R_3} + \cdots + \frac{1}{R_n} \right)$$

Reciprocal of total parallel resistance:

$$\frac{1}{R} = \left(\frac{1}{R_1} + \frac{1}{R_2} + \frac{1}{R_3} + \cdots + \frac{1}{R_n} \right)$$

Total parallel resistance:

$$R = \frac{1}{\dfrac{1}{R_1} + \dfrac{1}{R_2} + \dfrac{1}{R_3} + \cdots + \dfrac{1}{R_n}}$$

Total parallel conductance:

$$G = G_1 + G_2 + G_3 + \cdots + G_n$$

Two resistors in parallel:

$$R = \frac{R_1 \times R_2}{R_1 + R_2}$$

Current through R_1:

$$I_1 = I \left(\frac{R_2}{R_1 + R_2} \right)$$

Current through R_n:

$$I_n = I \frac{G_n}{G_1 + G_2 + G_3 + \cdots} = I \frac{R}{R_n}$$

Total power supplied:

$$P = P_1 + P_2 + P_3 + \cdots + P_n$$

REVIEW QUESTIONS

6-1 Sketch the circuit diagram for two parallel-connected resistors supplied by a battery. Show all voltage polarities and current directions and write expressions for the current through each resistor and for the total current drawn from the battery.

6-2 State Kirchhoff's current law. Also, explain the component voltage relationships in a parallel circuit.

6-3 Derive an equation for the equivalent resistance of n resistors connected in parallel. Repeat the derivation using conductances instead of resistances.

6-4 Derive the equations for the current through each of two parallel connected resistors, in terms of the total supply current. Also, show that for two parallel resistors,

$$R_1 \| R_2 = \frac{R_1 \times R_2}{R_1 + R_2}$$

6-5 Show that for conductances connected in parallel:

$$G = G_1 + G_2 + G_3 + \cdots G_n$$

6-6 For two resistors R_1 and R_2 connected in parallel, derive the equation for I_1 in terms of supply current and resistor values.

6-7 Discuss the effect that open-circuit and short-circuit conditions can have on parallel resistance circuits.

PROBLEMS

SECTION 6-1

6-1 For the circuit shown in Figure 6-10, determine the individual resistor currents and the current taken from the battery.

6-2 Four lamps are connected in parallel to a 120 V supply. Three of the lamp currents are $I_{L1} = 0.8$ A, $I_{L2} = 0.5$ A and $I_{L3} = 0.2$ A. If the total supply current is 2.75 A, calculate the resistance of each of the four lamps.

6-3 Calculate the current through each resistor and the total battery current for the circuit shown in Figure 6-11.

SECTION 6-2

6-4 Four parallel-connected resistors are supplied from a 25 V battery. Three of the resistors have values of $R_1 = 1$ kΩ, $R_2 = 12$ kΩ, and $R_3 = 8.2$ kΩ. If the battery current is measured as 36.5 mA, determine the value of the fourth resistor.

6-5 Determine the equivalent resistance for the two parallel-connected resistors in Figure 6-10. Also, draw the equivalent circuit and use it to calculate the total supply current.

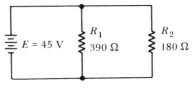

FIGURE 6-10.

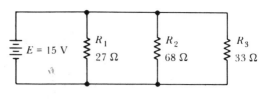

FIGURE 6-11.

6-6 Determine the equivalent resistance for the three parallel-connected resistors in Figure 6-11. Also, draw the equivalent circuit and use it to calculate the total supply current.

6-7 Determine the voltage E_2 in Figure 6-12.

6-8 If the battery voltage drops to 14.2 V in the circuit of Figure 6-11, calculate the new value for R_3 which will return the battery current to the level that flowed when $E = 15$ V.

SECTION 6-3

6-9 For the circuit in Figure 6-10, convert each resistance value into a conductance. Use the conductance values to calculate all current levels.

6-10 For the circuit in Figure 6-11, convert each resistance value into a conductance. Use the conductance values to calculate all current levels.

6-11 A 175 Ω resistor is connected in parallel with a 265 Ω resistor. Determine the value of a third parallel resistor if the total conductance of the three in parallel is to be 13.44 mS.

6-12 A 30 V supply provides 88 mA to a conductance $G_1 = 0.6$ mS in parallel with a conductance G_2. Calculate the value of G_2.

6-13 Calculate the value of R_4 in Figure 6-13.

SECTION 6-4

6-14 Two conductances connected in parallel take a total current of 750 mA. $G_1 = 14$ mS, $G_2 = 16$ mS. Calculate the current through each conductance.

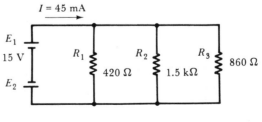

FIGURE 6-12.

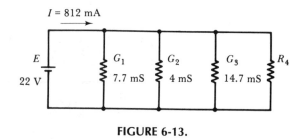

FIGURE 6-13.

6-15 Use the current divider rule to determine each branch current in the circuit of Figure 6-10.

6-16 Use the current divider rule to determine each branch current in the circuit of Figure 6-11.

6-17 Use the current divider rule to determine each branch current in the circuit of Figure 6-12.

6-18 Calculate the equivalent resistance and total supply current for the circuit in Figure 6-14. Use the current divider rule to determine each branch current.

SECTION 6-5

6-19 For the circuit in Figure 6-10, calculate the power dissipated in each resistor and the total power supplied by the battery.

6-20 For the circuit in Figure 6-11, calculate the power dissipated in each resistor and the total power supplied by the battery.

6-21 Four 115 V lamps are connected in parallel to a 115 V supply. The lamp ratings are $P_1 = 100$ W, $P_2 = 40$ W, $P_3 = 60$ W, and $P_4 = 25$ W. Determine the current that flows through each lamp and the total power delivered from the supply.

6-22 Three parallel-connected resistors, $R_1 = 1.2$ kΩ, $R_2 = 3.9$ kΩ, and $R_3 = 4.7$ kΩ, are supplied from a 5 V source. Calculate the equivalent resistance of the three in parallel. Also determine the individual resistor currents and the total power dissipated in the circuit.

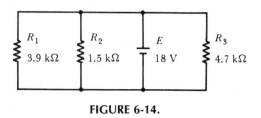

FIGURE 6-14.

Chap. 6   Parallel Resistive Circuits

6-23 Two resistors connected in parallel take a total of 75 mA from the supply. $R_1 = 620\ \Omega$ and $R_2 = 880\ \Omega$. Calculate I_1, I_2, and the level of the supply voltage. Also determine the power dissipated in each resistor.

6-24 The following parallel-connected lamps have a 120 V supply: $P_1 = 60$ W, $P_2 = 40$ W, $P_3 = 100$ W. If the supply current is not to exceed 3 A, determine the power of one additional lamp that may be connected.

6-25 For the circuit in Figure 6-12, calculate the power dissipated in each resistor and the total power supplied by the battery.

6-26 For the circuit in Figure 6-13, calculate the power dissipated in each resistor and the total power supplied by the battery.

SECTION 6-6

6-27 If E_2 in Figure 6-12 becomes short-circuited, calculate the value of the additional parallel resistance required to make $I = 100$ mA.

6-28 Repeat Problem 6-22 for the case of R_2 open-circuited.

6-29 If R_4 in Figure 6-13 becomes open-circuited, determine the new level of supply voltage required to keep the total current at 812 mA.

6-30 If the voltage source in Figure 6-14 has a resistance of 0.9 Ω, calculate the source current when R_2 becomes short-circuited.

6-31 Calculate each branch current in Figure 6-13 when one of the connections between G_3 and R_4 is open-circuited.

6-32 Calculate each resistor current in Figure 6-14 when the connection between the battery positive terminal and the junction of R_1 and R_2 is open-circuited.

COMPUTER PROBLEMS

6-33 Write a computer program to solve the type of problem presented in Example 6-6 for any level of supply current and any combination of resistor values.

6-34 Write a computer program to determine the individual lamp currents and the total supply current for a circuit consisting of six (differently rated) lamps connected in parallel to a 120 V supply.

ANSWERS TO PRACTICE PROBLEMS

6-1.1 667 μA, 1 mA, 625 μA, 2.29 mA

6-1.2 10.9 V

6-2.1 22.34 kΩ, 627 μA

6-2.2 37.5 kΩ

6-3.1   14.7 μS, 17.8 μS, 12.2 μS, 44.7 μS

6-3.2   133 μS, 200 μS, 125 μS, 458 μS

6-4.1   24.8 mA, 30.2 mA

6-4.2   728 μA, 511 μA, 890 μA

6-5.1   2.88 mW, 3.5 mW, 2.39 mW, 8.77 mW

6-5.2   418 mW, 511 mW, 571 mW

6-6.1   24 V, 1.4 mA, 24 V

7 SERIES-PARALLEL CIRCUITS

Objectives   You will be able to:

Indicate current directions and voltage polarities throughout series-parallel circuits.

Reduce a series-parallel circuit to a simple series or parallel equivalent circuit.

Solve problems involving equivalent resistance, current levels, voltage drops, and power dissipations in series-parallel resistive circuits.

Discuss the effects of open-circuits and short-circuits in series-parallel resistive circuits.

Introduction

Not all circuits are simple series or parallel arrangements. Many are combinations of parallel resistors connected in series with other resistors, or combined with other parallel groups. These can only be described as series-parallel circuits.

The simplest approach to analyzing a series-parallel circuit is to resolve each purely series group into its single equivalent resistance, and resolve each parallel group of resistors into its equivalent resistance. The process is repeated as many times as necessary.

As in all types of circuits, open-circuit and short-circuit conditions affect the currents and voltage drops throughout the circuit.

145

7-1
EQUIVALENT CIRCUIT OF A SERIES-PARALLEL CIRCUIT

In the circuit shown in Figure 7-1(a), resistors R_2 and R_3 are in parallel, and together they are in series with R_1. The level of current taken from the supply is easily calculated if R_2 and R_3 are first replaced with their equivalent resistance $(R_2 \| R_3)$ as illustrated in Figure 7-1(b). The circuit now becomes a simple two-resistor series circuit.

In Figure 7-2(a), another series-parallel resistor combination is shown. In this case the circuit is reduced to a simple parallel circuit when R_2 and R_3 are replaced by their equivalent resistance [Figure 7-2(b)].

EXAMPLE 7-1   Calculate the current drawn from the supply in the circuit shown in Figure 7-1(a).

SOLUTION

$$R_{eq} = R_2 \| R_3$$

From Equation (6-7),

$$R_{eq} = \frac{R_2 \times R_3}{R_2 + R_3} = \frac{20\ \Omega \times 30\ \Omega}{20\ \Omega + 30\ \Omega}$$

$$= 12\ \Omega$$

$$I = \frac{E}{R_1 + R_{eq}} = \frac{25\ V}{38\ \Omega + 12\ \Omega}$$

$$= 0.5\ A$$

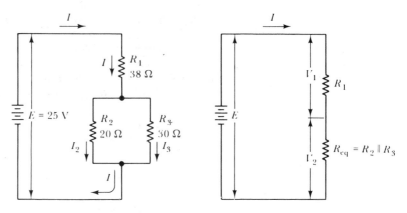

(a) Series-parallel resistance circuit   (b) Equivalent series resistance circuit

FIGURE 7-1. A series-parallel resistor circuit can be simplified by replacing each group of parallel-connected resistors with its equivalent resistance.

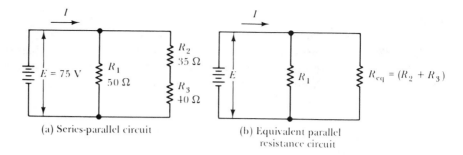

(a) Series-parallel circuit

(b) Equivalent parallel
resistance circuit

FIGURE 7-2. To simplify some series-parallel circuits, groups of series-connected resistors should first be replaced with their equivalent resistances.

EXAMPLE 7-2   Determine the level of battery current for the circuit shown in Figure 7-2(a).

SOLUTION

$$R_{eq} = R_2 + R_3$$

$$= 35 \ \Omega + 40 \ \Omega$$

$$= \textbf{75} \ \boldsymbol{\Omega}$$

R_1 and R_{eq} in parallel:

$$R = R_1 \| R_{eq}$$

From Equation (6-7),

$$R = \frac{R_1 \times R_{eq}}{R_1 + R_{eq}} = \frac{50 \ \Omega \times 75 \ \Omega}{50 \ \Omega + 75 \ \Omega}$$

$$= \textbf{30} \ \boldsymbol{\Omega}$$

$$I = \frac{E}{R} = \frac{75 \ \text{V}}{30 \ \Omega}$$

$$= \textbf{2.5 A}$$

PRACTICE PROBLEMS

7-1.1   A series-parallel circuit as in Figure 7-1(a) has a 5 V supply, and resistor values of $R_1 = 7.5 \ \text{k}\Omega$, $R_2 = 5 \ \text{k}\Omega$, and $R_3 = 8 \ \text{k}\Omega$. Calculate the supply current.

7-1.2   In the circuit in Problem 7-1.1 an additional resistor (R_4) is connected in parallel with R_1. If $R_4 = 3.3 \ \text{k}\Omega$, determine the new level of supply current.

7-2
CURRENTS IN A SERIES-PARALLEL CIRCUIT

The circuit of Figure 7-1(a) is reproduced in Figure 7-3 with the branch currents and voltages identified. It is seen that the battery current flows through resistor R_1, but that it splits up into I_2 and I_3 in order to flow through R_2 and R_3. Returning to the battery negative terminal, the current is once again I. Obviously,

$$I = I_2 + I_3$$

Similarly, the battery current splits up between the resistors in Figure 7-4, which is a reproduction of the circuit shown in Figure 7-2(a). Here, I_1 flows through R_1, while I_2 flows through R_2 and R_3, and the battery current is

$$I = I_1 + I_2$$

In each of the cases above, the current through the individual resistors can be calculated easily using the current divider rule.

EXAMPLE 7-3 Calculate the branch currents in the circuit of Figure 7-3, using the information available from Example 7-1.

SOLUTION

From Example 7-1,

$$I = 0.5 \text{ A}$$

Therefore, current through $R_1 = I = 0.5$ A

From Equation (6-8):

$$I_2 = I \left(\frac{R_3}{R_2 + R_3} \right)$$

$$= 0.5 \text{ A} \times \frac{30 \text{ } \Omega}{20 \text{ } \Omega + 30 \text{ } \Omega}$$

$$= \mathbf{0.3 \text{ A}}$$

and

$$I_3 = I \left(\frac{R_2}{R_2 + R_3} \right)$$

$$= 0.5 \text{ A} \times \frac{20 \text{ } \Omega}{20 \text{ } \Omega + 30 \text{ } \Omega}$$

$$= \mathbf{0.2 \text{ A}}$$

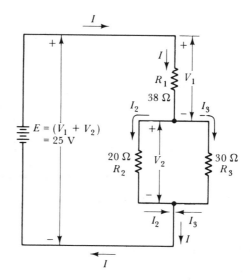

FIGURE 7-3. Currents and voltages in the series-parallel circuit of Figure 7-1(a). Supply current I flows through R_1, then splits into I_2 and I_3 to pass through R_2 and R_3 respectively.

EXAMPLE 7-4 Find the individual branch currents for the circuit of Figure 7-4, using the information given in Example 7-2.

SOLUTION

From Example 7-2:

$$I = 2.5 \text{ A}$$

and $$R_{eq} = 75 \ \Omega$$

From Equation (6-8):

$$I_1 = I \left(\frac{R_{eq}}{R_{eq} + R_1} \right)$$

$$= 2.5 \text{ A} \times \frac{75 \ \Omega}{75 \ \Omega + 50 \ \Omega}$$

$$= \mathbf{1.5 \text{ A}}$$

and $$I_2 = I \left(\frac{R_1}{R_{eq} + R_1} \right)$$

$$= 2.5 \text{ A} \times \frac{50 \ \Omega}{75 \ \Omega + 50 \ \Omega}$$

$$= \mathbf{1 \text{ A}}$$

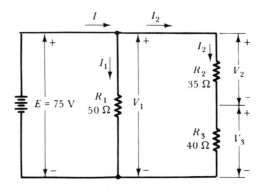

FIGURE 7-4. Currents and voltages in the series-parallel circuit of Figure 7-2(a). Supply current I splits into I_1 and I_2 to pass through R_1 and ($R_2 + R_3$) respectively.

PRACTICE PROBLEMS

7-2.1 Determine the individual resistor currents in the circuit in Problem 7-1.1

7-2.2 Calculate the individual resistor currents for the circuit in Problem 7-1.2.

7-3
VOLTAGE DROPS IN A SERIES-PARALLEL CIRCUIT

As always, the voltage drop across any resistor is the product of the resistance value and the current through the resistor. In Figure 7-3,

$$V_1 = IR_1$$

and

$$V_2 = I_2 R_2 = I_3 R_3$$

Also,

$$E = V_1 + V_2$$

Similarly, in Figure 7-4,

$$V_1 = I_1 R_1$$

$$V_2 = I_2 R_2$$

$$V_3 = I_2 R_3$$

$$E = V_1 = V_2 + V_3$$

Once the branch currents are known, the voltages across each resistor can be readily calculated. In some circumstances it may be more convenient to determine the resistor voltages first, then use these voltages to calculate the branch currents.

Chap. 7   Series-Parallel Circuits

EXAMPLE 7-5 Using the information available from Example 7-3, calculate the voltage drop across each resistor in Figure 7-3 as reproduced in Figure 7-5.

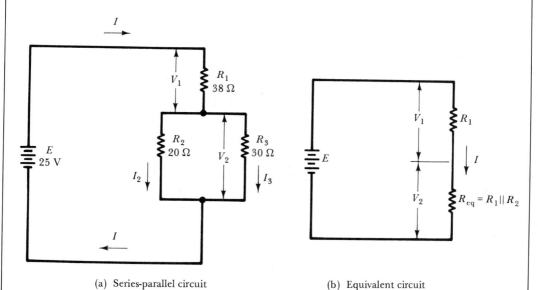

(a) Series-parallel circuit   (b) Equivalent circuit

FIGURE 7-5. Series-parallel circuit reproduced from Figure 7-1.

SOLUTION

From Examples 7-1 and 7-3,

$$I = 0.5 \text{ A}$$
$$I_2 = 0.3 \text{ A}$$

and

$$I_3 = 0.2 \text{ A}$$
$$V_1 = IR_1 = 0.5 \text{ A} \times 38 \text{ }\Omega$$
$$= \textbf{19 V}$$

$$V_2 = I_2 R_2 = 0.3 \text{ A} \times 20 \text{ }\Omega$$
$$= \textbf{6 V}$$

$$V_2 = I_3 R_3 = 0.2 \text{ A} \times 30 \text{ }\Omega$$
$$= \textbf{6 V}$$

EXAMPLE 7-6 Using the voltage divider theorem, analyze the circuit in Figure 7-5(a) to determine the resistor voltage drops and the branch currents.

SOLUTION

$$R_{eq} = R_2 \| R_3$$

$$= \frac{20 \ \Omega \times 30 \ \Omega}{20 \ \Omega + 30 \ \Omega}$$

$$R_{eq} = 12 \ \Omega$$

For voltage divider R_1 and R_{eq}, a shown in Figure 7-5(b):

$$V_2 = E \frac{R_{eq}}{R_1 + R_{eq}} = 25 \text{ V} \times \frac{12 \ \Omega}{38 \ \Omega + 12 \ \Omega}$$

$$= 6 \text{ V}$$

$$V_1 = E \frac{R_1}{R_1 + R_{eq}} = 25 \text{ V} \times \frac{38 \ \Omega}{38 \ \Omega + 12 \ \Omega}$$

$$= 19 \text{ V}$$

$$I = \frac{V_1}{R_1} = \frac{19 \text{ V}}{38 \ \Omega}$$

$$= 0.5 \text{ A}$$

$$I_2 = \frac{V_2}{R_2} = \frac{6 \text{ V}}{20 \ \Omega}$$

$$= 0.3 \text{ A}$$

$$I_3 = \frac{V_3}{R_3} = \frac{6 \text{ V}}{30 \ \Omega}$$

$$= 0.2 \text{ A}$$

EXAMPLE 7-7 Analyze the circuit in Figure 7-4, as reproduced in Figure 7-6, to determine resistor voltages and branch currents.

SOLUTION

$$V_1 = E = 75 \text{ V}$$

$$I_1 = \frac{V_1}{R_1} = \frac{75 \text{ V}}{50 \ \Omega}$$

$$= 1.5 \text{ A}$$

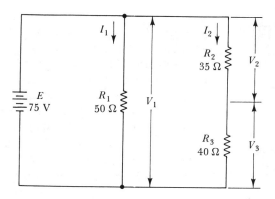

FIGURE 7-6. Series-parallel circuit reproduced from Figure 7-4.

For voltage divider R_2 and R_3:

$$V_3 = E \frac{R_3}{R_2 + R_3} = 75 \text{ V} \times \frac{40\ \Omega}{40\ \Omega + 35\ \Omega}$$

$$= 40 \text{ V}$$

$$V_2 = E \frac{R_2}{R_2 + R_3} = 75 \text{ V} \times \frac{35\ \Omega}{40\ \Omega + 35\ \Omega}$$

$$= 35 \text{ V}$$

$$I_2 = \frac{V_2}{R_2} = \frac{35 \text{ V}}{35\ \Omega}$$

$$= 1 \text{ A}$$

$$I_2 = \frac{V_3}{R_3} = \frac{40 \text{ V}}{40\ \Omega}$$

$$= 1 \text{ A}$$

PRACTICE PROBLEMS

7-3.1 Determine the voltage drops across each resistor in the circuit in Problem 7-1.1.

7-3.2 In the circuit in Problem 7-1.1 an additional resistor (R_4) is connected in series with R_3. If $R_4 = 3.3$ kΩ, determine the voltage drops across each resistor.

7-4

OPEN-CIRCUITS AND SHORT-CIRCUITS IN A SERIES-PARALLEL CIRCUIT

The effect of an open-circuit or short-circuit condition on a series-parallel circuit depends on just where in the circuit the fault occurs. Consider Figure 7-7(a), where an open-circuit is shown at one end of R_1. This has the same effect as an open circuit in the supply line, so that all current levels are zero. Also, because the currents are zero, there are no voltage drops across the resistors, and consequently all of the supply voltage E appears across the open-circuit.

In the case of an open-circuit at one end of one of the parallel resistors, as shown in Figure 7-7(b), I_3 goes to zero. The current through R_1 and R_2 is now equal to the battery current and is calculated as

$$I = \frac{E}{R_1 + R_2}$$

Also, since there is no current through R_3, there is no voltage drop across it, and the voltage at the open circuit is equal to V_2.

For the short-circuit condition shown in Figure 7-8(a), the resistance between the terminals of R_1 is effectively zero. Therefore, the battery voltage appears across R_2 and R_3 in parallel. This gives a battery current of

$$I = \frac{E}{R_2 \| R_3}$$

and the branch currents are

$$I_2 = \frac{E}{R_2}$$

and
$$I_3 = \frac{E}{R_3}$$

It is seen that the levels of current through R_2 and R_3 have been increased from the normal (i.e., before the short-circuit) condition. This could cause excessive power dissipation in the components if they have previously been operating near their maximum rating.

The short-circuit condition illustrated in Figure 7-8(b) effectively reduces I_2 and I_3 to zero and increases the battery current to

$$I = \frac{E}{R_1}$$

Obviously, the current through R_1 is now greater than normal, and again power dissipation might present a problem.

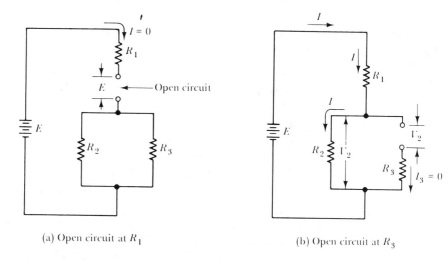

(a) Open circuit at R_1 (b) Open circuit at R_3

FIGURE 7-7. An open-circuit in one branch of a series-parallel circuit usually alters the current levels in several branches of the circuit.

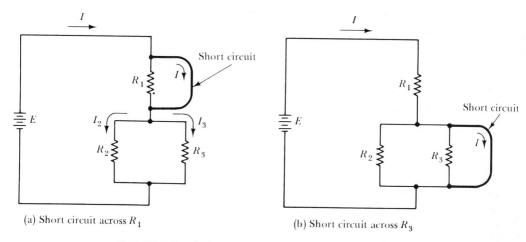

(a) Short circuit across R_1 (b) Short circuit across R_3

FIGURE 7-8. A short-circuit in one branch of a series-parallel circuit usually alters the current levels in several branches of the circuit.

PRACTICE PROBLEMS

7-4.1 If the circuit in Figure 7-1(a) has R_3 open-circuited, as in Figure 7-7(b), determine the new levels of resistor currents.

7-4.2 Determine the power dissipations in each resistor in the circuit of Figure 7-1(a). Then calculate the new power dissipations when R_1 is short-circuited as in Figure 7-8(a).

7-5

ANALYSIS OF A SERIES-PARALLEL CIRCUIT

ANALYSIS PROCEDURE FOR SERIES-PARALLEL RESISTOR CIRCUITS:

1. Draw a circuit diagram identifying all components by number and showing all currents and resistor voltage drops.

2. Convert all series branches of two or more resistors into a single equivalent resistance.

3. Convert all parallel combinations of two or more resistors into a single equivalent resistance.

4. Repeat procedures 2 and 3 until the desired level of simplification is achieved.

The final circuit should be a straightforward series or parallel circuit, which can be analyzed in the normal way. Once the current through each equivalent resistance, or the voltage across it, is known, the original circuit can be used to determine individual resistor currents and voltages.

EXAMPLE 7-8 Analyze the series-parallel circuit in Figure 7-9(a) to determine all resistor currents and voltages.

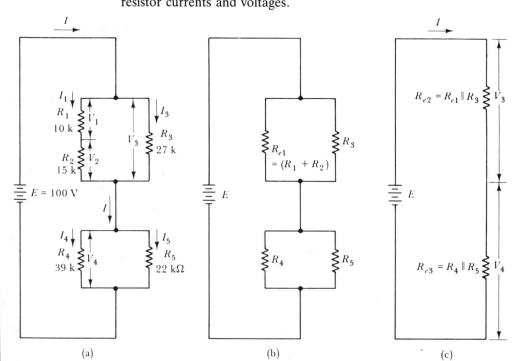

FIGURE 7-9. To analyze a series-parallel circuit, each group of series-connected resistors is first replaced with its equivalent resistance, then each parallel-connected group is replaced with its equivalent resistance. The process is repeated as necesary.

SOLUTION

$$R_{e1} = R_1 + R_2$$
$$= 10 \text{ k}\Omega + 15 \text{ k}\Omega$$
$$\boldsymbol{R_{e1} = 25 \text{ k}\Omega}$$

The circuit is now modified as shown in Figure 7-9(b), with R_{e1} replacing R_1 and R_2.

$$R_{e2} = R_{e1} \| R_3$$
$$= \frac{R_{e1} \times R_3}{R_{e1} + R_3} = \frac{25 \text{ k}\Omega \times 27 \text{ k}\Omega}{25 \text{ k}\Omega + 27 \text{ k}\Omega}$$
$$= \boldsymbol{12.98 \text{ k}\Omega}$$

$$R_{e3} = R_4 \| R_5$$
$$= \frac{R_4 \times R_5}{R_4 + R_5} = \frac{39 \text{ k}\Omega \times 22 \text{ k}\Omega}{39 \text{ k}\Omega + 22 \text{ k}\Omega}$$
$$= \boldsymbol{14.07 \text{ k}\Omega}$$

The circuit is now further modified as shown in Figure 7-9(c), with R_{e2} replacing $R_{e1} \| R_3$, and R_{e3} replacing $R_4 \| R_5$. Now,

$$I = \frac{E}{R_{e2} + R_{e3}} = \frac{100 \text{ V}}{12.98 \text{ k}\Omega + 14.07 \text{ k}\Omega}$$
$$\cong \boldsymbol{3.7 \text{ mA}}$$

$$V_3 = IR_{e2} = 3.7 \text{ mA} \times 12.98 \text{ k}\Omega$$
$$\cong \boldsymbol{48 \text{ V}}$$

$$V_4 = IR_{e3} = 3.7 \text{ mA} \times 14.07 \text{ k}\Omega$$
$$\cong \boldsymbol{52 \text{ V}}$$

$$I_1 = \frac{V_3}{R_1 + R_2} = \frac{48 \text{ V}}{25 \text{ k}\Omega}$$
$$\cong \boldsymbol{1.92 \text{ mA}}$$

$$V_1 = I_1 R_1 = 1.92 \text{ mA} \times 10 \text{ k}\Omega$$
$$\cong \boldsymbol{19.2 \text{ V}}$$

$$V_2 = I_1 \times R_2 = 1.92 \text{ mA} \times 15 \text{ k}\Omega$$
$$\cong \boldsymbol{28.8 \text{ V}}$$

$$I_3 = \frac{V_3}{R_3} = \frac{48 \text{ V}}{27 \text{ k}\Omega}$$

$$\cong \textbf{1.78 mA}$$

$$I_4 = \frac{V_4}{R_4} = \frac{52 \text{ V}}{39 \text{ k}\Omega}$$

$$\cong \textbf{1.33 mA}$$

$$I_5 = \frac{V_4}{R_5} = \frac{52 \text{ V}}{22 \text{ k}\Omega}$$

$$\cong \textbf{2.36 mA}$$

EXAMPLE 7-9 Determine the current through R_4 in the circuit shown in Figure 7-10. The supply voltage is $E = 15$ V, and the resistors are $R_1 = 2.7$ kΩ, $R_2 = 820$ Ω, $R_3 = 1.2$ kΩ, and $R_4 = 3.9$ kΩ.

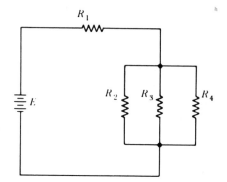

FIGURE 7-10. Circuit for Example 7-9.

SOLUTION

$$G_2 = \frac{1}{R_2} = \frac{1}{820 \text{ }\Omega}$$

$$= 1.22 \text{ mS}$$

$$G_3 = \frac{1}{R_3} = \frac{1}{1.2 \text{ k}\Omega}$$

$$= 0.83 \text{ mS}$$

$$G_4 = \frac{1}{R_4} = \frac{1}{3.9 \text{ k}\Omega}$$

$$= 0.26 \text{ mS}$$

For $R_2 \| R_3 \| R_4$,

$$G = G_2 + G_3 + G_4$$

$$= 1.22 \text{ mS} + 0.83 \text{ mS} + 0.26 \text{ mS}$$

$$= 2.31 \text{ mS}$$

$$R = \frac{1}{G} = \frac{1}{2.31 \text{ mS}}$$

$$= 433 \text{ }\Omega$$

$$I_1 = \frac{E}{R_1 + R}$$

$$= \frac{15 \text{ V}}{2.7 \text{ k}\Omega + 433 \text{ }\Omega}$$

$$= 4.79 \text{ mA}$$

From Equation (6-11):

$$I_4 = I_1 \frac{G_4}{G_2 + G_3 + G_4}$$

$$= 4.79 \text{ mA} \times \frac{0.26 \text{ mS}}{2.31 \text{ mS}}$$

$$= \mathbf{0.54 \text{ mA}}$$

PRACTICE PROBLEM

7-5.1 A series-parallel circuit consists of three parallel-connected resistors (R_1, R_2, and R_3), connected in series with resistors R_4 and R_5, which are also parallel-connected. The resistor values are $R_1 = 2.2$ kΩ, $R_2 = 3.9$ kΩ, $R_3 = 1.2$ kΩ, $R_4 = 1.5$ kΩ, and $R_5 = 1.8$ kΩ, and the supply voltage is $E = 45$ V. Analyze the circuit to determine all resistor voltages and currents.

PROBLEMS

SECTION 7-1

7-1 The circuit shown in Figure 7-1(a) has its components changed to $R_1 = 750$ Ω, $R_2 = 330$ Ω, $R_3 = 560$ Ω , and its supply voltage changed to 80 V. Determine the circuit equivalent resistance and the total battery current.

7-2 The series-parallel circuit in Figure 7-2(a) is changed as follows: $R_1 = 2.7$ kΩ, $R_2 = 8.2$ kΩ, $R_3 = 15$ kΩ, and $E = 50$ V. Determine the circuit equivalent resistance and the total battery current.

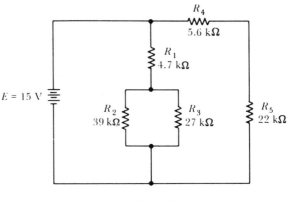

FIGURE 7-11.

7-3 Determine the equivalent resistance and the total supply current for the circuit in Figure 7-10 if $E = 9$ V, $R_1 = 3.3$ kΩ, $R_2 = 12$ kΩ, $R_3 = 6.8$ kΩ, and $R_4 = 18$ kΩ.

7-4 Determine the equivalent resistance and the total supply current for the circuit in Figure 7-11.

7-5 Determine the equivalent resistance and the total supply current for the circuit in Figure 7-12.

7-6 For the circuit in Figure 7-13, determine the value of R_1 that will give $I = 200$ mA.

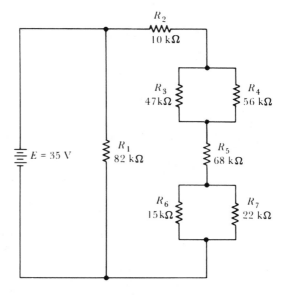

FIGURE 7-12.

Chap. 7   Series-Parallel Circuits

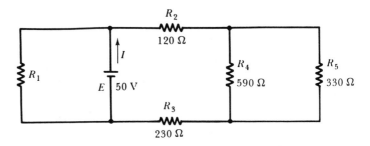

FIGURE 7-13.

SECTION 7-2

7-7 Calculate the branch currents for the circuit in Problem 7-1.

7-8 Calculate the branch currents for the circuit in Problem 7-2.

7-9 Calculate the branch currents for the circuit in Problem 7-3.

7-10 Calculate the branch currents for the circuit in Problem 7-4.

7-11 Calculate the branch currents for the circuit in Problem 7-5.

7-12 Calculate the branch currents for the circuit in Problem 7-6.

SECTION 7-3

7-13 Calculate the branch voltages for the circuit in Problem 7-1.

7-14 Calculate the branch voltages for the circuit in Problem 7-2.

7-15 Calculate the branch voltages for the circuit in Problem 7-3.

7-16 Calculate the branch voltages for the circuit in Problem 7-4.

7-17 Calculate the branch voltages for the circuit in Problem 7-5.

7-18 Calculate the branch voltages for the circuit in Problem 7-6.

SECTION 7-4

7-19 If resistor R_3 in Figure 7-12 becomes short-circuited, calculate the new levels of current and voltage throughout the circuit.

7-20 In the circuit of Figure 7-12, resistor R_7 becomes open-circuited. Determine the new levels of current and voltage throughout the circuit.

7-21 Determine the new level of battery current in the circuit of Figure 7-11 when R_2 is short-circuited.

7-22 Determine the new level of supply current in the circuit in Figure 7-11 when R_3 is open-circuited.

7-23 Calculate the new value of R_1 in Figure 7-13 to give $I = 200$ mA when R_5 is short-circuited.

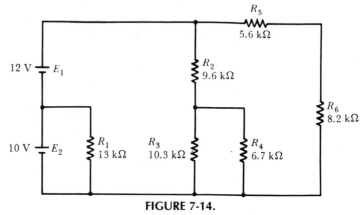

FIGURE 7-14.

SECTION 7-5

7-24 Determine all resistor currents and voltages in the circuit shown in Figure 7-14.

7-25 Repeat Problem 7-24 with E_2 polarity reversed.

7-26 In Figure 7-15, calculate the voltages at terminals A, B, and C with respect to ground.

7-27 Repeat Problem 7-26 with the polarity of E_1 reversed.

7-28 Repeat Problem 7-26 with R_4 short-circuited.

7-29 Repeat Problem 7-26 with R_6 open-circuited.

7-30 Calculate all resistor currents and voltages in the circuit shown in Figure 7-16.

7-31 Repeat Problem 7-30 with R_2 short-circuited.

7-32 For the circuit shown in Figure 7-16, calculate the new value of supply voltage that will cause 2.2 mW to be dissipated in R_2.

7-33 For the circuit shown in Figure 7-17, determine the value of R_2 that will give $I = 168$ μA.

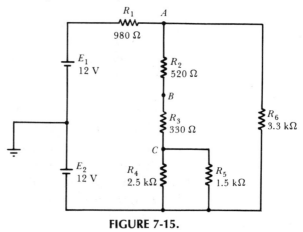

FIGURE 7-15.

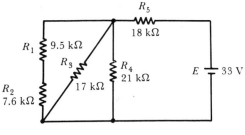

FIGURE 7-16.

7-34 For Problem 7-33, determine the new levels of resistor currents when R_3 becomes open-circuited.

COMPUTER PROBLEMS

7-35 Write a program to analyze the circuit in Figure 7-3, to determine resistor voltages, currents, power dissipations, and total supply current.

7-36 Write a program to analyze the circuit in Figure 7-12, to determine resistor voltages, currents, power dissipations, and total supply current.

ANSWERS TO PRACTICE PROBLEMS

7-1.1 473 μA

7-1.2 931 μA

7-2.1 473 μA, 291 μA, 182 μA

7-2.2 284 μA, 573 μA, 358 μA, 647 μA

7-3.1 3.55 V, 1.45 V, 1.45 V

7-3.2 3.42 V, 1.58 V, 1.12 V, 0.46 V

7-4.1 431 mA, 431 mA

7-4.2 9.5 W, 1.8 W, 1.2 W, 0, 31.25 W, 20.8 W

7-5.1 19.9 V, 19.9 V, 19.9 V, 25.1 V, 25.1 V, 9 mA, 5.1 mA, 16.6 mA, 16.7 mA, 13.9 mA

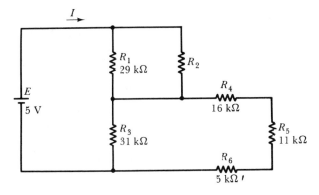

FIGURE 7-17.

8

NETWORK ANALYSIS TECHNIQUES

Objectives You will be able to:

Sketch the circuits of constant voltage sources and constant current sources.

Convert constant current sources to constant voltage sources, and vice versa.

Solve problems involving parallel and series operation of constant current sources and constant voltage sources.

Analyze resistive networks using Kirchhoff's laws.

Use loop equations to analyze resistive networks.

Solve resistive network problems by nodal analysis.

Apply delta-wye transformations to resistive networks.

Introduction

Some resistive networks are so complex that they cannot be analyzed by the simple series-parallel techniques described in Chapter 7. One example is a circuit that has more than one voltage source. Also, some circuits have current sources as well as voltage sources.

A network can sometimes be simplified by converting voltage sources to current sources, or vice versa. By direct application of Kirchhoff's voltage and current laws, equations can be derived that afford analysis of any circuit. *Loop equations, nodal analysis,* and *delta-wye transformations* are all useful techniques for determining the voltage and current levels in complex circuits.

8-1

VOLTAGE SOURCES AND CURRENT SOURCES

CONSTANT VOLTAGE SOURCE. Voltage cells and batteries can be represented by an *ideal cell* in series with the internal resistance of the cell or battery (see Section 10-2). The ideal cell is assumed to be a *constant voltage source,* and the output current produces a voltage drop across the internal resistance. Figure 8-1(a) shows such a constant voltage source, with voltage E, source resistance R_S, and an external load resistance R_L. Sometimes, instead of a cell, the symbol for a dc voltage generator is used to represent a constant voltage source [see Figure 8-1(b)]. Using the voltage divider rule (see Section 5-3), the output voltage developed across R_L from the constant voltage source can be quickly determined:

$$V_L = E \frac{R_L}{R_S + R_L}$$

(8-1)

If $R_S \ll R_L$,

$$V_L \cong E$$

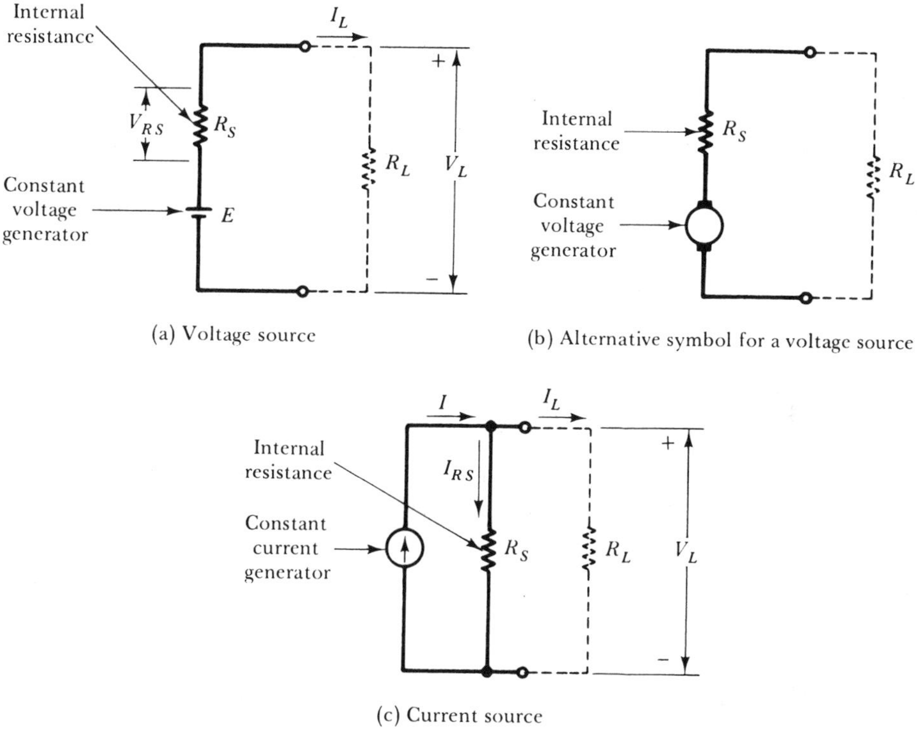

(a) Voltage source

(b) Alternative symbol for a voltage source

(c) Current source

FIGURE 8-1. A voltage source is represented by a constant voltage generator in series with its source resistance. A current source is represented by a constant current generator in parallel with its source resistance.

Where the load resistance is very much larger than the source resistance, the constant voltage source is assumed to have zero source resistance, and all of the source voltage is assumed to be applied to the load.

CONSTANT CURRENT SOURCE. Certain electronic devices can produce a current that tends to remain constant regardless of how the load resistance varies. Thus, it is possible to have a *constant current source.*

Figure 8-1(c) shows the circuit of a constant current source, together with its source resistance R_S and a load resistance R_L. Note that R_S is in parallel with the current source. Consequently, some of the source current flows through R_S and some flows through R_L. Using the current divider rule (see Section 6-4), the output current (or load current) from a constant current source can be determined in terms of R_L and R_S:

$$\boxed{I_L = I\,\frac{R_S}{R_L + R_S}}\qquad(8\text{-}2)$$

If $R_L \ll R_S$,

$$I_L \cong I$$

Referring to Figure 8-1(c) again, it is obvious that if R_S were very much larger than R_L, then I_{RS} would be very much smaller than I_L.

Where the load resistance is very much smaller than the source resistance, a constant current source is assumed to have an infinite source resistance, and all of the source current is assumed to flow through the load.

SOURCE CONVERSIONS. For the voltage source in Figure 8-1(a), the load current is calculated as

$$I_L = \frac{E}{R_S + R_L}$$

If E in this equation is multiplied by R_S/R_S, nothing is changed, because it is the same as multiplying by 1:

$$I_L = \frac{E\left(\dfrac{R_S}{R_S}\right)}{R_S + R_L}$$

or

$$\boxed{I_L = \frac{E}{R_S}\left(\frac{R_S}{R_S + R_L}\right)}\qquad(8\text{-}3)$$

In Equation (8-3), E/R_S is a current and $R_S/(R_S+R_L)$ is exactly the same as in the current divider equation, Equation (8-2). Thus, Equation (8-3) represents the output from a current source.

A voltage source having a voltage E and a source resistance R_S can be replaced by a current source with a current E/R_S and a source resistance R_S.

For the current source in Figure 8-1(c), the load voltage is

$$V_L = I_L R_L$$

I_L can be replaced by Equation (8-2). Thus,

$$V_L = \left(I \frac{R_S}{R_L + R_S} \right) R_L$$

or

$$\boxed{V_L = IR_S \left(\frac{R_L}{R_L + R_S} \right)} \tag{8-4}$$

In Equation (8-4), IR_S is a voltage and $R_L/(R_L + R_S)$ is exactly the same as in the voltage divider equation, Equation (8-1). Therefore, Equation (8-4) represents the output from a voltage source.

A current source having a current I and a source resistance R_S can be replaced by a voltage source with a voltage IR_S and a source resistance R_S.

EXAMPLE 8-1 The voltage source in Figure 8-1(a) has $E = 10$ V, $R_S = 1\ \Omega$, and $R_L = 1\ \Omega$. Determine I and R_S for the equivalent current source. Then calculate V_L and I_L for each type of source.

SOLUTION

For the current source,

$$I = \frac{E}{R_S}$$

$$= \frac{10\ \text{V}}{1\ \Omega} = \textbf{10 A}$$

current source R_S = voltage source R_S = **1 Ω**

Using the voltage source,

$$V_L = E \frac{R_S}{R_S + R_L} = 10\ \text{V} \times \frac{1\ \Omega}{1\ \Omega + 1\ \Omega}$$

$$V_L = 5 \text{ V}$$

$$I_L = \frac{E}{R_S + R_L} = \frac{10 \text{ V}}{1 \text{ }\Omega + 1 \text{ }\Omega}$$

$$= 5 \text{ A}$$

Using the current source,

$$I_L = I \times \frac{R_S}{R_L + R_S} = 10 \text{ A} \times \frac{1 \text{ }\Omega}{1 \text{ }\Omega + 1 \text{ }\Omega}$$

$$= 5 \text{ A}$$

$$V_L = I \times (R_L \| R_S)$$

$$= I \frac{R_L \times R_S}{R_L + R_S} = 10 \text{ A} \times \frac{1 \text{ }\Omega \times 1 \text{ }\Omega}{1 \text{ }\Omega + 1 \text{ }\Omega}$$

$$= 5 \text{ V}$$

PARALLEL AND SERIES OPERATION. *Voltage sources can be operated in series without difficulty* (see Section 10-4). For parallel operation, voltage sources must have closely equal terminal voltages (Section 10-5). *Parallel operation of voltage sources with differing terminal voltages is not possible,* because one source will tend to discharge the other.

Similar but converse rules apply to current sources. Consider Figure 8-2(a), in which two current sources are shown connected in parallel. The output current from the two is

$$I = I_{S1} + I_{S2}$$

$$= 2 \text{ A} + 3 \text{ A}$$

$$= 5 \text{ A}$$

and the internal resistance of the combination is

$$R = R_{S1} \| R_{S2}$$

$$= 2 \text{ }\Omega \| 2 \text{ }\Omega$$

$$= 1 \text{ }\Omega$$

The same two current sources are shown in Figure 8-2(b) with the direction of I_{S1} altered. In this case

$$I = I_{S2} - I_{S1}$$

$$= 3 \text{ A} - 2 \text{ A}$$

$$= 1 \text{ A}$$

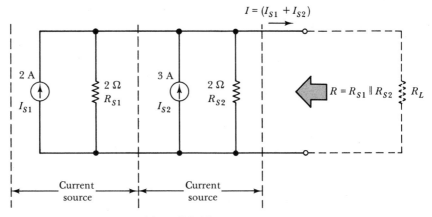

(a) Parallel-aiding current sources

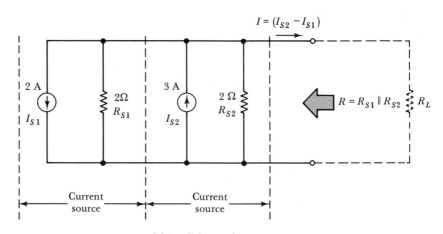

(b) Parallel-opposing current sources

FIGURE 8-2. Current sources may be operated parallel-aiding to give a total output current of $I_{s1} + I_{s2}$. It is wasteful to operate current sources parallel-opposing.

Apart from the fact that it it wasteful to connect two current sources with opposite polarity, it is seen that *current sources may be operated in parallel without difficulty.*

Now consider two current sources connected in series. Where they both generate exactly the same level of current, there is no problem. But suppose that one generates an output of 1 A and the other has an output of 2 A. The 2 A current would flow through the 1 A source, thus changing the 1 A output level. Or the resistance of the 1 A source might alter the level of the 2 A current. Consequently, *series operation of current sources with differing output currents it not possible.*

A current source may be connected in series or in parallel with a voltage source. However, the rules discussed above still apply. When a conversion is made, so that they are both voltage sources, parallel operation is possible only when the voltages are equal. When the conversion makes them both current sources, series operation is possible only when the currents are equal.

PRACTICE PROBLEMS

8-1.1 A voltage source has $E = 500$ mV, $R_S = 1$ kΩ, and $R_L = 82$ kΩ. Convert the voltage source into its equivalent current source and calculate the load current.

8-1.2 A constant current device has $I = 7.5$ mA, $R_S = 20$ kΩ, and $R_L = 2.7$ kΩ. Derive the equivalent voltage source and calculate the load voltage.

8-2
NETWORK ANALYSIS USING KIRCHHOFF'S LAWS

Kirchhoff's voltage law (see Section 5-2) and Kirchhoff's current law (see Section 6-1) readily lend themselves to the derivation of equations for solving complex circuits. Consider the simple series-parallel circuit in Figure 8-3. Applying Kirchhoff's voltage law to the closed path starting at point A and going through points B, E, and F and back to A again gives

$$E = V_1 + V_2$$

and since, $V_1 = I_1R_1$ and $V_2 = I_2R_2$,

$$E = I_1R_1 + I_2R_2 \qquad (1)$$

Summing the voltage drops around path $BCDEB$,

$$V_2 = V_3$$

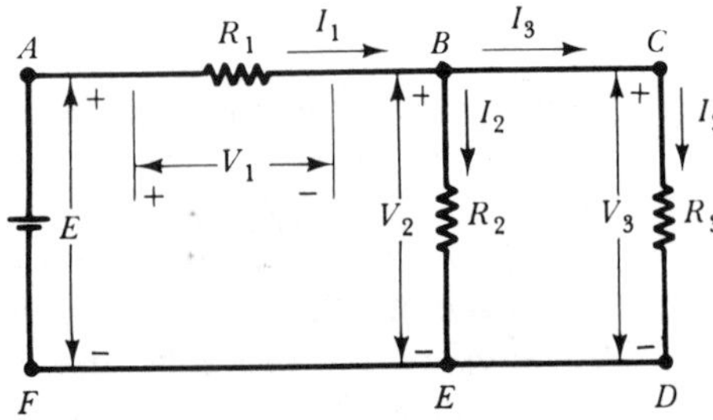

FIGURE 8-3. Kirchhoff's laws may be used to analyze a resistance network. For the circuit shown, Kirchhoff's voltage law gives $E = V_1 + V_2$, and Kirchhoff's current law gives $I_1 = I_2 + I_3$.

or
$$I_2 R_2 = I_3 R_3 \qquad (2)$$

For the path *ABCDEFA*,

$$E = V_1 + V_3$$

Therefore,
$$E = I_1 R_1 + I_3 R_3 \qquad (3)$$

And from Kirchhoff's current law, at point *B*,

$$I_1 = I_2 + I_3 \qquad (4)$$

Substituting the expression for I_1 from Equation (4) into Equation (1),

$$E = (I_2 + I_3) R_1 + I_2 R_2$$

Therefore,
$$E = I_2 R_1 + I_3 R_1 + I_2 R_2 \qquad (5)$$

and, from Equation (2),

$$I_3 = \frac{I_2 R_2}{R_3} \qquad (6)$$

Substituting for I_3 from Equation (6) into Equation (5),

$$E = I_2 R_1 + \frac{I_2 R_2}{R_3} R_1 + I_2 R_2$$

or
$$E = I_2 \left(R_1 + \frac{R_2 R_1}{R_3} + R_2 \right)$$

$$I_2 = \frac{E}{R_1 + R_2 + \dfrac{R_2 R_1}{R_3}} \qquad (7)$$

From Equation (7) the level of current I_2 can be calculated. Then I_2 can be substituted into Equation (6) to find I_3, and I_1 can be determined by substituting I_2 and I_3 into Equation (4). Once all the current levels are known, the voltage levels are easily calculated.

NETWORK ANALYSIS PROCEDURE USING KIRCHHOFF'S LAWS:

1. *Letter all junctions on the network A, B, C, etc.*
2. *Identify current directions and voltage polarities, and number them according to the resistor involved.*
3. *Identify each current path according to the lettered junctions, and applying Kirchhoff's voltage law, write the voltage equations for the paths.*

4. *Applying Kirchhoff's current law, write the equations for the currents entering and leaving all junctions where more than one current is involved.*
5. *Solve the equations by substitution to find the unknown currents.*

Note that in some circumstances currents and voltage polarities will turn out to be negative when the circuit is analyzed. This simply means that the assumed current directions and/or voltage polarities were incorrect.

EXAMPLE 8-2   Using Kirchhoff's laws, analyze the circuit in Figure 8-4 to determine I_1, I_2, and I_3.

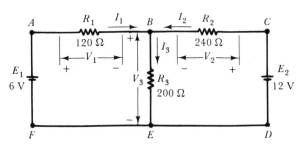

FIGURE 8-4.   A resistor network with two voltage sources can be analyzed by use of Kirchhoff's laws. For the circuit shown, $E_1 = V_1 + V_3$, $E_2 = V_2 + V_3$, and $I_3 = I_1 + I_2$.

SOLUTION

For the path ABEFA,

$$E_1 = V_1 + V_3$$

$$E_1 = I_1 R_1 + I_3 R_3$$

Therefore, $$6V = 120 I_1 + 200 I_3 \qquad (1)$$

For CBEDC,

$$E_2 = V_2 + V_3$$

$$E_2 = I_2 R_2 + I_3 R_3$$

$$12V = 240 I_2 + 200 I_3 \qquad (2)$$

For ABCDEFA,

$$E_1 = V_1 - V_2 + E_2$$

$$E_1 - E_2 = I_1 R_1 - I_2 R_2$$

$$6V - 12V = 120 I_1 - 240 I_2$$

$$-6V = 120 I_1 - 240 I_2 \qquad (3)$$

Chap. 8   Network Analysis Techniques

For point B,

$$I_3 = I_1 + I_2 \qquad (4)$$

Substituting for I_3 from Equation (4) *into* Equation (1),

$$6V = 120I_1 + 200I_1 + 200I_2$$

$$6V = 320I_1 + 200I_2 \qquad (5)$$

From Equation (3),

$$240I_2 = 120I_1 + 6V$$

Therefore,

$$I_2 = \frac{120}{240} I_1 + \frac{6V}{240}$$

$$I_2 = \frac{1}{2} I_1 + \frac{1V}{40} \qquad (6)$$

Substituting for I_2 from Equation (6) *into* Equation (5),

$$6V = 320I_1 + \frac{200}{2} I_1 + \frac{200V}{40}$$

Therefore,

$$6V = 420I_1 + 5V$$

$$I_1 = \frac{6V - 5V}{420 \ \Omega}$$

$$I_1 \cong \mathbf{2.38 \ mA}$$

Substituting for I_1 in Equation (1),

$$6V = (120 \times 2.38 \text{ mA}) + 200I_3$$

$$I_3 \cong \mathbf{28.57 \ mA}$$

Substituting for I_1 and I_3 in Equation (4),

$$28.57 \text{ mA} = 2.38 \text{ mA} + I_2$$

$$I_2 \cong \mathbf{26.19 \ mA}$$

PRACTICE PROBLEMS

8-2.1 In the circuit in Figure 7-4, the components are changed to $R_1 = 15$ kΩ, $R_2 = 2.2$ kΩ, $R_3 = 12$ kΩ, and $E_1 = 55$ V with $R_{S1} = 1.8$ kΩ. Use Kirchhoff's laws to determine I_2.

8-2.2 In the circuit described in Problem 8-2.1, resistors R_2 and R_3 are replaced by a voltage source with $E_2 = 35$ V and $R_{S2} = 3$ kΩ (+ terminal up). Use Kirchhoff's laws to determine I_1.

By the use of *loop equations,* also termed *mesh equations,* complex circuits can be solved with greater ease than by direct application of Kirchhoff's laws.

The circuit of Figure 8-4 is reproduced in Figure 8-5, and *loop currents* are shown in a clockwise direction. The loop current is simply the current that circulates in the closed current path. In Figure 8-5, loop current I_1 is exactly the same as the branch current that flows through R_1. However, I_1 is *not* the branch current in R_3, because loop current I_2 also flows through R_3 in an opposite direction to I_1. The branch current in R_3 is $I_3 = I_1 - I_2$. Loop current I_2 is the same as the branch current flowing in resistor R_2. A third loop could be drawn through E_1, R_1, R_2, E_2 and back to E_1. However, because all of the circuit voltage drops are included in the first two loops, the third loop is redundant.

The loop currents could actually be assigned an arbitrary direction, but assigning a clockwise direction simplifies the process of writing equations for the voltage drops around each loop. When the circuit analysis is complete, those branch currents that are in the same direction as the (clockwise) loop currents have a plus sign. Those that are in the opposite direction are given a minus sign. In Figure 8-5, note that the voltage drop $I_1 R_3$ is + at the top of R_3, while $I_2 R_3$ is + at the bottom of the resistor. Also note that the polarity of E_1 is opposite to the resistor voltage drops in loop 1, while in loop 2, E_2 has the same polarity as the resistor voltage drops.

DC CIRCUIT ANALYSIS PROCEDURE USING LOOP EQUATIONS:

1. *Convert all current sources to voltage sources.*
2. *Draw all loop currents in a clockwise direction and identify them by number.*
3. *Identify all resistor voltage drops as + to − in the direction of the loop current.*

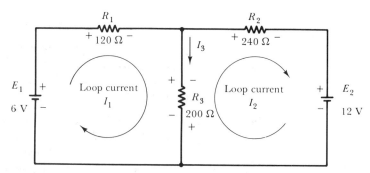

FIGURE 8-5. In the loop equation (or mesh equation) method of circuit analysis, the sum of the voltage drops around each circuit loop is equated to zero. For loop 1, $0 = I_1 R_1 + I_1 R_3 - I_2 R_3 - E_1$.

4. Identify all voltage sources according to their correct polarity.

5. Write the equations for the voltage drops around each loop in turn, by equating the sum of the voltage drops to zero.

6. Solve the equations to find the unknown currents.

EXAMPLE 8-3 Using loop equations, solve the resistor network shown in Figure 8-5 to determine the current through R_3.

SOLUTION

For loop 1,

$$0 = I_1 R_1 + I_1 R_3 - I_2 R_3 - E_1$$

Therefore, $E_1 = I_1 (R_1 + R_3) - I_2 R_3$

$$6V = I_1 (120\ \Omega + 200\ \Omega) - (I_2 \times 200\ \Omega)$$

$$6V = 320 I_1 - 200 I_2 \qquad (1)$$

For loop 2,

$$0 = I_2 R_2 + E_2 + I_2 R_3 - I_1 R_3$$

Therefore, $-E_2 = I_2 (R_2 + R_3) - I_1 R_3$

$$-12V = I_2 (240\ \Omega + 200\ \Omega) - (I_1 \times 200\ \Omega)$$

$$-12V = 440 I_2 - 200 I_1 \qquad (2)$$

Equations (1) *and* (2) *both contain two unknown quantities,* I_1 *and* I_2. *One of these two quantities must be eliminated before the other can be determined. Examining* Equations (1) *and* (2) *it is seen that if* Equation (1) *is multiplied by* 440/200, *then the multiple of* I_2 *becomes* 440, *as in* Equation (2). *This facilitates elimination of* I_2.

Equation (1) $\times \dfrac{440}{200}$:

$$13.2V = 704 I_1 - 440 I_2$$

Equation (2): $\quad -12V = -200 I_1 + 440 I_2$

Adding, $\qquad\quad 1.2V = 504 I_1 + 0$

Therefore, $\qquad\quad I_1 = \dfrac{1.2V}{504}$

$$I_1 \cong 2.38\ \text{mA}$$

Substituting for I_1 *in* Equation (2),

$$-12V = 440 I_2 - (200 \times 2.38\ \text{mA})$$

Therefore, $\qquad I_2 = \dfrac{-12V + (200 \times 2.38 \text{ mA})}{440}$

$$I_2 \cong -26.19 \text{ mA}$$

The negative sign indicates that the actual current direction for I_2 is opposite to the (clockwise) loop current.

$$I_3 = I_1 - I_2$$

$$\cong 2.38 \text{ mA} - (-26.19 \text{ mA})$$

$$\boldsymbol{I_3 \cong 28.57 \text{ mA}}$$

Compare these current levels to the answers obtained in Example 8-2.

PRACTICE PROBLEMS

8-3.1 If the circuit in Figure 8-3 has $R_1 = 100 \ \Omega$, $R_2 = 270 \ \Omega$, $R_3 = 180 \ \Omega$, and $E = 25$ V, use loop equations to calculate I_3.

8-3.2 The circuit in Problem 8-3.1 is altered by including a 12 V source in series with R_2 (+ terminal up). Use loop equations to calculate I_3.

8-4

**NODAL
ANALYSIS**

A *voltage node* is a junction in an electrical circuit at which a voltage can be measured with respect to another (reference) node. If one point in the circuit is grounded [see Figure 8-6(a)], that point is usually selected as the reference node. Otherwise, any convenient junction can be treated as a reference node.

From Figure 8-6(a), the current equation at node 1 is

$$I_3 = I_2 + I_1 \qquad\qquad (1)$$

An equation for each current can also be written in terms of the battery voltages and the node voltage:

$$I_1 = \frac{E_1 - V_1}{R_1} \qquad\qquad (2)$$

$$I_2 = \frac{E_2 - V_1}{R_2} \qquad\qquad (3)$$

$$I_3 = \frac{V_1}{R_3} \qquad\qquad (4)$$

Chap. 8   Network Analysis Techniques

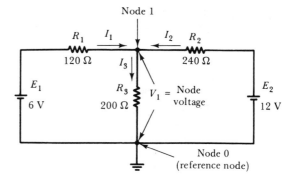

(a) Nodal analysis using voltage sources

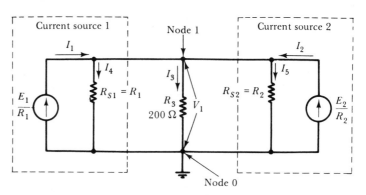

(b) Nodal analysis using current sources

FIGURE 8-6. For nodal analysis, the voltage at each node (or junction in the circuit) is identified with respect to ground (node zero). Then, circuit equations are written to determine the node voltages.

Substitution of Equations (2), (3), and (4) into Equation (1) gives

$$\frac{V_1}{R_3} = \frac{E_2 - V_1}{R_2} + \frac{E_1 - V_1}{R_1}$$

$$= \frac{E_2}{R_2} - \frac{V_1}{R_2} + \frac{E_1}{R_1} - \frac{V_1}{R_1}$$

or

$$V_1\left(\frac{1}{R_3} + \frac{1}{R_2} + \frac{1}{R_1}\right) = \frac{E_2}{R_2} + \frac{E_1}{R_1}$$

$$V_1 = \frac{\dfrac{E_1}{R_1} + \dfrac{E_2}{R_2}}{\dfrac{1}{R_3} + \dfrac{1}{R_1} + \dfrac{1}{R_2}} \tag{5}$$

Since V_1 is the only unknown quantity in Equation (5), the node voltage can be quickly determined. Then, using Equations (2), (3), and (4), the branch currents are readily calculated.

The above method is obviously simpler than either the Kirchhoff's law or loop-equation approach to network analysis. However, it can be simplified even further if the voltage sources and their series resistances are first converted to current sources, as illustrated in Figure 8-6(b).

NODAL ANALYSIS PROCEDURE:

1. *Convert all voltage sources to current sources, and redraw the circuit diagram.*
2. *Identify all nodes and choose a reference node.*
3. *Write the equations for the currents flowing into and out of each node, with the exception of the reference node.*
4. *Solve the equations to determine the node voltage and the required branch currents.*

EXAMPLE 8-4   For the circuit shown in Figure 8-6(a), convert the voltage sources to current sources, and use nodal analysis to determine the current through R_3.

SOLUTION

The circuit with voltage sources converted to current sources is shown in Figure 8-6(b). *Voltage source E_1 and series resistance R_1 convert to current source 1, where (from Section 8-1),*

$$I_1 = \frac{E_1}{R_1} = \frac{6\text{ V}}{120\text{ }\Omega}$$

and
$$R_{S1} = R_1 = 120\text{ }\Omega$$

For voltage source E_2 and series resistance R_2, current source 2 is

$$I_2 = \frac{E_2}{R_2} = \frac{12\text{ V}}{240\text{ }\Omega}$$

and
$$R_{S2} = R_2 = 240\text{ }\Omega$$

The modified circuit now shows three current paths, through R_{S1}, R_3, and R_{S2}.

Writing the current equation for node 1,

$$I_1 + I_2 = I_3 + I_4 + I_5 \qquad (1)$$

where
$$I_3 = \frac{V_1}{R_3}$$

Chap. 8   Network Analysis Techniques

$$I_4 = \frac{V_1}{R_{S1}}$$

$$I_5 = \frac{V_1}{R_{S2}}$$

Substituting into Equation (1),

$$\frac{E_1}{R_1} + \frac{E_2}{R_2} = \frac{V_1}{R_3} + \frac{V_1}{R_{S1}} + \frac{V_1}{R_{S2}} \qquad (2)$$

or

$$\frac{E_1}{R_1} + \frac{E_2}{R_2} = V_1 \left(\frac{1}{R_3} + \frac{1}{R_{S1}} + \frac{1}{R_{S2}} \right)$$

and

$$V_1 = \frac{\dfrac{E_1}{R_1} + \dfrac{E_2}{R_2}}{\dfrac{1}{R_3} + \dfrac{1}{R_{S1}} + \dfrac{1}{R_{S2}}}$$

$$V_1 = \frac{\dfrac{6\ \text{V}}{120\ \Omega} + \dfrac{12\ \text{V}}{240\ \Omega}}{\dfrac{1}{200\ \Omega} + \dfrac{1}{120\ \Omega} + \dfrac{1}{240\ \Omega}}$$

$$V_1 \cong 5.714\ \text{V}$$

$$I_3 = \frac{V_1}{R_3} \cong \frac{5.714\ \text{V}}{200\ \Omega}$$

$$\cong \textbf{28.57 mA}$$

EXAMPLE 8-5 Using nodal analysis, determine the current that flows through resistor R_3 in the circuit shown in Figure 8-7(a).

SOLUTION

Voltage source E_1 with series resistance R_1 becomes current source 1 in Figure 8-7(b), *with*

$$I_1 = \frac{E_1}{R_1} \quad \text{and} \quad R_S = R_1$$

Voltage source E_2 with series resistance R_2 becomes current source 2 in Figure 8-7(b), *with*

$$I_4 = \frac{E_2}{R_2} \quad \text{and} \quad R_S = R_2$$

The nodes are identified and numbered as 0, 1, *and* 2 [*see* Figure 8-7(b)].

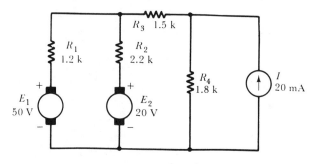

(a) Circuit to be analyzed

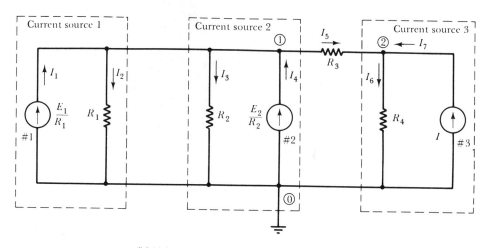

(b) Voltage sources replaced by current sources

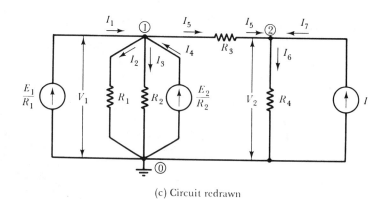

(c) Circuit redrawn

FIGURE 8-7. For nodal analysis, the voltage sources in (a) are first converted to current sources (b). Then, the circuit is redrawn in (c) to clearly show that there are only two nodes besides the reference node.

The circuit is redrawn in Figure 8-7(c) to more easily identify which currents flow into the nodes and which flow out.

At node 1:

$$I_1 + I_4 = I_2 + I_3 + I_5 \tag{1}$$

Therefore $\quad \dfrac{E_1}{R_1} + \dfrac{E_2}{R_2} = \dfrac{V_1}{R_1} + \dfrac{V_1}{R_2} + \dfrac{V_1 - V_2}{R_3}$

or,

$$\dfrac{E_1}{R_1} + \dfrac{E_2}{R_2} = \dfrac{V_1}{R_1} + \dfrac{V_1}{R_2} + \dfrac{V_1}{R_3} - \dfrac{V_2}{R_3}$$

$$\dfrac{50 \text{ V}}{1.2 \text{ k}\Omega} + \dfrac{20 \text{ V}}{2.2 \text{ k}\Omega} = V_1 \left(\dfrac{1}{1.2 \text{ k}\Omega} + \dfrac{1}{2.2 \text{ k}\Omega} + \dfrac{1}{1.5 \text{ k}\Omega} \right) - \dfrac{V_2}{1.5 \text{ k}\Omega}$$

$$50.76 \times 10^{-3} = 1.95 \times 10^{-3} \, V_1 - 0.667 \times 10^{-3} \, V_2 \tag{2}$$

At node 2:

$$I_6 = I_5 + I_7 \tag{3}$$

$$\dfrac{V_2}{R_4} = \dfrac{V_1 - V_2}{R_3} + I_7$$

$$\dfrac{V_2}{1.8 \text{ k}\Omega} = \dfrac{V_1}{1.5 \text{ k}\Omega} - \dfrac{V_2}{1.5 \text{ k}\Omega} + 20 \times 10^{-3}$$

or,

$$-20 \times 10^{-3} = \dfrac{V_1}{1.5 \text{ k}\Omega} - V_2 \left[\dfrac{1}{1.5 \text{ k}\Omega} + \dfrac{1}{1.8 \text{ k}\Omega} \right]$$

$$-20 \times 10^{-3} = 0.667 \times 10^{-3} \, V_1 - 1.22 \times 10^{-3} \, V_2 \tag{4}$$

Equation (2):

$$50.76 \times 10^{-3} = 1.95 \times 10^{-3} \, V_1 - 0.667 \times 10^{-3} \, V_2$$

Equation (4) $\times \dfrac{0.667}{1.22}$:

$$-10.93 \times 10^{-3} = 0.365 \times 10^{-3} \, V_1 - 0.667 \times 10^{-3} \, V_2 \tag{5}$$

Subtracting Equation (5) from Equation (2):

$$61.69 \times 10^{-3} = 1.585 \times 10^{-3} \, V_1$$

giving $\qquad V_1 \cong 38.92 \text{ V}$

From Equation (2),

$$50.76 \times 10^{-3} = (1.95 \times 10^{-3} \times 38.92 \text{ V}) - 0.667 \times 10^{-3} \, V_2$$

$$V_2 \cong 37.68 \text{ V}$$

$$I_3 = \frac{V_1 - V_2}{R_3}$$

$$\cong \frac{38.92 \text{ V} - 37.68 \text{ V}}{1.5 \text{ k}\Omega}$$

$$\cong \textbf{0.83 mA}$$

PRACTICE PROBLEMS

8-4.1 Use nodal analysis to determine I_2 in the circuit described in Problem 8-3.1.

8-4.2 Use nodal analysis to determine I_1 in the circuit described in Problem 8-3.2.

8-5
DELTA-WYE TRANS-FORMATIONS

Consider the circuit shown in Figure 8-8(a), in which resistors R_1, R_2, and R_3 are in the form of a *delta* (Δ) network. The circuit, as shown, could be analyzed by loop equations or by nodal analysis. However, the analysis can be considerably simplified if the Δ network is replaced by the *wye* (Y) network shown in Figure 8-8(b). If the Δ network is to be replaced by the Y network, both networks must have exactly the same effect on the rest of the circuit. Consequently, the resistance measured between any two of terminals *A, B,* and *C* must be exactly the same in each case. Figures 8-8(c) and (d) represent exactly the same circuits as in Figures 8-8(a) and (b), respectively.

For Δ–Y and Y–Δ transformation, equations must be derived to relate the resistors in the two networks. Consider Figures 8-9(a) and (b). *For the Y network, the resistance between terminals A and B is*

$$R_{AB} = R_a + R_b$$

and for the Δ network,

$$R_{AB} = R_{ab} \parallel (R_{ac} + R_{bc})$$

Therefore,
$$R_{AB} = \frac{R_{ab}(R_{ac} + R_{bc})}{R_{ab} + R_{ac} + R_{bc}}$$

$$R_{AB} = \frac{R_{ab}R_{ac} + R_{ab}R_{bc}}{R_{ab} + R_{ac} + R_{bc}}$$

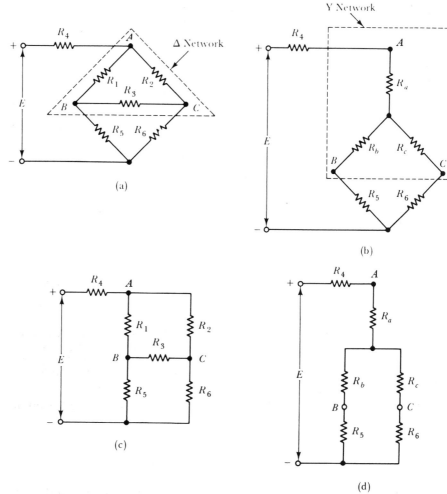

FIGURE 8-8. A delta network can be converted into an equivalent wye network, and vice versa, to facilitate circuit analysis.

For the same resistance value between terminals A and B on both Y and Δ networks,

$$R_a + R_b = \frac{R_{ab} R_{ac} + R_{ab} R_{bc}}{R_{ab} + R_{ac} + R_{bc}} \tag{1}$$

Similarly, it can be shown that

$$R_a + R_c = \frac{R_{ab} R_{ac} + R_{ac} R_{bc}}{R_{ab} + R_{ac} + R_{bc}} \tag{2}$$

and $$R_b + R_c = \frac{R_{ab} R_{bc} + R_{ac} R_{bc}}{R_{ab} + R_{ac} + R_{bc}} \tag{3}$$

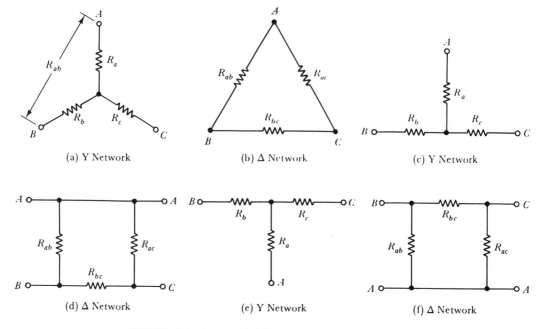

FIGURE 8-9. Wye and delta networks identified for convenience in converting between types. In the wye networks, each resistor is named for the terminal it is connected to—R_a, R_b, and R_c. For the delta network, each resistor is named for the two terminals it is connected across—R_{ab}, R_{bc}, and R_{ac}.

Adding Equations (1) *and* (2) *and subtracting* Equation (3) *gives*

$$R_a = \frac{R_{ab}R_{ac}}{R_{ab} + R_{ac} + R_{bc}} \qquad \text{(8-5)}$$

Also, Equations (1) + Equation (3) − Equation (2) *gives*

$$R_b = \frac{R_{ab}R_{bc}}{R_{ab} + R_{ac} + R_{bc}} \qquad \text{(8-6)}$$

and Equation (2) + Equation (3) − Equation (1) *gives*

$$R_c = \frac{R_{ac}R_{bc}}{R_{ab} + R_{ac} + R_{bc}} \qquad \text{(8-7)}$$

Equations (8-5), (8-6), and (8-7) can be used to convert from Δ to Y. Examination of Equation (8-5) shows that *to obtain the Y network resistor (R_a) connected to terminal A, the two Δ network resistors connected to A must be multiplied together, and the product divided by the sum of* *the three Δ network resistors.* A similar procedure applies for obtaining the Y network resistors connected to terminals *B* and *C*.

By further manipulation of Equations (1), (2), and (3), equations for conversion from Y to Δ can be obtained:

$$R_{ab} = \frac{R_a R_b + R_a R_c + R_b R_c}{R_c} \qquad (8\text{-}8)$$

$$R_{ac} = \frac{R_a R_b + R_a R_c + R_b R_c}{R_b} \qquad (8\text{-}9)$$

$$R_{bc} = \frac{R_a R_b + R_a R_c + R_b R_c}{R_a} \qquad (8\text{-}10)$$

Once again, examination of the three equations reveals a definite pattern. *To obtain the Δ resistor connected between any two terminals, divide the Y resistor connected to the terminal opposite those two (terminal C for the resistor between A and B) into the sum of the products of each pair of Y resistors.*

The Y network is also known as a *star network* or a *T network*. Figures 8-9(c) and (e) show Y networks redrawn in a slightly different form. Delta networks are also termed *mesh networks* or *π networks*. Figures 8-9(d) and (f) show delta networks redrawn in a different form.

Δ-*Y* and *Y*-Δ TRANSFORMATION PROCEDURE:

1. *When starting with a Δ network, draw a Y network.*
   When starting with a Y network, draw a Δ network.
2. *Identify the three corresponding terminals on each network as A, B, and C.*
3. *Identify the resistors on the Δ network:*
   resistor between terminals A and B as R_{ab}
   resistor between terminals A and C as R_{ac}
   resistor between terminals B and C as R_{bc}
4. *Identify the resistors on the Y network:*
   resistor connected to terminal A as R_a
   resistor connected to terminal B as R_b
   resistor connected to terminal C as R_c
5. *For Δ and Y transformation, substitute the Δ network resistor values into Equations (8-5), (8-6), and (8-7) to obtain the Y network resistor values.*
6. *For Y and Δ transformation, substitute the Y network resistor values into Equations (8-8), (8-9), and (8-10) to obtain the Δ network resistor values.*

EXAMPLE 8-6 Convert the Δ network shown in Figure 8-9(b) to a Y network. Then convert back again to prove the formula. Take $R_{ab} = 500\ \Omega$, $R_{ac} = 400\ \Omega$, and $R_{bc} = 300\ \Omega$.

SOLUTION

Equation (8-5):

$$R_a = \frac{R_{ab}\,R_{ac}}{R_{ab} + R_{ac} + R_{bc}}$$

$$= \frac{500\ \Omega \times 400\ \Omega}{500\ \Omega + 400\ \Omega + 300\ \Omega}$$

$$= \mathbf{166.7\ \Omega}$$

Equation (8-6):

$$R_b = \frac{R_{ab}\,R_{bc}}{R_{ab} + R_{ac} + R_{bc}}$$

$$= \frac{500\ \Omega \times 300\ \Omega}{500\ \Omega + 400\ \Omega + 300\ \Omega}$$

$$= \mathbf{125\ \Omega}$$

Equation (8-7):

$$R_c = \frac{R_{ac}\,R_{bc}}{R_{ab} + R_{ac} + R_{bc}}$$

$$= \frac{400\ \Omega \times 300\ \Omega}{500\ \Omega + 400\ \Omega + 300\ \Omega}$$

$$= \mathbf{100\ \Omega}$$

Converting back from Y to Δ:
Equation (8.8):

$$R_{ab} = \frac{R_a R_b + R_a R_c + R_b R_c}{R_c}$$

$$= \frac{(166.7\ \Omega \times 125\ \Omega) + (166.7\ \Omega \times 100\ \Omega) + (125\ \Omega \times 100\ \Omega)}{100\ \Omega}$$

$$= \frac{50\ 000\ \Omega}{100}$$

$$= \mathbf{500\ \Omega}$$

Equation (8-9):

$$R_{ac} = \frac{R_a R_b + R_a R_c + R_b R_c}{R_b}$$

$$= \frac{50\ 000\ \Omega}{125}$$

$$= \textbf{400}\ \boldsymbol{\Omega}$$

Equation (8-10):

$$R_{bc} = \frac{R_a R_b + R_a R_c + R_b R_c}{R_a}$$

$$= \frac{50\ 000\ \Omega}{166.7}$$

$$= \textbf{300}\ \boldsymbol{\Omega}$$

PRACTICE PROBLEMS

8-5.1 For the wye network in Figure 8-9(a), $R_a = 1.8\ \text{k}\Omega$, $R_b = 3.9\ \text{k}\Omega$, and $R_c = 2.7\ \text{k}\Omega$. Calculate the equivalent delta network component values.

8-5.2 The delta network in Figure 8-9(d) has $R_{ab} = 47\ \Omega$, $R_{bc} = 39\ \Omega$, and $R_{ac} = 56\ \Omega$. Calculate the equivalent wye network component values.

SUMMARY OF FORMULAS

Voltage-source to current-source conversion:

$$I = \frac{E}{R_s} \qquad R_s = R_s$$

Current-source to voltage-source conversion:

$$V = IR_s \qquad R_s = R_s$$

Δ-to-Y conversion:

$$R_a = \frac{R_{ab} R_{ac}}{R_{ab} + R_{ac} + R_{bc}}$$

$$R_b = \frac{R_{ab} R_{bc}}{R_{ab} + R_{ac} + R_{bc}}$$

$$R_c = \frac{R_{ac} R_{bc}}{R_{ab} + R_{ac} + R_{bc}}$$

Y-to-Δ conversion:

$$R_{ab} = \frac{R_a R_b + R_a R_c + R_b R_c}{R_c}$$

$$R_{ac} = \frac{R_a R_b + R_a R_c + R_b R_c}{R_b}$$

$$R_{bc} = \frac{R_a R_b + R_a R_c + R_b R_c}{R_a}$$

E.C.A.P. (Electronic Circuit Analysis Program)

E.C.A.P. is a powerful general purpose circuit analysis program that affords dc, ac, and transient analysis of electronic circuits. A very simple circuit is used to demonstrate the procedure for dc analysis by E.C.A.P.

Circuit to be analyzed

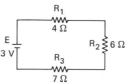

1. Draw an "E.C.A.P. equivalent circuit" simply by numbering the circuit nodes and branches. Also indicate the (expected) direction of current flow in each branch.

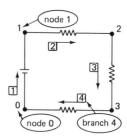

2. Write the E.C.A.P. program to describe the circuit:

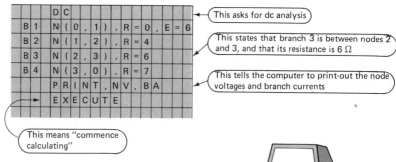

This asks for dc analysis

This states that branch 3 is between nodes 2 and 3, and that its resistance is 6 Ω

This tells the computer to print-out the node voltages and branch currents

This means "commence calculating"

3. Type the program into the terminal of a computer that can perform ECAP

4. The computer displays and prints the results of the analysis:

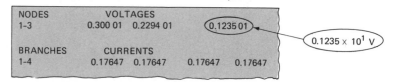

0.1235×10^1 V

NODES	VOLTAGES		
1-3	0.300 01	0.2294 01	0.1235 01
BRANCHES	CURRENTS			
1-4	0.17647	0.17647	0.17647	0.17647

8-1   Define constant voltage source, constant current source, loop equations, loop currents, and mesh questions.

8-2   Define voltage node, nodal analysis, delta network, π network, Y network, and T network.

8-3   Sketch circuit diagrams for constant voltage and constant current sources. Show all voltage polarities and current directions.

8-4   Derive the equations for converting between constant current and constant voltage sources. Discuss the effects of series and parallel operation of each type of source.

8-5   List the procedure for analyzing a complex network by use of Kirchhoff's laws.

8-6   List the procedure for network analysis using loop equations.

8-7   List the procedure for nodal analysis.

8-8   Write the equations for Δ–Y conversion and for Y–Δ conversion.

PROBLEMS

SECTION 8-1

8-1   A constant current source has $I = 15$ A and $R_s = 5\,\Omega$. Determine E and R_s for the equivalent constant voltage source. If $R_L = 25\,\Omega$, calculate the output voltage and current in each case.

8-2   For the circuit of Figure 8-10, convert the voltage sources (E_1, R_1) and (E_2, R_2) into their equivalent current sources.

8-3   For the circuit of Figure 8-11, convert (E_1, E_2, R_1) into the equivalent current source.

8-4   For the circuit of Figure 8-12, convert (E_1, R_2) into the equivalent current source.

8-5   For the circuit of Figure 8-17, convert (I_1, R_1) into the equivalent voltage source.

8-6   For the circuit of Figure 8-18, convert (I_1, R_1) into the equivalent voltage source.

8-7   For the circuit of Figure 8-19, convert (I_1, R_3) into the equivalent voltage source.

SECTION 8-2

8-8   Using Kirchhoff's laws, analyze the circuit in Figure 8-10 to determine the voltage drop across resistor R_5.

8-9   Using Kirchhoff's laws, determine I_2 in Figure 8-11.

8-10   Using Kirchhoff's laws, determine I_1 in Figure 8-12.

8-11   Using Kirchhoff's laws, determine I_3 in Figure 8-13.

8-12   Using Kirchhoff's laws, analyze the circuit in Figure 8-14 to determine the current through R_5.

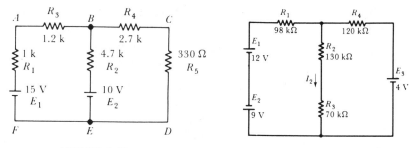

FIGURE 8-10.

FIGURE 8-11.

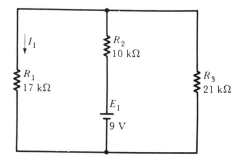

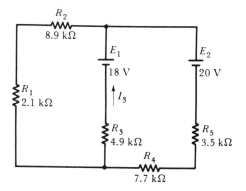

FIGURE 8-12.

FIGURE 8-13.

8-13 Using Kirchhoff's laws, analyze the circuit in Figure 8-15 to determine the voltage drop across R_5.

8-14 Using Kirchhoff's laws, analyze the circuit in Figure 8-16 to determine the current through R_3.

8-15 Using Kirchhoff's laws, analyze the circuit in Figure 8-17 to determine the current through resistor R_3.

8-16 Using Kirchhoff's laws, analyze the circuit in Figure 8-18 to determine the current through resistor R_5.

8-17 Using Kirchhoff's laws, analyze the circuit in Figure 8-19 to determine the current through resistor R_4.

190   Chap. 8   Network Analysis Techniques

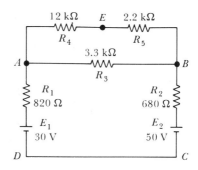

FIGURE 8-14.

SECTION 8-3

8-18   Repeat Problem 8-8 using loop equations.

8-19   Repeat Problem 8-9 using loop equations.

8-20   Repeat Problem 8-10 using loop equations.

8-21   Repeat Problem 8-11 using loop equations.

8-22   Repeat Problem 8-12 using loop equations.

8-23   Repeat Problem 8-13 using loop equations.

8-24   Repeat Problem 8-14 using loop equations.

8-25   Repeat Problem 8-15 using loop equations.

8-26   Repeat Problem 8-16 using loop equations.

8-27   Repeat Problem 8-17 using loop equations.

SECTION 8-4

8-28   Repeat Problem 8-8 using nodal analysis.

8-29   Repeat Problem 8-9 using nodal analysis.

8-30   Repeat Problem 8-10 using nodal analysis.

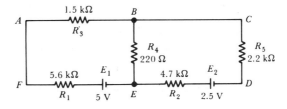

FIGURE 8-15.

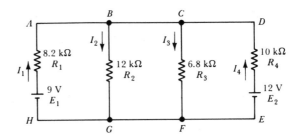

FIGURE 8-16.

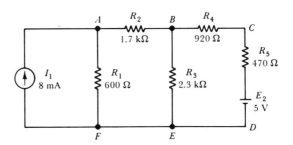

FIGURE 8-17.

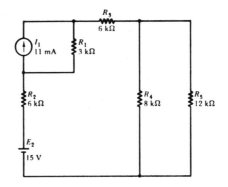

FIGURE 8-18.

8-31 Repeat Problem 8-11 using nodal analysis.

8-32 Repeat Problem 8-12 using nodal analysis.

8-33 Repeat Problem 8-13 using nodal analysis.

8-34 Repeat Problem 8-14 using nodal analysis.

8-35 Repeat Problem 8-15 using nodal analysis.

Chap. 8   Network Analysis Techniques

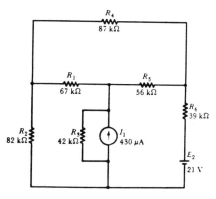

FIGURE 8-19.

8-36 Repeat Problem 8-16 using nodal analysis.

8-37 Repeat Problem 8-17 using nodal analysis.

SECTION 8-5

8-38 Apply wye-delta transformation to resistors R_1, R_2 and R_3 in the circuit in Figure 8-6(a), then determine the current taken from E_1.

8-39 Modify the circuit in Figure 8-11 to replace resistors R_1, R_2 and R_4 with the equivalent delta network.

8-40 Modify the circuit in Figure 8-19 to replace resistors R_1, R_4 and R_5 with the equivalent wye network.

COMPUTER PROBLEMS

8-41 Write a computer program to analyze the circuit in Figure 8-10 to determine the individual resistor voltages and currents.

8-42 Write a computer program to analyze the circuit in Figure 8-16 to determine the individual resistor voltages and currents.

ANSWERS TO PRACTICE PROBLEMS		
8-1.1	500 μA, 1 kΩ, 6.02 μA
8-1.2	150 V, 20 kΩ, 17.84 V
8-2.1	3.11 mA
8-2.2	2.95 mA
8-3.1	72.1 mA
8-3.2	84.9 mA
8-4.1	48.07 mA
8-4.2	97.1 mA
8-5.1	8.3 kΩ, 5.75 kΩ, 12.45 kΩ
8-5.2	18.54 kΩ, 12.91 kΩ, 15.38 kΩ

9

NETWORK
THEOREMS

Objectives   You will be able to:

State the superposition theorem and apply it to simplify the analysis of complex resistive networks.

State Thévenin's theorem and apply it to simplify the analysis of complex resistive networks.

State Norton's theorem and apply it to simplify problems involving complex resistive networks.

State Millman's theorem and apply it to simplify problems involving complex resistive networks.

State the maximum power transfer theorem and apply it to the solution of problems involving variations in load on a given source.

Introduction

Network analysis can be simplified by the use of *network theorems,* which state certain rules that may be applied in particular circumstances. The *superposition theorem,* for example, enables a circuit with several voltage sources and/or current sources to be treated as several circuits that each have only one source. *Thévenin's theorem* permits complex networks to be reduced to a single voltage source in series with a resistance. Thus simplified, the load current and voltage can be very easily determined for many different values of load resistance. *Norton's theorem* is just as powerful as Thévenin's theorem. In this case the network is reduced to a single current source and parallel resistance. How a circuit consisting of many sources in parallel may be treated as just one source is defined by *Millman's theorem,* while the *maximum power transfer theorem* predicts optimum load conditions.

THE SUPER-POSITION THEOREM

In a network containing more than one source of voltage or current, the current through any branch is the algebraic sum of the currents produced by each source acting independently.

Application of the superposition theorem simplifies the analysis of a network which has more than one source.

PROCEDURE FOR APPLYING THE SUPERPOSITION THEOREM:

1. *Select one source, and replace all other sources with their internal impedances.*
2. *Determine the level and direction of the current that flows through the desired branch as a result of the single source acting alone.*
3. *Repeat steps 1 and 2 using each source in turn until the branch current components have been calculated for all sources.*
4. *Algebraically sum the component currents to obtain the actual branch current.*

EXAMPLE 9-1 Use the superposition theorem to calculate the current I_3 in the circuit shown in Figure 9-1(a) (reproduced from Figure 8-4).

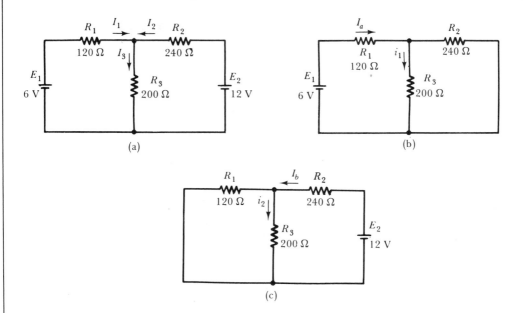

FIGURE 9-1. Application of the superposition theorem to determine the current I_3 through resistor R_3 in (a) above. The current i_1 produced by source E_1 when $E_2 = 0$ is first determined from circuit (b). Then, the current i_2 due to E_2 when $E_1 = 0$ is found from circuit (c). The result is $I_3 = i_1 + i_2$.

SOLUTION

In Figure 9-1(b) *voltage source* E_2 *is replaced with its internal resistance* (R_2), *and in* Figure 9-1(c) *voltage source* E_1 *is replaced with its internal resistance* (R_1).

For Figure 9-1(b):
The equivalent resistance of the circuit is

$$R_{eq1} = R_1 + R_2 \| R_3$$
$$= 120 \ \Omega + 240 \ \Omega \| 200 \ \Omega$$
$$\cong 229.09 \ \Omega$$

and
$$I_a = \frac{E_1}{R_{eq1}} \cong \frac{6 \text{ V}}{229.09 \ \Omega}$$
$$\cong 26.19 \text{ mA}$$

Using the current divider rule (*see* Section 6-4),

$$i_1 = I_a \times \frac{R_2}{R_3 + R_2}$$
$$\cong 26.19 \text{ mA} \times \frac{240 \ \Omega}{200 \ \Omega + 240 \ \Omega}$$
$$\cong 14.29 \text{ mA}$$

For Figure 9-1(c):

$$R_{eq2} = R_2 + R_1 \| R_3$$
$$= 240 \ \Omega + 120 \ \Omega \| 200 \ \Omega$$
$$= 315 \ \Omega$$
$$I_b = \frac{E_2}{R_{eq2}} = \frac{12 \text{ V}}{315 \ \Omega}$$
$$\cong 38.10 \text{ mA}$$

Using the current divider rule,

$$i_2 = I_b \times \frac{R_1}{R_1 + R_3}$$
$$\cong 38.1 \text{ mA} \times \frac{120 \ \Omega}{120 \ \Omega + 200 \ \Omega}$$
$$\cong 14.29 \text{ mA}$$

Total current,

$$I_3 = i_1 + i_2$$

$$\cong 14.29 \text{ mA} + 14.29 \text{ mA}$$

$$I_3 \cong \textbf{28.58 mA}$$

Note: The fact that i_1 and i_2 happen to be equal is peculiar to this particular circuit.

EXAMPLE 9-2   For the circuit of Figure 9-2(a) [reproduced from Figure 8-7(a)], determine the current through resistor R_3 using the superposition theorem.

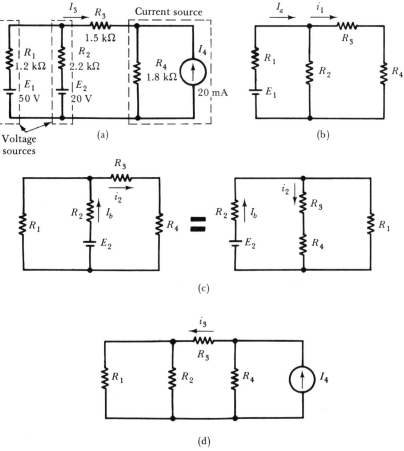

(a)

(b)

(c)

(d)

FIGURE 9-2.   Superposition theorem applied to determine the current I_3 in circuit (a). With $E_2 = 0$, and $I_7 = 0$, i_1 is calculated from circuit (b). Current i_2 is next found from circuit (c), in which $E_1 = 0$ and $I_7 = 0$. Finally, when E_1 and E_2 are both zero, i_3 is determined from circuit (d). Then $I_3 = i_1 + i_2 + i_3$.

SOLUTION

The circuits for each individual source acting independently are shown in Figure 9-2(b), (c), *and* (d).

In Figure 9-2(b),

$$I_a = \frac{E_1}{R_1 + R_2 \| (R_3 + R_4)}$$

$$= \frac{50 \text{ V}}{1.2 \text{ k}\Omega + 2.2 \text{ k}\Omega \| (1.5 \text{ k}\Omega + 1.8 \text{ k}\Omega)}$$

$$\cong 19.84 \text{ mA}$$

and using the current divider rule,

$$i_1 = I_a \times \frac{R_2}{R_2 + (R_3 + R_4)}$$

$$\cong 19.84 \text{ mA} \times \frac{2.2 \text{ k}\Omega}{2.2 \text{ k}\Omega + (1.5 \text{ k}\Omega + 1.8 \text{ k}\Omega)}$$

$$\cong 7.94 \text{ mA}$$

For Figure 9-2(c),

$$I_b = \frac{E_2}{R_2 + R_1 \| (R_3 + R_4)}$$

$$= \frac{20 \text{ V}}{2.2 \text{ k}\Omega + 1.2 \text{ k}\Omega \| (1.5 \text{ k}\Omega + 1.8 \text{ k}\Omega)}$$

$$\cong 6.49 \text{ mA}$$

Using the current divider rule,

$$i_2 = I_b \times \frac{R_1}{R_1 + (R_3 + R_4)}$$

$$\cong 6.49 \text{ mA} \times \frac{1.2 \text{ k}\Omega}{1.2 \text{ k}\Omega + (1.5 \text{ k}\Omega + 1.8 \text{ k}\Omega)}$$

$$\cong 1.73 \text{ mA}$$

For Figure 9-2(d):
Using the current divider rule,

$$i_3 = I_7 \times \frac{R_4}{R_4 + R_3 + (R_1 \| R_2)}$$

$$= 20 \text{ mA} \times \frac{1.8 \text{ k}\Omega}{1.8 \text{ k}\Omega + 1.5 \text{ k}\Omega + (1.2 \text{ k}\Omega \| 2.2 \text{ k}\Omega)}$$

$$\cong 8.83 \text{ mA}$$

Current through R_3,

$$I_3 = i_1 + i_2 - i_3$$
$$\cong 7.94 \text{ mA} + 1.73 \text{ mA} - 8.83 \text{ mA}$$

$I_3 \cong \textbf{0.84 mA}$

This compares with the result of Example 8-5.

PRACTICE PROBLEMS

9-1.1 Apply the superposition theorem to the circuit of Figure 8-10 to determine the current through resistor R_3.

9-1.2 Apply the superposition theorem to the circuit of Figure 8-11 to determine the current through resistor R_1.

9-2

THÉVENIN'S THEOREM

Any two-terminal network containing resistances and voltage sources and/or current sources may be replaced by a single voltage source in series with a single resistance. The emf of the voltage source is the open-circuit emf at the network terminals, and the series resistance is the resistance between the network terminals when all sources are replaced by their internal impedances.

By means of Thévenin's theorem, any one resistor in a network can be isolated. The entire remaining portion of the network can be replaced by a single source of emf and a single resistor. Then the current through the isolated resistor may be easily calculated for any value of resistance.

Consider the resistor network shown in Figure 9-3(a). Resistor R_L is variable. Consequently, a complete circuit analysis would seem to be necessary for each value of R_L. The circuit is simplified by replacing all of the network to the left of terminals A and B by its *Thévenin equivalent circuit*. This results in the circuit shown in Figure 9-3(b). Clearly, the calculation of load voltage and current can now be easily repeated for any number of values of R_L.

Determining the Thévenin equivalent circuit for a network is sometimes termed *thévenizing the circuit*.

THÉVENIZING PROCEDURE:

1. Calculate the open-circuit voltage (E_{TH}) at the network terminals.

2. Redraw the network with each source replaced by its internal resistance. (Note that when no source resistances are shown, voltage sources should be short-circuited and current sources should be open-circuited.)

3. Calculate the resistance (R_{TH}) of the redrawn network as seen from the output terminals.

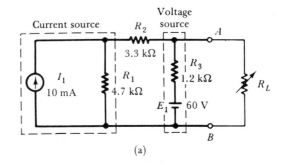

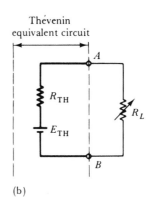

FIGURE 9-3. Application of Thévenin's theorem. The network to the left of terminals A and B in circuit (a) is replaced by its open-circuit output voltage E_{TH} and internal resistance R_{TH} to give circuit (b).

EXAMPLE 9-3 For the circuit in Figure 9-4(a) (reproduced from Figure 8-4), an external load R_L is to be connected across resistor R_3, as illustrated. Determine the Thévenin equivalent circuit for the network and calculate the load current when R_L is 330 Ω.

SOLUTION

The open-circuit voltage at terminals B and E must first be calculated. From Example 8-2,

$$I_3 \cong 28.57 \text{ mA}$$

$$V_{BE} = V_{R3} = I_3 R_3$$

$$\cong 28.57 \text{ mA} \times 200 \text{ Ω}$$

$$E_{TH} = V_{BE} \cong 5.71 \text{ V}$$

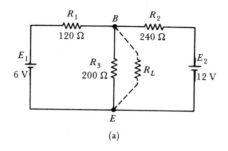

(a)

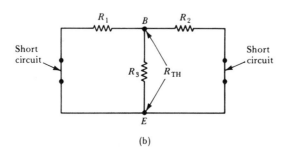

(b)

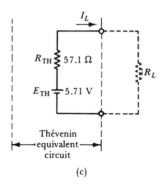

(c)

FIGURE 9-4. Application of Thévenin's theorem to determine the current through load R_L connected across terminals B and E in circuit (a). The open-circuit output voltage across B and E is first calculated to give E_{TH}. Then, the internal resistance R_{TH} is determined with E_1 and E_2 both equal zero, as in circuit (b).

Replacing E_1 and E_2 by short-circuits gives the circuit of Figure 9-4(b). The resistance seen looking into terminals B and E is

$$R_{TH} = R_3 \| R_1 \| R_2$$

$$= 200\ \Omega \| 120\ \Omega \| 240\ \Omega$$

$$R_{TH} \cong \mathbf{57.1\ \Omega}$$

The Thévenin equivalent circuit is now as shown in Figure 9-4(c).
The load current is calculated as

$$I_L = \frac{E_{TH}}{R_{TH} + R_L}$$

$$\cong \frac{5.71 \text{ V}}{57.1 \ \Omega + 330 \ \Omega}$$

$$\cong \mathbf{14.75 \ mA}$$

*Obviously, the load current can be readily recalculated for any number
of values of R_L.*

EXAMPLE 9-4   Derive the Thévenin equivalent circuit for the network shown in
Figure 9-3(a), and calculate the output voltage for $R_{L1} = 12 \ k\Omega$ and
for $R_{L2} = 5.6 \ k\Omega$.

SOLUTION

The circuit of Figure 9-3(a) *is reproduced in* Figure 9-5(a) *with the load
resistance removed. The open-circuit voltage at terminals A and B can
now be calculated by any of the methods already discussed.*
   *Using the superposition theorem, a current i_1 flows through R_3 due
to current source I_1, and a current i_2 flows in the opposite direction due
to voltage source E_1.*
   *For I_1 acting alone (voltage source E_1 replaced by R_3), using the
current divider rule:*

$$i_1 = I_1 \times \frac{R_1}{R_1 + (R_2 + R_3)}$$

$$= 10 \text{ mA} \times \frac{4.7 \ k\Omega}{4.7 \ k\Omega + 3.3 \ k\Omega + 1.2 \ k\Omega}$$

$$\cong 5.11 \text{ mA}$$

For E_1 acting alone (I_1 source open-circuited):

$$i_2 = \frac{E_1}{R_3 + R_2 + R_1}$$

$$= \frac{60 \text{ V}}{1.2 \ k\Omega + 3.3 \ k\Omega + 4.7 \ k\Omega}$$

$$\cong 6.52 \text{ mA}$$

Therefore, $I_3 = i_2 - i_1$

$$\cong 6.52 \text{ mA} - 5.11 \text{ mA}$$

$$\cong 1.41 \text{ mA}$$

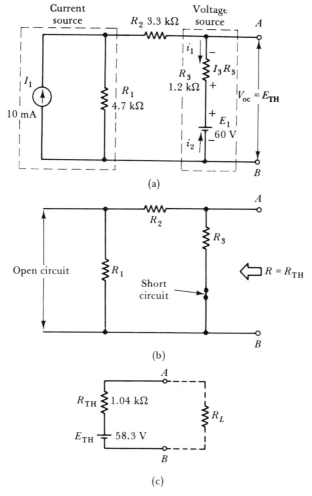

FIGURE 9-5. Application of Thévenin's theorem to the circuit of Figure 9-3(a). The open-circuit output voltage across A and B is first calculated to give E_{TH}. Then, the internal resistance R_{TH} is determined with I_1 open-circuited and E_1 short-circuited.

$$\textit{and} \qquad V_{AB} = E_1 - I_3 R_3 \qquad [\textit{see} \text{ Figure 9-5(a)}]$$

$$\cong 60 \text{ V} - (1.41 \text{ mA} \times 1.2 \text{ k}\Omega)$$

$$E_{TH} = V_{AB} \cong \mathbf{58.3 \text{ V}}$$

In Figure 9-5(b) the current source and voltage source are replaced by their internal resistances. From Figure 9-5(b),

$$R_{TH} = R_3 \| (R_2 + R_1)$$

$$= 1.2 \text{ k}\Omega \| (3.3 \text{ k}\Omega + 4.7 \text{ k}\Omega)$$

$$\cong \mathbf{1.04 \text{ k}\Omega}$$

The Thévenin equivalent circuit is now as shown in Figure 9-5(c). *The output voltage across R_L can be calculated using the voltage divider rule:*

$$V_L = E_{TH} \times \frac{R_L}{R_L + R_{TH}}$$

For $R_L = 12$ kΩ,

$$V_{L1} \cong 58.3 \text{ V} \times \frac{12 \text{ k}\Omega}{12 \text{ k}\Omega + 1.04 \text{ k}\Omega}$$

$$\cong \textbf{53.65 V}$$

For $R_L = 5.6$ kΩ,

$$V_{L2} \cong 58.3 \text{ V} \times \frac{5.6 \text{ k}\Omega}{5.6 \text{ k}\Omega + 1.04 \text{ k}\Omega}$$

$$\cong \textbf{49.17 V}$$

Example 9-4 illustrates the circuit simplification afforded by Thévenin's theorem. There would obviously be much more work involved if the circuit had to be completely analyzed each time a new value of R_L is substituted. Now look at Example 9-5, which further demonstrates how this useful circuit theorem tremendously cuts down on calculations. Were Thévenin's theorem not available, the circuit would have to be completely analyzed nine times for the nine different values of R_L listed.

EXAMPLE 9-5 For the circuit in Figure 9-5(a), calculate the load current for the following values of R_L (connected across terminals A and B): 12 kΩ, 5.6 kΩ, 3.3 kΩ, 10 kΩ, 6.8 kΩ, 8.2 kΩ, 4.7 kΩ, 13 kΩ, and 15 kΩ.

SOLUTION

Using the Thévenin equivalent circuit in Figure 9-5(c):

$$I_L = \frac{E_{TH}}{R_{TH} + R_L}$$

$$I_{L1} = \frac{58.3 \text{ V}}{1.04 \text{ k}\Omega + 12 \text{ k}\Omega} \cong \textbf{4.47 mA}$$

$$I_{L2} = \frac{58.3 \text{ V}}{1.04 \text{ k}\Omega + 5.6 \text{ k}\Omega} \cong \textbf{8.78 mA}$$

$$I_{L3} = \frac{58.3 \text{ V}}{1.04 \text{ k}\Omega + 3.3 \text{ k}\Omega} \cong \textbf{13.43 mA}$$

$$I_{L4} = \frac{58.3 \text{ V}}{1.04 \text{ k}\Omega + 10 \text{ k}\Omega} \cong \textbf{5.28 mA}$$

$$I_{L5} = \frac{58.3 \text{ V}}{1.04 \text{ k}\Omega + 6.8 \text{ k}\Omega} \cong \textbf{7.44 mA}$$

$$I_{L6} = \frac{58.3 \text{ V}}{1.04 \text{ k}\Omega + 8.2 \text{ k}\Omega} \cong \textbf{6.31 mA}$$

$$I_{L7} = \frac{58.3 \text{ V}}{1.04 \text{ k}\Omega + 4.7 \text{ k}\Omega} \cong \textbf{10.16 mA}$$

$$I_{L8} = \frac{58.3 \text{ V}}{1.04 \text{ k}\Omega + 13 \text{ k}\Omega} \cong \textbf{4.15 mA}$$

$$I_{L9} = \frac{58.3 \text{ V}}{1.04 \text{ k}\Omega + 15 \text{ k}\Omega} \cong \textbf{3.63 mA}$$

PRACTICE PROBLEMS

9-2.1   Use Thévenin's theorem to determine current I_3 in the circuit of Figure 7-1(a).

9-2.2   With R_1 open-circuited in Figure 7-13, apply Thévenin's theorem to calculate the voltage across resistor R_5.

9-3
NORTON'S THEOREM

In Section 8-1 it was shown that constant voltage sources can be converted to constant current sources, and vice versa. Therefore, a Thévenin equivalent circuit, which is a constant voltage source, can be converetd to a constant current source. Alternatively, instead of deriving the Thévenin constant voltage equivalent circuit for a complex network, a constant current equivalent circuit can be derived directly for the network. In this case the equivalent circuit is termed a *Norton equivalent circuit,* and the theory for its application is stated in *Norton's theorem:*

Norton's Theorem   **Any two-terminal network containing resistances and voltage sources and/or current sources may be replaced by a single current source in parallel with a single resistance. The output from the current source is the short-circuit current at the network terminals, and the parallel resistance is the resistance between the network terminals when all sources are replaced by their internal impedances.**

As with thévenizing, the determination of the Norton equivalent circuit is termed *nortonizing.*

NORTONIZING PROCEDURE:

1. Calculate the short-circuit current (I_N) at the network terminals.

2. Redraw the network with each source replaced by its internal re-

sistance. *(Note that when no source resistances are shown, voltage sources should be short-circuited and current sources should be open-circuited.)*

3. **Calculate the resistance (R_N) of the redrawn network as seen from the output terminals.**

Like the Thévenin theorem, application of the Norton theorem can save a lot of calculations when several values of load resistor are involved.

EXAMPLE 9-6   Derive the Norton equivalent circuit for the network shown in Figure 9-3(a) and calculate the load current for $R_L = 12$ kΩ and for $R_L = 5.6$ kΩ.

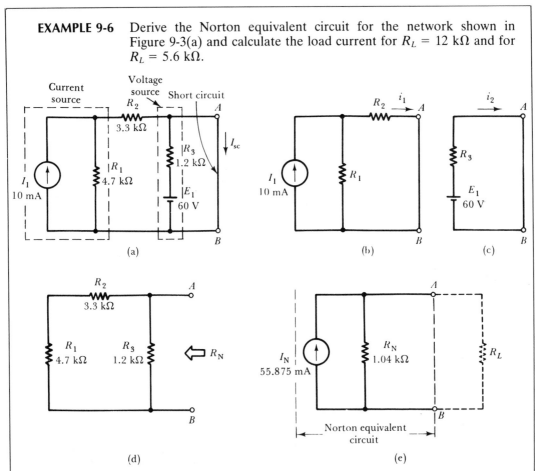

FIGURE 9-6. Application of Norton's theorem to the circuit of Figure 9-3(a). The output current when A and B are short-circuited is first calculated to give I_N. Then, the internal resistance R_N is determined with I_1 open-circuited and E_2 short-circuited.

SOLUTION

The circuit of Figure 9-3(a) *is reproduced in* Figure 9-6(a) *with R_L replaced by a short-circuit. Using the superposition theorem to determine I_{sc}:*

Chap. 9   Network Theorems

$$I_{sc} = i_1 + i_2 \quad [\textit{see} \text{ Figures 9-6(b) } \textit{and} \text{ (c)}]$$

$$i_1 = I_1 \times \frac{R_1}{R_1 + R_2} \quad \textit{by current divider rule}$$

$$= 10 \text{ mA} \times \frac{4.7 \text{ k}\Omega}{4.7 \text{ k}\Omega + 3.3 \text{ k}\Omega}$$

$$= 5.875 \text{ mA}$$

$$i_2 = \frac{E_1}{R_3} = \frac{60 \text{ V}}{1.2 \text{ k}\Omega}$$

$$= 50 \text{ mA}$$

$$I_{sc} = 5.875 \text{ mA} + 50 \text{ mA}$$

$$\boldsymbol{I_N = I_{sc} = 55.875 \text{ mA}}$$

In Figure 9-6(d), *the current source and voltage source are replaced by their internal resistances.*
      From Figure 9-6(d),

$$R_N = R_3 \| (R_2 + R_1)$$

$$= 1.2 \text{ k}\Omega \| (3.3 \text{ k}\Omega + 4.7 \text{ k}\Omega)$$

$$\boldsymbol{R_N \cong 1.04 \text{ k}\Omega}$$

The Norton equivalent circuit is now as shown in Figure 9-6(e).
      The output current through R_L can be calculated using the current divider rule:

$$I_L = I_N \times \frac{R_N}{R_N + R_L}$$

For R_L = 12 kΩ,

$$I_{L1} \cong 55.875 \text{ mA} \times \frac{1.04 \text{ k}\Omega}{1.04 \text{ k}\Omega + 12 \text{ k}\Omega}$$

$$\boldsymbol{I_{L1} \cong 4.46 \text{ mA}}$$

For R_L = 5.6 kΩ,

$$I_{L2} \cong 55.875 \text{ mA} \times \frac{1.04 \text{ k}\Omega}{1.04 \text{ k}\Omega + 5.6 \text{ k}\Omega}$$

$$\boldsymbol{I_{L2} \cong 8.75 \text{ mA}}$$

Note that because of approximations, I_{L1} and I_{L2} as calculated above are slightly different from the values obtained in Example 9-5.

9-3.1 Use Norton's theorem to determine current I_2 in the circuit of Figure 7-1(a).

9-3.2 For the circuit in Figure 7-12, apply Norton's theorem to find the current in resistor R_5.

9-4
MILLMAN'S THEOREM

Any number of current sources in parallel can be represented by a single current generator in which the generator current is the algebraic sum of the individual source currents, and the generator resistance is the parallel combination of the individual source resistances.

This definition does not give a complete idea of the applications of Millman's theorem, because it refers only to current sources. In Section 8-1 it is shown that voltage sources can be converted to current sources, and vice versa. Consequently, Millman's theorem can also be applied to voltage sources in parallel, or to a combination of voltage and current sources. Also, instead of ending up with a single current generator, the complete network can be represented as a single voltage generator.

One very important point is that *this theorem applies only to sources connected directly in parallel;* it does not apply where there are resistors between the sources. For example, Millman's theorem could not be applied directly to the two-generator circuit shown in Figure 9-3(a) because of the presence of resistor R_2.

PROCEDURE FOR APPLICATION OF MILLMAN'S THEOREM:

1. Convert all voltage sources to current sources.
2. Algebraically add all source currents to obtain the final generator current.
3. Determine the resistance value of the source resistances in parallel to obtain the final generator resistance.
4. If required, convert the current generator to a voltage generator.

EXAMPLE 9-7 Apply Millman's theorem to the circuit shown in Figure 9-7(a), to obtain the equivalent current generator circuit and the equivalent voltage generator circuit.

SOLUTION

Converting the voltage generator,

$$I_1 = \frac{E_1}{R_s} = \frac{5 \text{ V}}{100 \ \Omega}$$

$$= \textbf{50 mA} \quad [see \text{ Figure 9-7(b)}]$$

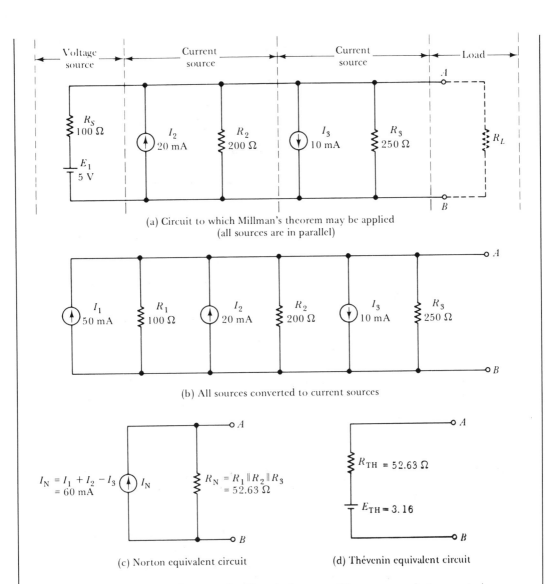

(a) Circuit to which Millman's theorem may be applied
(all sources are in parallel)

(b) All sources converted to current sources

$I_N = I_1 + I_2 - I_3$
$= 60$ mA

(c) Norton equivalent circuit

(d) Thévenin equivalent circuit

FIGURE 9-7. Application of Millman's theorem. All sources are first converted to current sources, and the total source current is summed to give I_N. Then, the internal resistance R_N is determined as the equivalent of all source resistances in parallel.

Current source resistance,

$$R_1 = R_s = \textbf{100 } \mathbf{\Omega} \qquad [see \text{ Figure 9-7(b)}]$$

For the final equivalent current generator,

$$I_N = I_1 + I_2 + I_3$$

$$= 50 \text{ mA} + 20 \text{ mA} + (-10 \text{ mA})$$

$$I_N = 60 \text{ mA}$$

$$R_N = R_1 \| R_2 \| R_3$$
$$= 100 \ \Omega \| 200 \ \Omega \| 250 \ \Omega$$
$$\cong \mathbf{52.63 \ \Omega} \quad [see \text{ Figure 9-7(c)}]$$

For the equivalent voltage generator,

$$E_{TH} = I_N R_N \cong 60 \text{ mA} \times 52.63 \ \Omega$$
$$\cong \mathbf{3.16 \ V}$$

and $\quad\quad\quad R_{TH} = R_N \cong \mathbf{52.63 \ \Omega} \quad [see \text{ Figure 9-7(d)}]$

PRACTICE PROBLEMS

9-4.1 In the circuit of Figure 8-7(a), resistor R_3 is shorted, and a new resistor R_5 is connected in parallel with R_4. If $R_5 = 3.3 \text{ k}\Omega$, apply Millman's theorem to determine the current through R_5.

9-4.2 For the circuit in Figure 8-10, apply Millman's theorem to find the current in resistor R_4.

9-5

MAXIMUM POWER TRANSFER THEOREM

Maximum output power is obtained from a network or source when the load resistance is equal to the output resistance of the network or source as seen from the terminals of the load.

The truth of the maximum power transfer theorem is easily tested by considering the Thévenin equivalent circuit in Figure 9-8(a), and calculating the power output for various values of load resistance. For the circuit in Figure 9-8(a), $R_{TH} = 50 \ \Omega$ and $E_{TH} = 100 \text{ V}$.

For $R_L = 50 \ \Omega$:

$$I_L = \frac{E_{TH}}{R_{TH} + R_L} = \frac{100 \text{ V}}{50 \ \Omega + 50 \ \Omega}$$

$$I_L = \mathbf{1 \ A}$$

$$V_L = I_L \times R_L = 1 \text{ A} \times 50 \ \Omega$$

$$V_L = \mathbf{50 \ V}$$

$$P_L = V_L \times I_L = 50 \text{ V} \times 1 \text{ A}$$

$$P_L = \mathbf{50 \ W}$$

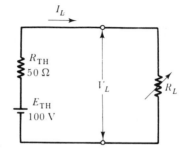

(a) Thévenin equivalent circuit with a variable load

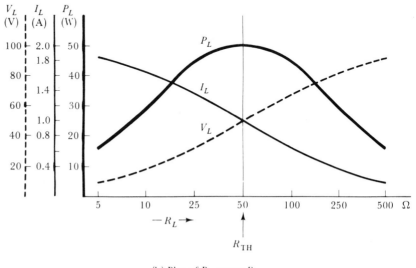

(b) Plot of P_L versus R_L

FIGURE 9-8. Illustration of the maximum power transfer theorem. Maximum output power occurs when $R_L = R_{TH}$.

Table 9-1 gives the values of load current, voltage, and power calculated for values of R_L ranging from 5 Ω to 500 Ω. The resistance values selected are all multiples of each other: 10 Ω = 2 × 5 Ω, 25 Ω = 5 × 5 Ω, etc. The reason for this becomes apparent when the graphs of V_L, I_L, and P_L are plotted versus R_L [Figure 9-8(b)]. The resistance scale is *logarithmic*

TABLE 9-1 I_L, V_L, and P_L for the Circuit of Figure 9-8(a) with Various Values of R_L

$R_L(\Omega)$	5	10	25	50	100	250	500
$V_L(V)$	9.09	16.7	33.3	50	66.7	83.3	90.9
$I_L(A)$	1.82	1.67	1.33	1	0.67	0.33	0.18
$P_L(W)$	16.5	27.8	44.4	50	44.4	27.8	16.5

rather than *linear,* and this gives the particular shape to the graph of P_L. Clearly the output power is at a peak value when $R_L = R_{TH}$. Also, note that the output current is greatest when R_L is very small, and the output voltage is greatest when R_L is very large.

PRACTICE PROBLEMS

9-5.1 For the circuit in Figure 8-10, a load resistor R_L is to be connected across terminals C and D. Calculate the resistance of R_L for maximum output power.

9-5.2 Calculate the new resistance for R_3 in Figure 8-11 to give maximum power dissipation in R_3. Calculate the power dissipated in R_3 after the change of value.

REVIEW QUESTIONS

9-1 State the superposition theorem and list the steps involved in applying it to the analysis of a resistor network.

9-2 State Thévenin's theorem and give the procedure for thévenizing a circuit. Explain the major advantages afforded by use of this theorem.

9-3 State Norton's theorem and list the steps for nortonizing a circuit. Compare the Norton equivalent circuit to the Thévenin equivalent circuit.

9-4 State Millman's theorem and list the procedure for its application to circuit analysis.

9-5 State the maximum power transfer theorem.

PROBLEMS

SECTION 9-1

9-1 Use the superposition theorem to determine the current through R_1 in the circuit in Figure 8-13.

9-2 Apply the superposition theorem to analyze the circuit in Figure 8-14 to determine the current through R_5.

9-3 Using the superposition theorem, analyze the circuit in Figure 8-15 to determine the current through R_5.

9-4 Apply the superposition theorem to analyze the circuit in Figure 8-16 to determine the current through R_3.

9-5 Using the superposition theorem, analyze the circuit in Figure 8-17 to determine the current through R_3.

9-6 For the circuit in Figure 8-19, use the superposition theorem to determine the current through R_2 when R_4 is open-circuited.

9-7 Use the superposition theorem to analyze the circuit in Figure 9-3(a) to determine the current through R_2 when R_L is open-circuited.

SECTION 9-2

9-8 Apply Thévenin's theorem to the circuit in Figure 7-10 to determine the current through resistor R_2.

9-9 For the circuit in Figure 7-11, use Thévenin's theorem to determine the current through resistor R_5.

9-10 Using Thévenin's theorem, determine the current through R_4 in Figure 7-13 when R_1 is open-circuited.

9-11 For the circuit in Figure 7-16, use Thévenin's theorem to determine the current through resistor R_3.

9-12 Using Thévenin's theorem, determine the current through R_3 in Figure 8-12.

9-13 Apply Thévenin's theorem to determine the current through R_5 in Figure 8-14.

9-14 Use Thévenin's theorem to determine the voltage across R_5 in Figure 8-15.

9-15 Apply Thévenin's theorem to the circuit in Figure 8-16 to determine the voltage across resistor R_3.

SECTION 9-3

9-16 Apply Norton's theorem to the circuit in Figure 7-10 to determine the current through resistor R_4.

9-17 Apply Norton's theorem to the circuit in Figure 7-11 to determine the voltage across resistor R_4.

9-18 Using Norton's theorem, determine the current through R_5 in Figure 7-13 when R_1 is open-circuited.

9-19 For the circuit in Figure 7-16, use Norton's theorem to determine the current through resistor R_4.

9-20 Using Norton's theorem, determine the new value of R_1 in Figure 8-12 which will give $I_1 = 0.4$ mA.

9-21 Apply Norton's theorem to the circuit in Figure 8-13 to determine the current through resistor R_1.

9-22 Apply Norton's theorem to determine the current through R_3 in Figure 8-14.

9-23 Use Norton's theorem to determine the voltage across R_3 in Figure 8-15.

9-24 Apply Norton's theorem to the circuit in Figure 8-16 to determine the voltage across resistor R_2.

SECTION 9-4

9-25 Using Millman's theorem, determine the current through resistor R_3 in Figure 8-16.

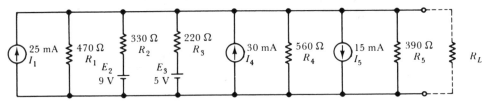

FIGURE 9-9.

9-26 Apply Millman's theorem to the circuit in Figure 8-4 to determine the voltage across resistor R_3.

9-27 Using Millman's theorem, reduce the voltage and current generators in Figure 9-9 to a single current generator. Then, calculate the load current for $R_L = 3.3$ kΩ, and for $R_L = 4.7$ kΩ.

9-28 Apply Millman's theorem to the circuit in Figure 8-11 to calculate I_2.

9-29 Using Millman's theorem, determine the current through resistor R_1 in the circuit shown in Figure 8-13.

SECTION 9-5

9-30 Using the current generator obtained in Problem 9-27, demonstrate the maximum power transfer theorem by calculating the load voltage, current, and power for various values of load resistance. Also, plot a graph of the quantities versus R_L.

9-31 Using the Thévenin circuit obtained in Problem 9-12, calculate the new value of R_3 that will give maximum power dissipation in R_3. Also, calculate the output power for resistor values equal to R_3, and to half and twice the value of R_3.

9-32 For the circuit in Figure 7-13, determine the value of R_5 that will give maximum power dissipation in R_5. Also, calculate the output power for resistor values equal to R_5, $R_5/2$ and $2R_5$.

9-33 For the circuit in Figure 7-16, use the maximum power transfer theorem to determine the resistance R_4 for maximum power dissipation.

COMPUTER PROBLEMS

9-34 Write a computer program which uses the superposition theorem to solve the type of problem presented in Example 9-1.

9-35 Write a computer program to solve the type of problem presented in Example 9-7.

ANSWERS TO PRACTICE PROBLEMS

9-1.1	2.73 mA
9-1.2	2.9 μA
9-2.1	0.2 A
9-2.2	18.8 V
9-3.1	0.3 A
9-3.2	311 μA
9-4.1	9.99 mA
9-4.2	2.96 mA
9-5.1	306 Ω
9-5.2	184 kΩ, 16.15 μW

10

VOLTAGE CELLS, BATTERIES, AND DC POWER SUPPLIES

Objectives You will be able to:

Sketch the basic construction and explain the operation of a simple voltage cell.

Discuss the performance of voltage cells, and solve problems involving internal resistance, ampere-hour rating, no-load output voltage, and performance under load.

Describe the construction, performance, and applications of zinc-carbon cells. Describe manganese-alkaline, mercury, and nickel-cadmium cells.

Show how cells may be connected in series, parallel, and series-parallel. Solve problems involving batteries of series- and parallel-connected cells.

Describe the construction and operation of a lead-acid battery. Discuss its performance and maintenance.

Solve problems involving charge and discharge of lead-acid batteries.

Describe the performance and use of laboratory-type dc power supplies.

Introduction

A voltage cell consists basically of two different metal plates immersed in an acid solution. The action of the acid causes electrons to be removed from one plate and to be accumulated on the other. In this way a potential difference is produced between the two plates, and a current can be made to flow through an external circuit.

Different types of voltage cells exist for different applications. Some can be recharged and used again many times, others can be used only once and are then discarded. One of the most important of the recharge-able cells is the *lead-acid cell,* which is the basic unit of the automobile battery.

Batteries of voltage cells can be operated in series or in parallel, or in series-parallel combinations. The performance of a voltage cell may be defined in terms of the maximum voltage and current it can supply, its *ampere-hour* rating, and the *cell equivalent circuit*.

Laboratory-type dc power supplies are usually designed to produce an adjustable output voltage and a specified maximum output current.

10-1

SIMPLE VOLTAGE CELL

OPERATION. The construction of a simple voltage cell is illustrated in Figure 10-1. The complete cell consists of copper and zinc *electrodes* partially immersed in a chemical solution known as the *electrolyte*. Diluted sulfuric acid is frequently used as the electrolyte, but any one of several other chemical compounds may be employed.

Sulfuric acid is a combination of *hydrogen* and *sulfate* in which each hydrogen atom has given up an electron to the sulfate molecules. The solution remains chemically stable until the electrodes are introduced. The surface of the zinc electrode readily dissolves in the sulfuric acid, and the zinc atoms combine with the sulfate to form *zinc sulfate*. In this process the zinc atoms leave electrons behind them on the zinc electrode [see Figure 10-2(a)]. This is because the sulfate already has an excess of electrons acquired from the hydrogen. As the zinc sulfate forms, hydrogen ions are released, and each hydrogen ion is an atom that is short of one electron (i.e., a positive ion). These ions travel to the copper electrode, where they acquire electrons, as illustrated.

In giving up electrons, the copper electrode becomes positively charged. Similarly, the zinc electrode becomes negatively charged as it accumulates excess electrons. Thus, a potential difference is created between the copper and zinc electrodes.

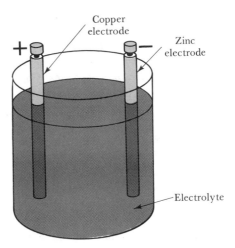

FIGURE 10-1. A simple voltage cell consists of copper and zinc rods partially immersed in a liquid electrolyte. The copper electrode is the positive terminal of the cell, and the zinc is the negative terminal.

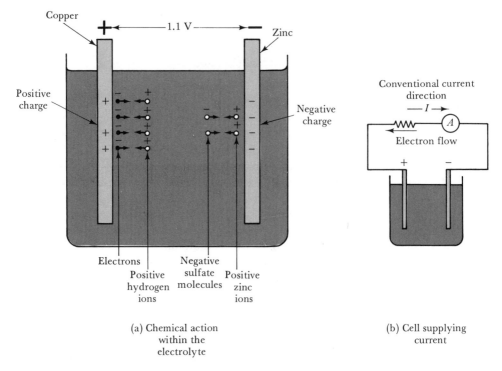

FIGURE 10-2. In the simple voltage cell, positive ions are removed from the zinc, creating a negative charge. Electrons are taken from the copper, leaving it positively charged. When a circuit is connected between the terminals, electrons flow from the negative terminal to the positive terminal.

If no current is drawn from the cell, the chemical action eventually ceases when the zinc becomes so negative that it repels the (negative) sulfate molecules, and the positive charge on the copper repels the (positive) hydrogen ions. In this condition the potential difference at the cell terminals is typically 1.1 V.

An external circuit connected to the cell terminals provides a path for electrons to flow from the (negative) zinc electrode to the (positive) copper electrode [Figure 10-2(b)]. Thus, the zinc electrode loses negative charges and the copper electrode gains negative charges (i.e., becomes less positive). Once the electrodes begin to lose some of their accumulated charges, the chemical action resumes, and the electrolyte continues to provide electrons to supply the output current from the cell.

POLARIZATION. The hydrogen released in the simple voltage cell frequently remains clinging to the copper electrode in the form of bubbles. When large areas of the electrode become coated with hydrogen bubbles, the active surface area is reduced and the cell's ability to supply current is diminished. The effect is known as *polarization*.

In more complex cells, several methods are employed to combat polarization. Materials used for this purpose are termed *depolarizers.*

OUTPUT VOLTAGE AND CURRENT. The potential difference produced at the terminals of a cell depends upon the electrode materials. For the zinc and copper electrodes of the simple voltage cell, the terminal voltage is approximately 1.1 V. This potential difference is produced regardless of the physical size of the electrodes or the quantity of the electrolyte.

The output current from a cell depends upon the output voltage and the external load resistance. As always,

$$I = \frac{E}{R}$$

where I is the output current, E is the cell terminal voltage, and R is the external load resistance. However, there is a limit to the maximum current that a given cell can supply. When this limit is approached, the cell's terminal voltage begins to fall, so that the output current cannot be increased. The maximum current that can be drawn from a cell is directly related to the physical size of the cell's components. Obviously, electrodes with large surface areas can release more atoms and generate more free electrons than is possible with small surface areas. Consequently, physically large voltage cells can sustain larger output currents than small cells.

The maximum output current from a voltage cell can be determined in terms of the cell's *internal resistance,* which is discussed in Section 10-2.

AMPERE-HOUR RATING. The amount of energy that can be supplied by any cell (or battery of cells) is defined in terms of its rating in *ampere-hours* (Ah).

$$\boxed{Ah\ rating = I \times t}$$ (10-1)

If a cell can supply a current of 5 A for a maximum time of 1 h, its rating is

$$I \times t = 5\ A \times 1\ h$$

$$= 5\ Ah$$

When the same cell is supplying a current of only 1 A, it can be expected to sustain this current for a time of

$$t = \frac{Ah\ rating}{I} = \frac{5\ Ah}{1\ A}$$

$$= 5\ h$$

For a time of 50 h, a cell with a rating of 5 Ah can be expected to supply a continuous current of

$$I = \frac{\text{Ah rating}}{t} = \frac{5 \text{ Ah}}{50 \text{ h}}$$

$$= 100 \text{ mA}$$

Usually a cell's Ah rating applies only to a certain range of load currents. For a given cell, the Ah rating when supplying a high current is not as great as that at a low current level. One reason for this is that polarization occurs more rapidly when the load current is high. Also, as already discussed, every cell has a limit to the maximum current it can supply.

EXAMPLE 10-1   A voltage cell supplies a load current of 0.5 A for a period of 20 h until its terminal voltage falls to an unacceptable level. Calculate the Ah rating of the cell and determine how long it could be expected to supply a current of 200 mA.

SOLUTION

From Equation (10-1):

$$\text{Ah } rating = I \times t = 0.5 \text{ A} \times 20 \text{ h}$$

$$= \textbf{10 Ah}$$

$$t = \frac{\text{Ah } rating}{I} = \frac{10 \text{ Ah}}{200 \text{ mA}}$$

$$= \textbf{50 h}$$

PRIMARY AND SECONDARY CELLS.   The simple voltage cell is classified as a *primary cell*. Primary cells can supply current until the negative electrode is completely dissolved, the electrolyte is exhausted, or perhaps until polarization renders the cell useless. A *secondary cell*, on the other hand, can be *recharged*. That is, the electrode surface can be re-formed and the electrolyte returned to its original condition. This is done by passing a current through the cell in an opposite direction to the current that normally flows from the cell.

The *lead-acid cell* discussed in Section 10-7 is a secondary cell, because it is rechargeable. Most *dry cells*, such as the *zinc-carbon cell* (see Section 10-3) cannot be recharged; consequently, they are primary cells.

10-1.1 A 9 V battery is to supply 100 mA for a period of 5 hours. Determine the required minimum Ah rating and calculate how long the same battery could be used to supply 35 mA.

10-1.2 A battery supplies the following currents before its voltage falls below a useful level: 10 mA for 30 minutes, 50 mA for 10 minutes, 25 mA for 4 h, 18 mA for 7 h. Calculate the total Ah rating of the battery.

10-2

CELL EQUIVALENT CIRCUIT

As already explained, when no current is being drawn from a voltage cell, charges are accumulated on the electrodes, and the chemical action ceases within the electrolyte. At this time, the potential difference beween the cell terminals is known as the *open-circuit output voltage* or the *no-load output voltage*.

When current flows from the cell, electrons are transferred externally from the negative electrode to the positive electrode. Thus, the potential difference between the electrodes falls, and the chemical action within the electrolyte commences to replace the lost charges. If the load current is very small, the terminal voltage under load may not be noticeably different from the open-circuit output voltage. When the load current is large, the terminal voltage falls below the open-circuit voltage level. If the load current is made too large, the cell terminal voltage falls to near zero, because the chemical action simply cannot occur fast enough to replace the charges removed from the electrodes. The maximum current that may be taken from a voltage cell without a substantial drop in output voltage is proportional to the surface area of the electrodes and to the composition of the electrolyte.

To account for the terminal voltage drop of a cell when supplying current, each cell is said to have a certain *internal resistance*. The cell can then be represented by an equivalent circuit consisting of its internal resistance together with an *ideal cell* which has a terminal voltage equal to the actual cell's open-circuit terminal voltage. An ideal cell is a theoretical cell that has a terminal voltage which is assumed to remain constant. The equivalent circuit of a voltage cell is illustrated in Figure 10-3.

When no current is flowing from the equivalent circuit, there is no voltage drop across the internal resistance (r_i) and the terminal voltage is equal to the ideal cell voltage. That is, the output voltage is the open circuit voltage E_c of the cell [see Figure 10-3(a)]. When a load resistance (R_L) is connected to the cell terminals [Figure 10-3(b)], the current that flows is

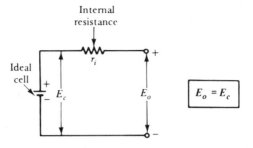

(a) Equivalent circuit of cell ($I_o = 0$)

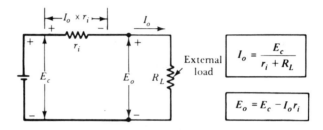

(b) Equivalent circuit of cell supplying a load current

FIGURE 10-3. Equivalent circuit of a voltage cell under no-load and load conditions. When a load current flows, the terminal voltage is reduced by the cell internal voltage drop.

$$I_0 = \frac{E_c}{r_i + R_L} \qquad (10\text{-}2)$$

When supplying current, a voltage drop occurs across r_i, and the output voltage becomes

$$E_0 = E_c - I_0 r_i \qquad (10\text{-}3)$$

Obviously, a cell that has a low internal resistance is capable of supplying a larger current than one that has a relatively high internal resistance.

EXAMPLE 10-2 A certain voltage cell has an open-circuit terminal voltage of $E_c = 2$ V and an internal resistance of $r_i = 0.1\ \Omega$. Determine the output voltage of the cell when the load current is 2 A. Also, determine the cell current when the terminals are short-circuited.

SOLUTION

When $I_0 = 2$ A:

From Equation (10-3):

$$E_0 = E_c - I_0 r_i$$
$$= 2\text{ V} - (2\text{ A} \times 0.1\text{ }\Omega)$$
$$= \textbf{1.8 V}$$

When the cell is short-circuited:

$$R_L = 0$$

From Equation (10-2):

$$I = \frac{E_c}{r_i + R_L} = \frac{2\text{ V}}{0.1\text{ }\Omega + 0\text{ }\Omega}$$
$$= \textbf{20 A}$$

PRACTICE PROBLEMS

10-2.1 The terminal voltage of a cell is measured as 1.55 V on open-circuit and as 1.47 V when supplying 50 mA. Calculate the cell internal resistance and the short-circuit current.

10-2.2 The terminal voltage of a 2.2 V cell is to drop by not more than 5% when supplying a load curent of 20 A. Determine the required maximum internal resistance of the cell.

10-3

DRY CELLS

ZINC-CARBON CELL. The *zinc-carbon* cell is a primary cell commonly used in electric flashlights and other portable equipment. As illustrated in Figure 10-4(a), this cell basically consists of a zinc can containing electrolyte with a carbon rod at the center. The zinc can is the negative electrode of the cell, and the carbon rod is the positive electrode. The electrolyte is in the form of a moist paste, so that the *dry cell* is, in fact, not dry at all. The cell becomes useless when the electolyte dries out because the necessary chemical reactions cannot occur.

The operation of the zinc-carbon cell is similar to that of the zinc-copper cell described in Section 10-1. Electrons accumulate on the zinc electrode, and electrons are removed from the carbon electrode as zinc atoms combine with the electrolyte. Thus, a potential difference occurs between the electrodes, with the zinc being negative and the carbon positive.

Figure 10-4(b) shows the construction of the zinc-carbon cell in more detail. The electrolyte consists of *sal ammoniac* (ammonium

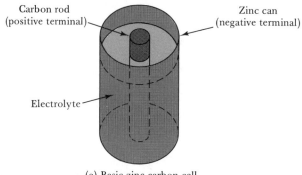

(a) Basic zinc-carbon cell

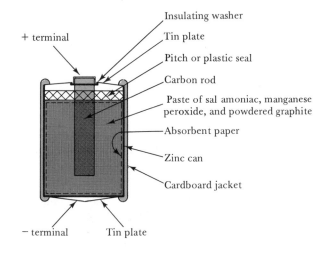

(b) Cross-section of zinc-carbon cell

FIGURE 10-4. Construction of a zinc-carbon cell. The zinc can is the negative terminal; the carbon rod is the positive terminal. The electrolyte is a paste of sal ammoniac, manganese peroxide, and powdered graphite.

chloride) mixed with *manganese peroxide* and powdered graphite. The manganese peroxide acts as a depolarizer by combining with the hydrogen, which appears at the positive electrode. The combination results in *manganous oxide* and water. This eliminates the polarizing effect of the hydrogen and helps to keep the electrolyte moist.

The electrolyte must be hermetically sealed in the can to prevent it from drying out quickly. A layer of pitch or plastic at the top of the can is the usual type of seal employed. The inside surface of the zinc is protected by a layer of absorbent paper [see Figure 10-4(b)] to help prevent the zinc from reacting directly with the carbon and other impurities in the electrolyte. Such reactions tend to set up little voltage cells at the surface of the zinc, and this is referred to as *local action*. Local action reduces the output voltage of the cell and shortens its life by using up the zinc. To combat local action due to impurities in the zinc, the zinc surface is usually coated with mercury, or else mercury is mixed with the zinc during the manufacturing process.

Most dry cells have a surrounding outer jacket of cardboard or tin plate. Also, tin-plate discs are used at the top and bottom, as illustrated. As well as providing protection for the cell, the outer jacket helps to protect battery-powered equipment from the corrosive action that occurs on the outside of the zinc can when dry cells are unused for a long time.

The typical output voltage from a zinc–carbon dry cell is 1.5 V, but it could range from 1.4 V to 1.6 V in a good cell. This terminal voltage is available no matter what the physical size of the cell. The very small *C cell* and the much larger *No. 6 cell* illustrated in Figure 10-5 each have a terminal voltage of approximately 1.5 V. For larger output voltages, several cells must be connected in series (see Section 10-4). In continuous use, the terminal voltage of the zinc–carbon cell tends to fall off rapidly unless the current is very small. Consequently, this type of cell is best applied where it will be used intermittently.

The output current that may be drawn from a zinc-carbon cell is very much dependent upon the physical size of the cell. The *No. 6 cell* can supply as much as 20 A for a short time. However, at this rate it would tend to discharge very quickly, and 1 A would be a more appro-

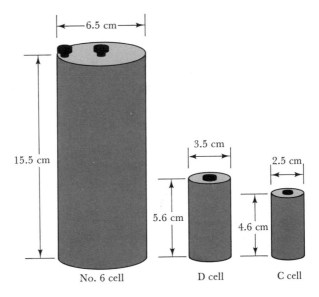

FIGURE 10-5. All zinc-carbon cells have a typical terminal voltage of 1.5 V, regardless of size. The largest cells are able to deliver the greatest currents. Alternatively, they may be used to supply a small current for a longer time than would be possible with a small cell.

priate maximum load. The *D cell*, which is commonly used in flashlights, cannot supply more than about 5 A. Again, this would be an excessive load, and a maximum of a few hundred milliamps is normal. The *C cells* supply even smaller maximum currents, corresponding to their smaller size.

MANGANESE–ALKALINE CELL. This is a rechargeable (or secondary) cell which can sustain a constant terminal voltage for heavy loads over a much larger period of time than the less expensive zinc-carbon cell. The electrolyte is *potassium hydroxide,* and the negative electrode is a *zinc–mercury* amalgam. The positive electrode is a layer of *manganese dioxide.* The cell is enclosed in a steel container. The open-circuit terminal voltage of this cell is typically 1.5 V.

MERCURY CELL. The mercury cell is a modification of the manganese–alkaline cell, in which *mercuric oxide* is used as the positive electrode (i.e., instead of manganese dioxide). The open-circuit terminal voltage of the mercury cell is usually around 1.4 V, falling to about 1.3 V under load conditions. The major advantage of this type of cell is that its terminal voltage is maintained constant for much longer time periods than either the manganese–alkaline or zinc–carbon cells. Figure 10-6 shows a graph of output voltage plotted versus time for comparable zinc–carbon, manganese–alkaline, and mercury cells. It is obvious from the graph that the terminal voltage of the zinc–carbon cell falls off relatively rapidly when supplying a constant load. The performance of the manganese-alkaline cell is better than that of the zinc-carbon cell, but the mercury cell gives by far the best performance.

NICKEL-CADMIUM CELL. The nickel–cadmium cell is a rechargeable cell widely used in portable electrical and electronic equipment. The initial cost of a set of nickel–cadmium cells and charger is much greater than the cost of a set of zinc–carbon cells. However, given the fact that

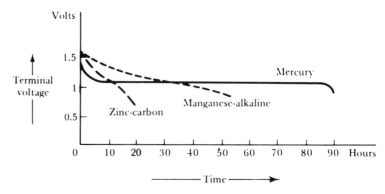

FIGURE 10-6. The graphs of terminal voltage versus hours of service for different types of primary cells show that mercury cells last longest.

the nickel cadmium cells could easily last 10 years at about two chargings per week, they are by far the least expensive in the long run.

10-4
VOLTAGE CELLS IN SERIES

When the positive terminal of one voltage cell is connected to the negative of another cell, the resultant output is the sum of the cell terminal voltages, and the cells are said to be *in series*.

OUTPUT VOLTAGE AND CURRENT. Figure 10-7(a) shows a battery of four cells connected in series. The negative terminal of each cell is connected to the positive terminal of the previous cell. With this arrangement, the cell voltages add together to give a relatively large battery terminal voltage. Thus, if each cell in Figure 10-7(a) has an emf of 1.5 V, the total output voltage is

$$E = E_1 + E_2 + E_3 + E_4 + \cdots$$

(10-4)

$$= 1.5 \text{ V} + 1.5 \text{ V} + 1.5 \text{ V} + 1.5 \text{ V}$$

$$= 6 \text{ V}$$

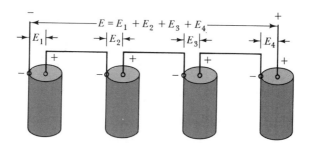

(a) Four cells correctly connected in series

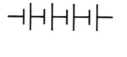

(b)   Circuit diagram for series-connected cells

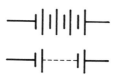

(c)   Alternative circuit diagrams for series-connected cells

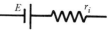

$$E = E_1 + E_2 + E_3 + E_4$$
$$r_i = r_1 + r_2 + r_3 + r_4$$

(d)   Equivalent circuit for series-connected cells

FIGURE 10-7. Series-connected cells with circuit diagrams and equivalent circuits. The output voltage is the sum of the individual cell voltages, and the battery internal resistance is the sum of the cell internal resistances.

Obviously, if the cell emf's were not all 1.5 V, the output voltage from the battery would still be the sum of the individual cell voltages.

Figure 10-7(b) shows the circuit diagram for the four cells in series. Alternative ways of graphically representing a battery of series-connected cells are shown in Figure 10-7(c).

The total internal resistance of the battery of series-connected cells is, of course, the sum of the internal resistances of the individual cells.

$$r_i = r_1 + r_2 + r_3 + \cdots \qquad (10\text{-}5)$$

Consequently, the equivalent circuit of the four-cell battery consists of a cell having an emf of $E = (E_1 + E_2 + E_3 + E_4)$ and an internal resistance of $r_i = (r_1 + r_2 + r_3 + r_4)$ [see Figure 10-7(d)].

When an external load resistance (R_L) is connected to a group of series-connected cells (Figure 10-8), the output current is

$$I = \frac{E}{r_i + R_L}$$

or

$$I = \frac{E_1 + E_2 + E_3 + E_4}{R_L + r_1 + r_2 + r_3 + r_4} \qquad (10\text{-}6)$$

Also

$$I = I_1 = I_2 = I_3 = I_4 \qquad (10\text{-}7)$$

Note from the circuit that the output current flows through each individual cell. Therefore, if each cell is capable of supplying a maximum current of 1 A, for example, the output current from the group of cells cannot exceed 1 A. The open-circuit output voltage and the output current from the battery can be calculated from Equations (10-4) and (10-6). The terminal voltage of the battery when supplying a load current is determined by use of Equation (10-3).

Voltage cells that are not identical can be connected in series, however, the maximum current that can be supplied by the battery of cells is limited to the maximum output of the lowest current cell.

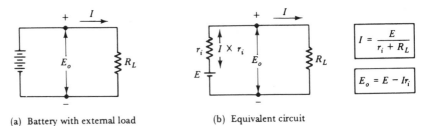

(a) Battery with external load   (b) Equivalent circuit

FIGURE 10-8. Circuit diagram and equivalent circuit of a battery of cells with an external load. The load current is $I = E/(r_i + R_L)$, and the battery terminal voltage is $E_0 = E - Ir_i$.

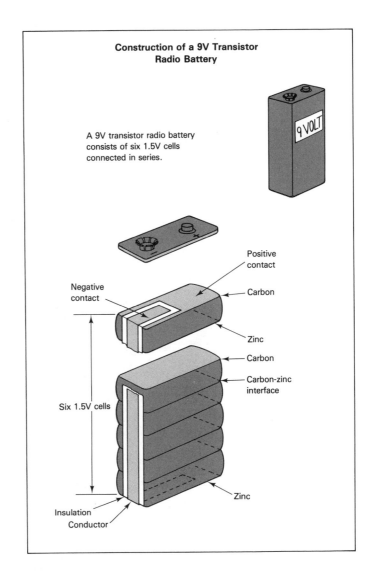

Construction of a 9V Transistor Radio Battery

A 9V transistor radio battery consists of six 1.5V cells connected in series.

9 VOLT

Positive contact

Negative contact

Carbon

Zinc

Carbon

Carbon-zinc interface

Six 1.5V cells

Zinc

Insulation

Conductor

Series-connected cells produce an output voltage equal to the sum of the individual cell voltages and can supply a maximum current equal to the maximum that can be taken from any one cell.

EXAMPLE 10-3 The battery shown in Figure 10-8(a) is made up of four individual cells each of which has an open-circuit terminal voltage of $E = 2$ V and an internal resistance of $r = 0.1$ Ω. Determine the terminal voltage of the battery for no-load and for $R_L = 9.6$ Ω. Also, determine the current that flows when the battery terminals are short-circuited.

SOLUTION

$$E = E_1 + E_2 + E_3 + E_4 = 4 \times 2 \text{ V}$$
$$= 8 \text{ V}$$
$$r_i = r_1 + r_2 + r_3 + r_4 = 4 \times 0.1 \text{ }\Omega$$
$$= 0.4 \text{ }\Omega$$

For no-load:

$$I_0 = 0$$

and
$$E_0 = E - I_0 r_i = 8 \text{ V} - 0$$
$$\mathbf{E_0 = 8 \text{ V}}$$

For $R_L = 9.6 \text{ }\Omega$:

$$I_0 = \frac{E}{r_i + R_L} = \frac{8 \text{ V}}{0.4 \text{ }\Omega + 9.6 \text{ }\Omega}$$
$$\mathbf{= 0.8 \text{ A}}$$
$$E_0 = E - I_0 r_i = 8 \text{ V} - (0.8 \text{ A} \times 0.4 \text{ }\Omega)$$
$$\mathbf{= 7.68 \text{ V}}$$

For short circuit:

$$R_L = 0$$
$$E_0 = \frac{E}{r_i + 0} = \frac{8 \text{ V}}{0.4 \text{ }\Omega + 0 \text{ }\Omega}$$
$$\mathbf{I_0 = 20 \text{ A}}$$

Note that short-circuiting a battery or cell terminals is to be avoided, because it rapidly discharges the cells. Also, the large current that flows can permanently damage rechargeable cells.

SERIES-AIDING AND SERIES-OPPOSING. The series cells illustrated in Figure 10-9(a) each have their positive terminal connected to the negative terminal of the adjacent cell. With this arrangement they assist each other to produce a current flow. The connection method is termed *series-aiding*. When cells are series-connected to produce currents in opposite directions, as shown in Figure 10-9(b), the connection is referred to as *series-opposing*. Series-opposing connections are normally avoided, and usually occur only as a result of error. The total output voltage is still the sum of the individual cell voltages, but the polarity

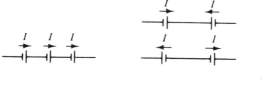

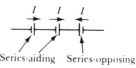

Series-aiding Series-opposing

| (a) Cells connected series-aiding | (b) Cells connected series-opposing | (c) One cell connected series-opposing |

FIGURE 10-9. When voltage cells are connected in series, they are always connected series-aiding so that their voltages add together. Cells connected series-opposing give a reduced output voltage.

of the voltages must be carefully considered. As illustrated in Figure 10-9(c), when one of three 1.5 V cells is connected series-opposing, the resultant terminal voltage of the battery is

$$E = E_1 + E_2 + E_3$$
$$= 1.5 \text{ V} + 1.5 \text{ V} + (-1.5 \text{ V})$$
$$= 1.5 \text{ V}$$

If a discharged cell is connected in series with fully charged cells, the discharged cell adds little or nothing to the total output voltage. However, it does increase the total internal resistance of the battery; therefore, a discharged cell should *not* be connected in series with good fully charged cells.

PLUS-MINUS SUPPLY. In Figure 10-10 two batteries, each having three series-connected cells are shown with their common terminal grounded (i.e., connected to *ground* or *earth*). (Note the graphic symbol for ground.) If each cell has an emf of 1.5 V, each battery has a terminal voltage of 4.5 V. Since the negative terminal of battery B_1 is grounded, its output voltage is $E_1 = 4.5$ V, that is, *positive* with respect to ground. The positive terminal of B_2 is grounded; therefore, the output voltage of B_2 is $E_2 = -4.5$ V, *negative* with respect to ground. The combined output of the two batteries is usually designated *±4.5 V.* Many electronics circuits use this *plus-minus* type of supply.

PRACTICE PROBLEMS

10-4.1 A battery consists of six series-connected 1.5 V cells which each have an internal resistance of 0.17 Ω. Determine the battery terminal voltage when supplying a 200 mA load.

10-4.2 A battery made up of three series-connected 2.2 V cells has a potential difference of 6 V at its terminals when supplying a 0.33 Ω load. Calculate its short-circuit current.

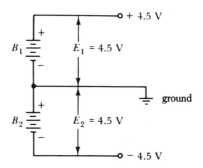

FIGURE 10-10. Two 4.5 V batteries may be connected in series with their common terminal grounded to provide a ±4.5 supply.

10-5

VOLTAGE CELLS IN PARALLEL

When the positive terminal of one voltage cell is connected to the positive of another cell, and the two negative terminals are also connected together, the cells are said to be *in parallel.*

Consider two voltage cells connected in parallel as illustrated in Figure 10-11(a). If each cell has a terminal voltage of exactly 1.5 V, the output voltage from the parallel combination is also 1.5 V.

$$E = E_1 = E_2 = E_3 = \cdots$$ (10-8)

If each cell is capable of supplying a maximum current of 1 A, the total output current that can be drawn from the two-cell parallel combination is 2 A. Similarly, for three such cells in parallel the maximum output current is three times the maximum current per cell. Of course, the actual output current still depends upon the output voltage and the load resistance. As shown by Figure 10-11(b), neglecting cell internal resistances the output current is

$$I_0 = \frac{E}{R_L}$$

But the output current I_0 is the sum of the individual currents from each cell. Therefore,

$$I_0 = I_1 + I_2 + I_3 + \cdots + I_n$$ (10-9)

For the case of two 1.5 V cells in parallel supplying a load resistor of 5 Ω:

$$I_0 = \frac{1.5 \text{ V}}{5 \text{ }\Omega} = 0.3 \text{ A} \quad \text{(neglecting the internal resistances)}$$

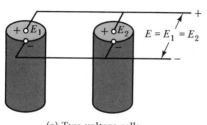

(a) Two voltage cells
connected in
parallel

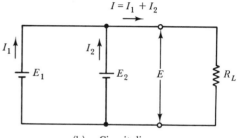

(b)   Circuit diagram
for parallel cells
with a load resistor

$$\boxed{E = E_1 = E_2}$$

$$\boxed{I = I_1 + I_2}$$

$$\boxed{I = \dfrac{E}{R_L}}$$

FIGURE 10-11. Parallel-connected cells give an output voltage equal to the terminal voltage of one cell, and a maximum output current equal to the sum of the individual cell maximum current levels.

and $\qquad I_1 = I_2 = 0.15$ A

giving $\qquad I_0 = 0.15$ A $+ 0.15$ A $= 0.3$ A

EXAMPLE 10-4 A 0.2 Ω resistance is to be supplied from a parallel-connected battery of 2 V cells. If each cell can supply a maximum current of 0.5 A, determine the number of cells that should be connected in parallel. Assume the cell internal resistances to be negligible.

SOLUTION

$$I_0 = \frac{E}{R_L} = \frac{2 \text{ V}}{0.2 \text{ } \Omega}$$

$$= 10 \text{ A}$$

$$number\ of\ parallel\ cells = \frac{I_0}{current\ per\ cell}$$

$$= \frac{10 \text{ A}}{0.5 \text{ A}}$$

number of parallel cells = 20

Each cell in a parallel combination of cells can be represented by its own equivalent circuit [Figure 10-12(a)]. The equivalent circuit of the battery

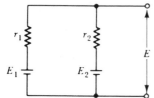

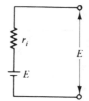

$$r_i = r_1 \| r_2$$
$$= \frac{r_1}{2} = \frac{r_2}{2} \text{ where } r_1 = r_2$$

$$E = E_1 = E_2$$

(a) Equivalent circuits
for parallel-connected
cells

(b) Single equivalent circuit
for parallel-connected
cells

FIGURE 10-12. The equivalent circuit for two parallel-connected cells shows that the internal resistance of the battery is $r_i = r_1 \| r_2$. Where $r_1 = r_2$, this gives $r_i = r_1/2$.

of cells can now be seen to be a voltage source equal to the terminal voltage of each cell, in series with an internal resistance which equals the parallel combination of the cell internal resistances. This is illustrated in Figure 10-12(b). For the case of two 1.5 V cells, each having an internal resistance of 0.1 Ω, the equivalent circuit has

$$E = E_1 = E_2 = 1.5 \text{ V}$$

and $\qquad r_i = r_1$ in parallel with r_2

or $\qquad r_i = r_1 \| r_2 = (0.1 \ \Omega)/2$

$$= 0.05 \ \Omega$$

If there were three identical cells in parallel, the internal resistance of one cell would be divided by 3 to find the internal resistance of the equivalent circuit, or

$$r_i = r_1 \| r_2 \| r_3 \qquad \text{(10-10)}$$

$$= r/3 \text{ for three identical cells}$$

Taking the cell internal resistance into consideration, the output current from a group of parallel cells supplying a load resistance R_L is

$$I_0 = \frac{E}{R_L + (r_1 \| r_2 \| r_3 \cdots)} \qquad \text{(10-11)}$$

In some cases cells connected in parallel may not have voltages which are exactly equal. For example, in Figure 10-11, E_1 may be 1.54 V while E_2 is 1.49 V. When this occurs, the cell with the largest terminal voltage will tend to discharge through the one with the smaller

Chap. 10 Voltage Cells, Batteries, and DC Power Supplies

terminal voltage, until both have equal voltages. If secondary (i.e., rechargeable) cells are used, one cell is charged from the other. *Cells that do not have closely equal terminal voltages should not be connected in parallel.*

EXAMPLE 10-5 Four cells connected in parallel each have terminal voltage of $E = 2$ V and internal resistances of $r_i = 0.1$ Ω. Each cell can normally supply a maximum load current of 500 mA for a time period of 4 h. Determine the open-circuit output voltage of the parallel combination and the maximum load current that can be supplied. Also, calculate the Ah rating and the internal resistance for the battery of cells.

SOLUTION

From Equation (10-8),

$$E = E_1 = E_2 = E_3 = E_4$$

$$= \mathbf{2\ V}$$

From Equation (10-9),

$$I_{0(max)} = I_1 + I_2 + I_3 + I_4$$

$$= 4 \times 500\ \text{mA}$$

$$= \mathbf{2\ A}$$

$$\text{Ah } \textit{rating of battery } = 4 \times (\text{Ah } \textit{rating per cell})$$

$$= 4 \times (500\ \text{mA} \times 4\ \text{h})$$

$$\text{Ah } \textit{rating of battery } = \mathbf{8\ Ah}$$

$$r_i = r_1 \| r_2 \| r_3 \| r_4$$

$$= \frac{0.1\ \Omega}{4}$$

$$= \mathbf{0.025\ \Omega}$$

Parallel-connected (similar) cells produce an output voltage equal to the terminal voltage of one cell and can supply a maximum output current equal to the sum of the maximum currents from each cell.

It is very important to ensure that parallel cells are correctly connected. Correct and incorrect ways of connecting cells in parallel are shown in Figure 10-13. All positive terminals should be connected together, and all negative terminals should be connected together, as shown in Figure 10-13(a). Where positive and negative terminals are

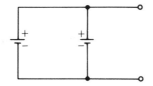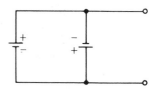

(a) Correctly connected
parallel cells

(b) Incorrectly
connected
parallel cells

FIGURE 10-13. Parallel-connected cells should always be connected positive-to-positive and negative-to-negative. Incorrectly connected cells discharge rapidly.

connected together as in Figure 10-13(b), *the cells short-circuit each other and discharge rapidly.*

PRACTICE PROBLEMS

10-5.1 Each cell in a battery of 100 parallel-connected cells has a 1.45 V terminal voltage when supplying a maximum current of 60 mA. Determine the minimum load resistance that can be connected to the battery.

10-5.2 A battery is made up of ten cells connected in parallel. Each cell has a 2.2 V open-circuit terminal voltage and an internal resistance of 0.015 Ω. Calculate the output voltage of the battery when supplying a 100 A load current.

10-5.3 A certain load connected to the battery in Problem 10-5.2 causes the terminal voltage to drop to 2 V. If this load can be supplied for a maximum of 3 hours, determine the Ah rating of the battery.

10-6

VOLTAGE CELLS IN SERIES-PARALLEL

Figure 10-14 shows the circuit diagram of two parallel-connected groups of cells, each group consisting of three 1.5 V cells in series. The terminal voltage of each group of series cells is obviously

$$E = E_1 + E_2 + E_3$$
$$= 1.5 \text{ V} + 1.5 \text{ V} + 1.5 \text{ V}$$
$$= 4.5 \text{ V}$$

Therefore, the terminal voltage of the entire battery is 4.5 V.

The currents flowing from each series group of cells in Figure 10-14 are identified as I_1 and I_2. If each cell is capable of supplying a maximum

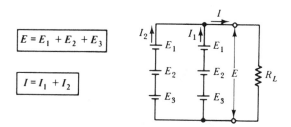

$$\boxed{E = E_1 + E_2 + E_3}$$

$$\boxed{I = I_1 + I_2}$$

FIGURE 10-14. Voltage cells can be connected in parallel combinations of series-connected groups. This arrangement gives a terminal voltage equal to that of each group and a maximum load current equal to the sum of the load currents from the groups.

of 1 A, the greatest output current that can be taken from each series group is also 1 A. For the parallel combination (or entire battery of cells) the maximum output current is

$$I = I_1 + I_2 = 1\text{ A} + 1\text{ A}$$
$$= 2\text{ A}$$

This kind of *series-parallel* arrangement of cells can be employed to give any desired combination of output voltage and maximum current. Where a greater output voltage is required, more cells are added to each series group. For larger maximum output currents, the number of parallel groups is increased.

PRACTICE PROBLEM

10-6.1   A battery of solar cells consists of 50 parallel-connected groups of 20 series-connected cells. Each cell has a 0.45 V terminal voltage when supplying a current of 50 mA. Calculate the maximum output power of the battery.

10-7
LEAD-ACID BATTERY

CONSTRUCTION.   The lead–acid battery is the most commonly used type of storage battery and is well known for its application in automobiles. The battery is made up of several cells, each of which consists of lead plates immersed in an electrolyte of dilute sulfuric acid. The voltage per cell is typically 2 V to 2.2 V. For a 6 V battery, three cells are connected in series, and for a 12 V battery six cells are series-connected.

The construction of a lead–acid automobile-type battery is illustrated in Figure 10-15. The electrodes are lead-antimony alloy plates with a pattern of recesses, so that they are in the form of grids [see Figure 10-15(a)]. Lead oxide (termed *active material*) is pressed into the re-

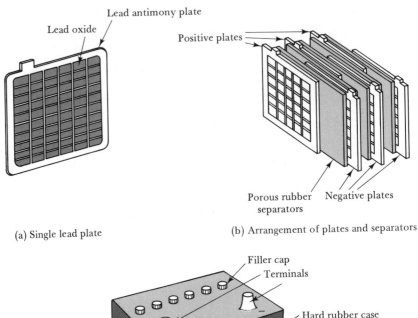

Lead antimony plate

Lead oxide

Positive plates

Porous rubber separators   Negative plates

(a) Single lead plate

(b) Arrangement of plates and separators

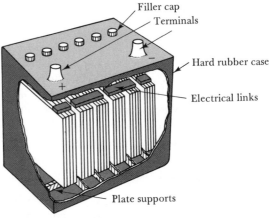

Filler cap

Terminals

Hard rubber case

Electrical links

Plate supports

(c) Complete 12 V battery assembly

FIGURE 10-15. A lead-acid battery consists of several series-connected lead-acid cells in a single container. Three cells are used for a 6 V battery; six cells are employed for a 12 V battery.

cesses of the plates. Each electrode consists of several plates connected in parallel with porous rubber separators in between, as illustrated in Figure 10-15(b). This arrangement, and the shape of the plates, gives the largest possible electrode surface area within the size limitations of the battery.

The complete 12 V battery, illustrated in Figure 10-15(c), has an outer case of hard rubber. The case is divided into six sections for the six separate cells. Projections are provided on the inside at the bottom of the case to support the plates. These projections ensure that the lower edges of the plates are normally well above the level of any active material that falls to the bottom of a cell. Such material can short out the positive and negative plates, and render a cell useless.

Every cell has a threaded *filler cap* with a small hole in its center. The filler caps provide access for adding electrolyte, and the holes allow gases to be vented to the atmosphere.

Thick electrical links connect the cells in series, and hefty battery terminals are provided. Since the battery may be required to supply a very heavy current, it is important that the resistance of all electrical connections be very low, to minimize voltage drops. A current of 250 A is not unusual for a battery driving an automobile starter.

OPERATION. When the lead–acid cell is charged, the lead oxide on the positive plates changes to *lead peroxide,* and that on the negative plates becomes *spongy* or *porous* lead. In this condition, the positive plates are brown in color and the negative plates are gray. When the battery is discharging (i.e., supplying a current), atoms from the spongy lead on the negative plates combine with sulfate molecules to form *lead sulfate* and *hydrogen.* As always, electrons are left behind on the negative plates so that they accumulate a negative potential. The hydrogen released in the electrolyte combines with the lead peroxide on the positive plate, removing electrons from the plate to give it a positive potential.

The combination of lead peroxide and hydrogen at the positive electrode produces water and lead sulfate. The water dilutes the electrolyte, making it a weaker solution, and the lead sulfate that is produced at both positive and negative plates tends to fill the pores of the active material. Both these effects (dilution of the electrolyte and formation of lead sulfate) render each cell less efficient and eventually cause the battery output voltage to fall. When the battery is recharged, a current (conventional direction) is made to flow into the positive electrode of each cell. This current causes the lead sulfate at the negative electrode to recombine with hydrogen ions, thus re-forming sulfuric acid in the electrolyte and spongy lead on the negative plates. Also, the lead sulfate on the positive electrodes recombines with water to regenerate lead peroxide on the positive plates and sulfuric acid in the electrolyte. The final result of charging the cell is that the electrodes are re-formed and the electrolyte is returned to its original strength. With proper care a lead-acid battery is capable of sustaining a great many cycles of charge and discharge, giving satisfactory service for a period of several years.

AMPERE-HOUR RATING OF LEAD-ACID BATTERY. Typical ampere-hour ratings for 12 V lead–acid automobile batteries range from 100 Ah to 300 Ah. This is usually specified for an 8 h discharge time, and it defines the amount of energy that can be drawn from the battery until the voltage drops to about 1.7 V per cell.

For a 240 Ah rating, the battery could be expected to supply 30 A for an 8 h period. With greater load currents, the discharge time is obviously shorter. However, the ampere-hour rating is also likely to be reduced for a shorter discharge time, because the battery is less efficient when supplying larger currents. Another method of rating a lead–acid battery is to define what its terminal voltage will be after about 5 s of supplying perhaps 250 A. This corresponds to the kind of load that a battery experiences in starting an automobile. It is important to avoid battery overloads that may demand excessive currents. Drawing a larger current than the battery is designed to supply may cause severe damage.

The rating of a battery is typically stated for temperatures around 25°C, and this must be revised for operation at lower temperatures. Since the chemical reactions occur more slowly at reduced temperatures, the available output current and voltage are less than at 25°C. Around −18°C a fully charged battery may be capable of delivering only 60% of its normal ampere-hour rating. As the cell is discharged and the electrolyte becomes weaker, freezing of the electrolyte becomes more likely. A fully charged cell is less susceptible to freezing, but even a fully charged cell may fail when its temperature falls to about −21°C.

EXAMPLE 10-6 A lead–acid battery has a rating of 300 Ah.
  a. Determine how long the battery might be employed to supply 25 A.
  b. If the battery rating is reduced to 100 Ah when supplying large currents, calculate how long it could be expected to supply 250 A.
  c. Under very cold conditions the battery supplies only 60% of its normal rating. Find the length of time that it might continue to supply 250 A to the starter motor of an automobile.

SOLUTION

a. *From* Equation (10-1),

$$Ah\ rating = I \times t$$

Therefore, $$t = \frac{Ah}{I} = \frac{300\ Ah}{25\ A}$$

$$= 12\ h$$

b. $$t = \frac{Ah}{I} = \frac{100\ Ah}{250\ A} = 0.4\ h$$

$$= 24\ min.$$

c. $$Ah = 60\%\ of\ 100\ Ah$$

$$= 60\ Ah$$

$$t = \frac{60\ Ah}{250\ A} = 0.24\ h$$

$$= 14.4\ min.$$

CHARGING. When a battery is to be charged, a dc charging voltage must be applied to its terminals. The polarity of the charging voltage must be such that it causes current to flow into the battery, in opposition to the normal direction of discharge current. This means that the positive output terminal of the *battery charger* must be connected to the positive terminal of the battery, and the charger negative terminal must be connected to the battery negative terminal. The arrangement is shown in Figure 10-16. The battery charger normally has a voltmeter and ammeter to monitor the charging voltage and current, and a control to adjust the rate of charge.

The output voltage of a battery charger must be greater than the battery voltage in order to cause current to flow into the battery positive terminal. The charging current depends upon the difference between the battery voltage and the charging voltage and upon the internal resistances of the battery. A very large charging current is to be avoided because it could cause the battery to overheat, possibly resulting in warping of the lead plates. The maximum safe charging current is frequently taken as the maximum output current from the battery when discharging at its 8 h rate.

EXAMPLE 10-7   A battery with a rating of 300 Ah is to be charged.
a. Determine a safe maximum charging current.
b. If the internal resistance of the battery is 0.008 Ω and its (discharged) terminal voltage is 11.5 V, calculate the initial output voltage level for the battery charger.

SOLUTION

a. Safe rate of charge at the 8 h *discharge rate:*

$$I = \frac{\text{Ah}}{t} = \frac{300 \text{ Ah}}{8 \text{ h}}$$

$$= \mathbf{37.5 \ A}$$

b. Charging current:

$$I = \frac{charger\ voltage - battery\ voltage}{r_i}$$

Therefore,   *charger voltage* $= (I \times r_i) + battery\ voltage$

$$= (37.5 \text{ A} \times 0.008 \ \Omega) + 11.5 \text{ V}$$

$$= \mathbf{11.8 \ V}$$

SPECIFIC GRAVITY.   When a lead–acid battery is in a nearly discharged condition, the electrolyte is in its weakest state. Conversely, the electrolyte is at its strongest (or greatest density) when the battery is fully

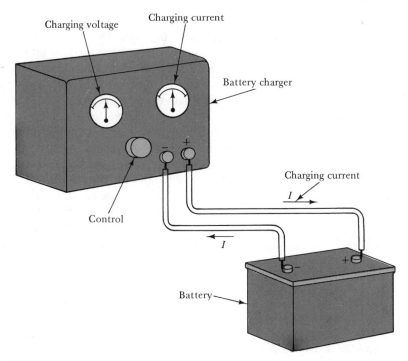

FIGURE 10-16. When a battery is being charged, current flows from the positive terminal of the charger into the positive terminal of the battery, and from the battery negative terminal into the charger negative terminal.

charged. The density of electrolyte related to the density of water is termed its *specific gravity*. The specific gravity of the electrolyte (measured by means of a hydrometer) is used as an indication of the state of charge of a lead–acid battery. Electrolyte with a specific gravity of 1100 to 1150 is 1.1 to 1.15 times as dense as water. At 1100 to 1150 the cell is completely discharged. When the specific gravity is 1280 to 1300, the cell may be assumed to be fully charged.

CARE OF LEAD–ACID BATTERIES

- The level of the electrolyte in each cell should be checked regularly, and distilled water added as necessary to keep the top of the plates covered by about 1 cm of liquid.
- Battery terminals should be kept clean and lightly coated with petroleum jelly to avoid corrosion.
- In cold weather, batteries should always be maintained in a nearly fully charged condition to avoid freezing.
- Lead–acid batteries should never be allowed to remain for long periods in a discharged state, because lead sulfate could harden and permanently clog the pores of the electrodes.

- Before storing for a long time the battery should be completely charged, then the electrolyte should be drained so that the battery is stored dry.

PRACTICE PROBLEMS

10-7.1 A 12 V lead–acid battery with a 400 Ah rating is to supply a 0.6 Ω load. Determine the maximum time period that the load may be connected. If the battery rating is reduced to 70% of normal at low temperatures, calculate the length of time that a 0.8 Ω load may be connected.

10-7.2 An automobile battery has an 80 Ah rating when supplying a 200 A starter current. If the rating is reduced to 65% at low temperatures, calculate the maximum time period for which it may keep the starter going.

10-7.3 A 400 Ah battery has an internal resistance of 0.01 Ω and a terminal voltage of 11.6 V when discharged. Determine the maximum safe charging current for the battery and the initial output voltage of the charger.

10-8

DC POWER SUPPLIES

Most electronics equipment uses a dc power supply rather than a battery. The power supply converts the usual domestic or industrial alternating voltage supply (see Chapter 17) into a direct voltage. In electronics laboratories dc power supplies are extensively used to supply experimental circuitry. Each laboratory-type power supply is designed to provide a particular output voltage, or range of voltages, and a specified maximum output current. When selecting a dc power supply for a particular application, the required output voltage and current must both be known.

Figure 10-17(a) shows the front panel of a typical laboratory-type dc power supply. As illustrated, there is an *on/off* switch for the ac input. A meter is provided for indicating output voltage or current, as selected by the *V/I* switch. Voltage and current controls are included for setting the output voltage and maximum output current to the desired level. Positive (+), negative (−), and *ground* terminals are available for connection purposes.

The output voltage may be taken from the + and − terminals and the ground terminal left unconnected. This gives a *floating* (i.e. ungrounded) output. Alternatively, either the − or + terminals may be connected to the ground terminal to give a voltage which is either positive or negative with respect to ground, as desired.

The voltage control is adjusted to set the output to the required voltage. Where there is no possibility of overload, the current control

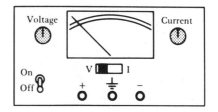

(a) Single output voltage dc power supply

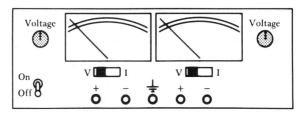

(b) Dual output voltage dc power supply

FIGURE 10-17. The front panel of a laboratory dc power supply usually has an ON/OFF switch, one or two meters, voltage and current controls, and a set of output terminals.

may be set to its maximum position. To limit the output current to a particular level, the output terminals must be short-circuited and the current control adjusted from minimum to the desired maximum current level.

Figure 10-17(b) shows the terminals of a dual output voltage supply with each output independently adjustable. The outputs may be series-connected to provide a greater voltage than can be obtained from one supply. Parallel connection for greater output current is also possible, but care must be taken to ensure that the load current is equally distributed between the two outputs. This type of supply is most frequently used with one + terminal and one − terminal grounded to give a *plus-minus* output voltage.

SUMMARY OF FORMULAS

Cell output current:

$$I_0 = \frac{E_c}{r_i + R_L}$$

Cell output voltage:

$$E_0 = E_c - I_0 r_i$$

For cells in series:

$$E = E_1 + E_2 + E_3 + \cdots$$

$$r_i = r_1 + r_2 + r_3 + \cdots$$

$$I = \frac{E_1 + E_2 + E_3 + \cdots}{R_L + r_1 + r_2 + r_3 + \cdots}$$

$$I = I_1 = I_2 = I_3 = I_4$$

For cells in parallel: $E = E_1 = E_2 = E_3 = \cdots$

$$r_i = r_1 \| r_2 \| r_3 \cdots$$

$$I = \frac{E}{R_L + (r_1 \| r_2 \| r_3 \cdots)}$$

REVIEW QUESTIONS

10-1 Draw a sketch to show the basic construction of a simple voltage cell. Explain the operation of the cell.

10-2 Discuss the output voltage and current that can be supplied by a simple voltage cell and define ampere-hour rating.

10-3 Define primary cell, secondary cell, open-circuit output voltage, no-load output voltage, internal resistance, and ideal cell.

10-4 Sketch the equivalent circuit for a voltage cell with an external load resistance (R_L). Write equations for the cell terminal voltage and output current. Explain briefly.

10-5 Draw a sketch to show the basic construction of a typical zinc–carbon dry cell. Explain the operation of the cell and discuss its ampere-hour rating.

10-6 Discuss the construction, terminal voltages, ampere-hour ratings, and application for the following: manganese–alkaline cell, mercury cell, and nickel–cadmium cell.

10-7 Sketch the circuit diagram for four voltage cells connected in series. Draw the equivalent circuit for the series combination of cells and write the equations for output voltage and current.

10-8 Draw circuit diagrams to show series-aiding and series-opposing connected cells and briefly explain. Also, show how a ±10 V supply could be constructed using 2 V cells.

10-9 Sketch the circuit diagram for four voltage cells connected in parallel. Draw the equivalent circuit for the parallel combination of cells and write the equations for output voltage and current.

10-10 Explain what occurs when cells having different terminal voltages are connected in parallel. Also, discuss the result of incorrectly connecting voltage cells in parallel.

10-11 Describe the construction of a 12 V lead–acid battery and explain the chemical process involved in charge and discharge within the individual cells.

10-12 Discuss typical ampere-hour ratings of a lead–acid battery with respect to supplying small currents, large currents, and low-temperature performance.

10-13 Explain how a lead–acid battery should be connected to a battery charger and discuss charging voltage and current levels.

10-14 Define specific gravity and state typical specific-gravity values for discharged and fully charged lead–acid batteries.

10-15 Discuss the care of lead–acid batteries.

PROBLEMS

SECTION 10-1

10-1   A voltage cell supplies a current of 0.75 A for 10 h. Then its terminal voltage drops to a low level. Calculate the Ah rating of the cell and determine how long it might be expected to supply a current of 250 mA.

10-2   An electronic circuit draws a constant current of 8 mA from a 9 V battery. If the battery has to be replaced every 30 days, calculate its ampere-hour rating.

10-3   A 3 V battery supplies current continously to a 2 Ω resistance. If the battery has a rating of 8 Ah, how long will it take to discharge?

10-4   A 1.5 V cell with a rating of 5 Ah is used to drive an electric clock. The cell has to be replaced approximately every 12 months. Determine the load resistance of the clock mechanism.

SECTION 10-2

10-5   The open-circuit voltage from a certain voltage cell is 1.65 V and its internal resistance is 0.05 Ω. Calculate the terminal voltage of the cell when the output current is 5 A. Also, determine the cell current when the terminals are short-circuited.

10-6   The cell in Problem 10-4 has a terminal voltage of 1.55 V when open-circuited and 1.5 V when connected to the clock. Calculate the internal resistance and short-circuit current for the cell.

10-7   A battery with an internal resistance of 3 Ω has an open-circuit voltage of 9.35 V. Calculate the minimum load resistance that can be connected to the battery if its terminal voltage is not to fall below 9 V.

10-8   A voltage cell has an open-circuit terminal voltage of 2.2 V and a short-circuit current of 1.47 mA. Determine the terminal voltage of the cell when supplying 100 μA.

SECTION 10-4

10-9   A battery is made up of five series-connected voltage cells, each of which has an open-circuit terminal voltage of 1.6 V and an internal resistance of 0.08 Ω. Calculate the battery terminal voltages for no-load condition and for $R_L = 6 \Omega$.

10-10   The terminal voltage of a battery consisting of 6 series-connected cells is measured as 9.5 V on open-circuit and 8.9 V when supplying 180 mA. Calculate the average internal resistance of one cell.

10-11   A 12 V supply is to be constructed using 1.5 V cells each of which has $r_i = 0.05 \Omega$. Determine the number of cells required and calculate the internal resistance of the battery. Also, calculate the battery terminal voltage when the load resistance is 10 Ω.

10-12 A plus-minus power supply constructed of eight 1.5 V cells has outputs of +6.5 V and −6.4 V on open-circuit. When supplying 500 mA, the outputs are +6.2 V and −6.15 V. Calculate the internal resistance of each half of the power supply.

10-13 A ±9 V supply is to be constructed using zinc-carbon D cells. Determine the total number of cells required and sketch the circuit diagrams for connecting the cells. If one of the cells is accidentally connected in reverse, what will be the output voltages from the supply?

10-14 Ten series-connected solar cells each have an open-circuit terminal voltage of 0.45 V and an internal resistance of 0.8 Ω. Determine the maximum current that can be drawn from the battery of cells if the terminal voltage is not to fall below 4 V.

SECTION 10-5

10-15 A parallel-connected battery of 1.5 V cells is to supply current to a 0.15 Ω resistor. If each cell can supply a maximum current of 1 A, determine the number of cells required. Neglect the internal resistances of each cell.

10-16 A 9 V power supply capable of sustaining a 1 A load current is to be constructed by parallel-connecting several of the type of batteries described in Problem 10-7. Determine the required number of batteries if the terminal voltage of the supply is not to fall below 9 V. Also, calculate the internal resistance and short-circuit current for the power supply.

10-17 Eight voltage cells connected in parallel each have an open-circuit terminal voltage of 1.7 V and an internal resistance of 0.07 Ω. Each cell can normally supply a maximum load current of 750 mA for a time period of 12 h. Determine the output voltage of the battery of cells and the maximum load current that can be supplied. Also, calculte the Ah rating and internal resistance for the parallel combination of cells.

10-18 If the ten solar cells described in Problem 10-14 are connected in parallel, determine the internal resistance and short-circuit current for the combination.

10-19 A 10 kW electric motor drives a vehicle at an average speed of 50 km/h. Ten parallel-connected 12 V, 100 Ah batteries supply the motor. Determine the maximum distance that the vehicle may travel before the batteries must be recharged.

10-20 A power supply made up of parallel-connected 1.5 V cells is to sustain a constant load current of 39 mA. The cells, which each have a rating of 5 Ah, are to be replaced every 14 days. Determine the number of cells required.

10-21 For the vehicle referred to in Problem 10-19, calculate the total Ah rating for batteries that will provide a range of 400 km.

SECTION 10-6

10-22 A 4.5 V power supply constructed of series-parallel-connected 1.5 V cells is to provide a constant current of 575 mA to an electronic circuit. The cells are each capable of supplying 100 mA without any serious drop in terminal voltage. Determine the number of cells required. If each cell has a 6 Ah rating, calculate the ampere-hour rating for the power supply.

10-23 A power supply to be constructed from voltage cells is required to supply a maximum current of 20 A with a terminal voltage of approximately 6 V. Several voltage cells are available, each of which has a terminal voltage of 1.5 V and each of which can supply a maximum current of 5 A. Determine the number of cells required and draw a circuit diagram to show how they should be connected.

10-24 A 100 m² roof area of a house is covered with solar cells which each occupy 5 cm². The terminal voltage of each cell is 0.46 V when supplying a current of 120 mA. How should the cells be arranged to produce a 115 V supply, and what is the maximum output current?

10-25 A solar cell supply with an output of approximately 13 V is to be constructed to recharge the 12 V batteries of a remote radio transmitter. The charging current is to be approximately 400 mA, and the available solar cells each have an output of 0.45 V and 65 mA. Determine the total number of cells required.

10-26 A 120 V power supply is to be constructed using 12 V batteries. Each battery can produce a continuous current of 10 A without a serious drop in terminal voltage. Determine the number of batteries required to produce an output current of 30 A. Also, determine the ampere-hour rating of the power supply if each battery is rated as 400 Ah.

SECTION 10-7

10-27 An automobile battery is to supply the starter motor with 150 A of current for a minimum time of 20 minutes. If the battery supplies only 65% of its normal rating under extremely cold conditions, calculate the minimum normal Ah rating for the battery.

10-28   The terminal voltage of a 12 V battery drops to 11.9 V when the load current is 10 A. Determine the short-circuit current for the battery. If the battery has a 300 Ah rating, which is reduced by 50% when overloaded, how long is the battery likely to sustain the short-circuit current?

10-29   A certain lead-acid battery has a rating of 400 Ah.
(a) Calculate how long the battery should be able to supply the current of 12 A.
(b) When supplying large currents the normal battery rating is reduced to 150 Ah. Determine how long it can be expected to supply 240 A to an automobile starter motor.
(c) If the battery rating is reduced to 60% of normal when very cold, calculate how long it might supply 240 A.

10-30   A 12 V battery becomes fully charged with a constant current of 20 A over a time period of 10 h. If the battery rating is reduced to 60% of normal when supplying high current levels, how long is it likely to sustain an 80 A load?

10-31   A 400 Ah lead-acid battery is to be charged.
(a) Calculate a maximum safe level of charging current.
(b) The internal resistance of the battery is 0.007 Ω and the terminal voltage when discharged is 11.4 V. Determine the initial output voltage level for the battery charger.

10-32   An automobile battery is almost completely discharged in a time of 8 h when the lights are left on with the engine off. If the lights draw a current of 20 A, estimate the ampere-hour rating of the battery.

COMPUTER PROBLEMS

10-33   Write a computer program to calculate the output voltage and current for a series-parallel battery of solar cells. The requested quantities should include cell voltage, cell current, number in series, and number of parallel-connected groups.

10-34   Write a computer program to determine the maximum safe charging current and initial charger voltage for a battery with a given ampere-hour rating. See Example 10-7.

ANSWERS TO
PRACTICE
PROBLEMS

10-1.1	0.5 Ah, 14.3 h
10-1.2	0.24 Ah
10-2.1	1.6 Ω, 0.97 A
10-2.2	0.0055 Ω
10-4.1	8.8 V

10-4.2 20 A

10-5.1 0.24 Ω

10-5.2 2.05 V

10-5.3 400 Ah

10-6.1 22.5 W

10-7.1 20 h, 18.7 h

10-7.2 15.6 min

10-7.3 50 A, 12.1 V

11 MAGNETISM

Objectives   You will be able to:

Discuss the behavior of bar magnets and sketch the magnetic force fields around bar magnets in various situations.

State the fundamental law of magnetism.

Describe magnetic lines of force and list their characteristics.

Define magnetic flux and flux density and solve problems involving these quantities in relationship to bar magnets.

Sketch the magnetic fields that occur around current-carrying conductors and relate the direction of the fields to the current directions.

Sketch the magnetic fields that occur around various shaped coils.

Solve problems involving the flux and flux density of the magnetic fields around coils.

Explain electromagnetic induction.

Explain the (atomic) theory of magnetism and classify various magnetic and nonmagnetic materials.

Define magnetomotive force and magnetic field strength and solve problems involving these quantities.

Calculate the force exerted on a current-carrying conductor situated in a magnetic field and define the direction of the force.

Calculate the torque exerted on a coil pivoted in a magnetic field and show how it occurs.

Introduction

Some two thousands years ago, Chinese mariners used a suspended bar of mineral known as *lodestone* as a crude compass. The mineral had magnetic properties, so its ends tended to point to the earth's magnetic poles. Today's magnetic compasses are much more sensitive than the lodestone, but they fulfill essentially the same purpose. As well as direction finding, there are a great many other applications of magnetism.

The magnetic force field around a bar magnet is easily plotted by means of iron filings or by use of compasses. It is found that the magnetic field passes through nonmagnetic material including air and vacuum. However, iron and steel are by far the best magnetic conductors, and in the presence of a magnetic field, pieces of iron and steel become magnetized.

A magnetic field also exists around a current-carrying conductor, and a current-carrying coil generates a field similar to that of a bar magnet.

11-1

MAGNETIC FIELD

NORTH AND SOUTH POLES. A bar magnet suspended on thread tends to align itself in relation to the earth in a north–south direction. This is also the case with the magnetic compass, which is simply a pivoted magnetized needle. The two ends of the bar magnet and the compass are identified as *north pole* (N) and *south pole* (S), in accordance with the direction that they tend to point.

Consider the effect of bringing the N pole of another bar magnet close to the N pole of the suspended magnet. As illustrated in Figure 11-1(a), the result is repulsion between the two magnets. A similar result occurs when the S pole of the hand-held magnet is brought close to the S pole of the suspended magnet. However, when N and S poles are brought together, the magnets are attracted to each other [see Figure 11-1(b)]. The same effects are demonstrated when one pole of a bar magnet is brought close to a magnetic compass [Figure 11-1(c)]. The needle of the compass always turns so that its S pole points to the N pole of the magnet, or so that the N pole points to the magnet's S pole. These results give rise to the *fundamental law of magnetism:*

Like poles repel; unlike poles attract.

This is similar to the fundamental law of electrification by friction (Section 1-1).

Because unlike poles attract, the pole of the magnetic compass that points to the earth's north pole [Figure 11-1(d)] must be *unlike* the earth's north pole. Consequently, the earth's north pole must actually be a south magnetic pole! Alternatively, it can be stated that the north pole of a compass is actually a *north-seeking pole* with respect to the earth.

MAGNETIC LINES OF FORCE. The fact that attraction or repulsion occurs when the poles of two magnets are brought together demonstrates that a *force field* exists around the magnets, and that the field tends to be concentrated at the magnet's poles. This force field around a magnet is termed a *magnetic field*, and it can be investigated further by means of iron filings sprinkled on a piece of cardboard placed on top of a magnet. When the cardboard is gently tapped, the iron filings align themselves into chains, as shown in Figure 11-2(a). The lines plotted by

Chap. 11   Magnetism

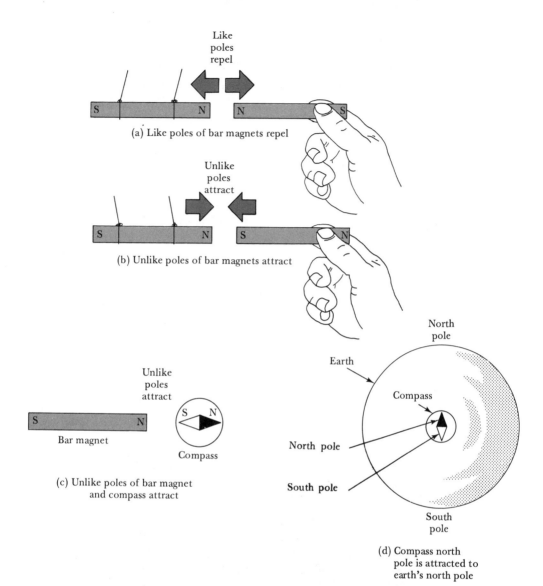

FIGURE 11-1. Like magnetic poles repel; unlike poles attract. Two north poles repel, two south poles repel, and a north and south pole attract. A compass north pole points to the earth's north pole which must be unlike the compass north pole!

the pattern of iron filings stretch from one pole to the other and are termed *magnetic lines of force*. Collectively, the magnetic lines of force are referred to as the *magnetic flux*.

The concentration of iron filings at the poles of the magnet shows that the magnetic force field has its greatest strength close to the magnet's poles. Obviously, the field becomes weaker at points farther away from the poles.

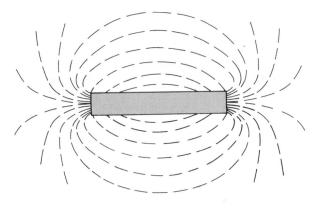

(a) Force field around a magnet, plotted by iron filings

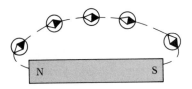

(b) Force field plotting using compasses

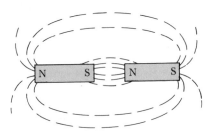

(c) Each half of a magnet has N and S poles

FIGURE 11-2. The magnetic force field that exists around a bar magnet can be plotted by the use of iron filings or by compasses.

The magnetic lines of force can also be plotted by the use of several compasses, as illustrated in Figure 11-2(b). The compass needles point along the lines of force in a direction from the magnet's north pole to its south pole. Because of this, it is assumed that the magnetic lines of force exit from the north pole of the magnet and enter the magnet at its south pole. The direction of the lines of force is defined as: *the direction in which an isolated north pole would move if it were placed in the magnetic field.* In fact, an *isolated north pole* could not exist because all magnets have north and south poles. However, it can be imagined that if one did exist, an isolated north pole would be repelled from the magnet's N pole and move along a line of force toward the S pole of the magnet.

When a bar magnet is cut into two pieces, as shown in Figure 11-2(c), it is found that each half is still a magnet with N and S poles. Also, the lines of force cross the gap between the two pieces of the magnet, as illustrated, demonstrating that the magnetic lines of force pass through the magnet itself. Thus, it is seen that *magnetic lines of force form closed loops.*

When a piece of soft iron is placed in a magnetic field, it is found that the lines of force bend away from their usual paths in order to pass through the iron [see Figure 11-3(a)]. It is also found that the iron has become magnetized temporarily and that a S pole can be identified at the end of the soft iron nearest the magnet's N pole, and a N pole can be identified at the other end. This effect is known as *induced magnetism.* No similar effect occurs when pieces of wood or glass or rubber or other

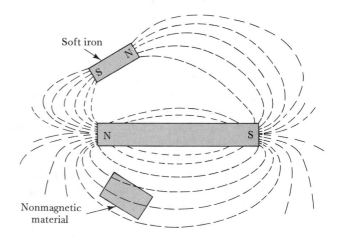

(a) Effects of magnetic and nonmagnetic material
on shape of magnetic force field

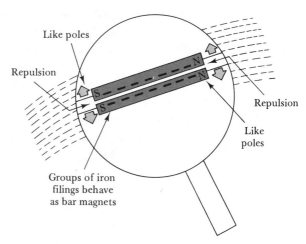

(b) "Magnified" view of lines of force
plotted by iron filings

FIGURE 11-3. Magnetic lines of force form closed loops and distort to pass through iron. Magnetic lines of force tend to repel each other, and two lines of force never intersect.

such *nonmagnetic materials* are placed in the magnetic field. Thus, it can be stated that *magnetic lines of force pass most easily through iron;* that is, iron is a better conductor of magnetic lines of force than such materials as air or wood or glass.

It can now be understood that the individual iron filings used to plot the magnetic lines of force must become magnetized, so that each has its own N and S poles. Figure 11-3(b) illustrates the situation and shows that the poles of each little magnet or group of magnets are likely to be adjacent to like poles of other magnetized iron filings alongside them. Thus repulsion occurs, and this demonstrates another piece of information about the magnetic lines of force; that is, *magnetic lines of force tend to repel each other.* This fact is also shown by the field pattern at the poles of the magnet in Figure 11-2(a). Since lines of force repel each other, *two lines of force cannot intersect.*

Figure 11-4(a) shows the pattern of the magnetic field when two N poles (or two S poles) are brought close together. It is clearly seen that the lines of force from each pole repel one another. Conversely, when a N and S pole are placed close together [Figure 11-4(b)], the lines of force run out of one pole into the other. In Figure 11-4(c), the field pattern at the poles of a horseshoe-shaped magnet is shown. Note that the magnetic lines of force take the shortest path between the poles, and this shows that *the lines of force are always in a state of tension.* Summarizing:

- Every magnet has two poles identified as a north pole (N) and a south pole (S).
- The N pole is actually a *north-seeking pole,* and the S pole is a *south-seeking pole,* with respect to the earth.
- The force field that exists around a magnet is concentrated at its poles.
- Like poles repel; unlike poles attract.
- Magnetic lines of force in a field around a magnet can be plotted by the use of iron filings.
- Lines of force form closed loops.
- Lines of force exit from a magnet's N pole and enter at its S pole.
- Lines of force travel most easily through soft iron.
- Lines of force repel each other.
- Lines of force cannot intersect.
- Lines of force from two like poles repel each other.
- Lines of force from two unlike poles run into each other.
- Lines of force are always in a state of tension.
- A piece of soft iron placed in a magnetic field is temporarily magnetized by induction.

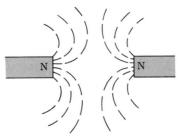

(a) Magnetic field at like poles

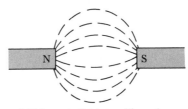

(b) Magnetic field at unlike poles

(c) Lines of force take
the shortest path

FIGURE 11-4. The magnetic force field plotted at two like poles shows repulsion, while that at two unlike poles demonstrates attraction.

11-2
ELECTRO-
MAGNETISM

MAGNETIC FIELD AROUND A CONDUCTOR. When an electric current flows in a conductor, a magnetic field is set up around the conductor. This is easily demonstrated by again using iron filings, as illustrated in Figure 11-5(a). A conductor is passed vertically through a hole in a horizontal piece of cardboard. Iron filings are sprinkled on the cardboard, and when the current through the conductor is zero, there is no evidence of a magnetic field. However, when a current flows through the conductor, gentle tapping of the cardboard causes the iron filings to settle in concentric rings around the conductor.

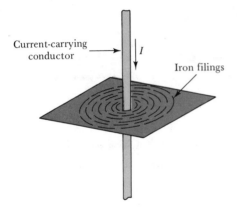

(a) Iron filings form
concentric rings around
a current-carrying conductor

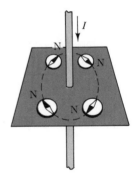

(b) Compasses point in
circle around a
current-carrying conductor

(c) Lines of force are in a
clockwise direction around a
conductor which is carrying
current away from the viewer

(d) Lines of force are in a
counterclockwise direction
around a conductor which is
carrying current toward the viewer

FIGURE 11-5. A magnetic force field in the form of concentric rings can be shown to exist around a current-carrying conductor.

As with other magnetic fields, the field around a current-carrying conductor can also be investigated by the use of magnetic compasses. Figure 11-5(b) shows the result of placing four compasses on the horizontal cardboard. For a current flowing in a (conventional) direction down through the cardboard, the compass needles point in a clockwise direction around the conductor. Since the direction of magnetic lines of force are defined in terms of the direction in which a free north pole would move, it can be stated that for the conditions illustrated, the magnetic lines of force are in a clockwise direction. When the current direction is reversed so that it flows up through the cardboard, the compass needles reverse and point in a counterclockwise direction.

Figure 11-5(c) and (d) further illustrate the direction of a magnetic field around a current-carrying conductor. The conductor is assumed to be directly opposite the viewer, so only its cross-sectional area is shown. In Figure 11-5(c), current is assumed to be flowing away from the viewer (i.e., into the page). This is indicated by the + sign, which represents the tail of an arrow. In this case the magnetic lines of force around the conductor have a clockwise direction. In Figure 11-5(d), the dot at the center of the conductor cross-section represents the point of an arrow. Therefore, current is flowing toward the viewer (i.e., out of the page). The magnetic lines of force now have a counterclockwise direction.

Two memory aids for determining the direction of the magnetic flux around a current-carrying conductor are shown in Figure 11-6. The *right-hand-screw rule* as illustrated in Figure 11-6(a) shows a wood screw being turned clockwise and progressing into a piece of wood. The horizontal direction of the screw is analogous to the direction of current in a conductor, and the circular motion of the screw shows the direction of magnetic flux around the conductor. In the *right-hand rule,* illustrated in Figure 11-6(b), a right hand is closed around a conductor with the thumb pointing in the (conventional) direction of current flow. The fingers point in the direction of the magnetic lines of force around the conductor.

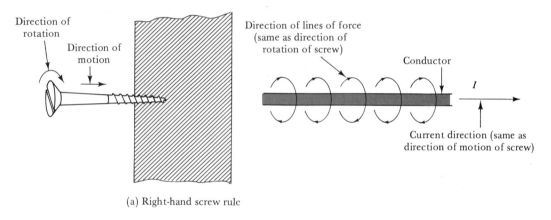

(a) Right-hand screw rule

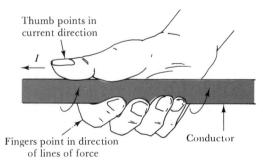

(b) Right-hand rule

FIGURE 11-6. The *right-hand-screw rule* and the *right-hand rule* can be used for determining the direction of the magnetic lines of force around a current-carrying conductor.

Since a current-carrying conductor has a magnetic field around it, when two current-carrying conductors are brought close together there will be interaction between the fields. Figure 11-7(a) shows the effect on the fields when two conductors carrying currents in opposite directions are adjacent. The directions of the magnetic fields round the conductors are in opposition, and this is similar to the situation when two like magnetic poles are brought close together. The fields exert a force of repulsion on each other, tending to push the conductors apart. When the adjacent conductors have currents flowing in the same direction

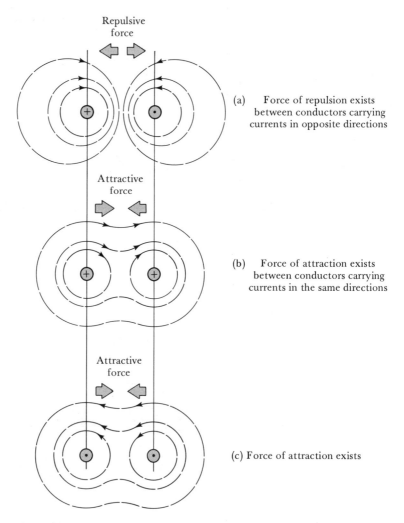

(a)　Force of repulsion exists between conductors carrying currents in opposite directions

(b)　Force of attraction exists between conductors carrying currents in the same directions

(c) Force of attraction exists

FIGURE 11-7. The magnetic field around two parallel current-carrying conductors shows that a force of repulsion occurs when the currents are in opposite directions, and that an attractive force exists when the currents are in the same direction.

[Figure 11-7(b) and (c)], the magnetic fields assist each other. Since the lines of force are always in tension, they are always trying to find the shortest path. Consequently, the fields exert a force that tends to pull the conductors together, as illustrated.

MAGNETIC FIELD AROUND A COIL.   Now consider the effect of passing a current through a one-turn coil of wire. Figures 11-8(a) and (b) show that all the magnetic flux generated by the electric current passes through the center of the coil. Therefore, the one-turn coil acts like a little magnet, and has a magnetic field with an identifiable N pole and S pole. Instead of a single turn, the coil may have many turns, as illustrated in Figure 11-8(c). In this case the flux generated by each of the individual current-carrying turns tends to link up and pass out of one end of the coil and back into the other end. This type of coil, known as a *solenoid,* obviously has a magnetic field pattern very similar to that of a bar magnet.

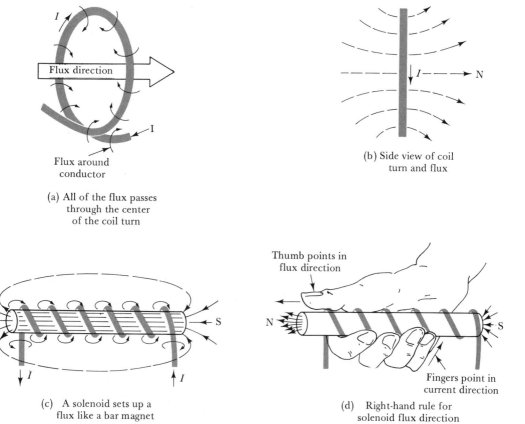

Flux direction

Flux around
conductor

(a) All of the flux passes
through the center
of the coil turn

(b) Side view of coil
turn and flux

Thumb points in
flux direction

Fingers point in
current direction

(c)   A solenoid sets up a
flux like a bar magnet

(d)   Right-hand rule for
solenoid flux direction

FIGURE 11-8.   In current-carrying coils, the magnetic lines of force around the conductors all pass through the center of the coil.

The right-hand rule for determining the direction of flux from a solenoid is illustrated in Figure 11-8(d). When the solenoid is gripped with the right hand such that the fingers are pointing in the direction of current flow in the coils, the thumb points in the direction of the flux (i.e., toward the N-pole end of the solenoid).

ELECTROMAGNETIC INDUCTION. It has been demonstrated that a magnetic flux is generated by an electric current flowing in a conductor. The converse is also possible; that is, a magnetic flux can produce a current flow in a conductor. Consider Figure 11-9(a), in which a hand-held bar magnet is shown being brought close to a coil of wire. As the bar magnet approaches the coil, the flux from the magnet *brushes across* the coil conductors, or *cuts* the conductors. This produces a current flow

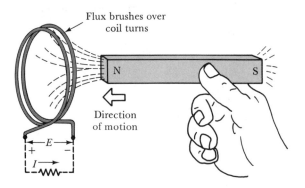

(a) emf induced in a coil by the motion
of the flux from the bar magnet

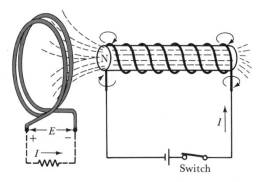

(b) emf induced in a coil by the motion
of the flux from the solenoid when
the current is switched on or off

FIGURE 11-9. An *electromotive force* (emf) is induced in a coil when the coil is *brushed* by a magnetic field. The magnetic field may be from a bar magnet or from a current-carrying coil.

in the conductors proportional to the total flux that *cuts* the coil. If the coil circuit is closed by a resistor (as shown broken in the figure), a current flows. Whether or not the circuit is closed, an *electromotive force* (emf) can be measured at the coil terminals. This effect is known as *electromagnetic induction.*

It is important to note that the emf within the coil (which produces the current flow) is generated only when the magnetic field is in motion with respect to the coil. When both the field and the coil are stationary, no emf is produced.

Now consider Figure 11-9(b), which shows a solenoid placed close to another coil. Both the solenoid and the other coil are stationary. The solenoid also has a battery and a switch that can be closed to pass current through the turns of the solenoid.

When the switch is open, there is obviously no current flow and therefore no flux to generate an emf in the second coil. However, with the switch closed, current flows through the solenoid, and the flux grows from zero to its maximum level. While the flux is growing, it is moving out from the solenoid and *brushing over* the coil. Consequently, an emf is generated in the coil. When the flux has reached its maximum level, it becomes stationary and is no longer able to generate the emf in the coil. If the switch is now opened, the flux falls to zero again, and in doing so it once more brushes over the conductors of the coil. Again, an emf is generated in the coil while the flux is in motion.

11-3
THEORY OF MAGNETISM

In relation to magnetism all materials can be categorized into four groups:

- *Nonmagnetic* materials have no more effect on a magnetic field than air or vacuum. This group includes such materials as wood and rubber.
- *Diamagnetic* materials exhibit a very slight opposition to magnetic lines of force. They tend to be repelled from both poles of a magnet, and if a bar of diamagnetic substance is suspended in a magnetic field, it tends to align itself at right angles to the field. The effect is so slight that these materials are usually classified as non-magnetic. Copper and silver are diamagnetic.
- *Paramagnetic* materials assist the passage of magnetic lines of force. Sometimes this term is used to include ferromagnetic materials. In general, however, it is applied only to material that shows a very slight magnetic effect. Aluminum and platinum are para-magnetic.
- *Ferromagnetic* materials tremendously assist the passage of magnetic lines of force. These are the materials used as permanent magnets and as electromagnets. Iron, nickel, cobalt, and a certain type of ceramic known as *ferrite* are all ferromagnetic materials.

It was shown in Section 11-2 that a current flowing through a conductor produces a magnetic field around the conductor, and that when the current-carrying conductor is bent in a circle, a magnetic flux having a definite direction is generated. Now recall that current flow is in fact electron motion. Thus, it follows that an electron in orbit around an atom can be thought of as a circular current flow, and that a magnetic flux is produced by the orbital motion of the electrons. This effect is referred to as the *magnetic moment* of the electron.

Figure 11-10(a) shows two electrons orbiting in the same plane and direction around adjacent atoms. Note that conventional current direction is opposite to that of the electron motion. The orbital motion of the two electrons produces magnetic fluxes in the same direction, and these fluxes reinforce each other. Now look at Figure 11-10(b). In this case the two electrons are shown orbiting in opposite directions. Thus, they generate magnetic fluxes that cancel each other, and the net result is zero flux.

Within nonmagnetic material, it can be assumed that all the magnetic effects of orbiting electrons add up to zero, and that the condition is unchanged by the influence of a magnetic field. Diamagnetic materials tend to be affected by a magnetic field, to the extent that they set up their own very weak magnetic flux that opposes the external magnetic field. The effect is almost undetectable, but it is fair to assume that the external magnetic field causes an alignment of some orbital electrons, and that the combined effect of these produces the opposing flux. With paramagnetic materials that exhibit very weak magnetization, it may be assumed that some orbiting electrons are aligned in such a way that they produce a magnetic flux in the same direction as the external flux.

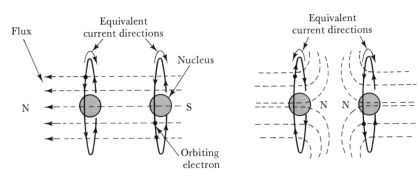

(a) Orbital magnetic moments assisting each other to establish a magnetic flux

(b) Orbital magnetic moments opposing each other; flux adds up to zero

FIGURE 11-10. The magnetic moment of electrons in adjacent atoms may assist each other to establish a magnetic flux or oppose and cancel each other's magnetic effect.

Chap. 11   Magnetism

In *ferromagnetic* materials, groups of atoms seem to act together. Within these groups, large numbers of orbital electrons are aligned in such a way that their magnetic fields reinforce each other. The groups of atoms are termed *magnetic domains,* and while the material is un-magnetized, the net effect of the magnetic domains is to cancel each other completely. When an external magnetic field is applied to ferro-magnetic material, the resulting alignment of the magnetic domains produces a tremendous increase in the magnetic flux. In the case of soft iron, the flux reduces almost to zero when the external field is removed. Thus, the magnetic domains appear to return (almost) to their original state. With hard steel and certain alloys of iron and nickel, the magnetic domains apparently remain realigned, and the steel continues to produce a magnetic flux even when the external field is removed. In this case the metal has become a *permanent magnet.*

11-4

MAGNETIC FLUX AND FLUX DENSITY

The total lines of force in a magnetic field are referred to as the *magnetic flux* (symbol Φ), and the flux per unit cross-sectional area of the field is termed the *flux density* (symbol B).

> *The **weber*** *(**Wb**) is the SI unit of magnetic flux. The **weber** is defined as the magnetic flux which, linking a single-turn coil, produces an emf of 1 V when the flux is reduced to zero at a constant rate in 1 s.*

> *The **tesla**[†] (**T**) is the SI unit of magnetic flux density. The **tesla** is the flux density in a magnetic field when 1 Wb of flux occurs in a plane of 1 m^2; that is, the **tesla** can be describd a 1 Wb/m^2.*

The relationship between flux and flux density is stated by the equation

$$B = \frac{\Phi}{A} \text{ teslas} \qquad (11\text{-}1)$$

where B is the flux density in teslas, Φ is the total flux in webers, and A is the cross-sectional area in m^2.

The flux density is frequently different from one point to another in the same magnetic field. For example, in the case of the bar magnet in Figure 11-2(a), the flux density is obviously greatest at points close to the poles of the magnet, or within the metal.

---

[*] Named after the German physicist Wilhelm Weber (1804–1890).

[†] Named for the Croatian-American researcher and inventor Nikola Tesla (1856–1943).

EXAMPLE 11-1 The total flux emitted from the pole of a bar magnet is 2×10^{-4} Wb (see Figure 11-11).

    a. If the magnet has a cross-sectional area of 1 cm^2, determine the flux density within the metal.

    b. If the flux spreads out so that at a certain distance from a pole it is distributed over an area of 2 cm by 2 cm, find the flux density at that point.

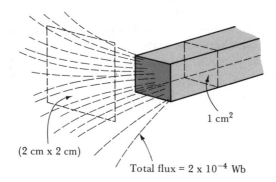

1 cm^2

(2 cm x 2 cm)

Total flux = 2 x 10^{-4} Wb

FIGURE 11-11. Because the magnetic flux spreads over a larger area as it is emitted from a pole of a bar magnet, the flux density is less than that within the magnet.

SOLUTION

a. Flux density within the metal:
Equation (11-1):

$$B = \frac{\Phi}{A} = \frac{2 \times 10^{-4} \text{ Wb}}{1 \times 10^{-4} \text{ m}^2}$$

$$= 2 \text{ T}$$

b. Flux density away from the pole:

$$B = \frac{\Phi}{A} = \frac{2 \times 10^{-4} \text{ Wb}}{(2 \times 10^{-2})^2 \text{ m}^2}$$

$$= 0.5 \text{ T}$$

PRACTICE PROBLEMS

11-4.1 Determine the flux density within a coil with a cross-sectional area of 9 cm^2 if the total flux is 18 μWb. Calculate the new flux density if the coil diameter is doubled and the total flux is unchanged.

11-4.2 The flux density of 1.7 T measured in the air gap between two magnetic poles is found to be constant over an area of 3 cm². If the poles each have a cross-sectional area 2.2 cm, determine the flux density within the poles.

11-5

MAGNETO-MOTIVE FORCE AND MAGNETIC FIELD STRENGTH

Just as an electric current is the result of an electromotive force (emf) acting on the electric circuit, so a magnetic flux is produced by a *magnetomotive force* (mmf), symbol F_m, acting upon the magnetic circuit. In the case of a solenoid or any other current-carrying coil, the magnetomotive force is the product of the current and the number of turns on the coil. Figure 11-12(a) shows a one-turn coil carrying a current of 1 A. All of the flux set up by the 1 A passes through the center of the coil. Therefore, the magnetomotive force is

$$F_m = 1 \text{ A} \times 1 \text{ turn}$$

$$= 1 \text{ A}$$

In the case of the four-turn coil with a current of 1 A [Figure 11-12(b)], the effect of the current is multiplied four times, because each turn of the coil generates a flux that passes through the center of the coil. Therefore, the magnetomotive force is

$$F_m = 1 \text{ A} \times 4 \text{ turns}$$

$$= 4 \text{ A}$$

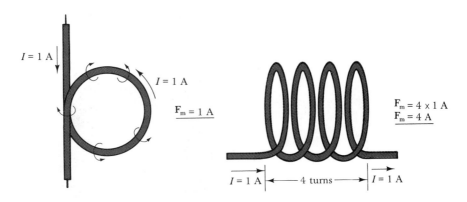

(a) Magnetomotive force of 1 A from a one-turn coil with $I = 1$ A

(b) Magnetomotive force of 4 A from a 4-turn coil with $I = 1$ A

FIGURE 11-12. The *magnetomotive force* (mmf) which generates a magnetic flux depends upon coil current and number of coil turns.

Sec. 11-5   Magnetomotive Force and Magnetic Field Strength

In general, for a coil of N turns carrying a current of I amperes, the magnetomotive force is

$$\boxed{\textit{magnetomotive force } \mathbf{F}_m = NI \textbf{ amperes}} \qquad \textbf{(11-2)}$$

The more descriptive term *ampere-turns* is sometimes used as the units of mmf. However, the approved SI unit of mmf is the ampere.

The quantity of flux that can be set up in a magnetic circuit is, of course, proportional to the mmf. It is also inversely proportional to the length of the magnetic circuit. Consider the *toroidal* coil shown in Figure 11-13. The length of the magnetic path is l, as illustrated. If l is very large, the mmf has to act over a long distance, and if l is very small, the mmf acts over a short distance. Obviously, the greatest intensity of magnetic flux occurs for the shortest magnetic path. This gives rise to the expression for *magnetic field strength:*

$$\textit{magnetic field strength, } H = \frac{\mathbf{F}_m}{l}$$

or

$$\boxed{H = \frac{NI}{l} \textbf{ amperes/meter}} \qquad \textbf{(11-3)}$$

When I is in amperes, N is number of turns, and l is in meters, the units of magnetic field strength are amperes per meter (A/m).

Equation (11-3) for magnetic field strength is analogous to the equation for electric field strength, Equation (4-1).

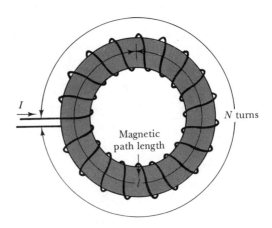

FIGURE 11-13. A toroidal coil wound on a circular former has a magnetic path through the center of the ring material.

EXAMPLE 11-2   The torodial coil shown in Figure 11-13 has 100 turns and carries a current of 0.5 A. If the length of the magnetic circuit is 10 cm, determine the mmf and the magnetic field strength.

SOLUTION

Equation (11-2):

$$F_m = NI = 100 \times 0.5 \text{ A}$$

$$= 50 \text{ A}$$

Equation (11-3):

$$H = \frac{NI}{l} = \frac{100 \times 0.5 \text{ A}}{10 \times 10^{-2} \text{ m}}$$

$$= 500 \text{ A/m}$$

PRACTICE PROBLEMS

11-5.1   A toroidal coil with 190 turns has a magnetic path length of 3.8 cm. Determine the current required to set up a magnetic field strength of 175 A/m.

11-5.2   A 10 mA current is to be passed through a toroidal coil which has a 4.5 cm magnetic path length. Calculate the number of coil turns required if the magnetic field strength is to be 120 A/m.

11-6
FORCE ON CURRENT-CARRYING CONDUCTORS IN A MAGNETIC FIELD

FORCE ON A CONDUCTOR.   Figure 11-14(a) shows an end view of a current-carrying conductor in a magnetic field. With the current direction as indicated by the + sign (i.e., away from the viewer), the flux set up by the current is in a clockwise direction around the conductor. When the field flux has the vertically down direction indicated, the flux from the conductor tends to assist the field flux on the right-hand side of the conductor. The conductor flux also opposes (or weakens) the field flux on the left-hand side of the conductor. Since the magnetic lines of force always tend to be in tension, those lines of force on the right-hand side of the conductor are tending to straighten out. The effect of this is to produce a force that pushes the conductor to the left.

When the current direction is reversed, as illustrated in Figure 11-14(b), the flux around the conductor is in a counterclockwise direction. Thus, the field flux is assisted on the left-hand side of the conductor and weakend on its right side. The result is a force that pushes the conductor to the right.

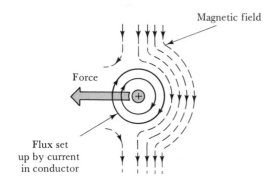

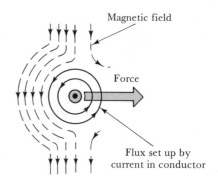

(a) Conductor current flowing
away from viewer

(b) Conductor current flowing
toward viewer

FIGURE 11-14. A current-carrying conductor situated in a magnetic field experiences a force due to the interaction of the conductor flux and the field flux. The conductor flux reinforces the field flux on one side of the conductor and weakens it on the other.

The effect is further illustrated in Figure 11-15, where the field flux is shown vertically downward. The current direction is such that the field set up around the conductor weakens the main field in front of the conductor and strengthens it behind the conductor. The result, as illustrated, is that a force acts to push the conductor forward.

Note that the conductor must be at right angles to the direction of the magnetic flux; otherwise, maximum force is not generated.

The magnitude of the force acting upon a current-carrying conductor in a magnetic field is proportional to the flux density of the field and to the density of the flux set up by the conductor. Since the flux set up around the conductor is directly proportional to the current, the force on the conductor is proportional to the current and to the flux density of the magnetic field. If a 2 m length of conductor is within the magnetic field, the force exerted upon it is obviously twice the force that would be exerted upon a 1 m length. The force on the conductor is thus seen to be proportional to the length of the conductor within the field, the field flux density, and the current carried by the conductor. Therefore,

$$(\text{force on conductor}) \propto (\text{flux density}) \times (\text{current}) \times (\text{length})$$

or

$$\boxed{F = BI\,\ell \text{ newtons}} \qquad (11\text{-}4)$$

*When B is in teslas, I is in amperes, and ℓ is in meters, the force F is given in **newtons**. Thus:*

EXAMPLE 11-3   A conductor carrying a current of 15 A is situated at right angles to a magnetic field with a flux density of 0.7 T (see Figure 11-15). If the length of the conductor within the field is 20 cm, determine the force on the conductor. Also calculate the new current level required to increase the force to 25 N.

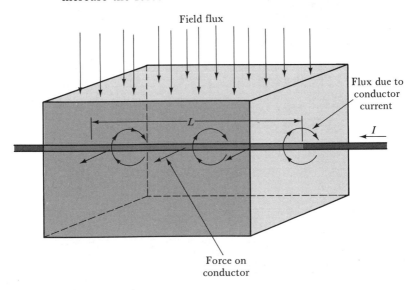

Field flux

Flux due to conductor current

I

L

Force on conductor

FIGURE 11-15. The force on a current-carrying conductor situated in a magnetic field is proportional to the current, the flux density of the field, and the length of the conductor within the field.

SOLUTION

Equation (11-4):

$$F = BI\ell \text{ newtons}$$
$$= 0.7 \text{ T} \times 15 \text{ A} \times 20 \times 10^{-2} \text{ m}$$
$$= \mathbf{2.1 \ N}$$
$$I = \frac{F}{B\ell} = \frac{25 \text{ N}}{0.7 \text{ T} \times 20 \times 10^{-2}}$$
$$\cong \mathbf{179 \ A}$$

FORCE ON A COIL.   In Figure 11-16(a), a single-turn coil is shown pivoted in the magnetic field set up by N and S poles. Consideration of the current direction through the coil shows that a downward force is

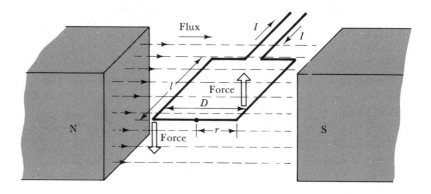

(a) Single-turn coil pivoted in a magnetic field

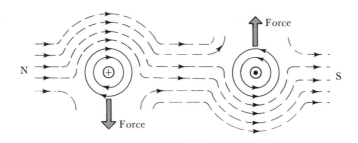

(b) Showing the force on each side of a single-turn
coil pivoted in a magnetic field

FIGURE 11-16. A force is exerted on each side of a current-carrying coil pivoted in a magnetic field. This force tends to cause the coil to rotate.

exerted on its left-hand side, while an upward force is exerted on its right-hand side [see Figure 11-16(b)]. Since there are two lengths (ℓ) of conductor within the magnetic field, the force exerted on the coil is

$$F = 2BI\ell \text{ newtons}$$

If the coil has N turns, the equation for force is

$$F = 2BI\ell N \text{ newtons}$$

and if the force acts at a radius r [see Figure 11-16(a)], the torque acting on the coil is

$$\text{torque} = 2BI\ell Nr \text{ newton meters}$$

Since the coil diameter D equals $2r$,

$$\boxed{torque = B\ell IND \text{ newton meters}} \qquad \text{(11-5)}$$

EXAMPLE 11-4 Calculate the torque acting on a 100 turn coil pivoted in a magnetic field having a flux density of 0.5 T. The current through the coil is 100 mA, its radius is 1 cm, and its axial length is 2 cm.

SOLUTION

Equation (11-5):

$$torque = B\ell IND \text{ newton meters}$$
$$= 0.5 \text{ T} \times 2 \times 10^{-2} \times 100 \text{ mA} \times 100 \times 2 \times 10^{-2}$$
$$\boldsymbol{torque = 2 \times 10^{-3} \text{ N} \cdot \text{m}}$$

The simple arrangement illustrated in Figure 11-16 is the basic moving part of a deflection-type measuring instrument. It also shows the principle behind the electric motor.

PRACTICE PROBLEMS

11-6.1 A conductor situated at right angles to a magnetic field experiences a force of 0.083 N. If the current is 286 mA and the conductor length within the field is 5.8 cm, calculate the flux density of the field.

11-6.2 A torque of 3.3 mN · m is to be exerted on a 75 turn coil pivoted in a magnetic field which has a flux density of 1.5 T. If the axial length of the coil is 1.65 cm and its diameter is 2 cm, determine the required current.

SUMMARY OF FORMULAS

Flux density:

$$B = \frac{\Phi}{A}$$

Magnetomotive force:

$$F_m = NI$$

Magnetic field strength:

$$H = \frac{NI}{l}$$

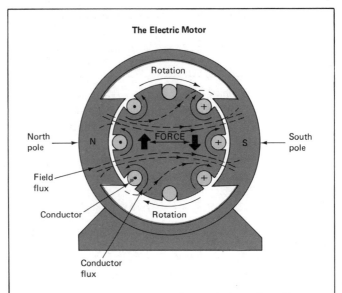

The Electric Motor

The cross-section of a basic electric motor is shown above. The outer shell of the motor is a ring-shaped magnet with inward pointing N and S poles. A cylindrical soft iron core is pivoted between the poles, and conductors are located in slots on the periphery of the core. Current flows through the conductors; in one direction when passing the N pole, and in the reverse direction when passing the S pole. In the illustration, the currents in those conductors adjacent to the N pole generate fluxes which weaken the magnetic field flux above the conductors and strengthen the field flux below the conductors. This produces an upward (or clockwise rotational force) on left side of the pivoted core. Similarly, the conductor fluxes adjacent to the S pole strengthen the field flux above the conductors, and weaken it below them. The result is a downward force on right side of the core. The two (left side and right side) forces result in clockwise rotation of the core.

Mechanical force:

$$F = BI\ell$$

Mechanical torque:

$$torque = B\ell IND$$

REVIEW QUESTIONS

11-1 Discuss the terms north-seeking pole and south-seeking pole in relation to a bar magnet and the earth's magnetic field. Also, explain what occurs when the ends of two bar magnets are brought close together.

11-2 Draw sketches to show the force field set up around a bar magnet and describe two methods of mapping the force field.

11-3 Define magnetic lines of force and magnetic flux, and list the characteristics of magnetic lines of force. Also, explain how their direction is defined.

11-4   Sketch the magnetic fields that occur at:
a. Two adjacent N poles of bar magnets.
b. Two adjacent S poles.
c. Adjacent N and S poles.
d. A bar magnet with a piece of soft iron in its field.

11-5   Sketch the shape of the magnetic field around a current-carrying conductor and show how the direction of the field is related to the direction of the current in the conductor.

11-6   Sketch the magnetic field that occurs around two adjacent current-carrying conductors:
a. For currents in opposite directions.
b. For currents in the same direction.
Explain how a force occurs between the conductors and indicate the direction of the force in each case.

11-7   Draw sketches to show the form of magnetic field that occurs with:
a. A single-turn coil with current flowing through it.
b. A long current-carrying solenoid.
Also, sketch a toroidal-shaped coil and show what its magnetic field should look like.

11-8   Discuss *electromagnetic induction* and draw sketches to explain the principle.

11-9   Define the following classifications of materials: nonmagnetic, diamagnetic, paramagnetic, and ferromagnetic.

11-10   Using illustrations, explain the theory of magnetism.

11-11   Define the weber and the tesla and state their relationship.

11-12   Define magnetomotive force and magnetic field strength and explain how the two are related.

11-13   Draw illustrations to show that a force is exerted on a current-carrying conductor situated in a magnetic field. Show how the direction of the force is related to the directions of the current and the magnetic field.

11-14   Discuss the relationships between the force on a current-carrying conductor in a magnetic field and the factors responsible for the force. Also, evolve the equation for the force and state the units for each factor.

11-15   Derive the expression for the torque on a pivoted coil situated within a magnetic field.

PROBLEMS

SECTION 11-4

11-1   The total flux emitted from the pole of a magnet is 0.5 μWb.
(a) If the magnet has a cross-sectional area of 1.5 cm², determine the flux density within the metal.

(b) Calculate the flux density a short distance from the pole if all of the flux is contained within an area 5 cm by 5 cm.

11-2 A circular magnetic pole with a diameter of 5 cm emits a total flux of 120 μWb. Calculate the flux density within the pole. Also, calculate the flux density a short distance from the pole if all of the flux is retained within an area 10% larger than the pole area.

11-3 The flux density in an air gap between two N and S poles is 2.5 T. The poles are circular with a diameter of 5.6 cm. Calculate the total flux crossing the air gap.

11-4 In Problem 11-3, allow for the flux spreading over an area 10% larger than the pole area. Recalculate the total flux and the flux density within the poles.

11-5 The cross-sectional area of a coil is 15 cm², and the flux density within it is 30 mT. Calculate the total flux.

11-6 If the diameter of the coil in Problem 11-5 is reduced by one third, determine the new level of flux density required to keep the total flux unchanged.

11-7 The flux density within a coil is 45 mT, and the total flux is 13 μWb. If the coil has a square cross-section, calculate its dimensions.

11-8 An iron ring has two air gaps, *a* and *b*. At gap *a*, the metal has a circular cross-section 1 cm in diameter. The cross-section at gap *b* is square with 1 cm sides. The flux density at gap *a* is measured as 20 mT. Calculate the flux density at gap *b*.

11-9 A magnetic pole with a cross-sectional area of 9 cm² has a flux density of 2 T. Calculate the total flux emitted from the pole.

11-10 For Problem 11-8, recalculate the flux density at gap *b*, if the flux spreads out at each gap over an area 10% larger than the pole face area.

11-11 A magnetic pole with a cross-sectional area 3 cm by 3 cm has a flux density of 1.11 mT. Calculate the flux density in an area 4 cm by 4 cm a short distance from the pole if 90% of the total flux passes through this area.

11-12 The flux density within a long coil is measured as 0.4 μT at a point close to one end, and as 0.48 μT at the middle of the coil. If the diameter of the coil is exactly 2.5 cm at the end, determine its diameter at the middle.

SECTION 11-5

11-13 A toroidal shaped coil with 280 turns and a magnetic path length of 50 cm has a current of 7.5 A. Calculate the mmf and magnetic field strength.

276   Chap. 11   Magnetism

11-14 A 500 turn toroidal coil with a current of 33 mA has a magnetic field strength of 110 A/m. Calculate the magnetic path length.

11-15 A toroidal coil has a magnetic path length of 33 cm and a magnetic field strength of 650 A/m. The coil current is 250 mA. Determine the total number of coil turns.

11-16 The coil current in Problem 11-15 is reduced to 180 mA. Calculate the new magnetic field strength.

11-17 Two toroidal coils (*a* and *b*) wound on a single coil former produce mmf is the same direction over a common magnetic path length of 45 cm. Coil *a* has 1500 turns and a current of 80 mA. Coil *b* has 1000 turns. If the magnetic field strength is 400 A/m, determine the current flowing in coil *b*.

11-18 For Problem 11-17, determine the new magnetic field strength if the current in coil *a* is reversed.

11-19 For Problem 11-17 with the current in coil *a* reversed, determine the level of coil *a* current that will completely cancel the effects of the current in coil *b*.

11-20 A toroidal coil with a 9 cm magnetic path length has a center-tapped coil with a total of 350 turns. One half of the coil has a current of 15 mA, and the other half has a 25 mA current. Both currents are in the same direction. Calculate the magnetic field strength.

SECTION 11-6

11-21 A conductor situated at right angles to a magnetic flux has a current of 5 A. The flux density is 1.5 T, and the conductor length within the magnetic field is 12 cm. Calculate the force on the conductor.

11-22 For Problem 11-21, determine the new flux density required to maintain the force unchanged when the current is reduced to 500 mA.

11-23 Two 25 cm conductors are situated at right angles to a magnetic flux with a flux density of 500 mT. Conductor *a* has a current of 100 A, and the current in conductor *b* is 39 A. Calculate the force exerted on each conductor.

11-24 For Problem 11-23, the length of conductor *b* within the magnetic field is increased to 33 cm. Calculate the new level of current that must flow in conductor *a* if the total force on the two conductors is to remain unchanged.

11-25 A 600 turn coil is pivoted within a magnetic field with a flux density of 2.2 T. The coil current is 50 μA, its radius is 0.75 cm, and its axial length is 1 cm. Calculate the torque acting on the coil.

11-26 For the coil in Problem 11-25, determine the level of coil current to produce a torque of 6 μN·m.

11-27 A 320 turn coil with a radius of 1.5 cm and an axial length of 3 cm is situated in a magnetic field with a flux density of 5 T. If the torque on the coil is to be 1.5 mN·m, determine the required coil current.

11-28 A 500 turn coil with a radius of 1 cm and an axial length of 2.5 cm is pivoted between the poles of a magnet at right angles to the magnetic flux. The coil current is 100 μA, and the torque on the coil is 50 μN·m. Calculate the flux density of the magnetic field.

11-29 A conductor with a current of 8 A is situated at right angles to the magnetic flux crossing a short air gap between two cylindrical magnetic poles. The diameter of each pole face is 7 cm, and the total flux is 125 μWb. Calculate the force on the conductor.

11-30 A torque of 33 mN·m is exerted on a rectangular coil pivoted within a magnetic field which has a flux density of 0.9 T. The coil current is 25 mA, and its dimensions are: axial length = 3 cm, and radius = 2 cm. Determine the total number of coil turns.

11-31 If the coil in Problem 11-27 has 75 of its turns wound in the opposite direction to the other turns, calculate the new level of coil current that must flow for the conditions described.

11-32 Two insulated conductors (*a* and *b*) are contained within a single electric cable. Conductor *a* has a current of 25 A, and the current in conductor *b* is 9.5 A. A 15 cm length of the cable is situated in a magnetic field with a flux density of 0.85 T. Calculate the force on the cable: (a) when the conductor currents are in the same direction, (b) when the currents are in opposite directions.

COMPUTER PROBLEMS

11-33 Write a computer program to determine the required current in a coil when the coil turns, magnetic path length, and field strength are given.

11-34 Write a computer program to determine the torque on a coil when the number of turns, flux density, current, and coil dimensions are given.

11-35 Modify the program written for Problem 11-34 to print the torque for each of ten different coil lengths.

ANSWERS TO PRACTICE PROBLEMS

11-4.1	20 mT, 5 mT
11-4.2	2.32 T
11-5.1	35 mA
11-5.2	540 turns
11-6.1	5 T
11-6.2	88.9 mA

12 MAGNETIC CIRCUITS

Objectives   You will be able to:

Define reluctance and permeability and explain the terms permeability of free space and relative permeability.

Calculate the magnetic field strength around a conductor when a current flows in the conductor.

Calculate the field strength, flux density, and total flux in an air-cored toroid.

Determine the mmf required to set up a given flux within an air gap.

Determine the field strength, flux density, and total flux in an air-cored solenoid and explain the result of introducing an iron core into the solenoid.

For various shaped magnetic cores, determine the coil current required to set up a given flux within the core or within an air gap in the core.

Calculate the energy stored in an air gap and determine the current that must flow in the coil of a given electromagnet in order to lift iron bars and plates.

Sketch and explain typical magnetization curves for ferromagnetic materials.

Sketch and explain typical hysteresis loops for ferromagnetic materials.

Explain the laminated construction of magnetic cores.

Introduction

Magnetic circuits are in many ways analogous to electrical circuits. *Magnetomotive force* (mmf), *flux*, and *reluctance* are the counterparts of electromotive force, current, and resistance, respectively. Electrical conductivity also has its magnetic analog in *permeability*. For all non-magnetic materials, the permeability has a very small fixed value, known as the *permeability of free space*. *Ferromagnetic* materials have relative permeability values which relate their magnetic properties to that of free space.

Magnetic circuits made up of cores of different thickness, and perhaps an air gap, are termed *composite magnetic circuits*. The mmf's required to set up a given flux are determined for each section of the circuit and then added together to find the total circuit mmf. Each type of ferromagnetic material has its own particular *magnetization curve*, the shape of which can be explained in terms of *magnetic domains*. Because of the *hysteresis* effect, the magnetization curves change to loops when the flux within a sample of ferromagnetic material is reversed several times. This gives rise to a loss of energy in magnetic cores.

12-1
RELUCTANCE AND PERMEABILITY

Recall from Section 4-4 that the resistance of a conductor can be calculated by use of Equation (4-2):

$$\text{resistance, } R = \frac{\rho \ell}{A}$$

where ρ is the resistivity of the conducting material, ℓ is the length of the conductor, and A is its cross-sectional area.

The magnetic circuit analog of the resistance of electrical circuits is termed *reluctance*. Therefore, *reluctance is a measure of the opposition offered by a magnetic circuit to the setting up of flux,* just as resistance is opposition to current flow in an electrical circuit. The corresponding equation for reluctance is

$$\boxed{\textit{reluctance, } \mathbf{R_m} = \frac{1}{\mu} \times \frac{l}{A}} \tag{12-1}$$

Here, l and A are length and cross-sectional area, respectively, of the magnetic circuit, and μ is the *permeability* of the material in the magnetic circuit. The reciprocal of permeability corresponds to the resistivity (ρ) of the electrical circuit. This means that magnetic permeability is the analog of electrical conductivity. Just as ρ is resistance of a cubic meter of electrical material, so $1/\mu$ can be termed *reluctance of a cubic meter of magnetic material.*

Ohm's law for an electrical circuit can be written:

$$\text{resistance, } R = \frac{\text{emf}}{\text{current}} = \frac{E}{I}$$

and

$$\text{conductance} = \frac{1}{R} = \frac{I}{E}$$

Similarly, for the magnetic circuit:

$$\text{reluctance} = \frac{\text{mmf}}{\text{flux}}$$

or

$$R_m = \frac{F_m}{\Phi}$$

(12-2)

The magnetic circuit analog of conductance is *permeance,* and since conductance is the reciprocal of resistance, so permeance is the reciprocal of reluctance:

$$permeance = \frac{1}{R_m} = \frac{\Phi}{F_m}$$

(12-3)

Substituting for R_m from Equation (12-1) into Equation (12-3) gives

$$\frac{1}{\frac{1}{\mu} \times \frac{l}{A}} = \frac{\Phi}{F_m}$$

or

$$\frac{\mu A}{l} = \frac{\Phi}{F_m}$$

Therefore,

$$\mu = \frac{\Phi}{A} \times \frac{l}{F_m}$$

from Equation (11-1),

$$\frac{\Phi}{A} = \text{flux density, } B$$

so,

$$\mu = B \times \frac{l}{F_m}$$

and from Equations (11-2) and (11-3),

$$\frac{F_m}{l} = \frac{NI}{l} = \text{magnetic field strength, } H$$

Therefore, $\text{permeability} = \dfrac{\text{flux density}}{\text{magnetic field strength}}$

or

$$\mu = \frac{B}{H}$$

(12-4)

The SI units for permeability are henrys/meter (H/m). The *henry* is the unit of inductance and is discussed in Chapter 14. However, as will be seen, μ is invariably used as a ratio.

12-2
PERMEABILITY OF FREE SPACE

A current-carrying conductor a is shown in Figure 12-1 with the magnetic flux around it in concentric circles. Another current-carrying conductor b is situated within the magnetic field generated by a, and therefore a force is exerted upon conductor b. The magnitude of the force can be determined using Equation (11-4):

$$F = BI\ell$$

or flux density,

$$B = \frac{F}{I\ell}$$

The definition of the ampere (see Appendix 2) states that the force per meter length exerted between two conductors 1 m apart when they are each carrying a current of 1 A is 2×10^{-7} N. Therefore, substituting: $F = 2 \times 10^{-7}$ N, $I = 1$ A, and $\ell = 1$ m:

$$B = 2 \times 10^{-7} \text{ teslas}$$

Thus, it is seen that *the flux density in air (or any nonmagnetic material) at 1 m from a long conductor carrying 1 A is 2×10^{-7} T.*

Returning to Figure 12-1, the length of the magnetic path at 1 m from conductor a is the circumference of a circle with a radius of 1 m. Therefore, the length of the magnetic path is 2π meters, and the mmf is

$$F_m = I \times N$$

$$= 1 \text{ A} \times (1 \text{ turn})$$

$$= 1 \text{ A}$$

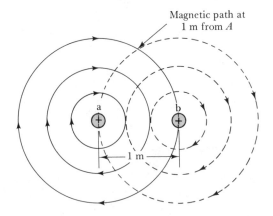

FIGURE 12-1. Cross-section of two long, current-carrying conductors situated 1 m apart in air. If the current flowing in the conductors is 1 A, the force between them is 2×10^{-7} N.

The magnetic field strength is calculated as

$$H = \frac{F_m}{l}$$

$$= \frac{1 \text{ A}}{2\pi \text{ m}}$$

$$= \frac{1}{2\pi} \text{ A/m}$$

From Equation (12-4), the *permeability of the material around the conductor is*

$$\mu = \frac{B}{H}$$

The material around the conductor was assumed to be air; however, it could be a vacuum or any nonmagnetic material. In this case, the permeability is a constant that is designated μ_0 and is referred to as the *permeability of free space.* Substituting for B and H gives

$$\mu_0 = \frac{2 \times 10^{-7} \text{ T}}{1/2\pi \text{ A/m}}$$

or   **permeability of free space, $\mu_0 = 4\pi \times 10^{-7}$**

The term μ_0 may now be employed in calculations relating to flux density and magnetic field strength.

EXAMPLE 12-1   An air-cored toroidal coil of the type shown in Figure 11-13 has 3000 turns and carries a current of 0.1 A. The cross-sectional area of the coil is 4 cm², and the length of the magnetic circuit is 15 cm. Determine the magnetic field strength, the flux density, and the total flux within the coil.

SOLUTION

Equation (11-3):

$$H = \frac{NI}{l} = \frac{3000 \times 0.1 \text{ A}}{15 \times 10^{-2} \text{ m}}$$

magnetic field strength, H = 2000 A/m

Equation (12-4):

$$\mu = \frac{B}{H}$$

Therefore,
$$B = \mu_0 H$$
$$= 4\pi \times 10^{-7} \times 2000 \text{ A/m}$$

flux density, $B = 2.5 \times 10^{-3}$ T

Equation (11-1):

$$B = \frac{\Phi}{A}$$

Therefore,
$$\Phi = BA$$
$$= 2.5 \times 10^{-3} \text{ T} \times 4 \times 10^{-4} \text{ m}^2$$

total flux, $\Phi = 1$ μWb

EXAMPLE 12-2   Determine the mmf required to generate a total flux of 100 μWb in an air gap 0.2 cm long. The cross-sectional area of the air gap is 25 cm^2.

SOLUTION

Equation (11-1):

$$B = \frac{\Phi}{A} = \frac{100 \times 10^{-6} \text{ Wb}}{25 \times 10^{-4} \text{ m}^2}$$
$$= 4 \times 10^{-2} \text{ T}$$

Equation (12-4):

$$\mu_0 = \frac{B}{H}$$

or
$$H = \frac{B}{\mu_0} = \frac{4 \times 10^{-2} \text{ T}}{4\pi \times 10^{-7}}$$
$$\cong 3.18 \times 10^4 \text{ A/m}$$

Equations (11-2) and (11-3):

$$H = \frac{F_m}{l}$$

Therefore,
$$\text{mmf} = F_m = H \times l$$
$$\cong 3.18 \times 10^4 \times 0.2 \times 10^{-2}$$

$F_m = 63.7$ A

PRACTICE PROBLEMS

12-2.1 Calculate the current required in a 263 turn coil to establish a flux density of 0.033 T in 0.01 cm air gap.

12-2.2 A 120 turn toroidal coil has a 1 mm diameter at its cross section and a coil length of 2 cm. Calculate the magnetic field strength, flux density, and total flux within the coil when a current of 1.5 A flows in the coil.

12-3

SOLENOID

A *solenoid* is a long, thin, air-cored coil, and when a current is passed through the coil, a magnetic field is set up, as shown in Figure 12-2(a). The path of the magnetic lines of flux is made up of two components: the length of the path within the coil (l_1), and the length of the path followed by the flux outside the coil (l_2). The length l_2 is longer than l_1, but not very much longer. Also, as illustrated in the figure, the cross-sectional area of the path outside the coil is very much larger than that inside the coil. Consequently, since they are both the same material (i.e., air), the reluctance of path l_2 is very much smaller than the reluctance of path l_1. Thus, the mmf required to set up a given total flux along path l_1 is very much larger than that required to set up the same flux along l_2. The total mmf required for the solenoid is the sum of the two components:

$$\text{total mmf} = (\text{mmf for } l_1) + (\text{mmf for } l_2)$$

but $\qquad$ $(\text{mmf for } l_1) \gg (\text{mmf for } l_2)$

Therefore, $\qquad$ $\text{total mmf} \cong \text{mmf for } l_1$

In Chapter 11 it was shown that magnetic lines of force are always in a state of tension, and always seeking to flow through the path of least reluctance. It was also seen that the reluctance of ferromagnetic materials is very much smaller than that of air. When a bar of ferromagnetic material is brought close to one end of a solenoid, the bar becomes magnetized by induction [see Figure 12-2(b)]. The induced magnetism produces poles as illustrated, so that the bar is attracted to the solenoid. If the bar is of suitable dimensions, it will be *sucked* right into the solenoid, and remain there. When this happens, the magnetic path inside the (iron-cored) solenoid has a very small reluctance, and it is the air path outside the solenoid that requires the greatest mmf to set up a given total flux.

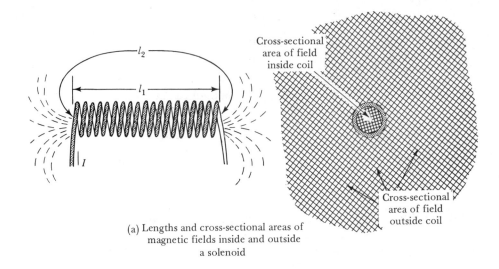

(a) Lengths and cross-sectional areas of
magnetic fields inside and outside
a solenoid

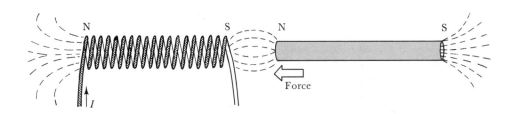

(b) Effect of bringing a soft iron
core close to a solenoid

FIGURE 12-2. The magnetic field of an air-cored solenoid passes through the solenoid and through the surrounding air. An iron bar located close to the solenoid will become magnetized and will experience a force tending to pull it into the solenoid.

EXAMPLE 12-3   An air-cored solenoid has length $l = 15$ cm and inside diameter $D = 1.5$ cm. If the coil has 900 turns, determine the total flux within the solenoid when the coil current is 100 mA.

SOLUTION

$$F_m = NI = 900 \times 100 \text{ mA}$$

$$= 90 \text{ A}$$

$$H = \frac{F_m}{l}$$

For the solenoid, $l \cong$ coil length. Therefore,

$$H \cong \frac{90 \text{ A}}{15 \times 10^{-2} \text{ m}}$$

$$= 600 \text{ A/m}$$

$$B = \mu_0 \times H$$

$$= 4\pi \times 10^{-7} \times 600 \text{ A/m}$$

$$= 24\pi \times 10^{-5} \text{ T}$$

$$\Phi = B \times A$$

$$= 24\pi \times 10^{-5} \times \pi \left(\frac{D}{2}\right)^2$$

$$= 24\pi \times 10^{-5} \times \pi \times \left(\frac{1.5 \times 10^{-2}}{2}\right)^2$$

$$\cong 1.33 \times 10^{-7} \text{ Wb}$$

PRACTICE PROBLEMS

12-3.1 The flux density within an air-cored solenoid is measured as 7.9 mT. If the solenoid has a length of 8 cm, a diameter of 1.25 cm, and 390 turns, calculate the coil current.

12-3.2 If a current of 0.85 A flows in the turns of the solenoid in Problem 12-3.1, determine the total flux.

12-4

RELATIVE PERMEABILITY

When iron or steel is inserted into a current-carrying coil, it is found that the total magnetic flux through the coil is increased a great many times. This occurs even though the mmf and magnetic field strength are not altered in any way. The reason (already discussed) is that iron and steel are much better conductors of magnetic flux than any other materials. Since permeability is the magnetic analog of conductivity, it is obvious that iron and steel have very much greater permeabilities than nonmagnetic materials.

The improvement in total flux and flux density when iron or steel is involved in a circuit is taken into account by assigning each material a *relative permeability* μ_r. Equation (12-4) is then modified to

$$\boxed{\mu_r \, \mu_0 = \frac{B}{H}} \qquad \text{(12-5)}$$

In the case of air and other nonmagnetic materials, $\mu_r = 1$. Depending upon the particular type of iron or steel, the relative permeability μ_r may range from 400 to perhaps 2500.

The solid lines in Figure 12-3 show plots of flux density (B) versus magnetic field strength (H) for various specimens of iron and steel. The measurements were made on ring-shaped specimens, so the magnetic circuit was a closed iron circuit with no air gaps. Thus, maximum possible flux densities were achieved. The magnetization curves can be used for direct determinations of flux density for a given value of magnetic field strength. Alternatively, the relative permeabilities of each specimen can be calculated and plotted versus H. The broken line in Figure 12-3 shows the plot of μ_r for sheet steel. This is derived simply by taking corresponding values of B and H from the permeability curve for sheet steel, and substituting them into Equation (12-5) to calculate the values of μ_r. Note that the relative permeability is by no means a constant quantity; instead, it is very much dependent upon the magnetic field strength H. From the plot of μ_r versus H for sheet steel, the greatest permeability is obviously achieved at a magnetic field strength of approximately 250 A/m.

It should be noted that all the B/H curves shown in Figure 12-3 are typical, and that for any given specimen of magnetic material, the

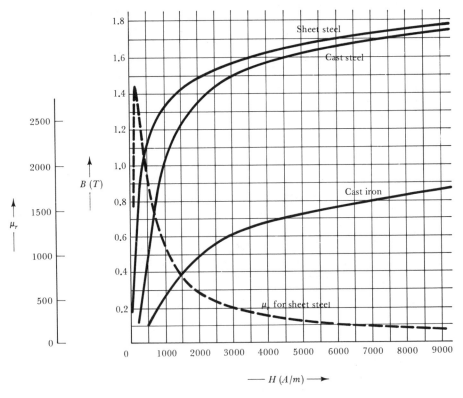

FIGURE 12-3. Magnetization curves (flux density B versus magnetic field strength H) for cast iron, cast steel, sheet steel, and the relative permeability curve for sheet steel.

actual *B/H relationships may be different from those illustrated.* This is because slight differences in the manufacturing process can significantly affect the material's magnetic properties. The shape of the *B/H* curve is considered further in Section 12-7.

EXAMPLE 12-4 Calculate the relative permeability of sheet steel at a magnetic field strength of 250 A/m.

SOLUTION

From Figure 12-3, *at H* = 250 A/m, *B* ≅ 0.8 T:

Equation (12-5): $\mu_r \, \mu_0 \cong \dfrac{B}{H}$

Therefore, $\mu_r = \dfrac{B}{\mu_0 \, H}$

$$\cong \frac{0.8 \text{ T}}{4\pi \times 10^{-7} \times 250 \text{ A/m}}$$

$$\cong \mathbf{2546}$$

EXAMPLE 12-5 The cast iron ring illustrated in Figure 12-4 has a 3000-turn coil that carries a current of 0.1 A. The cross-sectional area of the ring is 4 cm², and the length of the magnetic path is 15 cm. Determine the flux density and the total flux in the ring.

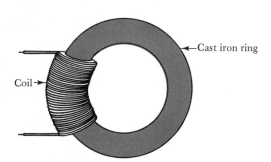

Coil→

←Cast iron ring

FIGURE 12-4. Cast iron ring for Example 12-5.

SOLUTION

Equation (11-3):

$$H = \frac{NI}{l} = \frac{3000 \times 0.1 \text{ A}}{15 \times 10^{-2} \text{ m}}$$

$$= 2000 \text{ A/m}$$

From Figure 12-3, *for cast iron at H* = 2000 A/m:

flux density, B ≃ 0.5 T

Equation (11-1):

$$B = \frac{\Phi}{A}$$

Therefore,

$$\Phi = BA$$
$$\cong 0.5 \text{ T} \times 4 \times 10^{-4} \text{ m}^2$$

total flux, Φ ≅ 200 μWb

Compare this to the total flux for the similar air-cored toroid in Example 12-1.

EXAMPLE 12-6　The cast steel core shown in Figure 12-5 has a relative permeability of 800, and the coil has 700 turns. If the total flux in the core is to be 4×10^{-4} Wb, determine the required current through the coil.

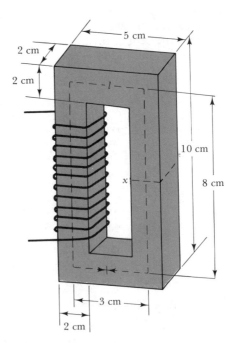

FIGURE 12-5.　Cast steel core for Examples 12-6 and 12-7.

Chap. 12　Magnetic Circuits

SOLUTION

$$B = \frac{\Phi}{A}$$

$$= \frac{4 \times 10^{-4} \text{ Wb}}{2 \times 10^{-2} \text{ m} \times 2 \times 10^{-2} \text{ m}}$$

$$= 1 \text{ T}$$

From Equation (12-5),

$$H = \frac{B}{\mu_r \mu_0}$$

$$= \frac{1 \text{ T}}{800 \times 4\pi \times 10^{-7}}$$

$$\cong 994.7 \text{ A/m}$$

The length of the magnetic circuit is l. As shown in Figure 12-5,

$$l = 2(3 \text{ cm} + 8 \text{ cm})$$

$$= 22 \text{ cm}$$

From Equations (11-2) *and* (11-3):

$$F_m = H \times l$$

$$\cong 994.7 \text{ A/m} \times 22 \times 10^{-2} \text{ m}$$

$$\cong 219 \text{ A}$$

and $$F_m = NI$$

Therefore, $$I \cong \frac{219 \text{ A}}{700 \text{ turns}}$$

$$\cong \mathbf{313 \text{ mA}}$$

PRACTICE PROBLEMS

12-4.1   Determine the relative permeability of sheet steel at 3500 A/m, cast steel at 4000 A/m, and cast iron at 8000 A/m.

12-4.2   A sheet steel magnetic core with a cross-sectional area of 3.3 cm^2 and a magnetic path length of 12 cm has a 2500 turn coil. If the coil current is 75 mA, calculate the total flux in the core.

12-4.3   If the flux density in the core described in Problem 12-4.2 is to be altered to 1.15 T, determine the required coil current.

12-5

COMPOSITE MAGNETIC CIRCUITS

Where a magnetic circuit has an air gap in its path, the air gap is *in series* with the rest of the magnetic circuit. The necessary magnetomotive forces must be calculated independently for the air gap and for the magnetic path, and then the two are added to determine the total required mmf. The same procedure applies when the cross-sectional area of a core is not constant along its entire length.

EXAMPLE 12-7   The core shown in Figure 12-5 has a 1 mm air gap at point *x*. $\mu_r = 800$ and $N = 700$, as described for Example 12-6. Calculate the new value of current through the coil to give a total core flux of 4×10^{-4} Wb.

SOLUTION

From Example 12-6,

$$B = 1 \text{ T}$$

and for the iron path,

$$F_m = 219 \text{ A}$$

For the air gap,

$$H = \frac{B}{\mu_0} = \frac{1 \text{ T}}{4\pi \times 10^{-7}}$$

$$\cong 7.96 \times 10^5 \text{ A/m}$$

and

$$l = 1 \text{ mm} = 1 \times 10^{-3} \text{ m}$$

$$F_m = H \times l$$

$$\cong 7.96 \times 10^5 \times 1 \times 10^{-3}$$

$$\cong 796 \text{ A}$$

Total mmf,

$$F_m \cong 219 \text{ A} + 796 \text{ A}$$

$$\cong 1015 \text{ A}$$

$$I = \frac{F_m}{N} \cong \frac{1015 \text{ A}}{700 \text{ turns}}$$

$$\cong \mathbf{1.45 \text{ A}}$$

(Note the significant increase in coil current required for a given flux density when a very short air gap is introduced into the core.)

EXAMPLE 12-8 The uneven ring-shaped core shown in Figure 12-6 has a relative permeability of 1000, and the flux density in the thickness section is to be 0.75 T. If the current through a coil wound on the core is to be 500 mA, determine the number of coil turns required.

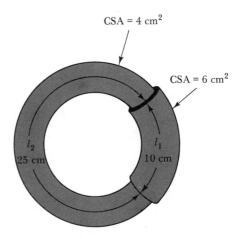

CSA = 4 cm^2

CSA = 6 cm^2

l_2
25 cm

l_1
10 cm

FIGURE 12-6. Composite magnetic circuit for Example 12-8.

SOLUTION

For the thick section,

$$H = \frac{B}{\mu_r\,\mu_0}$$

$$= \frac{0.75\,\mathrm{T}}{100 \times 4\pi \times 10^{-7}}$$

$$\cong 597\ \mathrm{A/m}$$

$$\mathrm{F_m} = H \times l$$

$$\cong 597\ \mathrm{A/m} \times 10 \times 10^{-2}\ \mathrm{m}$$

$$\mathrm{F_{m1}} \cong 59.7\ \mathrm{A}$$

Total flux in core,

$$\Phi = B \times A$$

$$= 0.75\ \mathrm{T} \times (A\ \text{of thick section})$$

$$= 0.75\ \mathrm{T} \times 6 \times 10^{-4}\ \mathrm{m^2}$$

$$\Phi = 4.5 \times 10^{-4}\ \mathrm{Wb}$$

For the thin section,

$$B = \frac{\Phi}{A} = \frac{4.5 \times 10^{-4} \text{ Wb}}{4 \times 10^{-4} \text{ m}^2}$$

$$= 1.125 \text{ T}$$

$$H = \frac{B}{\mu_r \mu_0} = \frac{1.125 \text{ T}}{1000 \times 4\pi \times 10^{-7}}$$

$$H_2 \cong 895 \text{ A/m}$$

$$F_m = H \times l$$

$$\cong 895 \text{ A/m} \times 25 \times 10^{-2} \text{ m}$$

$$F_{m2} \cong 224 \text{ A}$$

$$total \text{ mmf} = F_{m1} + F_{m2}$$

$$\cong 59.7 + 224$$

$$F_m \cong 283.7 \text{ A}$$

$$F_m = N \times I$$

Therefore,

$$N = \frac{F_m}{I} \cong \frac{283.7 \text{ A}}{500 \text{ mA}}$$

$$\cong \textbf{567 turns}$$

When crossing an air gap, the magnetic lines of force tend to bulge out as shown in Figure 12-7(a). This is because lines of force repel each other when passing through nonmagnetic material. The effect of this bulging, or *fringing* as it is termed, is to increase the cross-sectional area

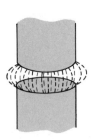

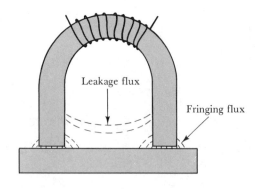

Leakage flux

Fringing flux

(a) Fringing flux at an air gap

(b) Electromagnet and iron bar

FIGURE 12-7. Fringing (bulging of magnetic fields) occurs at air gaps. Leakage flux may bypass part of a magnetic path at adjacent poles.

Chap. 12   Magnetic Circuits

of the magnetic field at the air gap and thus decrease its flux density. In a short air gap with a large cross-sectional area, the fringing may be insignificant. In other situations, 10% is typically added to the air gap's cross-sectional area to allow for fringing.

In Figure 12-7(b) an electromagnet is shown lifting an iron bar. In this case the uneven surfaces create air gaps at the junction between the magnet and the iron bar, and fringing occurs as illustrated. Some lines of force actually cross the gap between the magnet's poles without passing through the iron bar. This is termed *leakage flux,* and unlike the fringing flux, it does not assist in holding the iron bar to the magnet. Thus the leakage flux is wasted. However, the iron path around the magnetic circuit has a much lower reluctance than the air path taken by the leakage flux; consequently, leakage flux is usually small enough to be neglected.

EXAMPLE 12-9 The magnetic core shown in Figure 12-8 has the following dimensions: $l_1 = 10$ cm, $l_2 = l_3 = 18$ cm, cross-sectional area of l_1 path $= 6.25 \times 10^{-4}$ m², cross-sectional area of l_2 and l_3 paths $= 3 \times 10^{-4}$ m², length of air gap $(l_4) = 2$ mm. Determine the current that must be passed through the 600-turn coil to produce a total flux of 100 μWb in the air gap. Assume that the metal has a permeability of 800.

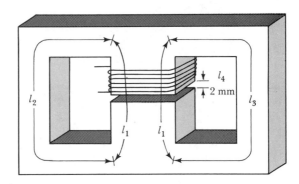

FIGURE 12-8. Composite magnetic core for Example 12-9.

SOLUTION

For the air gap:

$$B = \frac{\Phi}{A} = \frac{100 \times 10^{-6} \text{ Wb}}{6.25 \times 10^{-4} \text{ m}^2}$$

$$= 0.16 \text{ T}$$

$$H = \frac{B}{\mu_0} = \frac{0.16 \text{ T}}{4\pi \times 10^{-7}}$$

$$\cong 1.27 \times 10^5 \text{ A/m}$$

and
$$H = \frac{F_m}{l}$$

Therefore,
$$F_m = H \times l_4$$
$$\cong 1.27 \times 10^5 \text{ A/m} \times 2 \times 10^{-3} \text{ m}$$
$$F_{m4} \cong 254 \text{ A}$$

For path l_1:

$$B = 0.16 \text{ T}$$
$$H = \frac{B}{\mu_r \mu_0} = \frac{0.16 \text{ T}}{800 \times 4\pi \times 10^{-7}}$$
$$\cong 159 \text{ A/m}$$

and
$$F_m = H \times l_1 \cong 159 \text{ A/m} \times 10 \times 10^{-2} \text{ m}$$
$$F_{m1} \cong 15.9 \text{ A}$$

Since l_2 and l_3 are in parallel, and each has a cross-sectional area of $3 \times 10^{-4} \text{ m}^2$, they can be treated as a single path with a cross-sectional area of $6 \times 10^{-4} \text{ m}^2$.

For path l_2 and l_3:

$$B = \frac{\Phi}{A} = \frac{100 \times 10^{-6} \text{ Wb}}{6 \times 10^{-4} \text{ m}^2}$$
$$\cong 0.167 \text{ T}$$
$$H = \frac{B}{\mu_r \mu_0} \cong \frac{0.167 \text{ T}}{800 \times 4\pi \times 10^{-7}}$$
$$= 166 \text{ A/m}$$

and
$$F_{m2,3} = H \times l_2 \cong 166 \text{ A/m} \times 18 \times 10^{-2} \text{ m}$$
$$F_{m2,3} \cong 29.9 \text{ A}$$

$$total\ mmf \cong 254 \text{ A} + 15.9 \text{ A} + 29.9 \text{ A}$$
$$\cong 300 \text{ A}$$

and
$$F_m = NI$$

Therefore,
$$I = \frac{F_m}{N} \cong \frac{300}{600}$$
$$\cong \textbf{500 mA}$$

PRACTICE PROBLEMS

12-5.1 For Problem 12-4.2, calculate the required coil current to maintain the total core flux if two 0.1 mm air gaps are cut in the core.

12-5.2 A magnetic core with a relative permeability of 1800 has a cross-sectional area of 2.5 cm², a magnetic path length of 9 cm, and a 3900 turn coil. A 0.7 mm air gap in the core is to have a flux density of 0.9 T. Allowing 10% for fringing, calculate the required coil current.

12-6
FORCE BETWEEN TWO MAGNETIC SURFACES

The air gap in a magnetic circuit can be thought of as two magnetic poles of opposite polarity separated by air. Since opposite poles attract, there is a force of attraction tending to pull the two poles together. The magnitude of this force can be calculated in terms of the flux density and the cross-sectional area of the air gap. If the magnetic field is the result of current flowing in a coil wound on the core, energy must be supplied to the coil in order to set up the flux in the air gap. This energy input becomes energy stored in the air gap. Consider a ferromagnetic core with an air gap. The magnetic field strength in the air gap is

Equation (11-3):
$$H = \frac{IN}{l}$$

Magnetic tape recording

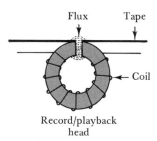

Flux   Tape

← Coil

Record/playback
head

A plastic tape with a surface track of magnetic material is passed over a record/playback head, which consists of an iron ring with an air gap and a coil. Alternating current in the coil produces an alternating flux in the air gap and through the magnetic track. Prior to recording, the particles in the track are like tiny magnets randomly arranged. The air gap flux aligns these magnets longditudinally in groups facing one way or the other. During playback, the groups of magnetized particles generate currents in the coil as the tape passes over the head.

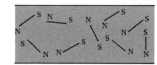

Randomly arranged particles
on unmagnetized track

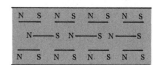

Longditudinally aligned particles
on magnetized track

Therefore,
$$I = \frac{H \times l}{N} \tag{1}$$

If I is increased from zero to its maximum level during the time Δt, the core flux changes from zero to a maximum of Φ. This flux change links with the coil and induces a voltage V at the coil terminals. From the definition of the weber (Section 11-4)—that *a flux change of* 1 *Wb in a time of* 1 *s induces* 1 *V in a one-turn coil*—the voltage induced in *each* turn of the coil is

$$V = \frac{\Phi}{\Delta t}$$

For a coil of N turns, the total induced voltage is

$$V = \frac{\Phi N}{\Delta t} \tag{2}$$

The coil supply voltage is (almost) equal in magnitude to the induced voltage, and the power input to the coil can be calculated as:

$$P = VI$$

However, P increased from zero to its maximum level during time Δt. Assuming that the power increased linearly, the average power input can be shown to be:

$$P_{av} = \frac{VI}{2}$$

Also, the energy supplied to the coil in a time of Δt is

$$W = P_{av} \times \Delta t$$

$$= \frac{V \times I \times \Delta t}{2} \quad \text{joules}$$

Substituting from equations *(1)* and *(2)* for V and I,

$$W = \frac{\Phi \times N}{\Delta t} \times \frac{H \times l}{N} \times \frac{\Delta t}{2}$$

Therefore,
$$W = \frac{\Phi \times H \times l}{2}$$

Since

$$\Phi = B \times A \quad \text{and} \quad H = B/\mu_0,$$

energy supplied, $\quad W = \dfrac{(B \times A) \times \left(\dfrac{B}{\mu_0}\right) \times l}{2}$

or

$$\boxed{\text{energy stored in air gap, } W = \dfrac{B^2 A l}{2\mu_0} \text{ joules}} \qquad (12\text{-}6)$$

When two magnetic surfaces are separated by a short distance, the mechanical energy involved in pulling them apart is

$$W = F \times d$$

Assuming that the surfaces are still attracted to each other, the electrical energy supplied to the air gap is equal to the mechanical energy supplied to pull the surfaces apart. Therefore,

$$F \times d = \dfrac{B^2 A l}{2\mu_0}$$

In this case d and l both represent the thickness of the air gap, and they cancel out, giving the force between surfaces as:

$$\boxed{F = \dfrac{B^2 A}{2\mu_0}} \qquad (12\text{-}7)$$

This expression includes the cross-sectional area (A) of the air gap, but does not include its length. When B is in teslas and A is in square meters, the force F is in newtons. Equation (12-7) can now be used to calculate the mechanical pull exerted by an electromagnet.

EXAMPLE 12-10 The electromagnet shown in Figure 12-7(b) has pole pieces which each have a cross-sectional area of 25 cm². The total flux crossing each pole is 250 μWb. Determine the maximum weight of iron plate that can be lifted by the magnet.

SOLUTION

$$B = \dfrac{\Phi}{A} = \dfrac{250 \ \mu\text{Wb}}{25 \times 10^{-4} \ \text{m}^2}$$

$$= 0.1 \ \text{T}$$

Equation (12-7):

$$F = \dfrac{B^2 A}{2\mu_0}$$

$$A = \textit{total cross-sectional area} = 2 \times 25 \times 10^{-4} \text{ m}^2$$

$$= 50 \times 10^{-4} \text{ m}^2$$

Therefore,
$$F = \frac{(0.1 \text{ T})^2 \times 50 \times 10^{-4} \text{ m}^2}{2 \times 4\pi \times 10^{-7}} \text{ newtons}$$

$$\cong 19.9 \text{ N}$$

and
$$F = m \times a$$

Therefore,
$$\text{mass } m = \frac{F}{a} \quad \begin{bmatrix} \textit{where a is the acceleration} \\ \textit{due to gravity, see Appendix 2} \end{bmatrix}$$

$$\cong \frac{19.9 \text{ N}}{9.81 \text{ m/s}^2}$$

weight that can be lifted $\cong$ 2.03 kg

PRACTICE PROBLEM

12-6.1 A magnet which has two poles, each with a cross-sectional area of 1.6 cm^2, is required to lift 1 kg bars of soft iron which completely cover the pole faces. Allowing 10% for fringing, determine the required flux density in the core.

12-7
MAGNETIZA-TION CURVES

In Figure 12-9 a typical B/H curve is shown to a slightly larger scale than those in Figure 12-3. Examination of the curve shows that the material is initially unmagnetized, because at zero magnetic field strength the flux density is zero. As the field strength is increased from zero, the flux density increases very slowly at first, giving a shallow slope, section o–a of the curve. Further increase in H causes B to increase progressively more rapidly, until there is a near-linear relationship between B and H over section a–b of the curve. With continued increase in H past point b, changes in B become less and less until the curve flattens out again to a very shallow slope beyond point c.

The shape of the curve can be explained in terms of the magnetic domains in ferromagnetic material (see Section 11-3). During portion o–a of the curve, the magnetic field strength is too weak to produce much realignment of the magnetic domains. Consequently, the increase in flux density is relatively small for these levels of field strength. Over portion a–b the increasing field strength is obviously causing more and more of the domains to be aligned in the same direction. Thus, B increases in a near-linear relationship to H. From point b to c, all of the domains are becoming completely aligned in the required direction for maximum magnetic effect. So, during the b–c portion of the curve, there

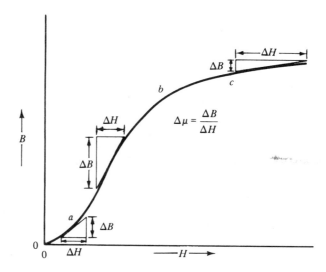

FIGURE 12-9. Permeability is the ratio B/H. When the flux density and magnetic field strength in a core vary by small amounts ΔB and ΔH, the incremental permeability $\Delta B/\Delta H$ can be important.

are fewer and fewer magnetic domains available to be aligned. Therefore, the increases in flux density in relation to increases in H are now becoming smaller. Finally, from point c on, it can be assumed that effectively all available magnetic domains are completely aligned in the direction enforced by the magnetic field strength. The magnetic material is said to be *saturated* at this point. The slope of the B/H curve beyond point c shows the same order of increase in B with respect to H as would be obtained with nonmagnetic material.

From Equation (12-5), the relationship between B and H is

$$permeability, \ \mu_r \, \mu_0 = \frac{B}{H}$$

Ideally, it is desirable to have the largest possible value of permeability, in order to achieve the greatest flux density with the smallest possible magnetic field strength. The plot of μ_r for sheet steel (Figure 12-3) shows that the largest possible value of μ_r is achieved around the center of the linear portion of the B/H curve. This direct relationship between constant values of B and H can be referred to as the *normal permeability* of the material. Another important relationship is that between changing values of *B and H*. Referring to Figure 12-9 again, the slope of the B/H curve at any point is known as the *incremental permeability*. This parameter defines the flux density change (ΔB) for a given magnetic field strength change (ΔH). Thus, the incremental permeability is

$$\Delta\mu = \frac{\Delta B}{\Delta H}$$

Reference to the B/H curve in Figure 12-9 reveals that (as with μ_r) $\Delta\mu$ is greatest at the center of the linear portion of the magnetization curve. Both the initial and final values of incremental permeability (see the illustration) are considerably smaller than that at the center of the curve.

The incremental permeability is important in applications where only small changes occur in the magnetic field strength and the largest possible changes in flux density are desired. One such application is the telephone earphone (or headphone) (see Figure 12-10), where audio signals are represented by small changes in coil current, which produce small changes in magnetic field strength. The resulting changes in flux density cause the metal disc (or *diaphragm*) to vibrate and re-create the original sound wave. To produce the largest possible changes in flux density the largest $\Delta\mu$ value is used; consequently, the magnetic field strength is usually set (or biased) at the center of the magnetization curve.

12-8
HYSTERESIS

The magnetization curve of Figure 12-9 is reproduced in Figure 12-11(a), in order to investigate what occurs when the magnetic field strength is reduced from its maximum level. As already discussed, portion o–a–b–c of the curve occurs when the material is initially unmagnetized and the field strength is increased from zero. Commencing at point c, when the field strength is decreased, the flux density does not

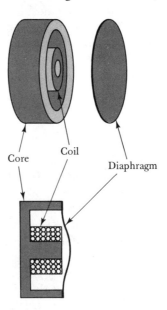

FIGURE 12-10. In a simple telephone earphone, small changes in coil current produce variations in magnetic field strength. These cause the diaphram to vibrate and thus create a sound wave.

Chap. 12   Magnetic Circuits

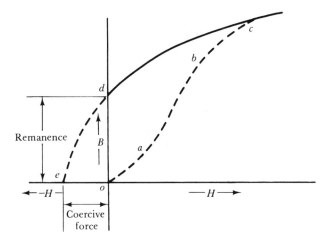

(a) Effect of reducing H from H_{max}

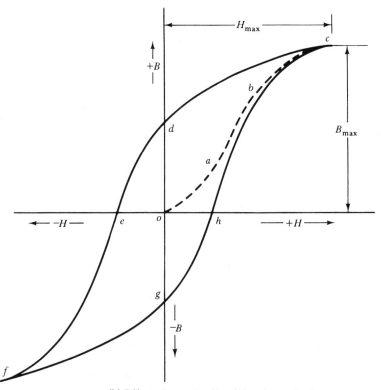

(b) Effect of reversing H and then increasing it
in a positive direction once again

FIGURE 12-11. When the magnetic field strength H in a sample of iron is decreased to zero, some flux density B may remain. If H is reversed, the residual magnetism is also reversed. Instead of giving a one line graph, a plot of B versus H is likely to produce a hysteresis loop.

move down the *c–b–a–o* curve, as might be expected. Instead, it falls along the curve *c–d*, as illustrated. The effect can be explained in terms of a kind of friction-force resisting movement of the magnetic domains. Thus, instead of returning to their original unmagnetized state, the magnetic domains remain partially aligned, and the material retains some magnetic flux, even though the field strength has been reduced to zero. The material has, in fact, been magnetized, and the retained flux density (*o–d*) is referred to as *remanence*, or *residual magnetism*. The ability of a given ferromagnetic material to retain residual magnetism is termed its *retentivity*.

To reduce the remanence to zero, a negative or reverse magnetic field strength must be applied (i.e., the current in the magnetizing coil must be reversed). As the reversed value of *H* is increased, the flux density moves down the *d–e* portion of the curve to a zero level of *B*. The magnetic field strength (*o–e*) required to reduce the remanence to zero is termed the *coercive force*.

When the reversed value of *H* is increased beyond the level needed to reduce *B* to zero, it is found that the flux density now increases in a negative direction. A new magnetization curve is generated, giving *e–f* in Figure 12-11(b). At point *f*, the reversed flux density has reached its saturation point, and this corresponds to point *c* on the positive half of the curve. Once again, when the reversed field strength is reduced to zero, the reversed flux density does not become zero. Instead, portion *f–g* of the *B/H* graph is generated, and *o–g* represents the retained flux density. Section *o–g* is exactly equal to *o–d*, although in the opposite direction. To reduce the flux density to zero once more, the field strength must be increased in a positive direction, giving portion *g–h* of the graph. This again represents a coercive force, and *o–h* is found to be equal in magnitude to *o–e*. Finally, when the magnetic field strength is further increased in a positive direction, the *B/H* relationship traces again out the *h–c* section of the graph.

It is seen that the *B/H* graph now forms a closed loop. The loop is symmetrical, with $+B_{(max)}$ and $-B_{(max)}$ equal in magnitude, and $+H_{max}$ equal in magnitude to $-H_{max}$. When a positive magnetic field strength (*o–h*) is applied to the specimen to which the graph refers, the flux density *B* is zero. Only when *H* is increased to H_{max} does the flux density become B_{max}. Similarly, when *H* is reduced from a positive level to zero, *B* remains equal to *o–d*. Continuing around the loop, it is seen that the changing levels of flux density lag behind the changes in magnetic field strength. This lagging effect is termed *hysteresis,* and the *B/H* graph is then referred to as a *hysteresis loop*.

The *B/H* loop demonstrates that some energy is absorbed into a magnetic core to overcome the *friction* involved in changing the alignment of the magnetic domains. Thus, a core that is subjected to repeated and rapid reversals of the magnetic field (as in the case of alternating currents) may absorb a lot of energy in this way. This energy results in heating of the core, and it is obviously wasted or lost energy. In fact, the area enclosed by the hysteresis loop can be shown to be proportional to the lost energy.

Figure 12-12 shows three typical hysteresis loops for three different types of ferromagnetic material. Loop (*a*) is the type of hysteresis loop obtained for soft iron. The fact that it is narrow means that the area enclosed by the loop is relatively small, and consequently the hysteresis losses in the core would be a minimum. For greatest efficiency, therefore, a soft iron core should be employed in situations where the magnetic field has to undergo a large number of reversals each second. Also, note that the residual magnetism is small for the soft iron core; consequently, soft iron is not suitable for permanent magnets.

The hysteresis loop shown in Figure 12-12(b) is typical of hard steel, and its large area obviously results in a large core loss. Thus, hard steel is not suitable where the magnetic field is being rapidly reversed. However, because its residual magnetism is large, hard steel is very suitable for permanent magnets.

The third hysteresis loop, Figure 12-12(c), is that of a material known as *ferrite*. This is a *ceramic core* made up of iron oxides. The shape of the loop suggests a large hysteresis loss. However, it also shows that B tends to remain constant in one direction until the value of H is increased almost to its maximum level in the opposite direction; then B rapidly reverses. This is exactly the characteristic required for a *magnetic memory*, and consequently, memories are one area of ferrite application.

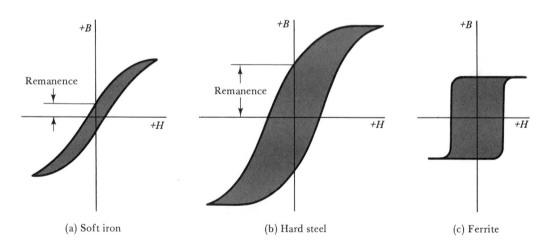

(a) Soft iron   (b) Hard steel   (c) Ferrite

FIGURE 12-12. The different types of hysteresis loops obtained with different ferromagnetic materials influence the application of the materials.

12-9

EDDY CURRENTS

In Section 11-2 it was explained that a changing magnetic field induces a voltage in a conductor situated within the field. When the current direction in a coil is continually reversing, a ferromagnetic core in that coil constitutes a conductor in a changing magnetic field. Consequently, voltages are induced in the core, and circulating currents (or *eddy currents*) are caused to flow, as illustrated in Figure 12-13(a).

Eddy currents cause heating of the core and add considerably to the total core losses. Even when the core is nonmagnetic, eddy currents are generated if the material is an electrical conductor. To combat the eddy current losses, magnetic cores used with alternating current are always made up of thin sheets, termed *laminations* [see Figure 12-13(b)]. The surfaces of the laminations are varnished or otherwise thinly insulated on either side so that they offer a high resistance to the flow of circulating eddy currents. By this method the eddy current core losses are rendered negligible without affecting the magnetic performance of the core. Note that the orientation of the laminations is such that they allow the setting up of the magnetic lines of force without difficulty. The varnished surface interface between laminations does not constitute an air gap in the magnetic path. The single negative effect of laminated cores is that the total cross-sectional area of the magnetic material is reduced by the total thickness of the insulation. This is usually taken into account by allowing an approximately 10% reduction in thickness of the core when making the magnetic calculations.

PRACTICE PROBLEM

12-9.1 If the core in Example 12-6 is laminated, calculate the new current level required in the coil.

SUMMARY OF FORMULAS

Reluctance:

$$R_m = \frac{1}{\mu} \times \frac{l}{A}$$

$$R_m = \frac{F_m}{\Phi}$$

Permeability:

$$\mu = \frac{B}{H}$$

$$\mu_0 = 4\pi \times 10^{-7}$$

$$\mu_r \mu_0 = \frac{B}{H}$$

$$\Delta\mu = \frac{\Delta B}{\Delta H}$$

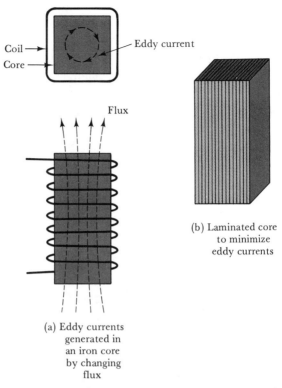

Coil → Core → ← Eddy current

Flux

(b) Laminated core to minimize eddy currents

(a) Eddy currents generated in an iron core by changing flux

FIGURE 12-13. Eddy currents induced in a solid magnetic core by a continuously reversing coil current can overheat the core. The effect is minimized by constructing the core of sheets, called laminations.

Energy stored in air gap:

$$W = \frac{B^2 A l}{2\mu_0}$$

Force between two magnetic surfaces:

$$F = \frac{B^2 A}{2\mu_0}$$

REVIEW QUESTIONS

12-1 Define reluctance and permeability and derive the equation relating the two quantities.

12-2 Show that the permeability of free space is $4\pi \times 10^{-7}$.

12-3 Using illustrations, explain why the length of a solenoid can be taken as the total length of the magnetic path in calculations of mmf, flux density, and so on. Also, discuss what occurs when a thin cylindrical iron core is brought close to one end of the solenoid.

12-4 Explain the term relative permeability and write the equation relating relative permeability to flux density and magnetic field strength.

12-5 Define fringing and leakage flux and discuss their effects on magnetic calculations.

12-6 Derive the equation for energy stored in an air gap and the equation for force between two magnetic surfaces.

12-7 Sketch a typical magnetization curve for an initially unmagnetized sample of ferromagnetic material and explain its shape. Also, show how the incremental permeability can be determined from the curve and discuss its importance in some applications.

12-8 Sketch a typical hysteresis loop for a sample of ferromagnetic material and explain its shape. Also, identify and define remanence and coercive force.

12-9 Sketch typical hysteresis loops for: (a) soft iron, (b) hard steel, and (c) ferrite. In each case state the applications that the material is most suitable for and explain why.

12-10 Explain the origin of eddy currents in a magnetic core and discuss their effects and the method employed to combat them.

PROBLEMS

SECTION 12-2

12-1 A toroidal coil has 1750 turns and carries a current a 5 A. The cross-sectional area of the coil is 2.25 cm^2 and the length of the magnetic path is 11 cm. Calculate the magnetic field strength, total flux, and flux density within the coil.

12-2 The coil described in Example 11-2 has an air core with a cross-sectional area of 3.2 cm^2. Calculate the total flux generated.

12-3 If the coil described in Problem 12-1 is to have a total flux of 0.2 μWb, determine the new level of current that must flow through the coil.

12-4 All of the mmf generated by the coil in Problem 12-1 is applied to produce a flux in a 1 mm air gap which has a cross-sectional area of 2.25 cm^2. Calculate the magnetic field strength, flux density, and total flux in the air gap.

12-5 A 0.5 mm air gap has a cross-sectional area of 7 cm^2. Calculate the mmf required to generate a total flux of 50 μWb in the air gap.

12-6 A coil with a magnetic path length of 28 cm and a cross-sectional area of 15 cm^2 produces a total flux of 45 μWb. If the coil current is to be 1.5 A, calculate the required number of turns.

SECTION 12-3

12-7 An air-cored solenoid is 18 cm long and has an inside diameter of 1.7 cm. The coil has 1400 turns. Calculate the current that must flow to give a total flux of 0.1 μWb within the coil.

12-8 Determine the total flux within the solenoid in Problem 12-7 when the coil current is 150 mA.

12-9 A short coil with a diameter of 1 cm is situated at the center of a 3300 turn solenoid. The dimensions of the solenoid are: length = 27 cm, diameter = 1.6 cm. Calculate the solenoid current that will give a total flux of 0.9 μWb within the small coil.

12-10 An air-cored solenoid is to be 25 cm long and 2.2 cm in diameter. The coil current of 100 mA has to produce a total flux of 0.3 μWb. Calculate the required number of turns.

SECTION 12-4

12-11 Referring to the magnetization curves shown in Figure 12-3, calculate the relative permeability of:
a. Cast iron at a magnetic field strength of 2000 A/m.
b. Cast steel at 1000 A/m.

12-12 Determine the relative permeability of sheet steel, cast steel, and cast iron when all three have a flux density of 0.7 T.

12-13 If the coil described in Problem 12-1 has a cast steel core, determine the value of total flux within the core when the current is 500 mA.

12-14 If the coil described in Problem 12-6 has a cast iron core, calculate the flux density when the coil current is 100 mA.

12-15 A cast iron ring has a cross-sectional area of 9 cm², and the length of the magnetic path is 25 cm. A coil of 5000 turns is wound on the ring and a current of 75 mA is passed through the coil. Calculate the magnetic field strength, the total flux, and the flux density within the core.

12-16 Recalculate the total core flux for Problem 12-15 when a sheet steel core is used.

SECTION 12-5

12-17 If the cast iron ring described in Problem 12-15 has a 2 mm air gap, calculate the new level of coil current required to establish the previously calculated level of flux density in the core.

12-18 The horeshoe-shaped magnet in Figure 12-7(b) has a magnetic path length of 50 cm and a cross-sectional area of 25 cm². The bar attached to the magnet's poles has a cross-sectional area of 35 cm² and the magnetic path through it is 15 cm long. Each air gap is approximately 5 mm thick. The magnet is cast iron and the bar is cast steel. Determine the current that must be passed through the 3000-turn coil to establish a total flux of 1500 μWb around the circuit.

12-19 The magnetic core shown in Figure 12-5 has four 0.5 mm air gaps. If the core material has $\mu_r = 800$ and the coil has 1400 turns, determine the current required to give a total core flux of 700 μWb.

12-20 Assume that the magnetic core shown in Figure 12-6 has a 1 mm air gap halfway along the thickest section. The relative permeability of the material is 1000 and the flux density in the air gap is to be 0.75 T. Determine the number of coil turns required if the coil current is to be 500 mA.

12-21 In Problem 12-17 fringing at the 2 mm air gap tends to increase the cross-sectional area of the air gap by 10%. Recalculate the coil current required to establish the previously calculated level of flux density in the air gap.

12-22 The magnetic core shown in Figure 12-14 is constructed of sheet steel, and the coil has 900 turns. Calculate the coil current required to produce a flux of 330 μWb in the air gap.

12-23 The magnetic core shown in Figure 12-8 is made of sheet steel and has the following dimensions: $l_1 = 8$ cm, $l_2 = l_3 = 15$ cm, cross-sectional area of l_1 path $= 5 \times 10^{-4}$ m², cross-sectional area of l_2 and l_3 paths $= 2 \times 10^{-4}$ m², and length of air gap $= 1.75$ mm. Calculate the number of coil turns required if the coil current is to be 5 mA and the flux density in the air gap is to be 1.2 T.

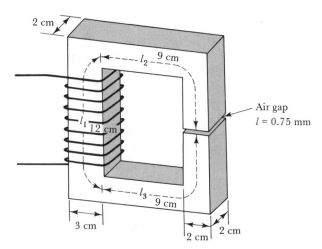

FIGURE 12-14.

12-24 An electromagnetic switch with a 1500 turn coil has a u-shaped sheet steel core with a cross-sectional area of 3 cm² and a magnetic path length of 10 cm. The sheet steel cross-bar which is lifted by the core has a magnetic path length of 4 cm, a cross sectional-area of 3 cm², and a weight of 80 gm. The two air gaps are 0.5 cm thick. Taking the relative permeability of sheet steel as 230, calculate the required coil current.

12-25 An electromagnet has two poles, each with a cross-sectional area of 35 cm². When the magnet is lifting a soft iron plate that completely covers the pole faces, the total flux around the magnetic circuit is 620 μWb. Determine the maximum weight of the iron plate that can be lifted.

12-26 For Problem 12-22, calculate the energy stored in the air gap.

12-27 For the electromagnet and steel bar described in Problem 12-18, calculate the maximum weight of steel bar that can be lifted.

SECTION 12-9

12-28 For Problem 12-24, recalculate the required coil current if the core and cross-bar are laminated so that the cross-sectional area is reduced by 10%.

12-29 For Problem 12-22, assume the core is constructed of *laminated* sheet steel so that the 2 cm thickness of the core actually contains only 1.8 cm of metal. Also assume that the fringing at the air gap increases its cross-sectional area by 10%. Recalculate the coil current required to produce the same flux density in the air gap.

12-30 For Problem 12-19, apply the same considerations (for a laminated core and air-gap, fringing) described in Problem 12-29. Recalculate the coil current.

12-31 An electromagnet used for lifting steel plates has two poles each of which have cross-sectional areas of 100 cm². The magnet is constructed of laminated sheet steel so that the actual cross-sectional area of the magnetic path is 0.9 times the measured area. The magnetic path length through the core is 35 cm. The path length through the sheet steel being lifted is 20 cm, and the cross-sectional area of the sheets is 0.02 m². The weight of each sheet is 15 kg and the air gaps are each 2 mm thick. Calculate the current that must flow in the 7000 turn coil. Take μ_r=250 for the core and for the steel plates.

COMPUTER PROBLEMS

12-32 Write a program to solve the type of problem presented in Example 12-1.

12-33 Write a program to determine the flux density within an air-cored solenoid for given dimensions, number of coil turns, and coil current.

12-34 Write a program to solve the type of problem presented in Example 12-7.

13 DC MEASURING INSTRUMENTS

Objectives   You will be able to:

Draw a sketch to show the construction of a permanent magnet moving-coil instrument and explain its operation.

Sketch the circuit of a dc ammeter and explain its operation.

Sketch a multirange dc ammeter circuit and calculate the resistance of ammeter shunts for various current ranges.

Sketch the circuit of a multirange dc voltmeter and explain its operation.

Calculate the resistance of voltmeter multipliers and solve problems involving voltmeter sensitivity.

Solve problems involving the accuracy of resistance measurement by ammeter and voltmeter.

Sketch the circuit and meter scale for a multirange ohmmeter. Explain the operation of the ohmmeter and solve problems involving ohmmeter measurement of resistance.

Explain the megohmmeter and discuss its use.

Sketch the basic construction of a dynamometer instrument, explain its operation, and show how it can be used as a wattmeter.

Sketch the circuit of a Wheatstone bridge and derive the balance equation.

Introduction

The *permanent magnet moving-coil* (PMMC) instrument is a basic deflection meter that may be connected to function as a *moving-coil ammeter*, a *moving-coil voltmeter*, or an *ohmmeter*. An understanding of the operation of the PMMC instrument is essential to the theory of more complex measurement methods.

It is possible to measure resistance by the use of an ammeter and voltmeter. However, certain errors can occur, and these dictate how the instruments should be connected. The *ohmmeter* provides a means of measuring medium-range resistance quickly, but not very accurately. For the measurement of very high resistance, the *megger* is applicable, and precise resistance measurement is made on a *Wheatstone bridge*.

The electrodynamic instrument, which has similarities to a PMMC instrument, is most frequently applied for *power measurement*.

13-1
PERMANENT MAGNET MOVING-COIL INSTRUMENT

A moving-coil instrument consists basically of a permanent magnet to provide a magnetic field and a small lightweight coil pivoted within the field. A soft iron core is included between the poles of the magnet, so that the coil rotates in the narrow air gap between the poles and the core. When a current is passed through the coil windings, a torque is exerted on the coil by the interaction of the magnet's field and the field set up by the current in the coil (see Section 11-6). The resulting deflection of the coil is indicated by a pointer that moves over a calibrated scale. Figure 13-1 shows the basic construction of a *permanent magnet moving-coil* (PMMC) instrument.

As well as a *deflecting force* provided by the coil current and the field from the permanent magnet, there is need for a *controlling force*. This is the force that returns the coil and pointer to the zero position when no current is flowing through the coil. The controlling force also *balances* the deflecting force, so that the pointer remains stationary for any constant level of current through the coil. The controlling force is usually provided by *spiral springs*, as illustrated in Figure 13-1. The springs are also employed as connecting leads for conducting current through the coil.

One other force, known as a *damping force*, is required for correct operation of a deflection instrument. When no damping force is present, the pointer swings above and below its final position on the scale for some time before settling down. In the case of the PMMC instrument, the damping force uses *eddy currents* (see Section 12-9). To facilitate this, the coil is wound on an aluminum frame or *coil former* in which eddy currents are generated by any rapid movement of the coil in the magnetic field. The eddy currents set up a magnetic flux that opposes the original movement that generates them. Thus, oscillations of the meter pointer are *damped out*.

The deflection of the pointer of a moving-coil instrument is directly proportional to the current flowing through the coil. As shown in the sketch, the scale might be calibrated to indicate a maximum current of 100 μA. When the coil current is 50 μA, the pointer is at the halfway mark between zero and 100 μA. Therefore, the scale is calibrated to read 50 μA at that point. Similarly, at ¼ of full-scale deflection (FSD),

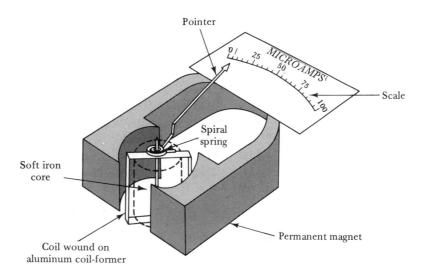

Pointer

MICROAMPS

Scale

Spiral
spring

Soft iron
core

Permanent magnet

Coil wound on
aluminum coil-former

FIGURE 13-1. In a permanent magnet moving-coil (PMMC) instrument, a light-weight coil is pivoted to move in the air gap between a soft iron core and the poles of a permanent magnet. Current flowing in the coil produces the deflecting force, and spiral springs provide a controlling force. A pointer attached to the coil indicates the amount of deflection.

the scale reads 25 μA, and at ¾ FSD the reading is 75 μA. The scale divisions for a given change of current are equal at all points on the scale, and therefore the scale is said to be *linear*. Certain other types of deflection instruments have scales that are not linearly divided.

In the PMMC instrument *it is important that the current flows through the coil in the correct direction.* When the current is in the wrong direction, the magnetic field set up around the coil reacts with the field from the magnet in a way that tends to cause the pointer to deflect to the left of the zero position. Therefore, for correct (i.e., positive) deflection over the scale of the instrument, there is only one direction for coil current flow. As already discussed in Chapter 2, the terminals of the instrument are identified as + and − to show the polarity for correctly connecting the meter into a circuit.

13-2
THE DC AMMETER

AMMETER CIRCUIT. Since the deflection of a PMMC instrument is directly proportional to the current through its coil, the instrument is essentially an *ammeter*. However, the meter can be employed directly as an ammeter only for very small current levels. Consequently, some modifications must be made where it is desired to have an ammeter that measures larger currents.

A dc ammeter is illustrated in Figure 13-2(a), where a resistor known as a *shunt* is shown connected in parallel with the PMMC instrument. Figure 13-2(b) shows the equivalent circuit of the complete instrument, including the coil resistance r_m. Some of the current to be measured passes through the instrument, and the rest of it passes through the shunt R_s. From a knowledge of the coil resistance and the FSD current of the instrument, the resistance of the shunt can be determined for any desired level of current to be measured. When the ammeter is designed to indicate a current of (for example) 100 A, the scale is recalibrated to read 100 A at FSD and proportional levels at other points. The procedure for determining the shunt resistance value is demonstrated by Example 13-1.

EXAMPLE 13-1 A PMMC instrument has a coil resistance of 200 Ω and gives full-scale deflection for a current of 750 μA. Determine the value of shunt resistance required if the instrument is to be employed as an ammeter with a FSD of 1 A.

SOLUTION

From Figure 13-2(b):

$$meter\ voltage = V_m = I_m \times r_m$$

at FSD,

$$I_m = 750\ \mu A$$

Therefore,

$$V_m = 750\ \mu A \times 200\ \Omega$$

$$= 150\ mV$$

and

$$meter\ voltage = shunt\ voltage = 150\ mV$$

Since

$$I = I_s + I_m$$

$$shunt\ current,\ I_s = I - I_m$$

$$= 1\ A - 750\ \mu A$$

$$= 999.25\ mA$$

and

$$shunt\ resistance,\ R_s = \frac{V_m}{I_s} = \frac{150\ mV}{999.25\ mA}$$

Therefore,

$$\mathbf{R_s \cong 0.15\ \Omega}$$

AMMETER RESISTANCE. Note that *it is very important for an ammeter to have a low resistance.* This is because the ammeter is always connected in series with the load that is to have its current measured (see Figure 2-1). If the ammeter resistance is not very much smaller than the load resistance, the load current can be substantially altered by the inclusion of the ammeter in the circuit. This is illustrated by Figure 13-3 and Example 13-2.

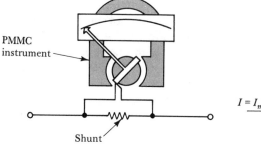
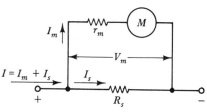

(a) Ammeter consisting
of PMMC instrument
and shunt

(b) Ammeter
equivalent
circuit

FIGURE 13-2. In an ammeter, a low resistance shunt causes most of the circuit current to be bypassed around the low-current PMMC instrument. The instrument measures a portion of the total current and indicates total current on its scale.

EXAMPLE 13-2   An ammeter having a resistance of 1 Ω is to be used to measure the current supplied to a 4 Ω resistor from a 100 V source. Calculate the current through the resistor before the ammeter is connected and after it is included in the circuit.

SOLUTION

Without the ammeter:

$$I = \frac{E}{R_x} = \frac{100\,\text{V}}{4\,\Omega}$$

$$= 25\,\text{A}$$

With the ammeter in circuit:

$$I = \frac{E}{R_x + R_a} = \frac{100\,\text{V}}{4\,\Omega + 1\,\Omega}$$

$$= 20\,\text{A}$$

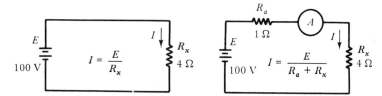

FIGURE 13-3. An ammeter should have a very low resistance, so that it does not add substantially to the total resistance in a circuit, and thus introduce an error in the measurement.

MULTIRANGE AMMETERS. A multirange ammeter can be constructed simply by employing several values of shunt resistor, with a rotary switch to select the desired range. Figure 13-4(a) shows the circuit arrangement. When an instrument is used in this fashion, care must be taken to ensure that the shunt does not become open-circuited, even for a brief instant. An ammeter is connected in series with the circuit in which current is to be measured. Consequently, if the shunt is open-circuited, a very large current may flow through the deflection instrument, possibly resulting in its destruction.

The *make-before-break* switch illustrated in Figure 13-4(b) protects an instrument from the possibility of the shunts becoming open-circuited in a multirange ammeter. The wide-ended moving contact connects to the next terminal to which it is being moved before it loses contact with the previous terminal. Thus, during the switching time there are two shunts in parallel with the instrument, and an open-circuited shunt is avoided.

AYRTON SHUNT. The *Ayrton shunt* shown in Figure 13-5(a) is another method employed to protect the deflection instrument in an ammeter from the possibility of excessive current flow. When the moving contact of the switch is connected to point B, the resistance of the shunt in parallel with the instrument is $(R_1 + R_2 + R_3)$. This is illustrated in Figure 13-5(b). When the moving contact is switched to point C [Figure 13-5(c)] the shunt becomes $(R_1 + R_2)$, and resistor R_3 is now in series with the meter. Finally, with the moving contact at point D [Figure 13-5(d)], R_1 is the shunt and $(R_2 + R_3)$ is in series with the meter. Note that there is a shunt in parallel with the instrument at all times. The shunt resistances

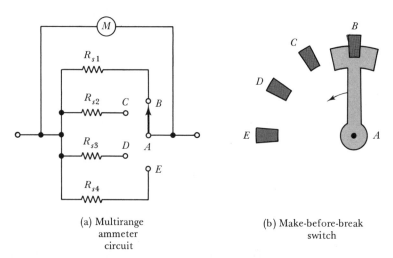

(a) Multirange ammeter circuit

(b) Make-before-break switch

FIGURE 13-4. In a multirange ammeter, any one of several shunts can be selected. A *make-before-break* switch must be used to keep the low current PMMC instrument shunted at all times.

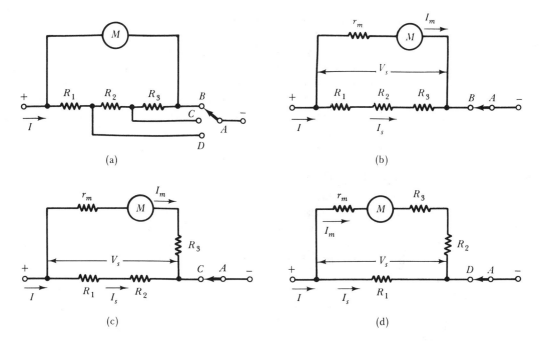

FIGURE 13-5. An Ayrton shunt has several series-connected shunts in parallel with the PMMC instrument. By selection of a connecting point at one of the junctions of the shunts, one or more shunts can be placed in parallel with the PMMC circuit, while the others are in series with the instrument.

are normally so much smaller than the instrument coil resistance that they have virtually no effect when they are in series with the instrument.

PRACTICE PROBLEMS

13-2.1 The ammeter described in Example 13-1 is to have its range changed to 100 mA and 10 mA. Calculate the new shunt resistance values.

13-2.2 The ammeter in Example 13-1 is used to measure the current supplied to a 1 Ω load. If the supply voltage is 1 V, determine the measured current and the measurement error.

13-2.3 An ammeter using an Ayrton shunt, as illustrated in Figure 13-5, has $R_1 = R_2 = R_3 = 0.0334\ \Omega$. The meter resistance is 50 Ω, and full-scale deflection current is 200 μA. Determine the ammeter full-scale range for each of the switched positions.

13-3
THE DC
VOLTMETER

VOLTMETER CIRCUIT. The deflection of a PMMC instrument is proportional to the current through its coil, and the current is, of course, proportional to the voltage across the coil resistance. Therefore, the scale of the instrument could be calibrated to indicate the applied voltage. Since the coil resistance is usually quite small, this arrangement would result in a voltmeter that could measure only very low voltage levels. To increase the voltage range the voltmeter resistance must be increased, and this is easily done by connecting a resistor (R) in series with the instrument [see Figure 13-6(a)]. Because the series resistor increases the range of the voltmeter, it is termed a *multiplier* resistor. As in the case of the ammeter, all calculations involving the voltmeter must include the meter resistance r_m [Figure 13-6(b)].

MULTIRANGE VOLTMETER. The equivalent circuit of a multirange voltmeter is shown in Figure 13-6(c). Any one of several multiplier resistors is selected by means of a rotary switch, as illustrated. Unlike the case of the ammeter, the rotary switch used with the voltmeter should be a *break-before-make* type; that is, the moving contact should disconnect from one terminal before connecting to the next terminal.

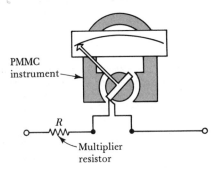

(a) Single-range dc voltmeter consisting of a deflection instrument and *multiplier* resistor

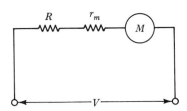

(b) The voltmeter equivalent circuit includes the meter resistance r_m

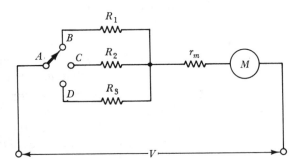

(c) Equivalent circuit of multirange voltmeter

FIGURE 13-6. A single-range voltmeter is constructed by connecting a suitable *multiplier* resistor in series with a PMMC instrument. For multirange operation, a switch is used to select one of several resistors.

Chap. 13   DC Measuring Instruments

EXAMPLE 13-3 A PMMC meter having a coil resistance of 100 Ω and an FSD current of 50 μA is to be used as a voltmeter having ranges of 100 V, 50 V, and 10 V. Determine the required values of multiplier resistor for each range.

SOLUTION

For the 100 V *range:*

$$I = \frac{E}{R_1 + r_m}$$

Therefore,

$$R_1 = \frac{E}{I} - r_m$$

$$= \frac{100 \text{ V}}{50 \text{ μA}} - 100 \text{ Ω}$$

$$= \mathbf{1.9999 \text{ MΩ}}$$

For the 50 V *range:*

$$R_2 = \frac{50 \text{ V}}{50 \text{ μA}} - 100 \text{ Ω}$$

$$= \mathbf{0.9999 \text{ MΩ}}$$

For the 10 V *range:*

$$R_3 = \frac{10 \text{ V}}{50 \text{ μA}} - 100 \text{ Ω}$$

$$= \mathbf{199.9 \text{ kΩ}}$$

VOLTMETER RESISTANCE. In Example 13-3, the total voltmeter resistance is

$$R_V = R \text{ of multiplier} + r_m$$

For the 100 V range this adds up to 2 MΩ. Dividing the 2 MΩ resistance by the 100 V range gives the *resistance per volt*, or *sensitivity*, of the voltmeter

$$sensitivity = \frac{2 \text{ MΩ}}{100 \text{ V}}$$

$$= 20 \text{ kΩ/V}$$

Calculating for the 50 V and 10 V ranges gives the same 20 kΩ/V constant. The voltmeter sensitivity is an important constant and is usually printed on the face of all voltmeters. To find the total voltmeter resistance, the resistance per volt is multiplied by the voltmeter range. For example, on a 50 V range with a sensitivity of 20 kΩ/V, the total voltmeter resistance is

$$R_V = (20 \text{ k}\Omega/\text{V}) \times (50 \text{ V})$$

$$= 1 \text{ M}\Omega$$

Unlike an ammeter, *a voltmeter should have a very high resistance.* This is because it is normally connected *in parallel* with the circuit where the voltage is to be measured (see Figure 2-5). To avoid altering the circuit conditions, the voltmeter operating current should be very small (i.e., its resistance should be high). Figure 13-7 and Example 13-4 illustrate the situation where a low-resistance voltmeter could substantially affect a circuit.

EXAMPLE 13-4   For the circuit shown in Figure 13-7, determine the voltage across resistance R_2:
a. Without the voltmeter in the circuit.
b. With the voltmeter connected.

SOLUTION

a. *Without the voltmeter:*

$$V_2 = \frac{E \times R_2}{R_1 + R_2}$$

$$= \frac{100 \text{ V} \times 1 \text{ k}\Omega}{100 \text{ k}\Omega + 1 \text{ k}\Omega}$$

$$\cong \mathbf{0.99 \text{ V}}$$

b. *With the voltmeter:*

voltmeter resistance $R_V = (\text{range}) \times (\text{sensitivity}) = (2 \text{ V}) \times (1 \text{ k}\Omega/\text{V})$

$$= 2 \text{ k}\Omega$$

$$R_V \| R_2 = 2 \text{ k}\Omega \| 1 \text{ k}\Omega$$

$$\cong 666.7 \text{ }\Omega$$

$$V_2 = \frac{E \times (R_V \| R_2)}{R_1 + (R_V \| R_2)}$$

$$\cong \frac{100 \text{ V} \times 666.7 \text{ }\Omega}{100 \text{ k}\Omega + 666.7 \text{ }\Omega}$$

$$\cong \mathbf{0.66 \text{ V}}$$

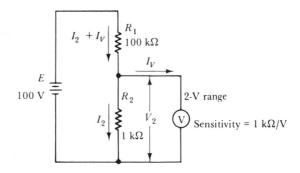

FIGURE 13-7. A voltmeter should have a very high resistance so that its current is extremely small. Voltmeter loading effect can change the conditions in a circuit and give a measurement error.

PRACTICE PROBLEMS

13-3.1 The voltmeter in Example 13-3 is to have additional ranges of 100 mV, 30 V, and 250 V. Calculate the resistance of the multiplier resistor for each range.

13-3.2 A voltmeter on a 10 V range is used to measure the output voltage from the potentiometer in Figure 5-7. If the voltmeter sensitivity is 2 kΩ/V, determine the measured maximum and minimum voltage.

13-4
MEASURING RESISTANCE BY AMMETER AND VOLTMETER

If the voltage across a resistor and the current flowing through it are measured, the resistance value can be calculated by applying Ohm's law. However, an error occurs depending on how the ammeter and voltmeter are connected, and this error may be insignificantly small or quite large.

Consider the arrangement shown in Figure 13-8(a). Since the voltmeter is connected directly across R_x, it measures the actual resistor voltage. However, the ammeter measures the resistor current I_x and the current I_V that flows through the voltmeter.

$$\text{measured resistance} = \frac{\text{voltmeter reading}}{\text{ammeter reading}}$$

$$\boxed{R = \frac{V_x}{I_x + I_V}} \tag{13-1}$$

The actual resistance of R_x is, $R_x = V_x/I_x$. Therefore, the presence of I_V in the equation introduces an error into the result. If I_V is very much smaller than I_x, the error may be insignificant. Obviously, this requires that I_x be a large current, which is the case when R_x is a small resistance.

Now consider the arrangement in Figure 13-8(b), where R_x and the ammeter are in series and the voltmeter is connected in parallel with the two of them. In this case the ammeter measures the actual current through R_x, but the voltmeter measures the voltage across the ammeter plus the voltage across R_x. Therefore,

$$\text{measured resistance} = \frac{\text{voltmeter reading}}{\text{ammeter reading}}$$

$$R = \frac{V_x + V_A}{I_x} \qquad\qquad \textbf{(13-2)}$$

Again, the actual resistance of R_x is $R_x = V_x/I_x$. Consequently, the presence of V_A in Equation (13-2) produces an error in the result. If V_A is very much smaller than V_x, the error could be insignificant. This requires that V_x be a large voltage, which means that R_x should be a large resistance.

It is seen that where R_x is a small resistance, the voltmeter should be connected directly across the resistance. Where R_x is a large resistance, the ammeter should be connected directly in series with the resistance for greatest accuracy. The correct arrangement can be easily determined by first connecting the ammeter in series with R_x, then observing the ammeter reading with the voltmeter temporarily connected directly across R_x [i.e., connected as in Figure 13-8(a)]. If the ammeter reading is not noticeably altered when the voltmeter is connected, the readings will give an accurate result. When the ammeter reading is noticeably changed by the voltmeter, the voltmeter should be moved to the other side of the ammeter.

EXAMPLE 13-5 An ammeter and voltmeter when connected as in Figure 13-8(a) indicate 10 A and 99 V, respectively. When the voltmeter is changed to reconnect the circuit as in Figure 13-8(b), the readings become 10 A and 100 V. The ammeter has a resistance of 0.1 Ω. The voltmeter is on its 100 V range and its sensitivity is 20 kΩ/V. Calculate the measured resistance for each case and determine which arrangements give the most accurate result.

SOLUTION

For the circuit of Figure 13-8(a):

$$R_x = \frac{V_x}{I_x + I_V} = \frac{99\ V}{10\ A}$$

$$R_x = 9.9\ \Omega$$

and   voltmeter resistance, $r_V = 20$ kΩ/V × *(voltmeter range)*

$$= 20\ \text{k}\Omega/\text{V} \times (100\ V)$$

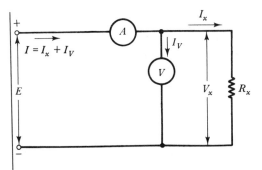

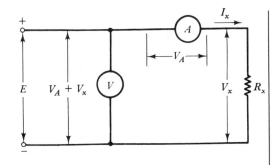

(a) Ammeter measures $I_x + I_V$ (b) Voltmeter measures $V_x + V_A$

FIGURE 13-8. Errors can occur when a resistance is measured by use of an ammeter and voltmeter. To minimize errors, the voltmeter should be connected directly across R_x when R_x is low, and the ammeter should be directly in series with R_x when R_x is high.

$$r_V = 2\ M\Omega$$

Therefore, $$I_V = \frac{V_x}{R_V} = \frac{99\ V}{2\ M\Omega}$$

$$I_V = 49.5\ \mu A$$

$$I_V \ll I_x$$

For the circuit of Figure 13-8(b):

$$R_x = \frac{V_x + V_A}{I_x} = \frac{100\ V}{10\ A}$$

$$\boldsymbol{R_x = 10\ \Omega}$$

and $$V_A = I_x + r_a$$

$$= 10\ A \times 0.1\ \Omega$$

$$V_A = 1\ V$$

In this case V_A is not very much smaller than V_x, and connection as in Figure 13-8(a) *gives the most accurate result.*

PRACTICE PROBLEM

13-4.1 The insulation resistance of an electrical cable is to be measured by use of a 1000 V source, a 1500 V voltmeter with a sensitivity of 10 kΩ/V, and a 500 μA ammeter with a resistance of 100 Ω. If the insulation resistance is 5 MΩ, determine the measured resistance using each of the two methods of connecting the instruments.

13-5

THE OHMMETER The *ohmmeter* provides a quick, but not very accurate, means of resistance measurement. The simplest direct reading ohmmeter is the basic *series ohmmeter* circuit illustrated in Figure 13-9(a). The circuit consists of a moving-coil instrument in series with a battery and adjustable resistor R_1, as shown.

With terminals A and B shorted together, R_1 is adjusted for full-scale deflection (FSD) on the instrument. A and B are the ohmmeter terminals and when they are short-circuited, the ohmmeter should read zero resistance. Thus, as shown in Figure 13-9(b), FSD is taken as an indication of zero ohms. When terminals A and B are open-circuited, the pointer should indicate infinity. Therefore, the zero deflection point on the dial is marked as infinite resistance [see Figure 13-9(b)].

When an unknown resistance R_x is connected to terminals A and B, some current flows through the meter, giving a reading between zero and infinity. As will be seen, for a given value of R_x the pointer position on the dial depends upon the value of R_1.

From Figure 13-9(a) it is seen that the meter current is

$$ I_m = \frac{E_b}{R_1 + R_x + r_m} \qquad (13\text{-}3) $$

where r_m is the coil resistance of the instrument. If $r_m \ll R_1$, the equation is simplified to

$$ I_m \cong \frac{E_b}{R_1 + R_x} \qquad (13\text{-}4) $$

EXAMPLE 13-6 The series ohmmeter shown in Figure 13-9 is made up of a 3 V battery, a 100 μA meter, and a resistance R_1 which has a fixed value of 30 kΩ. Determine the value of the unknown resistance R_x when the pointer indicates: *a.* ½ FSD, *b.* ¼ FSD, and *c.* ¾ FSD.

SOLUTION

When $R_x = 0$
Equation (13-4):

$$ I_m \cong \frac{E_b}{R_1 + R_x} = \frac{3 \text{ V}}{30 \text{ k}\Omega + 0} $$

$$ I_m = 100 \text{ μA} = \text{FSD} $$

a. At ½ FSD:

$$ I_m = \frac{100 \text{ μA}}{2} = 50 \text{ μA} $$

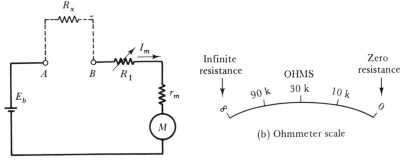

(a) Equivalent circuit of simple ohmmeter

(b) Ohmmeter scale

FIGURE 13-9. A simple ohmmeter consists of a PMMC instrument, a series resistor R_1, and a battery. When $R_x=0$, the pointer deflects to full scale. When R_x is open-circuited, pointer deflection is zero. At $R_x=R_1$, the pointer indicates half-scale deflection.

From Equation (13-4):

$$R_1 + R_x = \frac{E_b}{I_m} = \frac{3 \text{ V}}{50 \text{ μA}}$$

$$= 60 \text{ k}\Omega$$

and

$$R_x = 60 \text{ k}\Omega - R_1$$

$$= 60 \text{ k}\Omega - 30 \text{ k}\Omega$$

$$= \mathbf{30 \text{ k}\Omega}$$

b. *At ¼ FSD:*

$$I_m = \frac{100 \text{ μA}}{4} = 25 \text{ μA}$$

Therefore,

$$R_1 + R_x = \frac{3 \text{ V}}{25 \text{ μA}} = 120 \text{ k}\Omega$$

and

$$R_x = 120 \text{ k}\Omega - 30 \text{ k}\Omega$$

$$= \mathbf{90 \text{ k}\Omega}$$

c. *At ¾ FSD:*

$$I_m = \tfrac{3}{4} \times 100 \text{ μA} = 75 \text{ μA}$$

Therefore,

$$R_1 + R_x = \frac{3 \text{ V}}{75 \text{ μA}} = 40 \text{ k}\Omega$$

and

$$R_x = 40 \text{ k}\Omega - 30 \text{ k}\Omega$$

$$= \mathbf{10 \text{ k}\Omega}$$

The ohmmeter scale is now as marked in Figure 13-9(b).

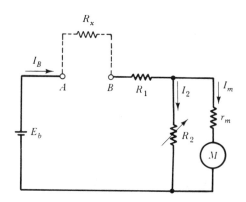

FIGURE 13-10. Series ohmmeter with zero adjust. To use the ohmmeter, the terminals are first short-circuited and R_2 is adjusted for zero ohms indication. Then R_x is connected, and the indicated resistance is read from the meter.

The results of Example 13-8 demonstrate that the ohmmeter scale is nonlinear. Also, the portion of the scale from ¾ FSD to FSD includes all resistance measurements from 10 kΩ to zero, while the portion from zero deflection to ¼ FSD includes all values from 90 kΩ to infinity. This shows that the useful range of the ohmmeter scale is approximately from ¼ FSD to ¾ FSD. The actual resistance values marked upon the scale depend on the value of R_1, which (instead of being variable) should be a fixed-value precision resistor. Where $R_1 = 30$ kΩ, as in Example 13-8, the resistance value at ½ FSD is 30 kΩ. If R_1 were changed to (for example) 100 kΩ, the scale must be marked 100 kΩ at ½ FSD. (E_b would also have to be changed.)

The circuit of Figure 13-9(a) relies upon the battery voltage remaining absolutely constant. When the battery terminal voltage falls (as they all do with use), the instrument scale is no longer accurate. Thus, some means of adjusting for battery voltage variations must be built into the circuit. Variable resistor R_2 shown in the circuit of Figure 13-10 provides the necessary adjustment. To measure a resistance R_x, the ohmmeter terminals are first short circuited, and R_2 is adjusted to give full scale deflection (zero ohms) on the meter. Then the terminals are open-circuited, and R_x is connected directly to the ohmmeter terminals. The parallel combination of R_x and r_m is usually very much smaller than R_1, so that normally $R_x = R_1$ when the meter indicates half-scale.

PRACTICE PROBLEMS

13-5.1 Determine the value of internal resistor R_1 for an ohmmeter which is to indicate 10 kΩ at half-scale. Also, calculate the measured resistance when the ohmmeter indicates ⅓ FSD and ⅔ FSD.

13-5.2 For the ohmmeter in Figure 13-10: $E_b = 3$ V, $R_1 = 30$ kΩ, $R_2 = 20$ Ω, $r_m = 20$ Ω, and the meter current is 50 μA at full scale. Determine the meter reading at $R_x = 0$ and $R_x = 30$ kΩ. Also, calculate the new resistance that R_2 must be adjusted to when E_b falls to 2.5 V.

13-6

THE MEGOHMMETER

The megohmmeter (or *megger*) is an instrument for measuring very high resistances, such as the insulation resistance of electrical cables. A high voltage source is required to pass a measurable current through such resistances. Thus, the megger is essentially an ohmmeter with a sensitive deflection instrument and a high voltage source. As illustrated in Figure 13-11, the voltage is usually produced by a hand-cranked generator. The generated voltage may be anything from 100 V to 2.5 kV.

As in the case of a low-resistance ohmmeter, the scale of the megger indicates *infinity* (∞) when measuring an open-circuit, zero on a short circuit, and half-scale when the unknown resistance equals a standard resistor inside the megohmmeter. At other points on the scale, the deflection is proportional to the ratio of the unknown and standard resistors. The range of the instrument can be altered by switching different values of standard resistor into the circuit.

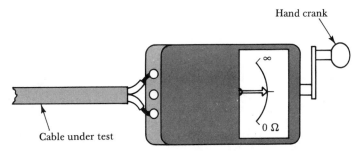

FIGURE 13-11. A megohmmeter (*megger*) is essentially an ohmmeter with its own hand-cranked high voltage generator. The high voltage source is required to pass a measurable current through the very high resistances measured by the megger.

13-7

POWER MEASUREMENT

In a dc circuit, the power supplied to a load can be determined by measuring the load voltage and current and multiplying them together: $P = EI$.

However, it is much more convenient to have an instrument that indicates power directly. The meter used for this purpose is called a *wattmeter*, and the instrument that can be applied as a wattmeter is known as a *dynamometer*, or sometimes as an *electrodynamic instrument*. The construction of a dynamometer instrument to some extent resembles the permanent magnet moving-coil instrument. The major difference from the PMMC construction is that the permanent magnet is replaced by two coils, as illustrated in Figure 13-12(a). (Compare this to Figure 13-1.) The magnetic field in which the lightweight moving coil is situated is generated by passing current through the stationary *field coils*. Then, when a current is passed through the lightweight moving coil, the moving coil and the meter pointer are deflected.

The deflection of the pointer of a dynamometer instrument is proportional to the current through the moving coil, but it is also proportional to the flux density of the magnetic field set up by the stationary coils. This means, of course, that deflection is also proportional to the current through the field coils.

Consider the arrangement shown in Figure 13-12(b). The moving coil of the instrument has a series resistor and is connected in parallel with the load that is to have its power measured. Consequently, current I_V through the moving coil is directly proportional to the load voltage. The field coils are connected in series with the load, so the current flowing through them is $(I_V + I_L)$, as shown. If $I_V \ll I_L$, I_V can be neglected and the field coil current assumed to be approximately equal to I_L. Because the meter deflection is proportional to the field coil current and to the moving-coil current,

$$\text{deflection} \propto V_L \times I_L$$

or

$$\text{deflection} \propto P$$

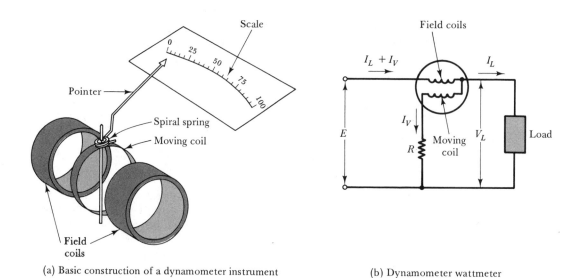

(a) Basic construction of a dynamometer instrument

(b) Dynamometer wattmeter

FIGURE 13-12. A *dynamometer*, or *electrodynamic instrument*, is similar to a PMMC instrument, except that the magnet is replaced by two field coils. When a load current is passed through the field coils and the load voltage is applied to the moving-coil circuit, the instrument indicates load power.

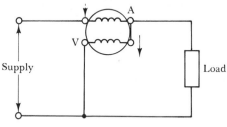

Wattmeter

The scale of the wattmeter illustrated is read directly only when the range switches are set to 1 A and 120 V. When set to 0.5 A and 60 V, full scale deflection represents: (0.5 A × 60 V) = 30 W.

It is important to avoid excessive current flow in the voltage and current coils of a wattmeter. When set to the 60 V range, the applied voltage should not exceed 60 V even when the pointer indication is less than full scale. The input terminal of each coil is identified as ↓. Sometimes a * or a ± sign is used. The marked terminal of the current coil should be connected to the supply, and the voltage coil marked terminal should be connected to the current coil, (see circuit).

The scale of the instrument can be calibrated to indicate watts, and thus it becomes a *dynamometer wattmeter*.

The dynamometer instrument can also be employed as a voltmeter or an ammeter. For use as a voltmeter, the field coils are connected in series with the moving coil and a multiplier resistor. For ammeter applications the field coils are connected in parallel with the moving coil, and in parallel with a shunt, as required.

Because fairly large currents are required to set up the necessary field flux, the dynamometer instrument is not as sensitive as a PMMC instrument. Consequently, its major application is as a wattmeter. One advantage that the dynamometer has over a PMMC instrument is that it can be used for both direct and alternating current/voltage measurements. This is discussed further in Chapter 25.

13-8
THE WHEATSTONE BRIDGE

Like the ohmmeter, the *Wheatstone bridge* is employed for measurement of resistance. But *unlike* the ohmmeter, the Wheatstone bridge measures resistance with a high degree of accuracy.

The circuit of a Wheatstone bridge is shown in Figure 13-13(a). R_1 and R_3 are precision resistors, R_4 is an adjustable precision resistor, R_2 is the unknown resistance to be measured, and G is a *galvanometer*. A galvanometer is essentially a PMMC instrument which has its pointer indicating zero at the center of its scale. It is also an extremely sensitive instrument, so very small currents through its coil cause it to deflect

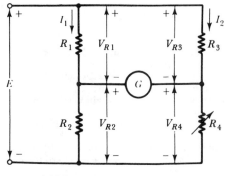

(a) Wheatstone bridge circuit

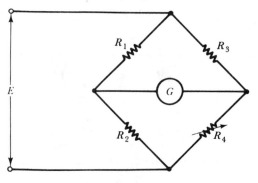

(b) Usual way to show a Wheatstone bridge circuit

FIGURE 13-13. In a Wheatstone bridge, a sensitive galvanometer is connected between the junctions of two potential dividers (R_1R_2) and (R_3R_4). When R_4 is adjusted to give zero deflection on the galvanometer, $R_2/R_1 = R_4/R_3$.

either to left or to right of zero. As applied in the Wheatstone bridge, the galvanometer is used to detect the zero-current condition. Consequently, it is usually termed a *null detector*.

In Figure 13-13(b), the Wheatstone bridge circuit is drawn in a slightly different form. This is the way the circuit is most frequently represented; however, although it may look different, it does not differ in any way from the circuit shown in Figure 13-13(a).

Referring again to Figure 13-13(a), when the galvanometer indi-, cates a null condition, the voltages on each side of the galvanometer must be equal. Also, since there is no current flowing through the galvanometer, current I_1 flows through R_1 and R_2, and current I_2 flows through R_3 and R_4. Therefore,

$$V_{R2} = V_{R4}$$

or
$$I_1R_2 = I_2R_4 \tag{1}$$

also
$$V_{R1} = V_{R3}$$

or
$$I_1R_1 = I_2R_3 \tag{2}$$

Dividing Equation (1) by Equation (2) gives

$$\frac{I_1R_2}{I_1R_1} = \frac{I_2R_4}{I_2R_3}$$

Therefore,
$$\frac{R_2}{R_1} = \frac{R_4}{R_3}$$

or

$$\boxed{\textit{unknown resistance, } R_2 = \frac{R_1 R_4}{R_3}} \qquad \textbf{(13-5)}$$

Since the precise values of R_1, R_3, and R_4 are known, the resistance R_2 can be accurately calculated.

EXAMPLE 13-7   A Wheatstone bridge has $R_1 = 5\ \text{k}\Omega$, $R_3 = 10\ \text{k}\Omega$, and the null condition is indicated on the galvanometer when $R_4 = 1994\ \Omega$. Determine the value of the unknown resistance R_2.

SOLUTION

Equation (13-5):

$$R_2 = \frac{R_1 R_4}{R_3}$$

$$= \frac{1994\ \Omega \times 5\ \text{k}\Omega}{10\ \text{k}\Omega}$$

$$= \mathbf{997\ \Omega}$$

PRACTICE PROBLEM

13-8.1   A Wheatstone bridge is to measure resistances ranging from 10 Ω to 100 Ω. If $R_1 = 250\ \Omega$ and $R_3 = 200\ \Omega$, calculate the required maximum and minimum resistances for R_4.

SUMMARY OF FORMULAS

For measuring resistance by ammeter and voltmeter:

$$R = \frac{V_L}{I_L + I_V}$$

$$R = \frac{V_L + V_A}{I_L}$$

Ohmmeter circuit:

$$I_m = \frac{E_b}{R_1 + R_x + r_m}$$

$$I_m \cong \frac{E_b}{R_1 + R_x}$$

Wheatstone bridge circuit:

$$R_2 = \frac{R_1 R_4}{R_3}$$

13-1 Draw a sketch to show the construction of a PMMC instrument. Define the three forces involved in operation of the instrument and explain briefly.

13-2 Sketch the circuit of an ammeter and briefly explain its operation. Comment on the resistance of an ammeter.

13-3 Draw a sketch to show the circuit of a multirange ammeter and discuss the type of switch required for range changing. Also, sketch the circuit of an ammeter using an Ayrton shunt and briefly explain.

13-4 Sketch the circuit of a multirange voltmeter and briefly explain its operation. Discuss any special requirement for the range-changing switch and comment on the resistance of a voltmeter in relation to its usual applications.

13-5 Sketch circuits to show the two possible arrangements for measuring resistance by use of an ammeter and voltmeter. Discuss the error sources with each arrangement.

13-6 Sketch the basic circuit of a series ohmmeter and explain its operation. Also, show how adjustments may be made for a falling battery voltage.

13-7 Discuss the measurement of insulation resistance by means of an ammeter and voltmeter. Sketch the circuitry involved.

13-8 Explain the basic operation of a megger.

13-9 Draw a sketch to show the basic construction of a dynamometer instrument and explain its operation. Also, show how this instrument can be employed as a wattmeter, and explain how it can give a direct indication of power.

13-10 Sketch the circuit of a Wheatstone bridge, explain its operation, and derive an expression from which the unknown resistance can be calculated.

PROBLEMS

SECTION 13-2

13-1 A PMMC instrument has a coil resistance of 270 Ω and gives FSD for a current of 100 μA. Determine the value of shunt resistance required to convert the instrument into a 100 mA ammeter.

13-2 An ammeter consists of a 37.5 μA meter in parallel with a 0.0018 Ω shunt. The meter coil resistance is 1.2 kΩ. Calculate the measured current at full scale.

13-3 Determine the new values of shunt resistor for the ammeter described in Problem 13-1 to change its range to
a. 1 A.   b. 10 A.

13-4 An ammeter which indicates full scale deflection at 300 mA has a 50 μA meter with a coil resistance of 900 Ω. Calculate the shunt resistance.

13-5 An ammeter having a resistance of 0.5 Ω is connected in series with a 20 V supply and a load that normally takes a current of 20 A from the supply. Calculate the current indicated by the ammeter.

13-6 The ammeter in Problem 13-2 indicates 15 A when connected in series with a load and a 5 V source. Calculate the current that flows when the ammeter is not connected in the circuit.

13-7 A PMMC instrument has a resistance of 100 Ω and FSD for a current of 100 μA. An Ayrton shunt is connected to the instrument to convert it to an ammeter. The Ayrton shunt has four resistors, each of which is 0.001 Ω. Determine the various ranges to which the ammeter may be switched.

13-8 The ammeter in Problem 13-7 is set on its lowest range and connected to measure the current flowing in a resistor which is precisely 0.9 Ω. If the supply voltage is exactly 2 V, determine the indicated current and the error introduced by the ammeter.

SECTION 13-3

13-9 A PMMC instrument with a resistance of 75 Ω and FSD current of 100 μA is to be used as a voltmeter with 250 V, 100 V, and 50 V ranges. Determine the required values of multiplier resistor for each range.

13-10 The voltmeter in Problem 13-9 is used to measure the voltage drop across each of two resistors connected as a potential divider as in Figure 13-7. The supply voltage is 50 V and the resistor values are $R_1 = 56$ kΩ and $R_2 = 68$ kΩ. If the voltmeter is set on its 50 V range, calculate the measured levels of V_{R1} and V_{R2}.

13-11 A voltmeter consists of a 150 μA meter with a 100 Ω coil resistance in series with three series-connected resistors. The resistor values are $R_1 = 166.6$ kΩ, $R_2 = 333.3$ kΩ, and $R_3 = 500$ kΩ. Determine the voltmeter range when the multiplier resistor is (a) R_1 alone, (b) $R_1 + R_2$, (c) $R_1 + R_2 + R_3$.

13-12 A voltmeter is to be constructed using a 37.5 μA meter with a resistance of 900 Ω, and three series-connected resistors as in Problem 13-11. The instrument ranges are to be 3 V, 30 V, and 300 V. Calculate the resistor values.

13-13 Two resistors are connected in series across a 250 V supply. The resistor values are $R_1 = 330$ kΩ and $R_2 = 220$ kΩ. The voltmeter described in Problem 13-9 is used to measure the voltage across each resistor. Determine the voltage indicated in each case.

13-14 Calculate the sensitivity of each of the voltmeters referred to in Problems 13-9, 13-11, and 13-12.

13-15 The voltmeter in Problem 13-9 is used on its lowest range to measure the voltages across R_1 and R_2 in the circuit in Figure 7-11. Determine the measured voltages.

13-16 The voltmeter in Problem 13-12 is used to measure the output of a potential divider circuit, as in Figure 5-7. The supply voltage is 12 V, and the resistor values are $R_1 = 27$ kΩ, $R_2 = 5$ kΩ, and $R_3 = 18$ kΩ. Calculate the measured maximum and minimum levels of V_0.

SECTION 13-4

13-17 An ammeter and voltmeter employed to measure resistance give readings of 196 μA and 240 V, respectively, when the voltmeter is connected directly in parallel with the resistor to be measured. When the ammeter is connected directly in series with the resistor, the readings are 100 μA and 240 V. The ammeter has a resistance of 50 Ω, and the voltmeter sensitivity is 10 kΩ/V. If the voltmeter is on a 250 V range, calculate the value of measured resistance for each case and determine which of the two gives the most accurate result.

13-18 The ammeter described in Problem 13-2 is used together with the voltmeter in Problem 13-12 to measure a resistance R_L. With the voltmeter on its lowest range connected directly in parallel with R_L, the meter readings are 2.75 V and 22 A. Determine the meter readings when the ammeter is directly in series with R_L.

13-19 A resistance R_x measured by voltmeter and ammeter gives readings of 500 V and 150 μA when the ammeter is directly in series with R_x. When the voltmeter is directly in parallel with R_x, the readings are 500 V and 200 μA. Determine the resistance of R_x and the sensitivity of the voltmeter.

13-20 Two voltmeters (V_1 and V_2), which each have a sensitivity of 20 kΩ/V, are employed to measure the insulation resistance of an electrical installation. V_1 is connected in parallel with the supply, and V_2 is in series with the supply and the resistance to be measured. Both voltmeters are on the 300 V range. If V_1 indicates 250 V and V_2 indicates 62.5 V, calculate the insulation resistance.

SECTION 13-5

13-21 A series ohmmeter uses a 200 μA meter and a 20 kΩ precision resistor. The supply is a battery with voltage $E = 4$ V. Determine the value of resistance measured by the ohmmeter at pointer deflections of ⅓ FSD, ½ FSD, and ⅔ FSD. Draw a sketch of the scale of the instrument, showing current levels and resistance values.

13-22 A 50 μA meter with $r_m = 50\ \Omega$ is to be used to construct a series ohmmeter which is to have a center scale indication of 15 kΩ. A shunt resistor R_2 is to be included across the meter, as in Figure 13-10. Determine a suitable resistance value for R_1 and R_2 and a suitable battery voltage.

13-23 A series ohmmeter has a meter with 100 μA FSD and a resistance of $r_m = 30\ \Omega$. The supply battery has a terminal voltage of $E_B = 4$ V, and the series resistor is 10 kΩ. An adjustable shunt resistor connected across the meter has a value of $R_2 = 30\ \Omega$. Determine the external resistance measured at ¼, ½, and ¾ of FSD.

13-24 For the ohmmeter in Problem 13-23, determine the new value that R_2 must be adjusted to when the battery voltage falls to 3 V.

13-25 For the ohmmeter in Problem 13-22, calculate the measured resistance at ⅓ and ⅔ of FSD.

13-26 For the ohmmeter in Problem 13-22, determine the new value that R_2 must be adjusted to when the battery voltage falls by 10%.

SECTION 13-8

13-27 A Wheatstone bridge has a 100 Ω precision resistor R_1 connected in series with the unknown resistor R_2. Another precision resistor R_3, having a value of 150 Ω, is connected in series with the variable resistor R_4. When the galvanometer indicates null, the value of R_4 is found to be 119.25 Ω. Calculate the value of the unknown resistance.

13-28 A Wheatstone bridge, as in Figure 13-13, is to be constructed to measure resistances in the range of 300 Ω to 30 kΩ. Precision resistors having values of 1 kΩ and 5 kΩ are available for use as R_1 and R_2 respectively. Calculate the required range of adjustment for R_4.

13-29 In a Wheatstone bridge circuit, as in Figure 13-13, $R_3 = 1$ kΩ, R_4 is adjustable from 100 Ω to 10 kΩ, and R_1 may be selected as 1 kΩ or 10 kΩ. Determine the two measurement ranges of the bridge.

COMPUTER PROBLEMS

13-30 Write a computer program to calculate the series resistor values for a multirange voltmeter when the required voltage ranges and meter characteristics are specified.

13-31 Write a computer program to calculate the Ayrton shunt resistor values for a multirange ammeter when the required current ranges and meter characteristics are specified.

14 INDUCTANCE

Introduction

Electromagnetic induction occurs when a magnetic flux in motion with respect to a single conductor or a coil *induces an emf* in the conductor or coil. Because the growth or decline of current through a coil generates a changing flux, an emf is induced in the coil by its own current change. The same effect can induce an emf in an adjacent coil. The level of emf induced in each case depends upon the *self-inductance* of the coil, or upon the *mutual inductance* between the two coils. In all cases the polarity of the induced emf is such that it opposes the original change that induced the emf.

Components called *inductors* or *chokes* are constructed to have specified values of inductance. *Energy* is *stored* in an inductive coil when a current is flowing in the coil windings. Inductors can be operated in series or in parallel. Even the shortest of conductors has an inductance. This is usually an unwanted quantity and is termed *stray inductance*.

14-1
ELECTRO-MAGNETIC INDUCTION

In Section 11-2 it is shown that when a magnetic flux cuts a conductor (or a coil), an emf is generated within the coil. This effect is termed *electromagnetic induction*. It is important to note that the emf is generated only when the flux is in motion with respect to the conductor or coil. When the flux and the conductor are stationary, no emf is generated.

Figure 14-1 illustrates electromagnetic induction in a single conductor in motion across a magnetic field. A center-zero galvanometer (see Section 13-8) is connected in series with the conductor, and it is found that the galvanometer deflects to the left when the conductor is moved up through the field [Figure 14-1(a)]. This indicates that the current induced in the conductor is flowing through the galvanometer from right to left, as illustrated. It also shows that the polarity of the induced emf in the conductor is + on the right-hand side of the conductor and − on the left-hand side, as shown in the figure. When the direction of the current induced in the conductor is considered, it is seen that the current sets up its own flux in a clockwise direction around the conductor [see Figure 14-1(a)]. The effect of this (conductor) flux is to strengthen the magnetic field above the conductor and to weaken it below the conductor, as shown in Figure 14-1(b). The strengthening of the magnetic field above the conductor and the weakening of the field below the conductor make it more difficult for the conductor to move upward. Thus it is seen that the magnetic flux set up by the current induced in the conductor actually opposes the direction of motion of the conductor.

A similar effect is produced when the conductor is moved down through the field, as illustrated in Figure 14-1(c) and (d). In this case, the induced emf is + on the conductor's left-hand side and − on its right-hand side. Galvanometer deflection is to the right, and the current direction is as shown in the figure. Figure 14-1(d) shows that the induced current in the conductor now generates a flux that strengthens the magnet's magnetic field below the conductor and weakens it above the conductor. Again, the field set up by the induced current opposes the conductor's direction of motion.

Chap. 14   Inductance

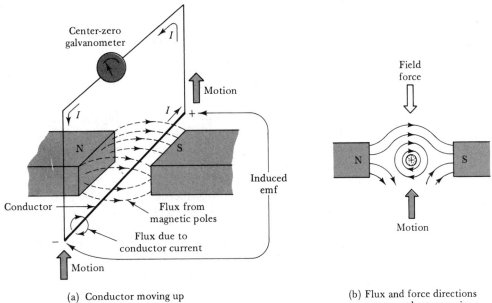

(a) Conductor moving up
through magnetic field

(b) Flux and force directions
on a conductor moving
up through field

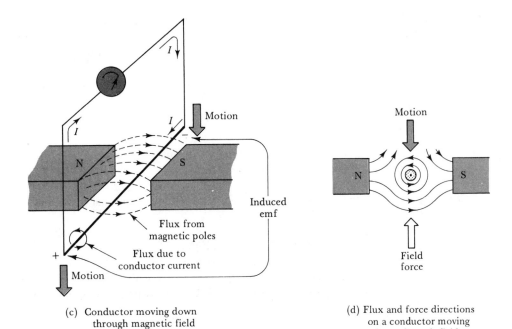

(c) Conductor moving down
through magnetic field

(d) Flux and force directions
on a conductor moving
down through field

FIGURE 14-1. Electromagnetic induction. An emf is generated in a conductor in motion through a magnetic field. A current flows when an external circuit is connected to the conductor.

The fact that the flux set up by the induced current always opposes the direction of motion is stated in *Lenz's* law*.

Lenz's Law **The induced current always develops a flux which opposes the motion or change producing the current.**

Lenz's law can be justified by considering the origin of the energy developed in the conductor. It is obviously generated as a result of the relative motion of the conductor and the magnetic field. Thus, work must be done in moving the conductor through the field, and for work to be done a force must oppose the motion of the conductor. This opposing force is the result of the flux set up by the induced current.

It should be noted that for electromagnetic induction to occur as illustrated in Figure 14-1, both the conductor and direction of motion must be at right angles to the magnetic field flux. If the conductor were to be moved axially, for example, or if it were oriented in a horizontal direction from the N pole to the S pole, no electromagnetic induction effect would occur.

14-2

INDUCED EMF AND CURRENT

In Section 11-4 the weber is defined as *the magnetic flux which, linking a single-turn coil, produces an emf of 1 V when the flux is reduced to zero at a constant rate in 1 s.* Therefore, the equation for induced emf can be written

$$e_L = \frac{\Delta\Phi}{\Delta t} \tag{14-1}$$

Here, e_L is in volts, $\Delta\Phi$ is in Wb, and Δt is in seconds. Equation (14-1) also originates from *Faraday's† law*.

Faraday's Law **The EMF induced in an electric circuit is proportional to the rate of change of flux linking the circuit.**

If the total field flux in Figure 14-1 were 1 Wb, and if the conductor were moved through the field in exactly 1 s, the emf measured at the conductor terminals would be exactly 1 V. If the conductor were to be moved through the field in 0.5 s, the voltage generated would be

$$e_L = \frac{1 \text{ Wb}}{0.5 \text{ s}}$$

$$= 2 \text{ V}$$

---

* Formulated by the Russian physicist Heinrich Lenz (1804–1865).

† Formulated by the English chemist and physicist Michael Faraday (1791–1867).

Similarly, if the flux were doubled, the generated voltage would be doubled.

Now consider Figure 14-2, which shows two coils on a ring-shaped iron core. Flux generated by a current flowing in the left-hand coil will pass through the iron core and link with the right-hand coil. Because the left-hand coil is the *input* coil and the right-hand coil is the *output*, the coils are identified as *primary* and *secondary*, respectively. While the current in the primary is constant, the core flux will not change and no emf will be induced in the secondary winding. When the primary current is increased or decreased (by adjustment of R), the core flux grows or declines, and in doing so it induces an emf in the secondary winding.

If N is the number of turns on the secondary winding, the induced emf is

$$e_L = \frac{\Delta \Phi \, N}{\Delta t}$$ (14-2)

EXAMPLE 14-1 In the circuit shown in Figure 14-2, the core flux is increased from zero to 0.05 Wb in a time of 4 s. Calculate the number of secondary turns required if the induced voltage is to be 1.5 V.

SOLUTION

Equation (14-2):

$$e_L = \frac{\Delta \Phi \, N}{\Delta t}$$

Therefore,

$$N = \frac{e_L \, \Delta t}{\Delta \Phi} = \frac{1.5 \text{ V} \times 4 \text{ s}}{0.05 \text{ Wb}}$$

$$= \textbf{120 turns}$$

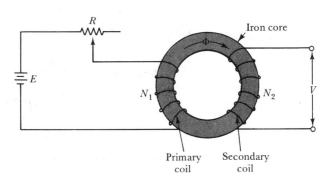

FIGURE 14-2. An iron core can be used to channel magnetic flux from one coil into another. An emf is induced in the secondary coil.

EXAMPLE 14-2 The dimensions of the magnetic core shown in Figure 14-2 are: cross-sectional area $A = 3$ cm^2, magnetic path length $l = 10$ cm, and the relative permeability is 250. The primary coil has $N_p = 100$ turns and the secondary coil has $N_s = 75$ turns. If the current is increased from zero to 5 A in 0.1 s, determine the emf induced in the secondary.

SOLUTION

Equation (11-2):

$$magnetomotive\ force,\ F_m = I \times N_p$$
$$= 5\ A \times 100$$
$$= 500\ A$$

Equations (11-2) and (11-3):

$$magnetic\ field\ strength,\ H = \frac{F_m}{l}$$
$$= \frac{500\ A}{10 \times 10^{-2}\ m}$$
$$= 5000\ A/m$$

Equation (12-5):

$$flux\ density,\ B = \mu_r\,\mu_0\,H$$
$$= 250 \times 4\pi \times 10^{-7} \times 5000\ A/m$$
$$= 1.57\ T$$

Equation (11-1):

$$total\ flux,\ \Phi = B \times A$$
$$= 1.57\ T \times 3 \times 10^{-4}\ m^2$$
$$= 471\ \mu Wb$$

Equation (14-2):

$$induced\ emf,\ e_L = \frac{\Delta\Phi\ N_s}{\Delta t}$$
$$= \frac{471\ \mu Wb \times 75}{0.1\ s}$$
$$\cong \mathbf{0.35\ V}$$

In Figure 14-3 an air-cored solenoid is shown, around which another (secondary) coil has been wound. When the switch is closed, the current through the primary coil causes the magnetic flux to grow from zero. All the flux from the solenoid cuts the secondary winding and induces an emf in the secondary coil. Once again, Equation (14-2) can be used to calculate the induced emf.

EXAMPLE 14-3 The solenoid shown in Figure 14-3 has 1000 turns, is 30 cm long, and has a cross sectional area of 5 cm². The secondary winding around it has 5 turns. Determine the emf induced in the secondary coil when the current through the solenoid is increased from zero to 20 A in a time period of 10 ms.

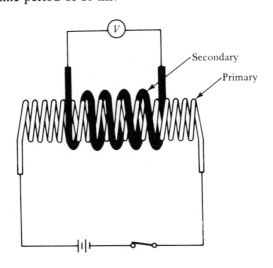

FIGURE 14-3. A secondary winding on top of an air-cored solenoid has an emf induced in it when a current increases or decreases in the solenoid.

SOLUTION

Equation (11-2):

$$\textit{magnetomotive force, } F_m = N_p I_p$$
$$= 1000 \times 20 \text{ A}$$
$$= 20\ 000 \text{ A}$$

Equations (11-2) and (11-3):

$$\textit{magnetic field strength, } H = \frac{F_m}{l}$$
$$= \frac{20\ 000 \text{ A}}{30 \times 10^{-2} \text{ m}}$$
$$\cong 6.67 \times 10^4 \text{ A/m}$$

Equation (12-4):

$$\textit{flux density, } B = H \times \mu_0$$

$$\cong 6.67 \times 10^4 \text{ A} \times 4\pi \times 10^{-7}$$

$$\cong 8.38 \times 10^{-2} \text{ T}$$

Equation (11-1):

$$\textit{total flux, } \Phi = B \times A$$

$$\cong 8.38 \times 10^{-2} \text{ T} \times 5 \times 10^{-4} \text{ m}^2$$

$$\cong 41.9 \ \mu\text{Wb}$$

Equation (14-2):

$$\textit{induced emf, } e_L = \frac{\Delta\Phi \ N_s}{\Delta t}$$

$$\cong \frac{41.9 \times 10^{-6} \text{ Wb} \times 5}{10 \times 10^{-3} \text{ s}}$$

$$\mathbf{e_L \cong 21 \text{ mV}}$$

PRACTICE PROBLEMS

14-2.1 Calculate the voltage induced in a 250 turn winding when the magnetic flux in its core changes from 0.3 μWb to 0.9 μWb in a time period of 4 ms.

14-2.2 A magnetic core with the same dimensions and number of coil turns as in Example 14-2 produces a secondary emf of 420 mV when the primary current is increased from zero to 1 A in 50 ms. Determine the relative permeability of the core.

14-2.3 A solenoid 20 cm long with an internal diameter of 2 cm has a 100 turn secondary winding. A secondary voltage of 33 mV is measured when the primary current changes by 500 mA in 2.5 ms. Determine the number of primary turns.

14-3

SELF-INDUCTANCE

COIL AND CONDUCTOR INDUCTANCE. It has been shown that a conductor moving through a magnetic field has an emf induced in it, and that the growth of current in a coil can induce an emf in another coil to which it is magnetically coupled. It is also possible for a coil to induce a voltage in itself as the current through it grows. This phenomenon is known as *self-inductance*, and the principle is illustrated in Figure 14-4.

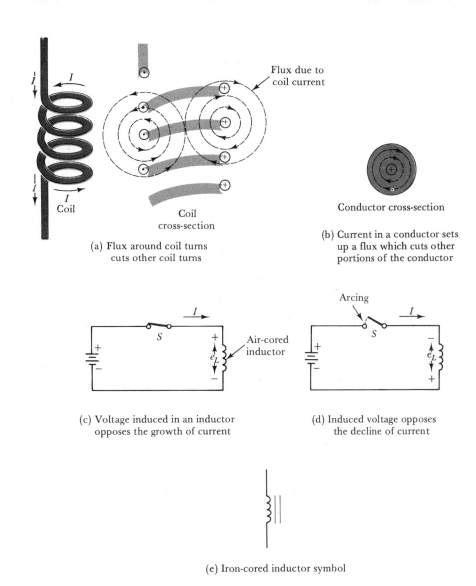

(a) Flux around coil turns cuts other coil turns

Flux due to coil current

Conductor cross-section

(b) Current in a conductor sets up a flux which cuts other portions of the conductor

(c) Voltage induced in an inductor opposes the growth of current

Air-cored inductor

(d) Induced voltage opposes the decline of current

Arcing

(e) Iron-cored inductor symbol

FIGURE 14-4. Magnetic flux growing outwards around the turns of a coil cuts (or brushes over) the other coil turns, and induces a emf in the coil. This *counter-emf* opposes the growth or decline of current in the coil. A similar effect occurs within a conductor.

A coil and its cross section area are shown in Figure 14-4(a), with arrow *tails* and *points* indicating the current directions in each turn. Every turn of the coil has a flux around it due to the current flowing through the coil. However, for convenience the illustration shows the growth of flux around only one turn on the coil. It is seen that as the current grows, the flux expands outward and cuts the other turns. This causes currents to be induced in the other turns, and the direction of the induced currents are such that they set up a flux which opposes the flux inducing them. Remembering that the current through the coil causes the flux to grow around all turns at once, it is seen that the flux from every turn induces a current that opposes it in every other turn.

To set up opposing fluxes, the induced current in a coil must be in opposition to the current flowing through the coil from the external source of supply. The induced current is, of course, the result of an induced emf. Thus, it is seen that the self-inductance of a coil sets up an induced emf which opposes the external emf that is driving current through the coil. Because this induced emf is in opposition to the supply voltage, it is usually termed the *counter-emf*, or *back-emf*. The counter-emf occurs only when the coil current is growing or declining. When the current has reached a constant level, the flux is no longer changing and no counter-emf is generated.

Even a single conductor has self-inductance. Figure 14-4(b) shows that when current is growing in a conductor, flux may grow outward from the center of the conductor. This flux cuts other portions of the conductor and induces a counter-emf.

In Figure 14-4(c) and (d), the polarity of the counter-emf induced in a coil is illustrated for a given supply voltage polarity. In Figure 14-4(c), the switch is closed and current I commences to grow from zero. The polarity of the counter-emf (e_L) is such that it opposes the growth of I, thus it is *series-opposing* with the supply voltage. When the switch is opened [Figure 14-4(d)], the current tends to fall to zero. But now the polarity of e_L is such that it opposes the decline of I. It is *series-aiding* with the supply voltage. In fact, e_L may cause arcing at the switch terminals as it attempts to maintain the flow of current.

The amplitude of the counter-emf induced in a coil by a given rate of change in current depends upon the coil's inductance.

*The SI unit of inductance is the **henry*** (H).*

*The inductance of a circuit is **one henry** (1 H), when an emf of 1 V is induced by the current changing at the rate of 1 A/s.*

Thus, the relationship among inductance, induced voltage, and rate of change of current is

$$L = \frac{e_L}{\Delta i / \Delta t}$$

(14-3)

where L is the inductance in henrys, e_L is the induced counter-emf in volts, and ($\Delta i / \Delta t$) is the rate of change of current in amperes/second. A negative sign is sometimes included in front of e_L to show that the induced emf

---

* Named for the American physicist Joseph Henry (1797–1878).

is in opposition to the applied emf. When $e_L = 1$ V and $\Delta i/\Delta t = 1$ A/s, $L = 1$ H. If the rate of change of current is 2 A/s and $e_L = 1$ V, the inductance is 0.5 H.

A coil that is constructed to have a certain inductance is usually referred to as an *inductor* or a *choke*. Note the graphic symbols for an inductor shown in Figure 14-4(c), (d), and (e).

EXAMPLE 14-4 The current in a coil grows linearly from 0 to 10 A in a time of 0.25 s. If the coil has an inductance of 0.75 H, calculate the induced counter-emf.

SOLUTION

From Equation (14-3):

$$e_L = L \left(\frac{\Delta i}{\Delta t} \right)$$

$$= 0.75 \text{ H} \times \left(\frac{10 \text{ A} - 0 \text{ A}}{0.25 \text{ s}} \right)$$

$$= \mathbf{30 \ V}$$

CALCULATION OF SELF-INDUCTANCE. An expression for inductance can also be derived involving the coil dimensions and the number of turns. From Equation (14-2),

$$e_L = \frac{\Delta \Phi \ N}{\Delta t}$$

Substituting for e_L into Equation (14-3) gives

$$L = \frac{\Delta \Phi \ N/\Delta t}{\Delta i/\Delta t}$$

or

$$\boxed{L = \frac{\Delta \Phi \ N}{\Delta i}} \qquad \text{(14-4)}$$

Also, $\qquad\qquad \Phi = B \times A$

and $\qquad\qquad B = \mu_r \mu_0 H$

$$= \mu_r \mu_0 \frac{IN}{l}$$

Therefore, $\qquad \Phi = \mu_r \mu_0 IN \frac{A}{l}$

Since I is a maximum current level, it also represents the change in current (Δi) from zero to the maximum level. Therefore, change in flux is

$$\Delta \Phi = \mu_r \, \mu_0 \, \Delta i \, N \frac{A}{l}$$ (14-5)

Substituting for $\Delta \Phi$ in Equation (14-4) gives

$$L = \frac{\left(\mu_r \, \mu_0 \, \Delta i \, N \frac{A}{l} \right) \times N}{\Delta i}$$

or

$$L = \mu_r \, \mu_0 \, N^2 \frac{A}{l}$$ (14-6)

Note that the inductance is proportional to the cross-sectional area of a coil and to the square of the number of turns. It is also inversely proportional to the coil length. Therefore, maximum inductance is obtained with a short coil with a large cross-sectional area and a large number of turns.

Equation (14-6) now affords a means of calculating the inductance of a coil of known dimensions. Alternatively, it can be used to determine the required dimensions for a coil that is to have a certain inductance. However, it is not so easily applied to iron-cored coils, because the permeability of ferromagnetic material changes when the flux density changes (see Figure 12-3). Consequently, the inductance of an iron-cored coil is constantly changing as the coil current increases and decreases.

EXAMPLE 14-5   A solenoid with 900 turns has a total flux of 1.33×10^{-7} Wb through its air core when the coil current is 100 mA. If the flux takes 75 ms to grow from zero to its maximum level, calculate the inductance of the coil. Also, determine the counter-emf induced in the coil during the flux growth.

SOLUTION

$$\Delta \Phi = 1.33 \times 10^{-7} \text{ Wb}$$

$$\Delta i = 100 \text{ mA}$$

$$\Delta t = 75 \text{ ms}$$

Equation (14-4):

$$L = \frac{\Delta\Phi \, N}{\Delta i}$$

$$= \frac{1.33 \times 10^{-7} \text{ Wb} \times 900}{100 \text{ mA}}$$

$$\cong 1.2 \text{ mH}$$

From Equation (14-2):

$$e_L = \frac{\Delta\Phi \, N}{\Delta t}$$

$$= \frac{1.33 \times 10^{-7} \text{ Wb} \times 900}{75 \text{ ms}}$$

$$\cong 1.6 \text{ mV}$$

EXAMPLE 14-6   The air-cored solenoid described in Example 14-5 has $l = 15$ cm and inside diameter $D = 1.5$ cm. Recalculate its inductance using the coil dimensions.

SOLUTION

$$\text{cross-sectional area, } A = \pi \left(\frac{D}{2}\right)^2$$

$$= \pi \times \left(\frac{1.5 \times 10^{-2} \text{ m}}{2}\right)^2$$

$$\cong 1.77 \times 10^{-4} \text{ m}^2$$

Equation (14-6):

$$L = \mu_r \, \mu_0 \, N^2 \frac{A}{l}$$

$$\cong 1 \times 4\pi \times 10^{-7} \times 900^2 \times \frac{1.77 \times 10^{-4} \text{ m}^2}{15 \times 10^{-2} \text{ m}}$$

$$\cong 1.2 \text{ mH}$$

This checks with the answer to Example 14-5. *Note that the solenoid dimensions employed are those used in* Example 12-3.

EXAMPLE 14-7   An air-cored coil is to be 2.5 cm long and to have an average cross-sectional area of 2 cm^2. Determine the number of turns required if the coil is to have an inductance of 100 μH.

SOLUTION

From Equation (14-6):

$$N^2 = \frac{Ll}{\mu_r \mu_0 A}$$

Therefore, $$N = \sqrt{\frac{Ll}{\mu_r \mu_0 A}}$$

$$= \sqrt{\frac{100 \times 10^{-6} \text{ H} \times 2.5 \times 10^{-2} \text{ m}}{4\pi \times 10^{-7} \times 2 \times 10^{-4} \text{ m}^2}}$$

$$\cong \textbf{100 turns}$$

EXAMPLE 14-8   A cast steel ring has a cross-sectional area of 4 cm^2 and a magnetic path length of 15 cm. A 300 turn coil is wound on the ring. Determine the coil inductance when the current change is: *a.* 2 A, *b.* 0.5 A.

SOLUTION

a. For I = 2 A,

$$H = \frac{NI}{l} = \frac{300 \times 2 \text{ A}}{15 \times 10^{-2} \text{ m}}$$

$$H_1 = 4000 \text{ A/m}$$

From Figure 12-3, at *H* = 4000 A/m, *B* ≅ 1.57 T *for cast steel,*

$$\mu_r \mu_0 = \frac{B}{H} \cong \frac{1.57 \text{ T}}{4000 \text{ A/m}}$$

$$\mu_r \mu_0 \cong 3.93 \times 10^{-4}$$

Equation (14-6):

$$L = \mu_r \mu_0 N^2 \frac{A}{l}$$

$$\cong \frac{3.93 \times 10^{-4} \times 300^2 \times 4 \times 10^{-4} \text{ m}^2}{15 \times 10^{-2} \text{ m}}$$

$$L_1 \cong \textbf{94.3 mH}$$

b. For $I = 0.5$ A,

$$H = \frac{300 \times 0.5 \text{ A}}{15 \times 10^{-2} \text{ m}}$$

$$H_2 = 1000 \text{ A/m}$$

From Figure 12-3, at $H = 1000$ A/m, $B \cong 1.05$ T *for cast steel,*

$$\mu_r \, \mu_0 \cong \frac{1.05 \text{ T}}{1000 \text{ A/m}}$$

$$\cong 1.05 \times 10^{-3}$$

Equation (14-6):

$$L = \mu_r \, \mu_0 \, N^2 \frac{A}{l}$$

$$\cong \frac{1.05 \times 10^{-3} \times 300^2 \times 4 \times 10^{-4} \text{ m}^2}{15 \times 10^{-2} \text{ m}}$$

$$L_2 \cong 252 \text{ mH}$$

INCREMENTAL INDUCTANCE. The steel-cored coil in Example 14-8 actually has *less inductance* for a current change of 2 A than for a current change of 0.5 A. This is because the 2 A current tends to drive the core into magnetic saturation, while with 0.5 A the core is operating on the steepest (near-linear) portion of the B/H curve. Therefore, when inductance is specified for a coil with a ferromagnetic core, the coil current should also be specified. Recall from Section 12-7 that the *incremental permeability* of ferromagnetic material varies along the length of the B/H curve. When the incremental permeability is used, the calculated inductance is known as the *incremental inductance.* This quantity applies in the case of an inductor in which the current level fluctuates while remaining substantially constant. As explained for incremental permeability in Section 12-7, a constant level of *bias* current is frequently employed to bias an iron core to the center of the steepest portion of its B/H characteristic. This gives the largest possible inductance value.

NONINDUCTIVE COIL. In many cases it is desired to have a noninductive coil; for example, precision resistors are usually noninductive. To construct such a coil, the coil winding is made double, as illustrated in Figure 14-5. Every coil turn has an adjacent turn which is carrying current in the opposite direction. The magnetic fields generated by adjacent turns cancel each other out. Therefore, no counter-emf is generated, and the coil is noninductive.

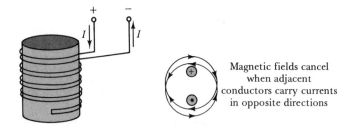

Magnetic fields cancel
when adjacent
conductors carry currents
in opposite directions

Noninductive coil with adjacent
conductors carrying currents
in opposite directions

FIGURE 14-5. A noninductive coil is constructed by winding a double set of coil turns and ensuring that adjacent turns carry currents in opposite directions.

PRACTICE PROBLEMS

14-3.1 Calculate the inductance of a 600 turn air-cored coil 5 cm long and 2.5 cm in diameter. Determine the counter-emf generated when the coil current collapses from 100 mA to zero in 5 ms.

14-3.2 Determine the number of turns required to create a 1.5 mH inductor if the dimensions of the coil are to be 4 cm long and 1.5 cm in diameter.

14-3.3 A ring made of sheet steel has a magnetic path length of 6.9 cm, a cross-sectional area of 2.7 cm², and a coil with 160 turns. Determine the coil inductance when the current level is 300 mA.

14-4
MUTUAL INDUCTANCE

When the flux from one coil cuts another adjacent (or magnetically coupled) coil, an emf is induced in the second coil. This was shown in Section 14-2. Following Lenz's law, the emf induced in the second coil sets up a flux that opposes the original flux from the first coil. Thus, the induced emf is again a counter-emf, and in this case the inductive effect is referred to as *mutual inductance*. Figure 14-6 shows the graphic symbols used for coils with mutual inductance, also termed *coupled coils*.

Like self-inductance, mutual inductance is meaured in henrys (H):

Two coils have a mutual inductance of 1 H when an emf of 1 V is induced in one coil by current changing at the rate of 1 A/s in the other coil.

This definition gives rise to the equation relating mutual inductance to induced voltage and rate of change of current:

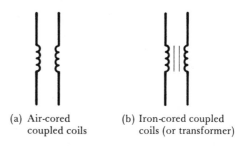

(a) Air-cored
coupled coils

(b) Iron-cored coupled
coils (or transformer)

FIGURE 14-6.   Circuit symbols for coupled coils.

$$M = \frac{e_L}{\Delta i / \Delta t} \qquad (14\text{-}7)$$

where M is the mutual inductance in henrys, e_L is the emf in volts induced in the secondary coil, and $(\Delta i / \Delta t)$ is the rate of change of current in the primary coil in amperes/second.

Recall from Section 14-2 that the coil through which a current is passed from an external source is termed the *primary,* and the coil that has an emf induced in it is referred to as the *secondary.*

An equation for the emf induced in the second coil can be written from Equation (14-2):

$$e_L = \frac{\Delta \Phi \, N_s}{\Delta t} \qquad (14\text{-}8)$$

Here $\Delta \Phi$ is the total change in flux linking with the secondary winding, N_s is the number of turns on the secondary winding, and Δt is the time required for the flux change.

Substituting for e_L from Equation (14-8) into Equation (14-7) gives

$$M = \frac{\Delta \Phi \, N_s / \Delta t}{\Delta i / \Delta t}$$

Therefore,

$$M = \frac{\Delta \Phi \, N_s}{\Delta i} \qquad (14\text{-}9)$$

Figure 14-7(a) illustrates the fact that when the two coils are wound on a single ferromagnetic core, effectively all of the flux generated by the primary coil links with the secondary coil. However, when the coils are air-cored, only a portion of the flux from the primary may link with the secondary [see Figure 14-7(b)]. Depending on how much of the primary flux cuts the secondary, the coils may be classified as *loosely coupled* or *tightly coupled.* One way to ensure tight coupling is shown in Figure 14-7(c), where each turn of the secondary winding is side by side with one turn of the primary winding. Coils wound in this fashion are said to be *bifilar.*

The amount of flux linkage from primary to secondary is also defined in terms of a *coefficient of coupling, k.* If all the primary flux links with the secondary, the coefficient of coupling is 1. When only 50 percent of the primary flux links with the secondary coil, the coefficient of coupling is 0.5. Thus,

$$k = \frac{\text{flux linkages between primary and secondary}}{\text{total flux produced by primary}}$$

Returning to Equation (14-9), when $\Delta\Phi$ is the total flux change in the primary coil, the flux linking with the secondary is $k\Delta\Phi$. Therefore, the equation for M becomes

$$M = k\,\frac{\Delta\Phi N_s}{\Delta i} \qquad \textbf{(14-10)}$$

Also, substituting for $\Delta\Phi$ from Equation (14-5) into Equation (14-10) gives

$$M = \frac{kN_s}{\Delta i}\,\mu_r\,\mu_0\,\Delta i\,N_p\,\frac{A}{l}$$

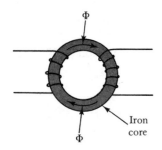

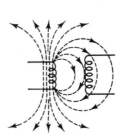

(a) Showing that with an iron core virtually all of the flux passes from one coil to the other

(b) Depending upon how closely air-cored coils are coupled, only a portion of the flux from the primary links with the secondary

(c) Bifilar winding for close coupling

FIGURE 14-7. The amount of flux from a primary winding that links with a secondary depends upon how closely the coils are coupled. The *coefficient of coupling* defines the linkage.

Chap. 14   Inductance

or

$$M = k\, N_p N_s\, \mu_r\, \mu_0\, \frac{A}{l} \qquad \text{(14-11)}$$

Each winding considered alone has a self-inductance that can be calculated from Equation (14-6). Thus, for the primary coil

$$L_1 = \mu_r\, \mu_0\, N_p^2\, \frac{A}{l}$$

and for the secondary

$$L_2 = \mu_r\, \mu_0\, N_s^2\, \frac{A}{l}$$

Assuming that the two windings share a common core (magnetic or nonmagnetic), the only difference in the expressions for L_1 and L_2 are the numbers of the coil turns. Therefore,

$$L_1 \times L_2 = N_p^2\, N_s^2 \left(\mu_r\, \mu_0\, \frac{A}{l} \right)^2$$

or

$$\sqrt{L_1 L_2} = N_p N_s\, \mu_r\, \mu_0\, \frac{A}{l} \qquad \text{(14-12)}$$

Comparing Equations (14-11) and (14-12), it is seen that

$$M = k\, \sqrt{L_1 L_2} \qquad \text{(14-13)}$$

EXAMPLE 14-9 Two identical coils are wound on a ring-shaped iron core that has a relative permeability of 500. Each coil has 100 turns, and the core dimensions are: cross-sectional area $A = 3$ cm^2 and magnetic path length $l = 20$ cm. Calculate the inductance of each coil and the mutual inductance between the coils.

SOLUTION

From Equation (14-6):

$$L_1 = L_2 = \mu_r\, \mu_0\, N^2\, \frac{A}{l}$$

$$= 500 \times 4\pi \times 10^{-7} \times 100^2 \times \frac{3 \times 10^{-4}\ \text{m}^2}{20 \times 10^{-2}\ \text{m}}$$

$$L \cong 9.42\ \text{mH}$$

Since the coils are wound on the same iron core, k = 1.

Equation (14-13): $M = k \sqrt{L_1 L_2}$

$$= \sqrt{9.42 \text{ mH} \times 9.42 \text{ mH}}$$

$$\cong \mathbf{9.42 \text{ mH}}$$

EXAMPLE 14-10   Two 100 turn end-to-end solenoids each have $l = 20$ cm and cross-sectional area = 3 cm². Calculate their coefficient of coupling when the mutual inductance between them is measured as 0.62 μH.

SOLUTION

$$L_1 = L_2 = \mu_r \, \mu_0 \, N^2 \frac{A}{l}$$

$$= 1 \times 4\pi \times 10^{-7} \times 100^2 \times \frac{3 \times 10^{-4} \text{ m}^2}{20 \times 10^{-2} \text{ m}}$$

$$\cong 18.8 \text{ μH}$$

$$M = k \sqrt{L_1 L_2}$$

Therefore, $$k = \frac{M}{\sqrt{L_1 L_2}}$$

$$\cong \frac{0.62 \text{ μH}}{\sqrt{(18.8 \text{ μH})^2}}$$

$$\cong \mathbf{0.033}$$

PRACTICE PROBLEMS

14-4.1   An iron ring with a relative permeability of 600 has a magnetic path length of 4 cm, a cross sectional area of 1 cm², an 80 turn coil, and a 190 turn coil. Calculate the mutual inductance between the coils.

14-4.2   A 50 turn secondary coil is wound directly on top of a 600 turn solenoid which is 10 cm long and 1.25 cm in diameter. The secondary is 10 cm long and 1.8 cm in diameter. Calculate the mutual inductance between the coils.

14-5

TYPES OF INDUCTORS

There are many different types of inductors, ranging from large high-current iron-cored chokes to tiny resistor-style low-current coils.

Chap. 14   Inductance

Automobile Ignition Coil

The automobile ignition coil consists of primary and secondary turns wound on an iron core. The supply voltage from the automobile battery to the primary, is rapidly switched *on* and *off* by a make-and-break switch (or by means of an electronic switch). The flux from the primary cuts the secondary, and an emf is induced in the secondary, first in one direction and then in the other. By selecting a suitable number of turns on each coil, a large output voltage can be derived from a 12 V battery.

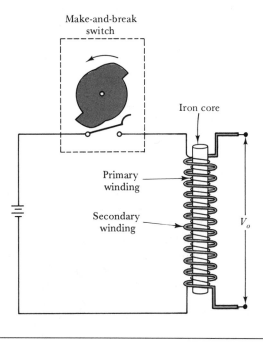

POWER SUPPLY INDUCTOR. Figure 14-8(a) shows the circuit of an inductor used with a dc power supply, and Figure 14-8(b) shows the appearance of a typical power supply inductor. In this application, the inductor is usually required to pass a direct current that has a fluctuating level. Since the inductor opposes any change in the current level through its windings, it tends to smooth out the fluctuations (see the waveforms illustrated). This is exactly why the inductor is employed in this particular application. Because of the presence of the direct current through the windings, it is the *incremental inductance* of this component that is important. Therefore, the inductance value must be specified at a given level of direct current. Typical values for such inductors range from 50 mH to 20 H, with direct currents up to about 10 A and insulation voltage ratings up to 1000 V.

HIGH-FREQUENCY INDUCTOR. In Figure 14-9, a low-current high-frequency type of inductor is illustrated. The core in this case is ferrite

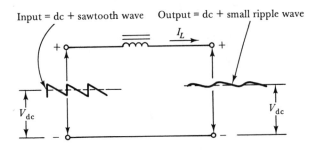

Input = dc + sawtooth wave   Output = dc + small ripple wave

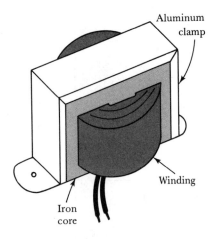

Aluminum clamp

Winding

Iron core

(a) Use of an inductor to smooth out an unwanted sawtooth waveform

(b) Typical power supply inductor

FIGURE 14-8. An inductor can be used with a dc power supply to smooth the input ripple voltage without affecting the dc voltage level.

material (see Section 12-8) in two mating sections known as a *pot core*. As well as increasing the coil's inductance, the pot core screens the coil to protect adjacent components against flux leakage and to protect the coil from external magnetic fields. The coil is wound on a bobbin, so its number of turns is easily modified.

Three different types of low-current inductors are illustrated in Figure 14-10. Figure 14-10(a) shows a type that is available either as an air-cored inductor or with a ferromagnetic core. With an air core, the inductance values range typically from 2.4 μH to 100 μH. With a ferromagnetic core, inductance values up to about 10 mH can be obtained. Depending on the thickness of wire used and the physical size of

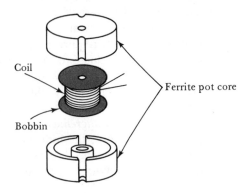

Coil

Ferrite pot core

Bobbin

FIGURE 14-9. Some low-current high-frequency inductors are wound on bobbins contained in a ferrite pot core. The ferrite core increases the winding inductance and screens the inductor.

(a) Inductor with air core or ferromagnetic core

(b) Circuit symbol for an inductor with an adjustable ferromagnetic core

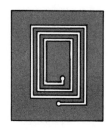

(c) Molded inductor

(d) Thin-film inductor

FIGURE 14-10. Small inductors may be wound on an insulating tube with an adjustable ferrite core, moulded like small resistors, or deposited as a conducting film on an insulating material.

the inductor, the maximum current can be anything from about 50 mA to 1 A. The core in such an inductor may be made adjustable so that it can be screwed into or partially out of the coil. Thus, the coil inductance is variable. Note the graphic symbol for an inductor with an adjustable core [Figure 14-10(b)].

MOLDED INDUCTORS. A *small molded inductor* is shown in Figure 14-10(c). Typical available values for this type range from 1.2 μH to 10 mH, with maximum currents of about 70 mA. The values of molded inductors are identified by a color code, similar to molded resistors. Figure 14-10(d) shows a tiny *thin-film* inductor used in certain types of electronic circuits. In this case the inductor is simply a thin metal film deposited in the form of a spiral on a ceramic base.

LABORATORY INDUCTORS. Laboratory-type variable inductors can be constructed in *decade box* format [see Figure 4-8(c)], in which precision inductors are switched into or out of a circuit by means of rotary switches. Alternatively, two coupled coils can be employed as a variable inductor. The coils may be connected in series or in parallel, and the total inductance is controlled by adjusting the position of one coil relative to the other.

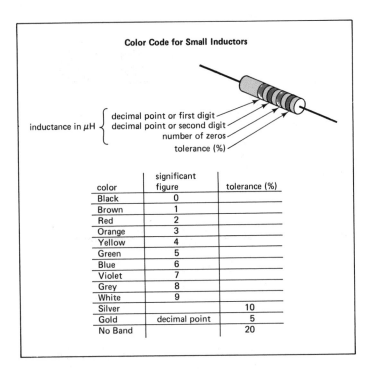

Color Code for Small Inductors

inductance in μH { decimal point or first digit
decimal point or second digit
number of zeros
tolerance (%)

color	significant figure	tolerance (%)
Black	0	
Brown	1	
Red	2	
Orange	3	
Yellow	4	
Green	5	
Blue	6	
Violet	7	
Grey	8	
White	9	
Silver		10
Gold	decimal point	5
No Band		20

14-6
ENERGY STORED IN AN INDUCTIVE CURRENT

From Equation (14-3), the emf induced in an inductance is

$$e_L = L \left(\frac{\Delta i}{\Delta t} \right)$$

If the current is controlled so that it grows linearly from zero to I in a time t, the average current is $0.5I$, the time change is $\Delta t = t$, and the current change is $\Delta i = I$. Thus the induced voltage e_L is a constant quantity.

$$e_L = L \left(\frac{I}{t} \right)$$

From Equations (3-3) and (3-7), the electrical energy input is

$$W = (\text{average voltage}) \times (\text{average current}) \times t$$

Therefore, the energy supplied to the inductive circuit is

$$W = e_L \times \tfrac{1}{2}I \times t$$

$$= L \left(\frac{I}{t} \right) \times \tfrac{1}{2}I \times t$$

Chap. 14   Inductance

or energy stored is

$$W = \tfrac{1}{2}LI^2$$ (14-14)

When L is in henrys and I is in amperes, W is given in joules.
From Equation (14-6),

$$L = \mu_r \mu_0 N^2 \frac{A}{l}$$

Substituting for L in Equation (14-14) gives

$$W = \tfrac{1}{2}\mu_r \mu_0 N^2 \frac{A}{l} I^2$$

This can be rewritten

$$W = \tfrac{1}{2}\mu_r \mu_0 A \left(\frac{N^2 I^2}{l^2}\right) l$$

$$= \tfrac{1}{2}\mu_r \mu_0 A H^2 l$$

and $$H = \frac{B}{\mu_r \mu_0}$$

Therefore, $$W = \tfrac{1}{2}\mu_r \mu_0 A l \left(\frac{B}{\mu_r \mu_0}\right)^2$$

or

$$W = \frac{B^2 A l}{2\mu_r \mu_0} \quad \text{joules}$$ (14-15)

When μ_r is replaced by 1 for an air core, this equation is exactly the same as Equation (12-6), which represents the energy stored in an air gap between two magnetic surfaces.

EXAMPLE 14-11　A 20 H choke with a resistance of 180 Ω has a 300 V supply. Calculate the energy stored.

SOLUTION

The steady-state current is

$$I = \frac{E}{R} = \frac{300 \text{ V}}{180 \text{ }\Omega}$$

$$\cong 1.67 \text{ A}$$

Equation (14-14):

$$W = \tfrac{1}{2}LI^2$$

$$\cong \tfrac{1}{2} \times 20 \text{ H} \times (1.67 \text{ A})^2$$

$$\cong 27.9 \text{ J}$$

PRACTICE PROBLEMS

14-6.1 Calculate the energy stored in the solenoid in Problem 14-3.1 when a constant current of 100 mA flows in the coil.

14-6.2 Determine the energy stored in the magnetic core in Example 12-6.

14-7
INDUCTORS IN SERIES AND IN PARALLEL

SERIES-CONNECTED INDUCTORS. When inductors are connected in series or parallel and *there is no mutual induction between them*, they can be treated exactly like resistors. For series-connected inductors, the same current flows through each, and the counter-emf generated in each is proportional to the rate of change of current. Therefore, as illustrated in Figure 14-11(a), the total counter-emf is

$$e_{L(\text{total})} = e_{L1} + e_{L2} + e_{L3} + \cdots$$

and since $\quad e_L = L(\Delta i/\Delta t),$

$$e_{L(\text{total})} = L_1\left(\frac{\Delta i}{\Delta t}\right) + L_2\left(\frac{\Delta i}{\Delta t}\right) + L_3\left(\frac{\Delta i}{\Delta t}\right) + \cdots$$

$$= \left(\frac{\Delta i}{\Delta t}\right) L_s, \text{ where } L_s \text{ is the total inductance.}$$

Therefore for series-connected inductors the total inductance is

$$\boxed{L_s = L_1 + L_2 + L_3 + \cdots} \qquad (14\text{-}16)$$

Note that Equation (14-16) is correct only when there is no mutual induction between the inductors.

PARALLEL-CONNECTED INDUCTORS. For parallel-connected inductors, the same voltage appears across each; consequently, the counter-emfs generated in each must be equal.

$$e_L = e_{L1} = e_{L2} = e_{L3} = \cdots$$

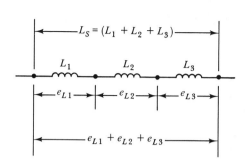

(a) Inductors in series

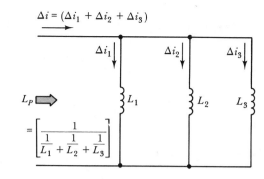

(b) Inductors in parallel

FIGURE 14-11. Inductors in series and in parallel behave similarly to resistors in series or parallel. Total series inductance is $L_S = L_1 + L_2 + L_3 + \cdots$. Total parallel inductance is $L_P = 1/(1/L_1 + 1/L_2 + 1/L_3 + \cdots)$.

As illustrated in Figure 14-11(b), the total current change is

$$\Delta i_{\text{total}} = \Delta i_1 + \Delta i_2 + \Delta i_3 + \cdots$$

and

$$e_L = L \frac{\Delta i}{\Delta t}$$

Therefore,

$$\Delta i = \frac{e_L \, \Delta t}{L}$$

and

$$\Delta i_{\text{total}} = \left(\frac{e_L \, \Delta t}{L_1} \right) + \left(\frac{e_L \, \Delta t}{L_2} \right) + \left(\frac{e_L \, \Delta t}{L_3} \right) + \cdots$$

or

$$\frac{\Delta i_{\text{total}}}{e_L \, \Delta t} = \frac{1}{L_1} + \frac{1}{L_2} + \frac{1}{L_3} + \cdots$$

$$= \frac{1}{L_p}, \text{ where } L_p \text{ is the total inductance.}$$

Therefore, for parallel-connected inductors the total inductance is determined from

$$\boxed{\frac{1}{L_p} = \frac{1}{L_1} + \frac{1}{L_2} + \frac{1}{L_3} + \cdots} \qquad \textbf{(14-17)}$$

Note that Equation (14-17) is correct only when there is no mutual induction between the inductors.

EXAMPLE 14-12   Three inductors have values of $L_1 = 10$ mH, $L_2 = 100$ μH, and $L_3 = 500$ μH. Determine the total inductance of the three:
a. When connected in series.
b. When connected in parallel.
Assume that there is no mutual induction between the inductors.

SOLUTION

a. *In series:*

$$L_s = L_1 + L_2 + L_3$$

$$= 10 \text{ mH} + 100 \text{ μH} + 500 \text{ μH}$$

$$= \textbf{10.6 mH}$$

b. *In parallel:*

$$L_p = \cfrac{1}{\cfrac{1}{L_1} + \cfrac{1}{L_2} + \cfrac{1}{L_3}}$$

$$= \cfrac{1}{\cfrac{1}{10 \text{ mH}} + \cfrac{1}{100 \text{ μH}} + \cfrac{1}{500 \text{ μH}}}$$

$$\cong \textbf{82.6 μH}$$

INDUCTORS WITH MUTUAL INDUCTANCE. Figure 14-12 illustrates the situation when mutual induction exists between two series-connected coils. In Figure 14-12(a) the coil connection is such that their fluxes assist each other, and the coils are said to be connected *series-aiding*. Figure 14-12(b) shows the graphic symbols used in circuit diagrams for mutually coupled coils which are connected series-aiding. The dots at the same ends of each coil indicate that the coils have fluxes in the same direction when a current flows.

The two coils shown in Figure 14-12(c) have their fluxes in opposite directions, and therefore they are said to be connected *series-opposing*. In Figure 14-12(d) the graphic symbols for series-opposing coils are illustrated. In this case the dots at opposite ends of each coil indicate that their fluxes are in opposition when a current flows.

For *series-aiding connections,* the total emf induced in L_1 is the sum of the emf due to the coil's self-inductance and the emf induced from L_2. Thus,

$$e_{L1} = L_1 \frac{\Delta i}{\Delta t} + M \frac{\Delta i}{\Delta t}$$

$$= \frac{\Delta i}{\Delta t} (L_1 + M)$$

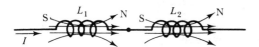

(a) Mutually coupled coils connected series-aiding

(b) Circuit symbols for mutually coupled coils connected series-aiding

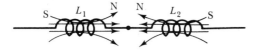

(c) Mutually coupled coils connected series-opposing

(d) Circuit symbols for mutually coupled coils connected series-opposing

FIGURE 14-12. As in the case of voltage cells, inductors can be connected series-aiding or series-opposing. Where the coils are mutually coupled, the total inductance is $L = L_1 + L_2 + 2M$ for a series-aiding connection, or $L = L_1 + L_2 - 2M$ for a series-opposing connection.

Similarly,
$$e_{L2} = \frac{\Delta i}{\Delta t}(L_2 + M)$$

and
$$e_{L(total)} = \frac{\Delta i}{\Delta t}(L_1 + M + L_2 + M)$$

also
$$\frac{e_{L(total)}}{\Delta i / \Delta t} = L_{total}$$

Therefore, for series-aiding connections,

$$\boxed{L = L_1 + L_2 + 2M} \qquad (14\text{-}18)$$

For *series-opposing connections*, the emf induced in L_1 from L_2 is in opposition to the self-induced emf. Therefore,

$$e_{L1} = L_1 \frac{\Delta i}{\Delta t} - M \frac{\Delta i}{\Delta t}$$

$$= \frac{\Delta i}{\Delta t}(L_1 - M)$$

Similarly,
$$e_{L2} = \frac{\Delta i}{\Delta t}(L_2 - M)$$

and
$$e_{L(total)} = \frac{\Delta i}{\Delta t}(L_1 - M + L_2 - M)$$

Therefore, for series-opposing connections,

$$\boxed{L = L_1 + L_2 - 2M} \qquad (14\text{-}19)$$

EXAMPLE 14-13 Two 500 μH coils have a mutual inductance of 200 μH. Determine the total inductance of the two coils:

a. When they are connected series-aiding.
b. When they are connected series-opposing.

SOLUTION

a. *Series-aiding:*

Equation (14-18) $L = L_1 + L_2 + 2M$

$= 500 \text{ μH} + 500 \text{ μH} + 2(200 \text{ μH})$

$= \mathbf{1.4 \text{ mH}}$

b. *Series-opposing:*

Equation (14-19) $L = L_1 + L_2 - 2M$

$= 500 \text{ μH} + 500 \text{ μH} - 2(200 \text{ μH})$

$= \mathbf{600 \text{ μH}}$

If the mutual inductance between two adjacent coils is not known, it can be determined by measuring the total inductance of the coils in series-aiding and series-opposing connections. Then,

$$L_a = L_1 + L_2 + 2M \quad \textit{for series-aiding}$$

and $$L_b = L_1 + L_2 - 2M \quad \textit{for series-opposing}$$

Subtracting, $L_a - L_b = 4M$

Therefore,

$$\boxed{M = \frac{L_a - L_b}{4}} \tag{14-20}$$

Recall that the mutual inductance between two coils is given by Equation (14-13) as

$$M = k\sqrt{L_1 L_2}$$

From these two equations, the coefficient of coupling of the two coils can be determined.

EXAMPLE 14-14 Use the results of Example 14-13 to determine the mutual inductance and coefficient of coupling for the two coils.

SOLUTION

$$L_a = 1.4 \text{ mH}$$

$$L_b = 600 \text{ μH}$$

Equation (14-20):

$$M = \frac{L_a - L_b}{4}$$

$$= \frac{1.4 \text{ mH} - 600 \text{ μH}}{4}$$

$$\mathbf{= 200 \text{ μH}}$$

From Equation (14-13),

$$k = \frac{M}{\sqrt{L_1 L_2}}$$

$$= \frac{200 \text{ μH}}{\sqrt{500 \text{ μH} \times 500 \text{ μH}}}$$

$$\mathbf{= 0.4}$$

PRACTICE PROBLEMS

14-7.1 Calculate the inductance that must be connected in parallel with a 100 μH inductor to give a total inductance of 70 μH. Assume no mutual inductance between the two.

14-7.2 Determine the total inductance of the two coils in Problem 14-4.2, (a) when connected series-aiding, (b) when connected series-opposing.

14-7.3 The two inductors in Problem 14-7.1 are connected in series, and the total inductance is measured as 220 μH. Calculate the coefficient of coupling between the two.

14-8
STRAY INDUCTANCE

Since inductance is (change in flux linkages)/(change in current), every current-carrying conductor has some self-inductance, and every pair of conductors has inductance. These *stray inductances* are usually unwanted, although they are sometimes used as components in a circuit design. In dc applications, stray inductance is normally unimportant, but in radio frequency ac circuits it can be a considerable nuisance. Stray inductance is normally minimized by keeping connecting wires as short as possible.

SUMMARY OF FORMULAS

Induced emf:

$$e_L = \frac{\Delta \Phi}{\Delta t}$$

Induced emf:

$$e_L = \frac{\Delta \Phi\, N}{\Delta t}$$

Inductance:

$$L = \frac{e_L}{\Delta i / \Delta t}$$

Inductance:

$$L = \frac{\Delta \Phi\, N}{\Delta i}$$

Flux change:

$$\Delta \Phi = \mu_r\, \mu_0\, \Delta i\, N \frac{A}{l}$$

Self-inductance:

$$L = \mu_r\, \mu_0\, N^2 \frac{A}{l}$$

Mutual inductance:

$$M = \frac{e_L}{\Delta i / \Delta t}$$

Induced emf:

$$e_L = \frac{\Delta \Phi\, N_s}{\Delta t}$$

Mutual inductance:

$$M = \frac{\Delta \Phi\, N_s}{\Delta i}$$

Mutual inductance:

$$M = k\, \frac{\Delta \Phi\, N_s}{\Delta i}$$

Mutual inductance:

$$M = k N_p\, N_s\, \mu_r\, \mu_0 \frac{A}{l}$$

Mutual inductance:

$$M = k\sqrt{L_1 L_2}$$

Energy stored:

$$W = \tfrac{1}{2} L I^2$$

Energy stored:

$$W = \frac{B^2 A l}{2\mu_r \mu_0}$$

Inductances in series:

$$L_s = L_1 + L_2 + L_3 + \cdots$$

Inductances in parallel:

$$\frac{1}{L_p} = \frac{1}{L_1} + \frac{1}{L_2} + \frac{1}{L_3} + \cdots$$

Total inductance (series-aiding):

$$L = L_1 + L_2 + 2M$$

Total inductance (series-opposing):

$$L = L_1 + L_2 - 2M$$

Mutual inductance:

$$M = \frac{L_a - L_b}{4}$$

REVIEW QUESTIONS

14-1 Draw sketches to show the direction of the induced current in a conductor in motion through a magnetic field. Also, show the field generated by the conductor. Briefly explain.

14-2 State Lenz's law and explain it in relation to the induced emf and current in a conductor moving through a magnetic field.

14-3 State Faraday's law, and define the weber. Use the definition of the weber to write an equation for induced emf.

14-4 Define electromagnetic induction, primary, secondary, self-inductance, mutual inductance, counter-emf, back-emf, and coefficient of coupling.

14-5 Draw sketches to show how the current through a coil induces a counter-emf in the coil. Explain carefully.

14-6 Define the henry and use the definition to write an equation for the inductance of a coil.

14-7 Derive an equation for inductance L in terms of flux linkages and current change. Also, derive the equation that relates L to the dimensions of a core. Discuss the difficulties that arise with ferromagnetic cored inductors in relation to coil current.

14-8 Show how a coil can be wound to be noninductive.

14-9 Explain mutual inductance and define the relationship between two coils when they have a mutual inductance of 1 H. Using the definition, write an equation for the mutual inductance between two coils.

14-10 Derive an equation for mutual inductance in terms of flux linkages and current change. Also explain coefficient of coupling and show how it affects the equation for M.

14-11 Derive the equation relating M to the dimensions of a core. Also, derive the equation relating M to the individual coil inductances.

14-12 Sketch the basic circuit of an automobile ignition coil and briefly explain how it operates.

14-13 Describe the various types of inductors and discuss their characteristics and important parameters.

14-14 Derive an expression for the energy stored in an inductive circuit in terms of the inductance value and the current through the coil. Also, derive an expression for the energy stored in terms of the flux density and core dimensions.

14-15 Derive the equations for the inductance of several inductors:
a. Connected in series.
b. Connected in parallel.
Assume that there is no mutual induction between the inductors.

14-16 Using illustrations, explain what occurs when mutual inductance exists between two series-connected inductors:
a. When they are connected *series-aiding*.
b. When they are connected *series-opposing*.
Derive equations for the total inductance in each case.

14-17 Explain how the mutual inductance between two coils can be determined and derive the appropriate equation.

PROBLEMS

SECTION 14-2

14-1 Write the equation for the emf induced in the secondary of two magnetically coupled coils. The flux linking the secondary of two coupled coils increases from zero to 0.12 Wb in a time of 0.3 s. Calculate the voltage induced in the secondary if it has 100 turns.

14-2 A 300 turn solenoid has a secondary winding with 250 turns. The solenoid is 20 cm long and 3 cm in diameter. Calculate the emf induced in the secondary when the solenoid current is increased from zero to 500 mA in a time of 5 ms.

14-3 A cast iron core with two coils has a closed magnetic path 20 cm long and a cross-sectional area of 5 cm^2. The primary winding has $N_p = 400$ turns and the secondary winding has $N_s = 250$ turns. If the current through the primary increases from zero to 1 A in a time of 10 ms, determine the emf induced in the secondary.

14-4 In Problem 14-3, another secondary winding having 200 turns is added on the core. Calculate the emf induced in each secondary when the primary current is increased from zero to 1 A in a time of 3.3 ms.

14-5 Two coils are wound on a ring-shaped cast steel core. The secondary of the two has $N_s = 250$ turns, and the output voltage from it is to be 25 V when the primary current increases from zero to its maximum level in a time of 8 ms. If the primary winding has 200 turns, determine the maximum level of the current that must flow through the primary. The core cross-sectional area is 10 cm² and the magnetic path length is 30 cm.

14-6 An additional 100 turns are added to the primary winding in Problem 14-5. Recalculate the required primary current.

14-7 A 5000 turn air-cored solenoid is 28 cm long and has an inside diameter of 2 cm. A secondary winding wound on top of the solenoid is to have 1 V induced in it when the solenoid current increases from zero to 15 A in a period of 2.5 ms. Determine the number of turns required on the secondary.

14-8 The secondary output voltage in Problem 14-7 is to be increased to 1.8 V without altering the primary current or the number of coil turns. How can the increase be achieved?

SECTION 14-3

14-9 A coil current grows linearly from zero to a maximum of 30 A in a time period of 15 ms. If the coil has an inductance of 5 mH, determine the level of the counter-emf.

14-10 Calculate the new maximum current level for Problem 14-9 if the counter-emf is not to exceed 100 mV.

14-11 A solenoid is to be constructed with a length of 10 cm and inside diameter of 1 cm. If the inductance of the coil is to be 500 μH, calculate the number of turns required.

14-12 Calculate the inductance of the solenoid (primary winding only) described in Problem 14-2. Also, determine the counter-emf induced during the current growth.

14-13 For the solenoid described in Problem 14-11, determine the total flux through the coil when the current is 25 mA. Also, calculate the counter-emf generated if the current increases linearly from zero in a time period of 50 ms.

14-14 Calculate the inductance of the solenoid (primary winding only) described in Problem 14-7. Also, determine the counter-emf induced during the current growth.

14-15 Calculate the inductance of a 2000 turn coil that is 12 cm in length and has an average inside diameter of 1.2 cm.

14-16 For Problem 14-15, recalculate the coil inductance (a) when its length is doubled, (b) when its diameter is doubled.

14-17 A cast iron ring has a cross-sectional area of 1 cm^2 and a magnetic path length of 10 cm. A 100 turn coil is wound on the ring. Determine the coil inductance when the current through it is a. $I = 1$ A, b. $I = 5$ A.

14-18 For Problem 14-17, recalculate the coil inductance for a 1 A current (a) when the core length is doubled, (b) when its diameter is doubled.

SECTION 14-4

14-19 Two 75 turn coils are wound on an iron core that has a closed magnetic path. The core dimensions are $l = 33$ cm and cross-sectional area = 9 cm^2. If the core has a relative permeability of 900, calculate the inductance of each coil and the mutual inductance between the coils.

14-20 Calculate the inductance of the secondary winding in Problem 14-2. Also, determine the mutual inductance between the two windings.

14-21 Two 150 turn solenoids each have $l = 22$ cm and cross-sectional area = 3.3 cm^2. The mutual inductance between the two is measured as 2 μH. Calculate their coefficient of coupling.

14-22 Calculate the inductance of the secondary winding in Problem 14-7. Also, determine the mutual inductance between the two windings.

14-23 Calculate the self inductance of each winding in Problem 14-3. Also, calculate the mutual inductance between the two windings. Assume that the core has a relative permeability of 500.

14-24 If the two coils in Problem 14-2 are separated so that their coefficient of coupling is 0.4, calculate the mutual inductance between them.

SECTION 14-6

14-25 Calculate the energy stored in a 5 H choke that has a resistance of 105 Ω. The applied voltage is 250 V.

14-26 Calculate the energy stored in the solenoid (primary winding only) in Problem 14-2 when the current remains constant at 500 mA.

14-27 When 500 V is applied to a choke with a resistance of 120 Ω, it is found that the energy stored is 33 J. Calculate the inductance of the choke.

14-28 Calculate the energy stored in the primary winding in Problem 14-5 when the current remains constant at 1 A.

14-29 Calculate the energy stored in the coil in Problem 14-9 when a constant current of 15 A flows.

14-30 For the coil in Problem 14-17, determine the energy stored (a) when the current is constant at 1 A, (b) when the current is constant at 5 A.

SECTION 14-7

14-31 Four inductors have values of $L_1 = 100$ mH, $L_2 = 50$ mH, $L_3 = 750$ μH, and $L_4 = 1$ H. Determine the total inductance of the four:
a. When connected in series.
b. When connected in parallel.
Assume that there is no mutual induction between the inductors.

14-32 Two inductors which have values of 500 μH and 800 μH are connected in parallel. Calculate the value of a third inductor which when connected in parallel with the other two will give a total inductance of 250 μH.

14-33 A 600 μH coil and a 400 μH coil have a mutual inductance between them of 100 μH. Determine the total inductance of the two:
a. When they are connected series-aiding.
b. When connected series-opposing.

14-34 Use the results of Problem 14-33 to determine the coefficient of coupling between the two coils.

14-35 Two identical coils when connected series-aiding give a total inductance of 850 μH. When connected series-opposing, their total inductance is 250 μH. Determine the mutual inductance between the two. Also, determine the values of individual inductances.

COMPUTER PROBLEMS

14-36 Write a computer program to solve the type of problem set in Example 14-3 for three different cross-sectional areas.

14-37 Write a computer program to solve the type of problem set in Example 14-6 for five different coil lengths.

ANSWERS TO PRACTICE PROBLEMS

14-2.1	37.5 mV
14-2.2	742
14-2.3	836
14-3.1	4.44 mH, 88.8 mV
14-3.2	519
14-3.3	173 mH
14-4.1	28.6 mH
14-4.2	66.6 μH
14-6.1	22.2 μJ
14-6.2	43.9 mJ
14-7.1	233 μH
14-7.2	696 μH, 430 μH
14-7.3	0.3

15 CAPACITANCE

Objectives	You will be able to:

Draw sketches to illustrate the process of electronic charge storage in two metal plates separated by a layer of insulating material.

Solve problems involving electric flux, electric flux density, and electric field strength in the dielectric of a capacitor.

Calculate the capacitance of a capacitor with given dimensions and given type of dielectric.

Describe the various types of capacitors and list the characteristics of each.

Calculate the equivalent capacitance of several capacitors connected in series or in parallel.

Calculate the energy stored in a charged capacitor.

Introduction

A capacitor consists of a layer of insulating material sandwiched between two metal plates. The electric field in a capacitor has many parallels to a magnetic field, including a parameter known as the *permittivity*, which is analogous to magnetic permeability. The capacitance value of a capacitor is a measure of the amount of electric charge that can be stored in the device. The capacitance can be calculated from a knowledge of the capacitor dimensions and the permittivity of the insulating material. There are many different types of capacitors, each with its own particular characteristics and application. Parallel-connected and series-connected capacitors *cannot* be treated in the same way as parallel and series resistance circuits.

15-1
ELECTRIC CHARGE STORAGE

Figure 15-1(a) shows a device that consists of a layer of insulating material sandwiched between two metal plates. This device is known as a *capacitor* and, as will be shown, it has the ability to store an electric charge. The figure also shows a battery and switch S_1 for connecting the battery across the plates of the capacitor. Initially, there is no potential difference between the plates, but when the switch is closed, the plates assume the same potentials as the battery terminals. So a potential difference now exists between the capacitor plates [see Figure 15-1(b)].

Recall that the positive electrode of a battery is positive because many of its (*negative*) electrons have been removed. Also, the battery's negative electrode has an accumulation of electrons. For the capacitor plates in the illustration to have the same potential difference as the battery terminals, electrons must be removed from the top plate to make it positive, and the removed electrons must be delivered to the bottom plate to make it negative. This means that electrons flow out of the top plate and into the bottom plate, as illustrated in Figure 15-1(b). The flow of electrons constitutes a *conventional direction* current flow into the top plate and out of the bottom plate. The current flow lasts only a brief time (less than milliseconds) and ceases just as soon as the plates are at the same potentials as the battery terminals.

If the switch is now opened, the capacitor plates are found to retain their potential difference [see Figure 15-1(c)]. It can be shown that the capacitor has, in fact, stored an electric charge, and that the quantity of charge stored depends on the dimension of the plates and insulating material as well as on the battery voltage. The capacitor was *charged* by the brief pulse of current that flowed when switch S_1 was closed. If *short-circuiting switch* S_2 in Figure 15-1(d) is now closed, the surplus of electrons on the bottom plate returns to the top plate, as shown. Again, this can be described as a conventional direction current flow from the positive plate to the negative plate. In any case, closing S_2 *discharges* the capacitor and returns the plates to the same potential.

The layer of insulating material between the capacitor plates is known as the *dielectric*. The dielectric has to be insulating material, because if it were conducting material, it would simply short-circuit the plates. Typical dielectric materials are rubber, mica, paper, and air. The capacitor plates must be conducting material in order to have large quantities of electrons that can be transferred between plates. The graphic symbols for fixed and variable capacitors are shown in Figure 15-1(e). The straight line identifies the terminal of the capacitor that should be always made positive.

15-2
ELECTRIC FIELD

ELECTRIC LINES OF FORCE. In Chapter 1 it was explained that a force exists between electrically charged bodies, and that bodies with like charges repel each other whereas those with unlike charges experience

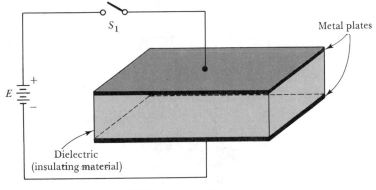

(a) Capacitor supplied
from a battery

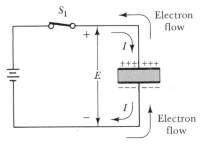

(b) Capacitor becomes charged to
battery voltage when S_1 is closed

(c) Capacitor retains its
charge when S_1 is opened

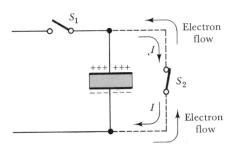

(d) Capacitor is discharged by
short-circuiting switch S_2

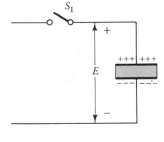

(e) Capacitor circuit symbols

FIGURE 15-1. A capacitor basically consists of two conducting plates separated by a layer of insulation. A capacitor connected to a voltage source assumes the same terminal voltage as the source, and this remains constant when the source is disconnected; the capacitor retains a charge.

a force of attraction. These forces are similar in many ways to the forces that occur between magnetic poles. As with magnetism, a *force field* exists around electrically charged bodies. *Electric lines of force* are spoken of, although their existence cannot be demonstrated as easily as in the case of magnetism. The *electric field strength, electric flux*, and *electric flux density* can all be calculated.

In Figure 15-2(a), the bottom plate of a capacitor is shown grounded, and the top plate has a potential of $+E$ *with respect to* (w.r.t.) ground. In the space between the two plates, the potential at the halfway point becomes $+E/2$ w.r.t. ground. Similarly, at $\frac{1}{4}d$ and $\frac{3}{4}d$, the potentials are $+E/4$ and $(+3/4)E$, respectively, w.r.t. ground. At each of these points a line exists along which the potential is a constant quantity. Appropriately, the lines are termed *equipotential lines*.

In the space between the capacitor plates, the equipotential lines are found to be horizontal. However, at the edges of the plates, they

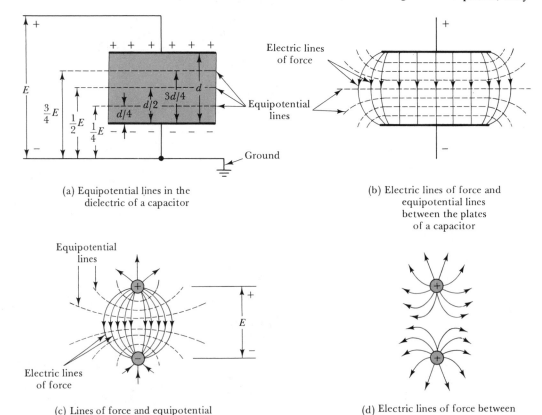

(a) Equipotential lines in the
dielectric of a capacitor

(b) Electric lines of force and
equipotential lines
between the plates
of a capacitor

(c) Lines of force and equipotential
lines between two oppositely
charged bodies

(d) Electric lines of force between
two similarly charged bodies

FIGURE 15-2. An electric force field exists around charged bodies. Electric lines of force and equipotential lines can be plotted. Electric field strength and flux density can be calculated.

curve outward, as illustrated in Figure 15-2(b). It is also found that the electric lines of force always cross the equipotential lines at right angles. Thus, the lines of force are vertical in the space between the plates and bulge outward at the plate edges, as shown in the figure.

The direction of the electric lines of force is assumed to be the direction in which a free, positively charged particle would move if it were placed in the electric field. Such a particle would obviously be repelled from the positive plate of the capacitor and attracted to the negative plate. Therefore, the electric lines of force are said to flow from positive to negative. Figure 15-2(c) shows the electric field pattern that exists between two round bodies that are oppositely charged. The lines of force from each are seen to join together. In Figure 15-2(d), the lines of force from the two like-charged bodies are shown repelling each other. As with magnetic lines of force, it is found that *electric lines of force repel each other, never intersect, and are always in a state of tension.*

ELECTRIC CHARGE, FLUX, AND FLUX DENSITY. Electric charge is measured in coulombs (see Appendix 2), and the coulomb is also the unit of electric flux. Therefore, a body that is charged to Q coulombs emits a total electric flux of Q coulombs. Similarly, a capacitor that has Q coulombs of charge has a total electric flux of Q coulombs between its plates.

$$\boxed{\textit{electric flux, } \psi = Q} \qquad \textbf{(15-1)}$$

where ψ is the electric flux in coulombs, and Q is the charge in coulombs.
The electric flux density is simply the flux per unit area. Therefore,

$$\boxed{\textit{electric flux density, } D = \frac{Q}{A}} \qquad \textbf{(15-2)}$$

where D is electric flux density in C/m^2, Q is total charge in coulombs, and A is the area in m^2.
The electric field strength in the space between the plates of a capacitor depends on the applied voltage and on the distance between the plates:

$$\boxed{\textit{electric field strength } = \frac{E}{d}} \qquad \textbf{(15-3)}$$

where E is the voltage difference between the plates in volts, d is the distance between the plates in meters, and the electric field strength is in V/m [see Section 4-2 and Equation (4-1)].

DIELECTRIC EFFECTS. The insulating dielectric material between the plates of a capacitor is subjected to electric stress; that is, the applied voltage (electric pressure) is pressing for a current flow. Where the voltage between the plates is low, the electric stress on the dielectric is low. When the voltage is high, the stress is greater, and if the stress becomes high enough, the material may break down. This is discussed in Section 4-2. The stress on the dielectric is measured in volts/meter, and it is, in fact, the electric field strength. The *dielectric strength* of a given material is the voltage per unit thickness at which the material may break down (see Table 4-1).

Since the dielectric is an insulating material, the electrons within the dielectric are never detached from their atoms (except in a break-down situation). However, a certain amount of distortion of the atoms occurs, as illustrated in Figure 15-3. When the capacitor plates are at the same potential there is no electric stress on the dielectric, and in the dielectric atoms the electrons are orbiting normally [see Figure 15-3(a)]. When a potential difference exists between the plates, the orbiting electrons experience a force attracting them toward the positive plate, while the atomic nucleus is attracted toward the negative plate. The result is that the electrons tend to go into a distorted orbit around the nucleus, as illustrated in Figure 15-3(b). In this situation the atoms are said to be *polarized*.

When a capacitor is discharged by short-circuiting its plates, the polarized atoms may be expected to return to their normal state. With some dielectric materials, it is found that the polarized atoms do not return completely to their normal state. Consequently, when the short circuit is removed from the capacitor plates, a small voltage can again be measured across the plates. The plate voltage is, of course, the result of

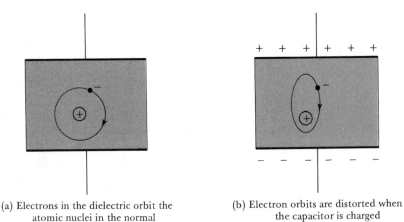

(a) Electrons in the dielectric orbit the atomic nuclei in the normal symmetrical way when the capacitor is uncharged

(b) Electron orbits are distorted when the capacitor is charged

FIGURE 15-3. Electric stress occurs on the dielectric of charged capacitors. This distorts the orbits of electrons within the dielectric.

Chap. 15   Capacitance

residual polarization within the dielectric. This phenomenon of energy being retained by the dielectric is referred to as *dielectric absorption*.

EXAMPLE 15-1 A capacitor has a plate area of 400 cm² and a dielectric thickness of 1 mm. If the capacitor has a charge of 10×10^{-3} C when the voltage between its plates is 25 V, determine the electric field strength and electric flux density within the dielectric.

SOLUTION

Equation (15-3):

$$electric\ field\ strength = \frac{E}{d}$$

$$= \frac{25\ V}{1 \times 10^{-3} m}$$

$$= \mathbf{25\ 000\ V/m}$$

Equation (15-2):

$$electric\ flux\ density,\ D = \frac{Q}{A}$$

$$= \frac{10 \times 10^{-3}\ C}{400 \times 10^{-4} m^2}$$

$$= \mathbf{0.25\ C/m^2}$$

PRACTICE PROBLEMS

15-2.1 Two conductors in a cable are separated by 0.3 cm of insulation. Calculate the electric field strength within the insulation when the conductor voltage difference is 120 V. Also, determine the voltage at which the insulation may break down if its dielectric strength is 270 kV/cm.

15-2.2 A capacitor to contain a charge of 40 μC is to have a maximum electric flux density of 1.6 mC/m². Calculate its minimum plate area.

15-3

CAPACITANCE AND CAPACITOR DIMENSIONS

PLATE AREA AND DIELECTRIC THICKNESS. The quantity of charge that can be stored by a capacitor with a given terminal voltage is termed its *capacitance*. The capacitance of a capacitor has a definite relationship to the area of the plates and to the thickness of the dielectric.

Refer again to Figure 15-1(b) and recall that electrons are attracted to a positive potential. Therefore, the presence of the positive potential on the top plate causes electrons to be attracted (from the negative terminal of the battery) into the adjacent bottom plate of the capacitor. The electrons cannot pass through the insulating dielectric to the positive top plate. Instead, they get as close to the top plate as possible by accumulating at the surface of the bottom plate close to the dielectric. Similarly, the negative potential on the bottom plate causes electrons to be driven out of the adjacent top plate. Consequently, a positive charge (i.e., a lack of electrons) is accumulated at the surface of the top plate close to the dielectric.

Now look at Figure 15-4(a), which shows a capacitor consisting of two metal plates with a very thick dielectric between them. With such a thick piece of insulating material between the two plates, the potential on each plate has very little influence on the electrons in the other plate. Therefore, only a relatively small charge can be stored. In Figure 15-4(b), a capacitor with a very thin dielectric is shown. In this case the potential on each plate has a strong influence on the very close adjacent plate. The result is, of course, that a relatively large number of electrons is accumulated on the bottom plate, and an equally large number of electrons is driven out of the top plate, constituting a relatively large charge on the capacitor. This suggests (and it can be experimentally demonstrated) that the amount of charge stored in a capacitor is *inversely proportional* to the thickness of the dielectric.

Figure 15-4(c) shows two identical capacitors connected in parallel. If each capacitor has a charge of Q coulombs, the total charge stored by the two is $2Q$ coulombs. Paralleling the two similar capacitors means that the plate area for one capacitor has been doubled. Thus, a capacitor with twice the plate area would also store twice the charge. Now it can be written

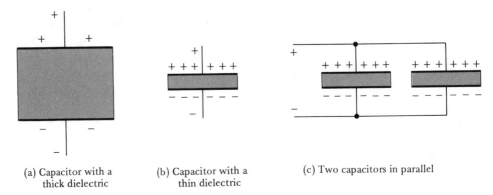

(a) Capacitor with a thick dielectric

(b) Capacitor with a thin dielectric

(c) Two capacitors in parallel

FIGURE 15-4. A capacitor with a thin dielectric can store more charge than a similar capacitor with a thick dielectric. A capacitor with a large plate area has a greater capacitance than one with a small area. Capacitance is proportional to plate area and inversely proportional to dielectric thickness.

Chap. 15   Capacitance

$$\text{capacitance} \propto \frac{\text{plate area}}{\text{distance between plates}}$$

or
$$C \propto \frac{A}{d}$$

Refer once more to Figure 15-4(b). It can be shown that for a given applied voltage, a definite number of electrons is attracted into the bottom plate, and the same number of electrons is driven out of the top plate of the capacitor. If the applied voltage is now doubled, twice as many electrons are displaced from the positive plate to the negative plate, and therefore the accumulated charge is doubled. Thus, the charge on a capacitor is directly proportional to the applied voltage as well as to the capacitance of the capacitor:

$$Q \propto CE$$

The *farad** (*F*) is the SI unit of capacitance.

The *farad* is the capacitance of a capacitor that contains a charge of 1 coulomb when the potential difference between its terminals is 1 volt.

It is found that the farad is a very large unit; consequently, microfarads (μF), nanofarads (nF), and picofarads (pF) are most frequently used for expressing capacitance.

From the definition of the farad, the equation for charge becomes

$$\boxed{Q = CE} \qquad \text{(15-4)}$$

where Q is charge in coulombs, C is capacitance in farads, and E is potential difference between the plates in volts.

PERMITTIVITY. Before proceeding further, another factor, known as the *permittivity* must be introduced. Electric permittivity is analogous to magnetic permeability. It specifies the ease with which electric flux is *permitted* to pass through a given dielectric material. Like permeability, permittivity is subdivided into the *permittivity of free space* ϵ_0, and the *relative permittivity* ϵ_r, which is also known as the *dielectric constant*. Table 15-1 lists the relative permittivites of various dielectric materials. Note that the permittivity of air is 1.0006, and therefore it is usually taken as 1 (i.e., the same as for vacuum).

---

*Named for the English chemist and physicist Michael Faraday (1791–1867).

TABLE 15-1

MATERIAL	TYPICAL DIELECTRIC CONSTANT (ϵ_r)
Vacuum	1
Air	1.0006
Ceramic (low loss)	6–20
Ceramic (high ϵ_r)	>1000
Glass	5–100
Mica	3–7
Mylar	3
Oxide film	5–25
Paper	4–6
Polystyrene	2.5
Teflon	2

$$\textit{permittivity of free space, } \epsilon_0 = \frac{1}{36\pi \times 10^9}$$

or

$$\epsilon_0 \cong 8.84 \times 10^{-12}$$

It is interesting to note that

$$\frac{1}{\sqrt{\mu_0 \epsilon_0}} = \frac{1}{\sqrt{4\pi \times 10^{-7} \times \dfrac{1}{36\pi \times 10^9}}}$$

$$= 3 \times 10^8$$

or

$$\frac{1}{\sqrt{\mu_0 \epsilon_0}} = \begin{cases} \text{velocity of light and other electromagnetic} \\ \text{waves, in meters per second*} \end{cases}$$

The dielectric permittivity has a similar relationship to field strength and flux density as does the magnetic permeability:

$$\boxed{\epsilon_r \epsilon_0 = \frac{D}{\textbf{electric field strength}}} \qquad \textbf{(15-5)}$$

where electric field strength is in V/m and D is electric flux density in C/m².

CAPACITANCE EQUATION. From Equation (15-2),

$$D = \frac{Q}{A}$$

---

*This relationship was discovered in 1865 by the Scottish physicist James C. Maxwell (1831–1879). From it he was able to predict the existence of electromagnetic waves many years before they were demonstrated experimentally.

and from Equation (15-3),

$$\text{electric field strength} = \frac{E}{d}$$

Therefore,

$$\epsilon_r \epsilon_0 = \frac{Q}{A} \times \frac{d}{E}$$

From Equation (15-4),

$$Q = CE$$

Therefore,

$$\epsilon_r \epsilon_0 = \frac{Cd}{A}$$

or

$$\boxed{C = \frac{\epsilon_r \epsilon_0 A}{d}} \tag{15-6}$$

When A is plate area in m^2, d is dielectric thickness in meters, and C is the capacitance in farads.

EXAMPLE 15-2 Calculate the capacitance of a capacitor with a plate area of 400 cm² and a dielectric thickness of 1 mm:

  a. When the dielectric is air, and
  b. When the dielectric is mica with a relative permittivity of 5.
  c. Determine the charge on the capacitor in the case of (a) and (b) when the applied voltage is 25 V.

SOLUTION

  a. From Equation (15-6),

$$C = \frac{\epsilon_r \epsilon_0 A}{d}$$

$$= \frac{8.84 \times 10^{-12} \times 400 \times 10^{-4} \text{ m}^2}{1 \times 10^{-3} \text{ m}}$$

$$\cong \mathbf{354\ pF}$$

  b.

$$C = \frac{5 \times 8.84 \times 10^{-12} \times 400 \times 10^{-4} \text{ m}^2}{1 \times 10^{-3} \text{ m}}$$

$$\cong \mathbf{1.77\ nF}$$

  c. From Equation (15-4),

$$Q = CE$$

For $C = 354$ pF,

$$Q = 354 \text{ pF} \times 25 \text{ V}$$
$$= 8.85 \times 10^{-9} \text{ C}$$

For $C = 1.77$ nF,

$$Q = 1.77 \text{ nF} \times 25 \text{ V}$$
$$= 4.425 \times 10^{-8} \text{ C}$$

EXAMPLE 15-3   A 1 μF capacitor is to be constructed from rolled-up sheets of aluminum foil separated by a layer of paper 0.1 mm thick. Calculate the required area for each sheet of foil if the relative permittivity of the paper is 6.

SOLUTION

From Equation (15-6),

$$A = \frac{Cd}{\epsilon_r \epsilon_0}$$

$$= \frac{1 \text{ μF} \times 0.1 \times 10^{-3} \text{ m}}{6 \times 8.84 \times 10^{-12}}$$

$$\cong 1.885 \text{ m}^2$$

PRACTICE PROBLEMS

15-3.1   Two conductors in a cable are 0.5 cm in diameter, and the average thickness of the insulation between the conductors is 0.25 cm. Calculate the cable capacitance per meter if the insulation has a dielectric constant of 2.2.

15-3.2   Determine the charge contained on 100 m of the cable in Problem 15-3.1 when the conductor voltage difference is 120 V.

15-3.3   Two copper films on opposite sides of a printed circuit board are separated by 2 mm of insulation with a dielectric constant of 2. If each plate is 2 cm by 1 cm, calculate the capacitance.

15-4

CAPACITOR TYPES AND CHARAC- TERISTICS

In addition to the actual value of capacitance, there are many other important characteristics that must be considered in selecting a capacitor for a particular application.

WORKING VOLTAGE.   The maximum voltage that may be safely applied to a capacitor is usually expressed in terms of its *dc working voltage*. A *maximum rms ac voltage* (see Chapter 17) may also be specified, and this is usually little more than half the maximum dc voltage. These limits should never be exceeded; otherwise, dielectric breakdown may occur.

CAPACITOR TOLERANCE.   This is simply the accuracy with which the capacitance value is specified. Tolerances of ±5% to ±20% are normal for most small-value capacitors, but more precise capacitors can be purchased at increased cost. In the case of larger-value capacitors, the capacitance has to be a certain minimum value in many applications. Consequently, manufacturers tend to specify the tolerance as −10% +150%. This means, for example, that a 100 μF capacitor could have a value as low as 90 μF or as high as 250 μF.

TEMPERATURE EFFECTS.   Every capacitor type has an operating temperature range specified by the manufacturer. Typical ranges are −20°C to +65°C, −40°C to +65°C, and −55°C to +125°C. Obviously, no capacitor should be employed in an environment where the temperatures may be beyond its range. The capacitance value is also likely to change slightly over the temperature range. The maximum change that can occur is specified by the manufacturer either in *parts per million per degree celsius* (ppm/°C), or as a percentage change for the temperature extremes. Sometimes a graph is provided of capacitance percentage change versus temperature.

LEAKAGE CURRENT.   Despite the fact that the dielectric is an insulator, small leakage currents flow between the plates of a capacitor. The actual level of leakage current depends on the insulation resistance of the dielectric. *Plastic film* capacitors, for example, may have insulation resistances higher than 100 000 MΩ. At the other extreme, an *electrolytic* capacitor may have several microamps of leakage current, with only 10 V applied to its terminals.

POLARIZATION.   Electrolytic capacitors normally have one terminal identified as the most positive connection. Thus, they are said to be *polarized*. This usually limits their application to situations where the polarity of the applied voltage will not change.

CAPACITOR EQUIVALENT CIRCUIT.   An *ideal capacitor* has a dielectric that has an infinite resistance and plates that have zero resistance. However, an ideal capacitor does not exist, since all dielectrics have some leakage current and all capacitor plates have some resistance. The complete equivalent circuit for a capacitor [shown in Figure 15-5(a)] consists of an ideal capacitor C in series with a resistance R_D representing the resistance of the plates, and in parallel with a resistance R_L representing

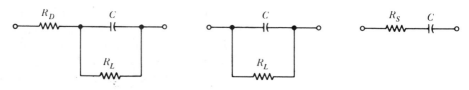

(a) Complete equivalent circuit   (b) Parallel equivalent circuit   (c) Series equivalent circuit

FIGURE 15-5. A capacitor equivalent circuit consists of the capacitance C, the leakage resistance R_L in parallel with C, and the plate resistance R_D in series with C and R_L.

the leakage resistance of the dielectric. Usually, the plate resistance can be completely neglected, and the equivalent circuit becomes that shown in Figure 15-5(b). With capacitors that have a very high leakage resistance (e.g., mica and plastic film capacitors), the parallel resistor is frequently omitted in the equivalent circuit, and the capacitor is then treated as an ideal capacitor. This *cannot* normally be done for electrolytic capacitors, for example, which have relatively low leakage resistances. In Section 20-6 it is shown that the circuit of Figure 15-5(b) has an equivalent series circuit as shown in Figure 15-5(c).

Because a capacitor's dielectric is largely responsible for determining its most important characteristics, capacitors are usually identified by the type of dielectric used.

AIR CAPACITORS. A typical capacitor using air as a dielectric is illustrated in Figure 15-6. The capacitance is variable, as is the case with virtually all air capacitors. There are two sets of metal plates, one set fixed and one movable. The movable plates can be adjusted into or out of the spaces between the fixed plates by means of the rotatable shaft. Thus, the area of the plates opposite each other is increased or decreased, and the capacitance value is altered.

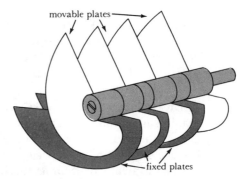

movable plates

fixed plates

FIGURE 15-6. A variable air capacitor is made up of a set of movable plates and a set of fixed plates separated by air.

PAPER CAPACITORS. In its simplest form, a paper capacitor consists of a layer of paper between two layers of metal foil. The metal foil and paper are rolled up, as illustrated in Figure 15-7(a); external connections are brought out from the foil layers and the complete assembly is dipped in wax or plastic. A variation of this is the *metalized paper* construction, in which the foil is replaced by thin films of metal deposited on the surface of the paper. One end of the capacitor sometimes has a band around it [see Figure 15-7(b)]. This does not mean that the device is polarized, but simply identifies the terminal that connects to the outside metal film, so that when required it can be grounded to avoid pickup of unwanted signals.

Paper capacitors are available in value ranging from about 500 pF to 50 μF, and in dc working voltages up to about 600 V. They are among the lowest-cost capacitors for a given capacitance value but are physically larger than several other types having the same capacitance value.

PLASTIC FILM CAPACITORS. The construction of plastic film capacitors is similar to that of paper capacitors, except that the paper is replaced by a thin film that is typically polystyrene or Mylar. This type of dielectric gives insulation resistances greater than 100 000 MΩ. Working voltages are as high as 600 V, with the capacitor surviving 1500 V surges for a brief period. Capacitance tolerances of ±2.5% are typical, as are temperature coefficients of 60 to 150 ppm/°C.

Plastic film capacitors are physically smaller but more expensive than paper capacitors. They are typically available in values ranging from 5 pF to 0.47 μF.

MICA CAPACITORS. As illustrated in Figure 15-8(a), mica capacitors consist of layers of mica alternated with layers of metal foil. Connections are made to the metal foil for capacitor leads, and the entire assembly is dipped in plastic or encapsulated in a molded plastic jacket. Typical

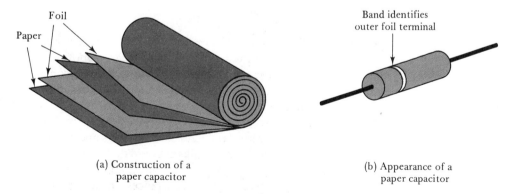

Foil

Paper

(a) Construction of a
paper capacitor

Band identifies
outer foil terminal

(b) Appearance of a
paper capacitor

FIGURE 15-7. In a paper capacitor, two sheets of metal foil separated by a sheet of paper are rolled up together. External connections are made to the foil sheets.

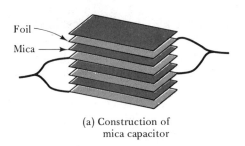

(a) Construction of mica capacitor

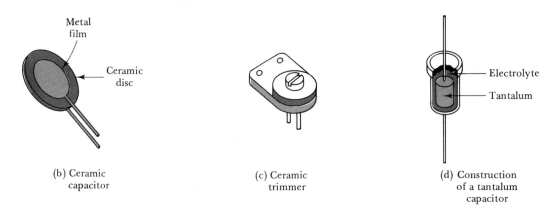

(b) Ceramic capacitor

(c) Ceramic trimmer

(d) Construction of a tantalum capacitor

FIGURE 15-8. Mica capacitors consist of sheets of mica interleaved with foil. A ceramic disc silvered on each side makes a ceramic capacitor; in a ceramic trimmer, the plate area is screwdriver adjustable. A tantalum capacitor has a relatively large capacitance in a small volume.

capacitance values range from 1 pF to 0.1 μF, and voltage ratings as high as 35 000 V are possible. Precise capacitance values and wide operating temperatures are obtainable with mica capacitors. In a variation of the process, *silvered mica* capacitors use films of silver deposited on the mica layers instead of metal foil.

CERAMIC CAPACITORS. The construction of a typical ceramic capacitor is illustrated in Figure 15-8(b). Films of metal are deposited on each side of a thin ceramic disc, and copper wire terminals are connected to the metal. The entire unit is then encapsulated in a protective coating of plastic. Two different types of ceramic are used, one of which has extremely high relative permittivity. This gives capacitors that are much smaller than paper or mica capacitors having the same capacitance value. One disadvantage of this particular ceramic dielectric is that its leakage resistance is not as high as with other types. Another type of ceramic is also used, and this gives leakage resistances on the order of 7500 MΩ. Because it has a lower permittivity, this type of ceramic produces capacitors that are relatively large for a given value of capacitance.

The range of capacitance values available with ceramic capacitors is typically 1 pF to 0.1 μF, with dc working voltages up to 1000 V.

Figure 15-8(c) shows a variable ceramic capacitor known as a *trimmer*. By means of a screwdriver, the area of plate on each side of a dielectric can be adjusted to alter the capacitance value. Typical ranges of adjustment available are 1.5 pF to 3 pF and 7 pF to 45 pF.

ELECTROLYTIC CAPACITORS. The most important feature of electrolytic capacitors is their very large capacitance value in a physically small container. For example, a capacitance of 5000 μF can be obtained in a cylindrical package approximately 5 cm long by 2 cm in diameter. In this case the dc working voltage is only 10 V. Similarly, a 1 F capacitor is available in a 22 cm by 7.5 cm cylinder, with a working voltage of only 3 V. Typical values for electrolytic capacitors range from 1 μF through 100 000 μF.

The construction of an electrolytic capacitor is basically the same as that of a paper capacitor (see Figure 15-7). Two sheets of aluminum foil separated by a fine gauze soaked in electrolyte are rolled up and encased in an aluminum cylinder for protection. When assembled, a direct voltage is applied to the capacitor terminals, and this causes a thin layer of aluminum oxide to form on the surface of the positive plate next to the electrolyte. The aluminum oxide is the dielectric, and the electrolyte and positive sheet of foil are the capacitor plates. The extremely thin oxide dielectric gives the very large value of capacitance.

It is very important that electrolytic capacitors be connected with the correct polarity. When incorrectly connected, gas forms within the electrolyte and the capacitor may *explode*! Such an explosion blows the capacitor apart and spreads its contents around. This could have *tragic consequences* for the eyes of an experimenter who happens to be closely examining the circuit when the explosion occurs. Nonpolarized electrolytic capacitors can be obtained. They consist essentially of two capacitors in one package connected *back to back*, so one of the oxide films is always correctly biased.

Electrolytic capacitors are available with dc working voltages greater than 400 V, but in this case capacitance values do not exceed 100 μF. In addition to their low working voltage and polarized operation, another disadvantage of electrolytic capacitors is their relatively high leakage current.

TANTALUM CAPACITORS. This is essentially another type of electrolytic capacitor. Powdered tantalum is *sintered* (or baked), typically into a cylindrical shape. The resulting solid is quite porous, so that when immersed in a container of electrolyte, the electrolyte is absorbed into the tantalum. The tantalum then has a large surface area in contact with the electrolyte [see Figure 15-8(d)]. When a dc *forming voltage* is applied, a thin oxide film is formed throughout the electrolyte-tantalum

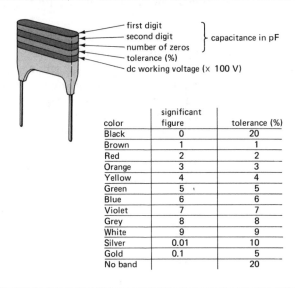

Capacitor Color Codes

Physically large capacitors usually have their capacitance value, tolerance and dc working voltage printed on the side of the case. Small capacitors (like small resistors) use a code of colored bands (or sometimes colored dots) to indicate the component parameters.

There are several capacitor color codes in current use. Here is one of the most common:

first digit
second digit } capacitance in pF
number of zeros
tolerance (%)
dc working voltage (× 100 V)

color	significant figure	tolerance (%)
Black	0	20
Brown	1	1
Red	2	2
Orange	3	3
Yellow	4	4
Green	5	5
Blue	6	6
Violet	7	7
Grey	8	8
White	9	9
Silver	0.01	10
Gold	0.1	5
No band		20

contact area. The result, again, is a large capacitance value in a small volume.

A typical tantalum capacitor in a cylindrical shape 2 cm by 1 cm might have a capacitance of 100 μF and a dc working voltage of 20 V. Other types are available with a working voltage up to 630 V, but with capacitance values on the order of 3.5 μF. Like aluminum-foil electrolytic capacitors, tantalum capacitors must be connected with the correct polarity.

15-5
CAPACITORS IN SERIES AND IN PARALLEL

PARALLEL-CONNECTED CAPACITORS. When capacitors are connected in *parallel*, the result is the same as increasing the total plate area. From Equation (15-6), the capacitance of a capacitor is given by

$$C = \frac{\epsilon_r \epsilon_0 A}{d}$$

Therefore, when three capacitors with plate areas of A_1, A_2, and A_3 and dialectric thickness d are connected in parallel as shown in Figure 15-9(a), the total capacitance is

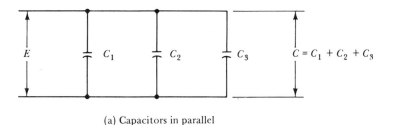

(a) Capacitors in parallel

(b) Capacitors in series

FIGURE 15-9. Parallel-connected capacitors have a total capacitance of $C_p = (C_1 + C_2 + C_3 + \cdots)$. With series-connected capacitors, $C_s = 1/(1/C_1 + 1/C_2 + 1/C_3 + \cdots)$.

$$C_p = \frac{\epsilon_r \epsilon_0}{d} (A_1 + A_2 + A_3)$$

$$= \frac{\epsilon_r \epsilon_0 A_1}{d} + \frac{\epsilon_r \epsilon_0 A_2}{d} + \frac{\epsilon_r \epsilon_0 A_3}{d}$$

or

$$\boxed{C_p = C_1 + C_2 + C_3 + \cdots} \qquad (15\text{-}7)$$

Thus, the total capacitance of capacitors in parallel is the sum of the individual capacitance values.

For the parallel capacitors shown in Figure 15-9(a), the applied voltage is obviously common to all three. From Equation (15-4), the charge on a capacitor is

$$Q = CE$$

Therefore, the charge on each capacitor is

$$Q_1 = C_1 E, \quad Q_2 = C_2 E, \quad \text{and} \quad Q_3 = C_3 E$$

so that the total charge is

$$Q = Q_1 + Q_2 + Q_3$$
$$= C_1 E + C_2 E + C_3 E$$

or
$$Q = C_p E$$

EXAMPLE 15-4 Three capacitors have values $C_1 = 1 \ \mu F$, $C_2 = 2 \ \mu F$, and $C_3 = 3 \ \mu F$. Determine the total capacitance and the charge on each capacitor when the three are connected in parallel across 100 V supply.

SOLUTION

Equation (15-7):

$$C_p = C_1 + C_2 + C_3$$
$$= 1 \ \mu F + 2 \ \mu F + 3 \ \mu F$$
$$= \mathbf{6 \ \mu F}$$

Equation (15-4): $Q = CE$

Therefore, $Q_1 = 1 \ \mu F \times 100 \ V$
$$= \mathbf{100 \ \mu C}$$

$Q_2 = 2 \ \mu F \times 100 \ V$
$$= \mathbf{200 \ \mu C}$$

$Q_3 = 3 \ \mu F \times 100 \ V$
$$= \mathbf{300 \ \mu C}$$

SERIES-CONNECTED CAPACITORS.   Three series-connected capacitors are shown in Figure 15-9(b). It is seen that connecting capacitors in series amounts to increasing the total thickness of the dielectric between the two outer plates. Equation (15-6) gives the capacitance as

$$C = \frac{\epsilon_r \epsilon_0 A}{d}$$

Therefore, when three capacitors with similar plate areas and with dielectric thickness of d_1, d_2, and d_3 are connected in series, as shown in Figure 15-9(b), the total capacitance is

$$C_s = \frac{\epsilon_r \epsilon_0 A}{d_1 + d_2 + d_3}$$

giving
$$\frac{1}{C_s} = \frac{d_1 + d_2 + d_3}{\epsilon_r \epsilon_0 A}$$

$$= \frac{d_1}{\epsilon_r \epsilon_0 A} + \frac{d_2}{\epsilon_r \epsilon_0 A} + \frac{d_3}{\epsilon_r \epsilon_0 A}$$

or

$$\boxed{\frac{1}{C_s} = \frac{1}{C_1} + \frac{1}{C_2} + \frac{1}{C_3} + \cdots} \qquad \text{(15-8)}$$

Therefore, the total capacitance of capacitors connected in series is the reciprocal of the sum of the reciprocals of the individual capacitances.

Note that the formulas for total capacitance with parallel-connected and series-connected capacitors are the reverse of the formulas for similar resistance circuits.

For the three series-connected capacitors in Figure 15-9(b), the charging current must flow through all three capacitors. If a charging current *I* flows for a time *t,* then *the charge supplied to each capacitor is equal to the charge supplied to all three capacitors:*

$$Q = Q_1 = Q_2 = Q_3$$
$$Q = It$$

and, from Equation (15-4),

$$E = \frac{Q}{C}$$

Therefore, the voltage across the individual capacitors can be calculated as

$$E_1 = \frac{Q}{C_1}, \quad E_2 = \frac{Q}{C_2}, \quad \text{and} \quad E_3 = \frac{Q}{C_3}$$

Note that the largest capacitor has the smallest terminal voltage. The total applied voltage is the sum of the individual capacitor voltages [see Figure 15-9(b)]:

$$E = E_1 + E_2 + E_3$$

EXAMPLE 15-5   If the three capacitors in Example 15-4 are connected in series across the 100 V supply, determine the total capacitance and the voltage across each capacitor.

SOLUTION

Equation (15-8):

$$\frac{1}{C_s} = \frac{1}{C_1} + \frac{1}{C_2} + \frac{1}{C_3}$$

$$= \frac{1}{1\ \mu F} + \frac{1}{2\ \mu F} + \frac{1}{3\ \mu F}$$

Therefore, $C_s \cong \mathbf{0.545\ \mu F}$

Equation (15-4):

$$Q = CE \cong 0.545\ \mu F \times 100\ V$$

$$\cong 54.5\ \mu C$$

and $Q = Q_1 = Q_2 = Q_3$

Therefore, $E_1 = \dfrac{Q}{C_1} \cong \dfrac{54.5\ \mu C}{1\ \mu F}$

$$\cong \mathbf{54.5\ V}$$

$$E_2 = \frac{Q}{C_2} \cong \frac{54.5\ \mu C}{2\ \mu F}$$

$$\cong \mathbf{27.3\ V}$$

$$E_3 = \frac{Q}{C_3} \cong \frac{54.5\ \mu C}{3\ \mu F}$$

$$\cong \mathbf{18.2\ V}$$

$$E = E_1 + E_2 + E_3$$

$$= 54.5\ V + 27.3\ V + 18.2\ V$$

$$= \mathbf{100\ V}$$

PRACTICE PROBLEMS

15-5.1   Calculate the total capacitance of a 100 pF capacitor C_1 in parallel with two series-connected capacitors: $C_2 = 40$ pF and $C_3 = 1000$ pF.

15-5.2 If a 15 V supply is connected to the capacitor circuit in Problem 15-5.1, calculate the total charge and the voltage across each capacitor.

15-6

ENERGY STORED IN CHARGED CAPACITOR

From Equation (15-4):

$$C = \frac{Q}{E}$$

and

$$Q = It$$

Therefore,

$$C = \frac{It}{E}$$

or

$$\boxed{I = \frac{CE}{t}}$$ (15-9)

When a capacitor is charged from a constant current source for a time t seconds, the voltage across it grows linearly from zero to E volts. The constant level of input current is given by Equation (15-9), and the average input voltage is $\frac{1}{2}E$.

From Equations (3-3) and (3-7), the electrical energy is

$$W = \text{(average voltage)} \times \text{(average current)} \times t$$

Therefore, the energy supplied to the capacitive circuit is

$$W = \tfrac{1}{2}E \times I \times t$$

$$W = \tfrac{1}{2}E \times \left(\frac{CE}{t}\right) \times t$$

or energy stored is

$$\boxed{W = \tfrac{1}{2}CE^2}$$ (15-10)

When C is in farads and E is in volts, W is given in joules. Note the similarity between Equations (15-10) and (14-14).

EXAMPLE 15-6 Calculate the energy stored in each of the three series-connected capacitors referred to in Example 15-5. Also, calculate the total energy stored.

SOLUTION

For C_1:

$$W_1 = \tfrac{1}{2} C_1 E_1^2$$
$$= \tfrac{1}{2} \times 1 \ \mu F \times (54.5 \ V)^2$$
$$\cong 1.49 \times 10^{-3} \ J$$

For C_2:

$$W_2 = \tfrac{1}{2} C_2 E_2^2$$
$$= \tfrac{1}{2} \times 2 \ \mu F \times (27.3 \ V)^2$$
$$\cong 745 \ \mu J$$

For C_3:

$$W_3 = \tfrac{1}{2} C_3 E_3^2$$
$$= \tfrac{1}{2} \times 3 \ \mu F \times (18.2 \ V)^2$$
$$\cong 497 \ \mu J$$

total energy stored, $W = \tfrac{1}{2} C E^2$
$$= \tfrac{1}{2} \times 0.545 \ \mu F \times (100 \ V)^2$$
$$\cong 2.73 \ mJ$$

and total energy stored, $W = W_1 + W_2 + W_3$
$$= (1.49 + 0.745 + 0.497) \times 10^{-3} \ J$$
$$\cong 2.73 \ mJ$$

PRACTICE PROBLEMS

15-6.1 Determine the energy stored in each of the capacitors in Problem 15-5.2.

15-6.2 Calculate the voltage that must be applied to a 100 μF capacitor to store 33 mJ of energy.

15-7

STRAY CAPACITANCE

Capacitance exists where any conductors are separated by an insulator. Thus, all pairs of adjacent conductors have capacitance between them, and every conductor has a capacitance to ground. This unwanted capacitance is termed *stray capacitance,* and because capacitance is inversely proportional to dielectric thickness, stray capacitance is most easily minimized by keeping conductors as far apart as possible. In dc applications, stray capacitance may be absolutely negligible, but in alternating current circuits, particularly at radio frequencies, the stray capacitance (like stray inductance) can be a severe problem.

SUMMARY OF FORMULAS

Electric flux density:

$$D = \frac{Q}{A}$$

Electric field strength:

$$electric\ field\ strength = \frac{E}{d}$$

Electric charge:

$$Q = CE$$

Permittivity:

$$\epsilon_0 = 8.84 \times 10^{-12}$$

$$= \frac{1}{36\pi \times 10^9}$$

$$\epsilon_0 \epsilon_r = \frac{D}{E}$$

Capacitance:

$$C = \frac{\epsilon_r \epsilon_0 A}{d}$$

Capacitors in parallel:
   Total capacitance:

$$C_p = C_1 + C_2 + C_3 + \cdots$$

Capacitors in series:
   Total capacitance:

$$\frac{1}{C_s} = \frac{1}{C_1} + \frac{1}{C_2} + \frac{1}{C_3} + \cdots$$

Total charge:

$$Q = Q_1 = Q_2 = Q_3 = \cdots$$

Energy stored:

$$W = \tfrac{1}{2}CE^2 \text{ joules}$$

REVIEW QUESTIONS

15-1 Using illustrations, explain the process of electric charge storage in two metal plates separated by a thin layer of insulating material.

15-2 Discuss the electric field, comparing it to a magnetic field. Write the equations for electric field strength and electric flux density and explain each.

15-3 Explain electric lines of force, equipotential lines, dielectric strength, dielectric absorption, permittivity, and relative permittivity.

15-4 State the definition of the farad and from it write an equation for the charge stored in a capacitor.

15-5 Explain why the capacitance of a capacitor is proportional to the plate area and inversely proportional to dielectric thickness.

15-6 Derive an equation for the capacitance of a parallel-plate capacitor.

15-7 Discuss the following characteristics of capacitors: working voltage, tolerance, temperature effects, leakage current, and polarization.

15-8 Sketch the equivalent circuit of a capacitor and briefly explain.

15-9 Describe the construction of air capacitors, paper capacitors, plastic film capacitors, and mica capacitors. Also, explain the most important parameter for each type of capacitor.

15-10 Repeat Review Question 15-9 for ceramic, electrolytic, and tantalum capacitors.

15-11 Derive equations for total capacitance of capacitors connected in parallel, and for the total charge.

15-12 Derive equations for the total capacitance of capacitors connected in series and for the terminal voltage on each capacitor.

15-13 Derive an equation for the energy stored in a charged capacitor.

15-14 Define stray capacitance and explain the problems associated with it.

Chap. 15   Capacitance

PROBLEMS

SECTION 15-2

15-1   A capacitor with a plate area of 50 cm² and a dielectric thickness of 0.5 mm has a charge of 10×10^{-9} C when the applied voltage is 20 V. Calculate the electric field strength and flux density.

15-2   Calculate the plate area for a capacitor which is to store a charge of 750 μC with an electric flux density of 0.3 C/m².

15-3   If the dielectric strength for the capacitor described in Problem 15-1 is 100 000 V/m, calculate the applied voltage at which the dielectric is likely to break down.

15-4   Calculate the electric field strength in a capacitor with 60 V applied to its terminals (a) when the dielectric thickness is 0.3 mm, (b) when the dielectric is 0.025 mm thick.

SECTION 15-3

15-5   Determine the capacitance of a capacitor with a plate area of 50 cm² and a dielectric thickness of 0.5 mm:
a. When the dielectric is air.
b. When the dielectric is glass with a relative permittivity of 75.

15-6   For Problem 15-5, determine the necessary applied voltage in each case to store a charge of 3.3 nC.

15-7   A 20 μF capacitor is to be constructed using metal foil with a plastic film dielectric. If the dielectric is 0.05 mm thick and has a relative permittivity of 3, calculate the area of the metal foil plates.

15-8   Two conductors which are 2 mm in diameter are separated by an insulation with an average thickness of 0.3 cm. The dielectric constant of the insulation is 3.5. Calculate the capacitance per meter of the cable.

15-9   A 50 μF electrolytic capacitor has aluminum-foil plates that are 10 cm × 75 cm. If the oxide dielectric has a relative permittivity of 20, calculate the thickness of the dielectric.

15-10 Calculate the capacitance of two 0.5 m² sheets of foil which are separated by 0.2 mm of paper with a dielectric constant of 4.7.

15-11 A 10 km electric cable has two conductors, each of which is 1 cm thick. The average distance between the two conductors is 0.5 cm, and the insulation material is oil-soaked paper with a relative permittivity of 5. The potential difference between the conductors is 500 V. Determine the capacitance of the cable.

15-12 Two copper conductors each of which are 10 cm long and 3 mm wide are printed on opposite sides of a circuit board. The board material is 1.5 mm thick and has a dielectric constant of 1.8. Calculate the capacitance.

15-13 Two sheets of aluminum foil 1 m in width and 15 m long are rolled up with two sheets of plastic. The plastic is 0.5 mm thick and has a relative permittivity of 3. Calculate the capacitance of the roll. Also, calculate its charge when the terminal voltage is 12 V.

15-14 Calculate the charge on the printed circuit conductors in Problem 15-12 if their potential difference is (a) 33 V, (b) 5 V.

15-15 A printed circuit board has a total conductor area of 25 cm². The opposite side of the board is completely coated with copper. The board is made of teflon and is 1 mm thick with a dielectric constant of 2. Determine the total capacitance.

15-16 Calculate the applied voltage for the capacitor in Problem 15-10 to store a charge of (a) 15 μC, (b) 1.5 nC.

15-17 The charge taken by a capacitor is 0.13 μC when its terminal voltage is 50 V. The plate area of the capacitor is 39 cm², and its dielectric thickness is 0.9 mm. Calculate the relative permittivity of the dielectric.

15-18 The dielectric in the capacitor in Problem 15-17 is replaced with paper which has a dielectric constant of 5.5. Calculate the new capacitance value and the new level of charge when the terminal voltage is 50 V.

15-19 Two circular metal plates 30 cm in diameter are separated by air. The plates take a charge of 0.01 μC when the potential difference between them is 10 V. Calculate the distance between the plates.

15-20 Determine the capacitance of the two plates in Problem 15-19 when the distance between the plates is (a) 0.3 mm, (b) 1 mm.

SECTION 15-5

15-21 For the capacitor described in Problem 15-1, determine the dielectric constant and list several possible dielectric materials.

15-22 A 5 μF capacitor and a 3 μF capacitor are connected in parallel. The parallel combination is connected in series with a 7 μF capacitor and a 10 μF capacitor, and 55 V is applied to the terminals of the circuit. Calculate the total capacitance, the voltage across each capacitor, and the charge on each capacitor.

15-23 Four capacitors, having values of 100 μF, 50 μF, 25 μF, and 10 μF, are connected in series to a 25 V supply. Calculate the total capacitance, the voltage across each capacitor, and the charge on each capacitor.

15-24 Repeat the calculation in Problem 15-23 for the case of all capacitors connected in parallel.

15-25 A 100 μF capacitor is connected in parallel with a 50 μF capacitor, then the two are connected in series with a 25 µF capacitor. Calculate the total capacitance.

15-26 For Problem 15-25, calculate the voltage across each capacitor when a 25 V supply is applied to the terminals of the circuit.

SECTION 15-6

15-27 Calculate the energy stored in each of the capacitors in Problems 15-25 and 15-26.

15-28 Determine the energy stored in the cable referred to in Problem 15-11.

15-29 Calculate the energy stored in each capacitor in Problem 15-23.

15-30 Calculate the energy stored in each capacitor in Problem 15-25.

15-31 The printed circuit board in Problem 15-15 has half the conductor area at 12 V with respect to the copper sheet on the other side, and half at 3 V with respect to the copper sheet. Calculate the energy stored in each half.

15-32 Calculate the voltage that must be applied to two series-connected capacitors to store 700 µJ of energy if the capacitances are $C_1 = 33$ µF and $C_2 = 15$ µF. Also, determine the energy stored in each capacitor.

COMPUTER PROBLEMS

15-33 Write a computer program to calculate the capacitance of each of several different capacitors when their plate areas, dielectric thicknesses, and relative permittivities are specified.

15-34 Write a computer program for series-connected capacitors with a given applied voltage, to determine the total capacitance, voltage on each capacitor, and charge on each capacitor.

ANSWERS TO PRACTICE PROBLEMS

15-2.1	40 kV/m, 81 kV
15-2.2	25×10^{-3} m^2
15-3.1	38.9 pF
15-3.2	0.47 µC
15-3.3	1.8 pF
15-5.1	138.5 pF
15-5.2	2 nC, 15 V, 14.4 V, 0.6 V
15-6.1	11.25 nJ, 4.15 nJ, 180 pJ
15-6.2	25.7 V

16 INDUCTANCE AND CAPACITANCE IN DC CIRCUITS

Objectives   You will be able to:

Sketch a graph showing the growth of current with time in an inductive-resistive circuit.

Calculate the instantaneous levels of current and voltage in an *LR* circuit at any given time.

Define the time constant of an *LR* circuit and explain its special relationship to the performance of the circuit.

Solve problems involving open-circuiting of inductors in which energy is stored.

Sketch the various inductor and resistor voltage waveforms that result when a square wave is applied to an *LR* circuit.

Sketch a graph showing the growth of a capacitor voltage with time in a capacitive-resistive circuit.

Calculate the instantaneous levels of voltage and current in a *CR* circuit at any given time.

Define the time constant of a *CR* circuit and explain its special relationship to the performance of the circuit.

Solve problems involving discharge of charged capacitors.

Sketch the various capacitor and resistor voltage waveforms that result when a square wave is applied to a *CR* circuit.

Introduction

When an inductive-resistive series circuit has its supply voltage switched on, the inductance produces an initial maximum level of counter-emf that gradually falls to zero. The circuit current is zero initially and grows gradually to its maximum level. The behavior of an *LR* circuit is most easily understood by plotting the graphs of instantaneous current versus time and instantaneous inductor voltage versus time. Care must be taken in open-circuiting an *LR* circuit, to avoid an extremely high induced voltage.

In the case of a capacitive-resistive series circuit, when the supply is first switched on the charging current is initially at its maximum level, then it gradually falls to zero. The capacitor voltage is zero at first and grows gradually to its maximum level. As with the *LR* circuit, the behavior of a *CR* circuit can be represented graphically by plotting instantaneous current and voltage versus time. Because energy is stored in a charged capacitor, a large current can flow when the capacitor terminals are short-circuited.

16-1
LR CIRCUIT OPERATION

When switch S_1 is closed in the simple battery and resistor circuit of Figure 16-1(a), the current tends to jump instantaneously to its maximum level of $I = E/R$. This is illustrated by the graph of i versus t in Figure 16-1(b), and it assumes that R is purely resistive. (Actually, there can be no such thing as *pure resistance*, and as will be seen later, there is always some *rise time* involved in the current going from zero to its maximum level.)

Now consider the circuit shown in Figure 16-2(a), and assume that L is a pure inductance. Recall from Equation (14-3) that when the current changes through an inductor, a counter-emf is generated that has the value

$$e_L = L \frac{\Delta i}{\Delta t} \quad \text{volts}$$

The counter-emf opposes the current change that generates it; thus, its polarity is as shown in the circuit. Using Kirchhoff's voltage law, the instantaneous level of current is

$$i = \frac{E - e_L}{R} \qquad \text{(16-1)}$$

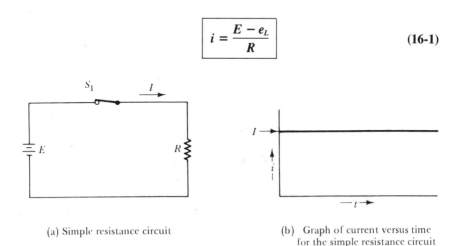

(a) Simple resistance circuit

(b) Graph of current versus time for the simple resistance circuit

FIGURE 16-1. In a simple resistance circuit, the current tends to jump to its final level and remain constant.

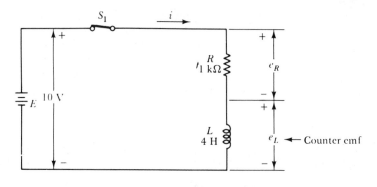

(a) Series LR circuit

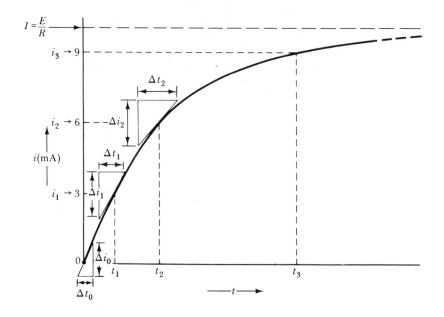

(b) Growth of current in LR circuit

FIGURE 16-2. The counter-emf generated in an inductor when the current level changes causes the current in an LR circuit to grow slowly to its final level when the supply voltage is switched *on*.

Although the actual current level is very small when S_1 is first closed, its *rate of growth* is large. The initial rapid rate of growth of current causes the counter-emf e_L to be virtually equal to E. Putting $e_L = E$ into Equation (16-1) and using the voltage and component values shown in the circuit gives

$$i = \frac{10\,\text{V} - 10\,\text{V}}{1\,\text{k}\Omega} = 0$$

Chap. 16   Inductance and Capacitance in DC Circuits

Therefore, on the graph of i versus t, Figure 16-2(b), i is zero at $t = 0$. At this instant the rate of change of current ($\Delta i_0/\Delta t_0$ on the graph) is at its maximum value. Consequently, because $e_L = L\,\Delta i/\Delta t$, e_L also has its maximum level at $t = 0$.

Some time t_1 after S_1 was closed, the rate of change of current has decreased to a value ($\Delta i_1/\Delta t_1$ on the graph) that gives $e_L = 7$ V. Therefore, using Equation (16-1), at time t_1 the instantaneous current level is

$$i_1 = \frac{10\,\text{V} - 7\,\text{V}}{1\,\text{k}\Omega}$$

$$= 3\,\text{mA}$$

On the graph of i versus t, $i_1 = 3$ mA is plotted at time t_1.

Because e_L is falling, the current level is increasing. As the current level grows, it gets closer to its maximum level of $I = E/R$, and in doing so the rate of change of current $\Delta i/\Delta t$ continues to decrease. This decrease in $\Delta i/\Delta t$ is the cause of the decreasing level of counter-emf $[e_L = L\,\Delta i/\Delta t]$.

After some time t_2, $\Delta i/\Delta t$ has fallen to a level that gives $e_L = 4$ V. Thus,

$$i_2 = \frac{10\,\text{V} - 4\,\text{V}}{1\,\text{k}\Omega} = 6\,\text{mA}$$

and at time t_3, $\Delta i/\Delta t$ is such that $e_L = 1$ V, giving

$$i_3 = \frac{10\,\text{V} - 1\,\text{V}}{1\,\text{k}\Omega} = 9\,\text{mA}$$

The values of i_2 and i_3 are plotted on the graph at times t_2 and t_3, respectively. The final current level when e_L becomes zero is, of course, $I = E/R$.

Because the counter-emf is initially large and falls off smoothly to zero, the instantaneous current grows continuously from zero to its maximum level. As the current approaches maximum, the rate of change of current keeps decreasing, and this produces the decreasing level of e_L.

16-2

INSTANTANE-OUS CURRENT AND VOLTAGE IN *LR* CIRCUITS

CURRENT EQUATION. By means of differential calculus it can be shown that the equation for instantaneous current in an inductive-resistive circuit is

$$\boxed{i = \frac{E}{R}\left(1 - \epsilon^{-t(R/L)}\right)} \qquad \text{(16-2)}$$

where

> i = *instantaneous current, in amperes, at time t*

> E = *supply voltage*

> R = *series resistance, in ohms (including the winding resistance of the inductor)*

> ϵ = *exponential constant* = 2.718

> t = *time, in seconds, from current commencement*

> L = *inductance of the inductor, in henrys*

Using Equation (16-2), the instantaneous current levels can be calculated for several different time intervals from $t = 0$ for a given circuit. The corresponding values of i and t can be used to plot an accurate graph of i versus t for the circuit. The equation can also be manipulated to determine the required value of t, R, or L for a given current level:

$$i = \frac{E}{R}\left(1 - \epsilon^{-t(R/L)}\right)$$

$$\frac{iR}{E} = 1 - \epsilon^{-t(R/L)}$$

$$\epsilon^{-t(R/L)} = 1 - \frac{iR}{E}$$

$$= \frac{E - iR}{E}$$

so that

$$\epsilon^{t(R/L)} = \frac{E}{E - iR}$$

Taking the natural logarithm of both sides,

$$\boxed{t\,\frac{R}{L} = \ln\left(\frac{E}{E - iR}\right)} \tag{16-3}$$

EXAMPLE 16-1 Determine the instantaneous values of current at 2 ms intervals from $t = 0$ for the circuit of Figure 16-2(a).

SOLUTION

Equation (16-2):

$$i = \frac{E}{R}\left(1 - \epsilon^{-t(R/L)}\right)$$

$$At\ t = 0, \qquad i = 0 \qquad\qquad\qquad\qquad\qquad\qquad\text{\textit{point 1 in}}$$
$$\text{\textit{Figure 16-3}}$$

$$At\ t = 2\ ms, \quad i = \frac{10\ V}{1\ k\Omega}\ (1 - \epsilon^{-2\ ms \times (1\ k\Omega/4\ H)}) \cong 3.93\ mA \qquad \text{\textit{point 2}}$$

$$At\ t = 4\ ms, \quad i = \frac{10\ V}{1\ k\Omega}\ (1 - \epsilon^{-4\ ms \times (1\ k\Omega/4\ H)}) \cong 6.32\ mA \qquad \text{\textit{point 3}}$$

At t = 6 ms, *i = 7.77 mA* *point 4*

At t = 8 ms, *i = 8.65 mA* *point 5*

At t = 10 ms, *i = 9.18 mA* *point 6*

At t = 12 ms, *i = 9.5 mA* *point 7*

At t = 14 ms, *i = 9.7 mA* *point 8*

At t = 16 ms, *i = 9.82 mA* *point 9*

At t = ∞, *i = 10 mA = maximum current level*

The corresponding values of t and i are plotted in Figure 16-3

TIME CONSTANT. Referring to the graph of i versus t plotted in Figure 16-3, it is seen that when $t = 4$ ms, i is 6.32 mA. Also, 6.32 mA is 63.2 percent of the maximum current level and

$$t = \frac{L}{R} = \frac{4\ H}{1\ k\Omega} = 4\ ms$$

Therefore, when $t = L/R$, the instantaneous current level is always 63.2 percent of E/R. The quantity L/R is termed the *time constant* of an inductive-resistive circuit, and this is a very important quantity in determining the behavior of the circuit. Sometimes the greek letter τ is used as the symbol for the time constant. It can be shown that after a time of $t = 5L/R$, the current is 99.3 percent of its maximum level (see the graph). By drawing a straight line from the origin at a tangent to the graph of i/t, it is seen that if the initial rate of change of current were maintained, the current would reach its maximum level in a time of $t = L/R$ (see Figure 16-3).

The graph of e_L versus t in Figure 16-4 shows how the counter-emf e_L changes with time. At $t = 0$, e_L is equal to the supply voltage E. At $t = L/R$, e_L has fallen by 63.2 percent of E, while at $t = 5L/R$, e_L has fallen through 99.3 percent of its initial level. Comparing the graphs of i versus t and e_L versus t, it is seen that inductor terminal voltage drops as the current increases.

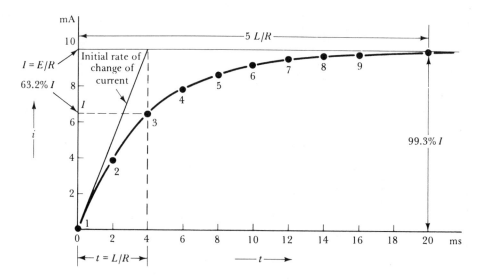

FIGURE 16-3. Graph of current i versus time t for a series LR circuit. The current increases to 63.2% of its maximum level at $t = L/R$ and to 99.3% of its maximum at $t = 5L/R$.

EXAMPLE 16-2 Calculate the level of the counter-emf for the circuit in Figure 16-2(a) at $t = L/R$ and at $t = 5L/R$.

SOLUTION

Equation (16-2):

$$i = \frac{E}{R}\left(1 - \epsilon^{-t(R/L)}\right)$$

Therefore, $\qquad iR = E - E\epsilon^{-t(R/L)}$

and from Equation (16-1):

$$iR = E - e_L$$

Therefore, $\qquad E - e_L = E - E\epsilon^{-t(R/L)}$

giving $\qquad e_L = E\epsilon^{-t(R/L)}$

At $t = L/R$,

$$e_L = E\epsilon^{-1}$$
$$= 10 \text{ V} \times \epsilon^{-1}$$
$$e_L \cong \textbf{3.68 V} \qquad \textit{point 1 on Figure 16-4}$$

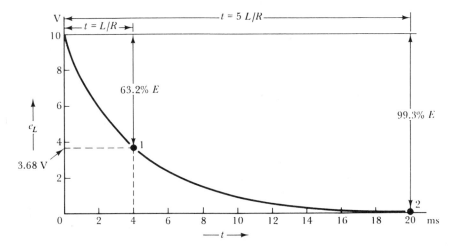

FIGURE 16-4. Graph of inductor voltage e_L versus time t for a series LR circuit. The voltage falls by 63.2% of its maximum level at $t = L/R$ and by 99.3% of its maximum at $t = 5L/R$.

At $t = 5L/R$,

$$e_L = 10 \text{ V} \times \epsilon^{-5}$$

$$e_L \cong \mathbf{0.067 \text{ V}} \qquad \textit{point 2 on Figure 16-4}$$

PRACTICE PROBLEMS

16-2.1 A series LR circuit has $L = 33$ mH, $R = 3.3$ kΩ, and $E = 12$ V. Determine the instantaneous current levels at five evenly spaced time intervals from the supply is switched *on* until the current is at its maximum level.

16-2.2 An inductor is included in series with a 100 Ω resistor to limit the rate of current growth when the supply is switched *on*. If the current is to reach 80% of its maximum level in a time of 3 ms, determine the required inductance.

16-3

OPEN-CIRCUITING AN INDUCTIVE CIRCUIT

COUNTER-EMF. From Equation (14-14), the energy stored in an inductive circuit is

$$W = \tfrac{1}{2}LI^2 \text{ joules}$$

The equation shows that when a steady-state current is flowing through an inductor, energy is stored in the inductor. Also, when the current goes to zero, there is obviously no energy stored.

Now consider what occurs when the switch is opened in an inductive circuit, such as that shown in Figure 16-5. Before the switch is opened, a constant current is flowing through the inductor, and energy is stored. When the switch is opened, the current must go to zero; consequently, the energy contained in the inductor must also go to zero. This means that the energy must somehow flow out of the inductor at the instant that the switch is opened. Now recall that the counter-emf generated in an inductor opposes the current change that produces it. Therefore, since the current is falling when the switch is opened, the counter-emf generated is such that it tends to oppose the fall in current; that is, its polarity is such that it tries to maintain the current flow. This is illustrated in Figure 16-5, where the counter-emf is + at the bottom of the inductor and − at the top. This polarity is opposite to that of the counter-emf when the switch was first closed [see Figure 16-2(a)].

The effect of the counter-emf trying to maintain current flow when the switch is opened is that a spark jumps across the open switch contacts. The arcing across the contacts continues for the brief period necessary to discharge the energy contained in the inductor. To produce such an arc across open contacts requires a very high voltage. The high voltage is, of course, the counter-emf. It is not surprising that a very high voltage is generated when it is remembered that the equation for counter-emf is,

$$e_L = L \frac{\Delta i}{\Delta t}$$

Because the open-circuit tries to make the current go to zero instantaneously, the rate of change of current $\Delta i / \Delta t$ is a very large quantity, and therefore $L(\Delta i / \Delta t)$ is also very large.

To get some idea of the voltage generated when the switch is opened in an inductive circuit, consider again the circuit in Figure 16-2(a). The steady-state current level while the switch is closed is

$$I = \frac{E}{R} = \frac{10 \text{ V}}{1 \text{ k}\Omega} = 10 \text{ mA}$$

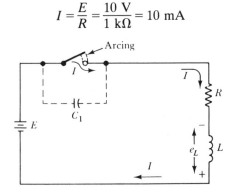

FIGURE 16-5. When an inductive circuit is open-circuited, the energy stored in the inductor tends to keep current flowing in the circuit; arcing may occur at a switch.

At the instant that the switch is opened, the counter-emf is such that for a brief time period, this 10 mA continues to flow across the air gap between the switch contacts. Since the resistance of the air gap could very easily be much greater than 1 MΩ, the voltage across the air gap is at least

$$e = IR$$

$$= 10 \text{ mA} \times 1 \text{ M}\Omega$$

$$= 10\ 000 \text{ V}$$

This 10 000 V is the generated counter-emf.

CONTROLLING THE COUNTER-EMF. It is seen that when a switch is opened in an inductive circuit, the high level of counter-emf may damage the circuit insulation. Obviously, some means of discharging the energy in the inductor must be provided. A capacitor connected across the switch, as shown broken in Figure 16-5, is sometimes employed for this purpose. The capacitor initially behaves as a short-circuit when the switch is opened, and this allows a current flow to continue until the capacitor is charged to the level of the supply voltage E.

Another method of protecting an inductive circuit against high levels of counter-emf is shown in Figure 16-6. Resistor R_2, in parallel with R_1 and L, has no effect on the steady-state current through R_1 and L while the switch remains closed. When the switch is opened, the total resistance in series with L is $(R_1 + R_2)$. Therefore, the maximum level of counter-emf that must be generated to maintain the initial level of I is

$$e_L = I\,(R_1 + R_2)$$

For $R_2 = 5$ kΩ connected in the circuit of Figure 16-2(a), the counter-emf is

$$e_L = 10 \text{ mA}(5 \text{ k}\Omega + 1 \text{ k}\Omega)$$

$$= 60 \text{ V}$$

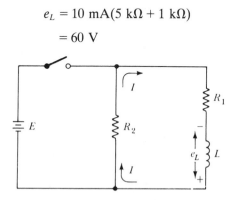

FIGURE 16-6. One method of protecting an inductive circuit from the effects of high counter-emf levels when the switch is open-circuited. The energy stored in the inductor is discharged via resistor R_2.

EXAMPLE 16-3   A circuit connected as shown in Figure 16-6 has $R_1 = 500 \ \Omega$, $L = 500$ mH, and $E = 100$ V. Determine how long it takes the inductor current to get to 1.7 mA after the switch is closed. Also, calculate the required value of R_2 if the counter-emf is not to exceed 300 V when the switch is opened.

SOLUTION

From Equation (16-3),

$$t = \frac{L}{R_1} \ln \left(\frac{E}{E - iR_1} \right)$$

$$= \left(\frac{500 \text{ mH}}{500 \ \Omega} \right) \ln \left(\frac{100 \text{ V}}{100 \text{ V} - (1.7 \text{ mA} \times 500 \ \Omega)} \right)$$

$$= 8.5 \ \mu s$$

$$\text{maximum current level, } I = \frac{E}{R_1} = \frac{100 \text{ V}}{500 \ \Omega}$$

$$I = 200 \text{ mA}$$

$$\text{counter-emf, } e_L = I \, (R_1 + R_2)$$

Therefore,

$$\frac{e_L}{I} = R_1 + R_2$$

and

$$R_2 = \frac{e_L}{I} - R_1$$

$$= \frac{300 \text{ V}}{200 \text{ mA}} - 500 \ \Omega$$

$$= 1 \text{ k}\Omega$$

PRACTICE PROBLEM

16-3.1   For the *LR* circuit in Problem 16-2.1 determine the required parallel resistance (as in Figure 16-6) to limit the counter-emf to 18 V at switch-*off*.

16-4

LR CIRCUIT WAVEFORMS

Consider the *LR* circuit once again, as illustrated in Figure 16-7(a). Suppose that switch S_1 is closed for an accurately measured time t_1, then opened again for an exactly equal time t_2, and that the sequence repeated over and over. The resulting voltage applied to the circuit would be as shown in Figure 16-7(b). For obvious reasons the shape is termed a *square wave* or *pulse wave* and the time t_1 is referred to as the *pulse width*. It would not really be practical to attempt to manually open and

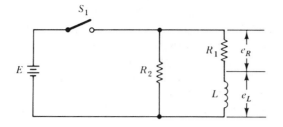

(a) Circuit for generating waveforms below

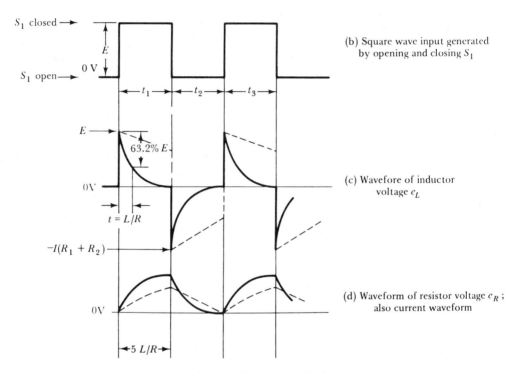

(b) Square wave input generated by opening and closing S_1

(c) Wavefore of inductor voltage e_L

(d) Waveform of resistor voltage e_R; also current waveform

FIGURE 16-7. When a voltage is switched *on* to an *LR* circuit, e_L jumps to the level of E then decreases exponentially, and I increases exponentially from zero. When the voltage is switched *off*, e_L jumps to $-I(R_1 + R_2)$ then rises exponentially to zero, and I decreases exponentially from its maximum level.

close a switch in order to generate a square wave; suitable electronic instruments are available for the purpose. However, going along with the assumption that a square wave is applied as illustrated, the resulting voltages across L and R can be investigated.

When E is switched *on* to the circuit, the counter-emf e_L immediately jumps to the level of the applied voltage. Then as the current grows, e_L slowly falls to zero, as shown in Figure 16-7(c). The counter-emf remains at zero until the switch is opened, and as already discussed, it now immediately jumps to $-I(R_1 + R_2)$, from which it again slowly goes to zero, as illustrated. The positive and negative step and decline of e_L is repeated over and over as the switch is closed and opened, giving the *spike waveform* illustrated in Figure 16-7(c).

The current through R_1 and L is zero initially when S_1 is closed, and rises to its maximum level of $I = E/R_1$ when e_L has fallen to zero. When S_1 is opened, the current falls slowly as e_L moves from $-I(R_1 + R_2)$ back to zero. The resulting current waveform is shown in Figure 16-7(d). Since the voltage across R_1 is simply $I \times R_1$, the waveshape of e_R is exactly the same as the current waveshape.

Recalling that in a time $t = 5(L/R)$ the counter-emf e_L falls by 99.3 percent of its maximum level, it is seen from Figure 16-7(c) that $t_1 = t_2 \cong 5(L/R)$. In this case, where the time constant L/R is short by comparison to the pulse width t, the circuit is said to have a *short time constant*. When L/R is greater than t, the typical shape of the waveforms are as shown by the broken lines in Figure 16-7(c) and (d), and the circuit is said to have a *long time constant*.

16-5
CR CIRCUIT OPERATION

A capacitor and resistor are shown connected in series in Figure 16-8(a), together with a supply voltage E and a switch S_1. At any instant the voltage across R is

$$e_R = E - e_C$$

The charging current is

$$i = \frac{e_R}{R}$$

or

$$\boxed{i = \frac{E - e_C}{R}} \qquad \textbf{(16-4)}$$

If the charge on the capacitor is zero at the instant the switch is closed, then $e_C = 0$, and for the voltage and component values shown in Figure 16-8(a),

$$i = \frac{10 \text{ V} - 0}{1 \text{ k}\Omega}$$

$$= 10 \text{ mA}$$

The current flow causes the capacitor to charge with the polarity illustrated in the figure. After a time t_1, the capacitor voltage might be 3 V [see Figure 16-8(b)]. Then the charging current becomes

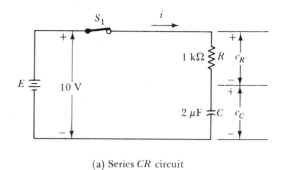

(a) Series CR circuit

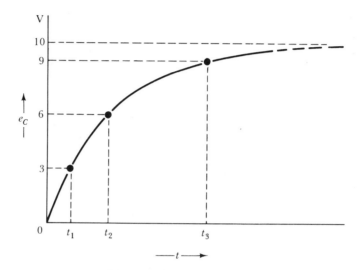

(b) Growth of capacitor voltage in
series CR circuit

FIGURE 16-8. The capacitor voltage in a series CR circuit tends to grow slowly from zero to its final level when the supply voltage is first switched *on*.

$$i_1 = \frac{10\,\text{V} - 3\,\text{V}}{1\,\text{k}\Omega}$$

$$= 7\,\text{mA}$$

It is seen that because C has accumulated some charge, the voltage across R is reduced, and consequently the charging current is reduced from 10 mA to 7 mA. Because the charging current has been reduced, the capacitor voltage is now growing at a slower rate than before.

After time t_2, e_C has grown to 6 V, and the charging current becomes

$$i_2 = \frac{10 \text{ V} - 6 \text{ V}}{1 \text{ k}\Omega} = 4 \text{ mA}$$

The charging current has now been further reduced (from 7 mA to 4 mA), so the capacitor is charging at an even slower rate than before. An even longer time is now required for the capacitor voltage to grow by another 3 V. Because e_C is continuously increasing, the voltage across R is continuously decreasing, and therefore the charging current is continuously decreasing. This means that C is charged at a rapid rate initially, then the rate decreases as the capacitor voltage grows.

16-6

INSTANTANE- OUS CURRENT AND VOLTAGE IN *CR* CIRCUITS

VOLTAGE EQUATION. The equation for the instantaneous voltage on a capacitor in a resistive-capacitive circuit can be derived by differential calculus:

$$\boxed{e_C = E - (E - E_0)\epsilon^{-t/(CR)}} \tag{16-5}$$

where

> e_C = *capacitor voltage at time t*
>
> E = *supply voltage*
>
> E_0 = *initial level of capacitor voltage*
>
> ϵ = *exponential constant* = 2.718
>
> t = *time, in seconds, from commencement of charge*
>
> C = *capacitance value, in farads*
>
> R = *charging resistance, in ohms*

Using Equation (16-5), the instantaneous levels of capacitor voltage can be calculated for several different time intervals from $t = 0$ for a given circuit. The corresponding values of e_C and t can then be plotted to give an accurate graph of e_C versus t for the circuit. The equation can also be manipulated to determine expressions for t, C, and R for a given capacitor voltage level.

When the capacitor is initially uncharged,

$$E_0 = 0$$

and

$$e_C = E - E\epsilon^{-t/(CR)}$$

or

$$\boxed{e_C = E(1 - \epsilon^{-t/(CR)})} \tag{16-6}$$

Also,
$$E\epsilon^{-t/(CR)} = E - e_C$$

$$\epsilon^{-t/(CR)} = \frac{E - e_C}{E}$$

$$\epsilon^{t/(CR)} = \frac{E}{E - e_C}$$

Taking the natural logarithm of both sides,

$$\boxed{\frac{t}{CR} = \ln\left(\frac{E}{E - e_C}\right)} \qquad \textbf{(16-7)}$$

EXAMPLE 16-4 Determine the instantaneous values of capacitor voltage at 1 ms intervals from $t = 0$ for the circuit in Figure 16-8(a).

SOLUTION

Assuming that the initial level of capacitor voltage is zero, Equation (16-6) can be used:

$$e_C = E\left(1 - \epsilon^{-t/(CR)}\right)$$

At $t = 0$, $e_C = 0$ V   *point 1 in Figure 16-9*

At $t = 1$ ms, $e_C = 10$ V$(1 - \epsilon^{-1\ ms/(2\ \mu F \times 1\ k\Omega)})$

                 $e_C = 3.93$ V   *point 2*

At $t = 2$ ms, $e_C \cong 6.32$ V   *point 3*

At $t = 3$ ms, $e_C \cong 7.77$ V   *point 4*

At $t = 4$ ms, $e_C \cong 8.65$ V   *point 5*

At $t = 5$ ms, $e_C \cong 9.18$ V   *point 6*

At $t = 6$ ms, $e_C \cong 9.5$ V   *point 7*

At $t = 7$ ms, $e_C \cong 9.7$ V   *point 8*

At $t = 8$ ms, $e_C \cong 9.82$ V   *point 9*

At $t = \infty$, $e_C = 10$ V $=$ *maximum voltage level*

TIME CONSTANT.   Referring to the graph of e_C versus t plotted in Figure 16-9, it is seen that when $t = 2$ ms, e_C is 6.32 V. Also, 6.32 V is 63.2 percent of the maximum voltage level and

$$t = CR = 2\ \mu F \times 1\ k\Omega = 2\ ms$$

Therefore, when $t = CR$, the instantaneous capacitor voltage level is always 63.2 percent of E. The quantity CR is the *time constant* of a capacitive-resistive circuit, and, as in the case of an *LR* circuit, the time constant largely determines the behavior of the circuit. After a time period of 5 *CR*, the capacitor voltage is 99.3 percent of its maximum level. By drawing a straight line at a tangent to the graph of e_C/t, it can be shown that if the initial rate of charge were maintained, the capacitor voltage would reach its maximum level in a time of $t = CR$ (see Figure 16-9).

The graph of i_C versus t in Figure 16-10 shows how the charging current changes with time. At $t = 0$, $i_C = E/R$. At $t = CR$, i_C has fallen by 63.2 percent of E/R. And at $t = 5\,CR$, i_C has fallen through 99.3 percent of its initial level.

EXAMPLE 16-5   Calculate the value of capacitor charging current for the circuit in Figure 16-8(a) at $t = CR$ and $t = 5\,CR$.

SOLUTION

Equation (16-6):

$$e_C = E\,(1 - \epsilon^{-t/(CR)})$$

and from Equation (16-4)

$$i = \frac{E - e_C}{R}$$

Therefore,
$$i = \frac{E - E\,(1 - \epsilon^{-t/(CR)})}{R}$$

$$= \frac{E\epsilon^{-t/(CR)}}{R}$$

At $t = CR$,

$$i = \frac{E\epsilon^{-1}}{R} = \frac{10\text{ V} \times \epsilon^{-1}}{1\text{ k}\Omega}$$

$$\cong \textbf{3.68 mA} \qquad point\ 1\ on\ Figure\ 16\text{-}10$$

At $t = 5\,CR$,

$$i = \frac{10\text{ V} \times \epsilon^{-5}}{1\text{ k}\Omega}$$

$$\cong \textbf{67 μA} \qquad point\ 2\ on\ Figure\ 16\text{-}10$$

PRACTICE PROBLEMS

16-6.1   A series *CR* circuit has $C = 10$ μF, $R = 3.3$ kΩ, and $E = 12$ V. Determine the instantaneous current levels at five evenly

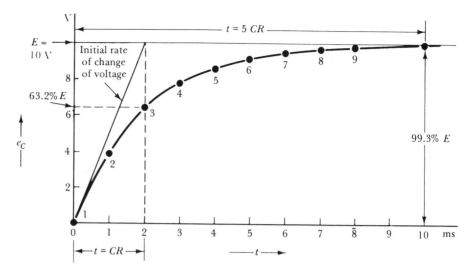

FIGURE 16-9. Graph of capacitor voltage e_C versus time t for a series CR circuit. The voltage increases to 63.2% of its maximum level at $t = CR$, and to 99.3% of its maximum at $t = 5CR$.

spaced time intervals from the supply is switched *on* until the current is at its maximum level.

16-6.2 Two series-connected resistors $R_1 = 4.7$ kΩ and $R_2 = 22$ kΩ are connected via a switch to a 5 V supply. A 100 pF capacitor is connected in parallel with R_1. Determine the circuit current level at switch-*on*, and the time taken for the capacitor to discharge to zero when the supply is open-circuited.

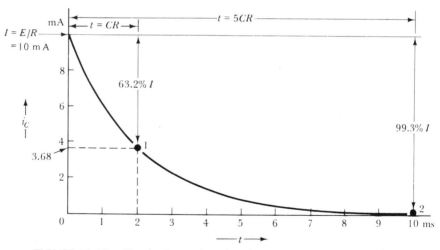

FIGURE 16-10. Graph of capacitor charging current i_C versus time t for a series CR circuit. The current falls by 63.2% of its maximum level at $t = CR$, and by 99.3% of its maximum at $t = 5CR$.

16-7
DISCHARGING A CAPACITOR

CAPACITOR VOLTAGE. From Equation (15-11), the energy stored in a charged capacitor is

$$W = \tfrac{1}{2}CE^2 \text{ joules}$$

From the equation, it is seen that when a steady-state voltage level exists across the capacitor terminals, there is a charge stored in the capacitor.

When the switch is opened in a series CR circuit such as that shown in Figure 16-8(a), there is no effect on the charge stored in the capacitor. The capacitor is able to hold its charge as a static quantity, and its terminal voltage is constant. This contrasts with the case of an inductor, where the charge is a dynamic quantity that cannot be maintained when the current is interrupted. The capacitor can hold its charge indefinitely, and *if it is charged to a high voltage it can be **dangerous** to someone working on a circuit believed to be safely disconnected from the supply.*

SAFE DISCHARGING. Sometimes the switch employed in a CR circuit has two contacts, as illustrated in Figure 16-11(a). Then, when S_1 is switched to position 2, after being in position 1 for some time, C_1 is discharged via R_1. In this case the capacitor has already been charged up to E volts, with the polarity shown. When S_1 is in position 2, the capacitor acts like a voltage source and causes current to flow through R_1, as illustrated. This current discharges the capacitor afer a time approximately equal to $5\,CR$. Note that the direction of the discharging current through R_1 is opposite to the charging current direction when S_1 was in position 1. Consequently, the voltage drop e_R across R_1 now also has reversed polarity.

During discharge, the capacitor voltage at any instant can be determined from Equation (16-5). With S_1 in position 2 in the circuit of Figure

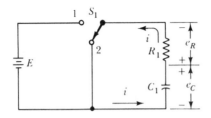

(a) Discharge of C_1 through R_1

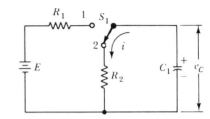

(b) Discharge of C_1 through R_2

FIGURE 16-11. Method of discharging a charged capacitor. When the supply is switched *off*, C_1 discharges through R_1 or R_2.

16-11(a), the supply voltage to the CR circuit is zero. Therefore, putting $E = 0$ in Equation (16-5) gives

$$e_C = E_0 \epsilon^{-t/(CR)} \qquad \text{(16-8)}$$

Figure 16-11(b) shows a slightly different discharge circuit arrangement. When S_1 is in position 1, C_1 is charged from the battery via R_1, as before. When S_1 goes to position 2, R_1 is taken out of the circuit and the capacitor is discharged via resistor R_2. By selecting R_2 larger than R_1, the discharge time for the capacitor may be made longer than the charge time. Alternatively, R_2 may be selected smaller than R_1, so the capacitor discharge time is shorter than the charging time. In this circumstance, the initial level of discharge current can be very large. For example, if e_C is 10 V and $R_2 = 0.1\ \Omega$, the initial level of discharge current is

$$i = \frac{10\ \text{V}}{0.1\ \Omega} = 100\ \text{A}$$

EXAMPLE 16-6   A CR circuit connected as shown in Figure 16-11(b) has $E = 100$ V and $C_1 = 1\ \mu\text{F}$. Determine the value of R_1 if C_1 is to become charged to 50 V in 20 ms. Also, calculate the value of R_2 that will limit the initial level of discharge current to 1 mA.

SOLUTION

From Equation (16-7),

$$R_1 = \frac{t}{C\ \ln[E/(E - e_C)]}$$

$$= \frac{20\ \text{ms}}{1\ \mu\text{F}\ \ln[100\ \text{V}/(100\ \text{V} - 50\ \text{V})]}$$

$$\cong \mathbf{28.9\ k\Omega}$$

The maximum level of e_C is

$$e_C = E = 100\ \text{V}$$

The initial level of discharge current is

$$i = \frac{e_C}{R_2}$$

Therefore, $$R_2 = \frac{e_C}{i} = \frac{100\ \text{V}}{1\ \text{mA}}$$

$$= \mathbf{100\ k\Omega}$$

16-7.1 The circuit in Figure 16-11(a) has $R_1 = 470 \, \Omega$, $C_1 = 2.7 \, \mu\text{F}$, and $E = 18 \, \text{V}$. Calculate the time for the capacitor to reach 80% of maximum charge, and determine the initial level of discharge current at switch-*off*.

16-8

CR CIRCUIT WAVEFORMS

As explained in Section 16-4, a *square wave* can theoretically be generated by opening and closing a switch connected between a circuit and the supply battery. Refer to the circuit of Figure 16-12(a), and assume that switch S_1 is held in position 1 for an accurately timed period t_1, and then moved to position 2 for an equal time t_2. During t_1, capacitor C_1 is charged from the battery via R_1, and during t_2, the capacitor discharges via R_1. The input waveform is as shown in Figure 16-12(b).

When the switch first goes to position 1, the capacitor voltage starts to grow slowly from 0 toward E, as illustrated in Figure 16-12(c). When e_C finally reaches E, it remains constant until S_1 is moved to position 2. Then e_C immediately begins to fall slowly toward zero, at which level it remains until the switch is moved over again. This growth and decline of e_C is repeated over and over as the switch is moved from one position to the other, resulting in the waveform shown.

When S_1 is switched to position 1 and e_C is zero, the charging current is a maximum of $I = E/R$. Then as the capacitor voltage grows, the current falls off to zero. This is illustrated in Figure 16-12(d). Similarly, when S_1 is moved to position 2, the current becomes a discharge current and it has an initial maximum level of $I = -E/R$. Thus, as shown in Figure 16-12(d), the current is a negative quantity falling off to zero during t_2. Once again, the sequence is repeated as S_1 is moved from one position to the other, and the illustrated *spike waveform* is generated. The resistor voltage e_R is, of course, $i \times R_1$; consequently, the waveform of e_R is identical in shape to the current waveform.

The waveforms shown in Figure 16-12 are those that are typically obtained when $t_1 = t_2 \cong 5CR$ [see Figure 16-12(c)]. CR is therefore thought of as a *short time constant* by comparison to the charge and discharge times of the capacitor. When CR is several times greater than the pulse width, the circuit is said to have a *long time constant*. In this case the typical waveforms would be as shown by the broken lines in Figure 16-12(c) and (d).

SUMMARY OF FORMULAS

Current in a series LR circuit:

$$i = \frac{E - e_L}{R}$$

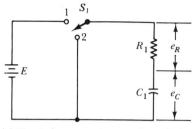

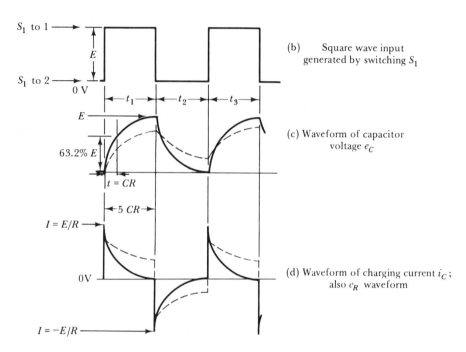

(a) Circuit for generating waveforms below

(b) Square wave input generated by switching S_1

(c) Waveform of capacitor voltage e_C

(d) Waveform of charging current i_C; also e_R waveform

FIGURE 16-12. When a voltage is switched *on* to a series CR circuit, e_c increases exponentially toward the level of E over a time of $5CR$. The current jumps to $I = E/R$ then decreases exponentially to zero. When the voltage is switched *off*, e_C decreases exponentially from its maximum level, and I jumps to $-E/R$ then rises exponentially to zero.

Instantaneous current in an LR series circuit:

$$i = \frac{E}{R}\left(1 - \epsilon^{-t(R/L)}\right)$$

R, L, and t values in an LR series circuit:

$$t\frac{R}{L} = \ln\left(\frac{E}{E - iR}\right)$$

Current in a series CR circuit:

$$i = \frac{E - e_C}{R}$$

Summary of Formulas

Instantaneous voltage in a series CR circuit:

$$e_C = E - (E - E_0)\epsilon^{-t/(CR)}$$

When $E_0 = 0$:

$$e_C = E(1 - \epsilon^{-t/(CR)})$$

R, C, and t values in a series CR circuit:

$$\frac{t}{CR} = \ln\left(\frac{E}{E - e_C}\right)$$

Instantaneous discharging voltage in a series CR circuit:

$$e_C = E_0\epsilon^{-t/(CR)}$$

REVIEW QUESTIONS

16-1 Sketch an approximate representation of the current growth in a series inductive-resistive circuit. Carefully explain how the current and the counter-emf vary with time.

16-2 Define the *time constant* of an *LR* circuit and explain the special relationships between the time constant and the time at which the inductor's instantaneous current and voltage reach certain levels.

16-3 Explain what occurs when the supply switch in a series *LR* circuit is opened. Discuss the danger to the circuit that occurs and methods used to avoid damage.

16-4 Sketch the voltage and current waveforms that result when a square wave is applied to a series *LR* circuit:
a. For a circuit with long time constant.
b. For a circuit with short time constant.
Explain briefly.

16-5 Sketch an approximate representation of the voltage growth across a capacitor being charged via a series resistor. Carefully explain how the charging current and voltage vary with time.

16-6 Define the *time constant* of a *CR* circuit and explain the special relationship between the time constant and the time at which the capacitor's instantaneous voltage and charging current reach certain levels.

16-7 Discuss what occurs when the supply to a *CR* circuit is interrupted, and when a fully charged capacitor is short-circuited. Explain any danger that results, and the methods used to avoid damage.

16-8 Sketch the voltage and current waveforms that result when a square wave is applied to a series *CR* circuit:
a. For a circuit with long time constant.
b. For a circuit with short time constant.
Explain briefly.

SECTION 16-2

16-1   A circuit consisting of a 100 mH inductor in series with a 500 Ω resistor has a 50 V supply. Calculate the instantaneous levels of circuit current at 0.2 ms intervals up to $t = 1$ ms and plot the graph of i versus t.

16-2   A 500 mH inductor with a coil resistance of 280 Ω is to be connected to a 12 V supply via a resistor. The inductor current is to be approximately 63% of its maximum level 0.39 ms after switch-*on*. Select a suitable resistor value and determine the time taken for the current to reach its approximate maximum level.

16-3   For the circuit described in Problem 16-1, calculate the instantaneous levels of inductor voltage at 0.25 ms intervals and plot the graph of e_L versus t.

16-4   For the circuit in Problem 16-2, determine the inductor current and voltage at 0.1 ms, 0.2 ms, and 0.3 ms from the instant of supply switch-*on*.

16-5   A series LR circuit with a 35 V supply is to have an instantaneous current of 5 mA at a time of 22 ms from switch-*on*. If $R = 4.7$ kΩ, determine the required value of L.

16-6   For the circuit in Problem 16-5, calculate the approximate times from switch-*on* for the inductor voltage to reach 13 V and zero.

16-7   A 600 mH inductor with a winding resistance of 250 Ω is connected in series with a 330 Ω resistor and a 24 V supply. Determine the lengths of time that elapse from the instant the supply is switched *on* until the inductor voltage reaches 5 V and 17 V.

16-8   An inductor is included in series with an electric lamp filament to limit the current growth at switch-*on*. The filament has a resistance of 1 Ω, and the current is to be approximately 30% of maximum 20 ms after switch-*on*. Determine a suitable inductance value.

16-9   A 24 V supply is connected to an inductor via a 1.2 kΩ resistor. A time of 270 μs is to elapse from the instant of switch-*on* until the resistor voltage reaches 5 V. Determine the required value of inductance.

16-10   For the circuit described in Problem 16-9, calculate the times from switch-*on* for the resistor voltage to reach 3 V, 6 V, 12 V, and approximately 24 V.

SECTION 16-3

16-11   For the series circuit described in Problem 16-7, show how an additional resistor should be connected to limit the inductor emf when the supply is switched *off*. Determine the required resistor value to limit the inductor emf to a maximum of 75 V.

16-12 Calculate the counter-emf generated in the inductor in Problem 16-5 when the supply is replaced with a short circuit. Also, determine the possible counter-emf generated when the supply is switched *off*, if the open-circuited switch resistance is 10 MΩ.

16-13 A 500 mH inductor is connected to a 30 V supply via a 12 kΩ resistor. If the switch is open-circuited, and the resistance between its terminals is 25 MΩ, calculate the counter-emf generated by the inductor. Also, determine a suitable value of resistor to be connected in parallel with the inductor to limit its counter-emf to a maximum of 60 V.

16-14 The circuit in Figure 16-6 has $R_1 = 1.5$ kΩ, $R_2 = 2.7$ kΩ, $L = 300$ mH, and $E = 15$ V. Calculate the time required from switch-*on* for the inductor voltage to fall to 3 V. Also determine the counter-emf generated at switch-*off*.

SECTION 16-6

16-15 A series circuit consisting of a 10 μF capacitor and a 2.2 kΩ resistor is connected to a 50 V supply. Calculate the instantaneous levels of capacitor voltage at 15 ms intervals from supply switch-*on* up to $t = 105$ ms. Plot the graph of e_C versus t.

16-16 A 200 μF capacitor is to be connected to a 15 V supply via a resistor. The capacitor voltage is to be approximately 63% of maximum 0.78 s after switch-*on*. Select a suitable resistor value and calculate the approximate time for the capacitor to become fully charged.

16-17 For the circuit described in Problem 16-15, calculate the instantaneous charging current levels at 20 ms intervals from switch-on and plot the graph of i versus t.

16-18 Calculate the charging current and capacitor voltage for the circuit in Problem 16-16 at time intervals of 0.2 s, 0.4 s, 0.8 s, and 1.6 s from switch-*on*.

16-19 A series CR circuit with a 30 V supply is to have an instantaneous capacitor voltage of 22 V at a time of 65 μs from supply switch-*on*. If $C = 0.1$ μF, determine the required resistance of R.

16-20 For the circuit described in Problem 16-19, calculate the times from switch-*on* for the charging current to reach 30 mA and approximately zero.

16-21 A 47 μF capacitor is connected in series with a 4.7 kΩ resistor and a 5 V supply. Determine how long it takes from the instant of switch-*on* for the capacitor voltage to reach 2.2 V.

16-22 A printed circuit board has a 150 pF capacitance between conductors and ground. If a dc voltage is applied to the circuit board conductors via a 33 kΩ resistance, how long will it take for the conductor voltage to reach 90% of maximum?

16-23 A 1500 μF capacitor is charged from a 200 V supply and then connected in parallel with an uncharged 1200 μF capacitor. Determine the terminal voltage of the two capacitors in parallel.

16-24 A 120 km cable with a capacitance of 50 pF per meter has a dc voltage applied from a 6 kΩ source. Calculate the approximate time for the cable to become fully charged.

16-25 The terminal voltage of a capacitor is to take 750 μs from the instant of supply switch-*on* to arrive at 3.3 V. Calculate the required capacitor value if it is to be connected to a 5 V supply via a 1.5 kΩ resistor.

16-26 For the circuit described in Problem 16-25, calculate the times from switch-*on* for the charging current to reach 200 μA, 400 μA, 800 μA, 1.6 mA, and 3.2 mA.

SECTION 16-7

16-27 The 10 km cable described in Problem 15-11 has an insulation resistance of 10 MΩ between the two conductors. Calculate the time required for the potential difference between the conductors to fall from 500 V to 10 V when the supply is disconnected.

16-28 Calculate the time required for the voltage of the fully charged capacitor in Problem 16-21 to fall to 3 V when the supply is replaced with a short circuit.

16-29 Sketch the series circuit described in Problem 16-21 and show how an additional resistor and switch should be included for safely discharging the capacitor. Determine the value of the additional resistor if the capacitor discharge current is not to exceed 1 mA. Also, calculate approximately how long it will take for the capacitor to become completely discharged.

16-30 Determine the time required for a 100 μF capacitor to discharge from 120 V to 20 V if the discharge occurs thorugh a 100 MΩ insulation resistance. Calculate the value of a resistor through which the capacitor can be safely discharged with a maximum current of 10 mA and determine the new time required for discharge from 120 V to 20 V.

COMPUTER PROBLEMS

16-31 Write a computer program to determine the intantaneous levels of current in a series *LR* circuit at a selectable number of equally spaced time intervals from the instant of supply switch-*on*.

16-32 Write a computer program to determine the instantaneous levels of discharge current in a *CR* circuit at a selectable number of equally spaced time intervals from maximum charge.

ANSWERS TO PRACTICE PROBLEMS

16-2.1	2.3 mA, 3.15 mA, 3.46 mA, 3.57 mA, 3.62 mA
16-2.2	186 mH
16-3.1	1.65 kΩ
16-6.1	1.34 mA, 0.49 mA, 181 μA, 66.7 μA, 25 μA
16-6.2	227 μA, 2.35 μs
16-7.1	2.04 ms, 38.3 mA

17

ALTERNATING CURRENT AND VOLTAGE

Objectives   You will be able to:

Draw a sketch of a simple alternating current generator. Explain how the generator produces an ac output and sketch the output waveform.

Calculate the instantaneous output voltage from an ac generator at any given time.

For a given sinusoidal waveform, determine peak value, rms value, average value, and frequency.

Calculate the time period and wavelength of a given sine wave.

Determine the phase relationship between two sine waves.

Calculate the instantaneous, peak, average, and rms levels of current in a resistor when a sinusoidal voltage is applied. Also, determine the power dissipated in the resistor.

Identify the controls on the front panel of a cathode ray oscilloscope and explain the function of each control.

Use an oscillosocpe to measure voltage, frequency, and phase difference.

Introduction

Alternating current (ac) is current that is continuously reversing direction, alternately flowing in one direction and then in the other. Alternating voltage can be similarly described. Usually ac voltages and currents are sinusoidal, and definite relationships exist among the *peak*, *average*, and *effective* values. The effective value, or *rms value*, of an ac quantity is the normally quoted value and is the value normally applied in Ohm's law and power calculations. The *frequency* and *phase angle* of an ac quantity can also be important, and the *wavelength* can be calculated from a knowledge of the frequency and the velocity of the wave. The cathode ray oscilloscope is the instrument used for the study of alternating waveforms.

17-1

GENERATION OF ALTERNATING VOLTAGE

Electromagnetic induction has been discussed in Sections 11-2 and 14-1. In Section 14-1 it was shown how an emf is induced in a conductor in motion through a magnetic field. Now consider two series-connected conductors rotating in a magnetic field, as shown in Figure 17-1. Conductors 1 and 2 form a loop situated within the field set up between two magnetic poles. The conductors are mechanically fastened to an axle equipped with a crank for hand turning. Each conductor is connected to a *slip ring*, and the slip rings have *brushes* for electrically connecting to external terminals. When the loop is rotated, the conductors cut the magnetic flux, and an emf is generated in each conductor.

Figure 17-2 illustrates the conditions that exist at each of several instants while the loop is being rotated in the magnetic field. These can be explained as follows:

- *Figure 17-2(a)*: Here the loop is horizontal, and conductor 1 is moving up while conductor 2 is moving down. Because the motion of each conductor is parallel to the direction of the magnetic field at this instant, no flux is being cut by the conductors and so no emf is generated. Therefore, as illustrated by point 1 on the graph of voltage e versus time t, the voltage at output terminals A and B is zero, and no current flows through the external load resistor (R_L in Figure 17-1).

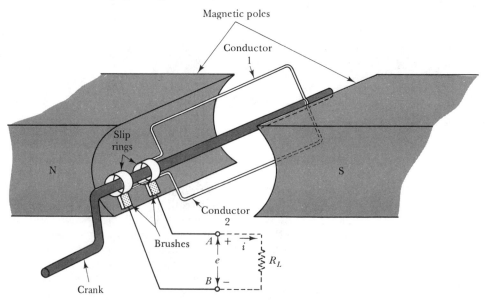

FIGURE 17-1. Simple alternating voltage generator consisting of a conducting loop rotated in a magnetic field. An emf is induced in each conductor as it is swept past a magnetic pole. The emf reverses polarity when the conductors move from one pole to the other.

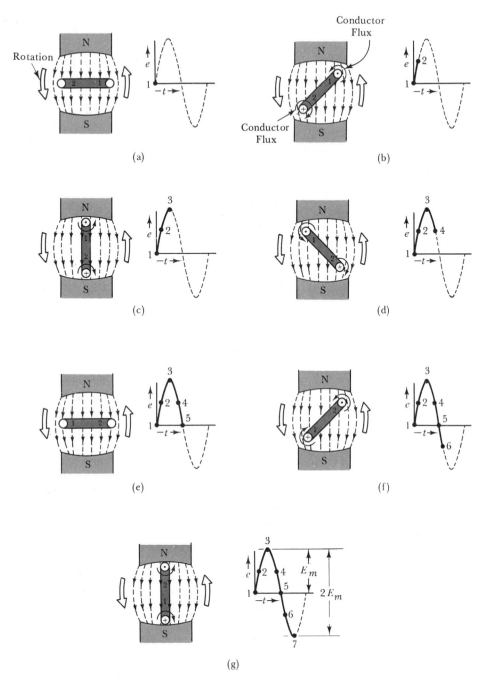

FIGURE 17-2. A sinusoidal voltage waveform is generated at the terminals of a conducting loop rotated in a magnetic field. Peak output voltage is produced when the conductors are moving perpendicular to the field [see (c) and (g) above]. Zero output is generated at the instants when the conductors are moving parallel to the field [see (a) and (e) above].

- *Figure 17-2(b)*: The conductors are now moving at an angle across the field, and so each is cutting some flux, and an emf is induced in each conductor. According to Lenz's law, the direction of the induced emf must be such that it opposes the motion that causes the emf. Thus, the emf induced in conductor 1 causes a current to flow towards the viewer, as illustrated. This current produces a counter-clockwise magnetic flux around the conductor, which strengthens the magnetic field in front of the conductor and weakens it behind (i.e., it opposes the conductor's motion). Similarly, the induced emf in conductor 2 causes a current in the direction shown, away from the viewer. A clockwise flux is set up around conductor 2, and this again weakens the flux behind conductor 2 and strenghens it ahead of the conductor, again opposing the motion of the conductor. It is seen that the current directions in the two conductors are such that they assist each other (i.e., the current is flowing around the loop in a single direction). Consequently, the output voltage is positive at terminal A and negative at terminal B in Figure 17-1. (Conductor 1 is connected to A, and conductor 2 is connected to B.) A current flows in the direction shown (from A and B) through the external load R_L (see Figure 17-1). When the growth of the terminal voltage e is plotted versus time t, the result is as illustrated by the graph in Figure 17-2(b). Point 1 represents the zero output voltage obtained for the conditions in Figure 17-2(a), and point 2 is the output voltage some time later when the conductor positions are as shown in Figure 17-2(b).
- *Figure 17-2(c)*: The conductors are now moving perpendicular to the magnetic field. Thus, they are cutting maximum flux and generating maximum emf. Again, the current directions are such that current flows around the loop, and terminal A remains positive while terminal B is negative. The output current through R_L is now greater than before, and as shown on the graph of e versus t, the output voltage has reached a peak value, point 3.
- *Figure 17-2(d)*: Now the conductors are again moving at an angle less than 90° with respect to the direction of the magnetic field. Therefore, the amount of flux being cut is reduced, and consequently the generated emf is less than for the conditions in Figure 17-2(c). The current direction and output voltage polarity are the same as before. However, the output voltage has now dropped below the peak level and is as shown at point 4 on the e versus t graph.
- *Figure 17-2(e)*: At this time the conductors are again moving parallel to the direction of the magnetic field. No flux is being cut, so no emf is generated. The output voltage is now zero and is plotted at point 5 on the voltage versus time graph.
- *Figure 17-2(f)*: The conductors are once more moving at an angle with respect to the direction of the magnetic field. However, both

conductors now have the induced current directions reversed. The current direction in conductor 1 is flowing away from the viewer, while that in conductor 2 is toward the viewer. It is seen that the currents still assist each other, so that the current around the loop still has a single (but reversed) direction. The output voltage is now such that terminal B is positive, while terminal A is negative. The instantaneous output voltage e plotted versus time t gives point 6 on the graph.

- *Figure 17-2(g)*: The motion of the conductors is once again perpendicular to the magnetic field, and maximum flux is being cut. Thus, maximum emf is again generated. The output voltage is still a negative quantity and has reached its peak value, plotted at point 7 on the graph of e versus t. It takes only a little imagination to see that as the loop rotation continues, the output voltage goes to zero once more, then reverses again, and the cycle is repeated over and over.

Since the voltage generated by the rotating loop is alternately positive and negative, it is referred to as an *alternating voltage*. The current produced in a load supplied by an alternating voltage flows first in one direction and then in the other. Thus, it is an *alternating current (ac)*. The designation *ac* is normally applied to both current and voltage. The voltage/time graph constructed in Figure 17-2 is spoken of as an *ac waveform*.

17-2
SINE WAVE

SINUSOIDAL WAVEFORM. The waveform traced out on the graph of e versus t in Figure 17-2 has a shape termed *sinusoidal*, so the wave is referred to as a *sine wave*. As will be shown, these terms arise because the instantaneous voltage is found to be dependent on the trigonometrical *sine of an angle*.

In order to derive an equation for the instantaneous emf induced in the rotating loop, recall that the induced emf depends on the amount of flux cut per second. From Equation (14-1), the emf induced in a single conductor is

$$ e = \frac{\Delta\Phi}{\Delta t} $$

where $\Delta\Phi$ is the total flux cut by the conductor during time Δt. Since $\Phi = B \times A$, the equation can be rewritten

$$ e = B \times \frac{\Delta A}{\Delta t} $$

where B is the flux density between the magnetic poles and ΔA is the area swept by one conductor during time Δt.

Consider the situation illustrated in Figure 17-2(c). If the total length of the two conductors within the magnetic field is ℓ and if the conductors move through a distance Δd during time Δt, then the area swept by the conductors during Δt is

$$\Delta A = \ell \times \Delta d$$

and the equation for emf induced in the loop becomes

$$e = B\ell\,\frac{\Delta d}{\Delta t} \qquad \text{(17-1)}$$

Because maximum flux is being cut at the instant shown in Figure 17-2(c), the peak or maximum value of emf (E_m) is being generated. Also, since $\Delta d/\Delta t$ is meters/second (i.e., a velocity v), the equation can be written

$$E_m = B\ell v \qquad \text{(17-2)}$$

where E_m is in volts, B is in teslas, ℓ is in meters, and v is in m/s. In this case v is the actual linear velocity of the conductors. It can also be described as the *peripheral velocity* with respect to the circle traced out by each conductor. It is important to note that Equation (17-2) is an expression for the maximum or *peak value E_m* of the voltage generated [i.e., the voltage represented by point 3 on the e/t graph in Figure 17-2].

INSTANTANEOUS VALUE. Now consider Figure 17-3, which illustrates the movement of conductor 1 from its position in Figure 17-2(a) toward that in Figure 17-2(b). It is seen that as the conductor moves through a distance Δd, the distance that it moves perpendicular to the flux is $\Delta d'$; and

$$\Delta d' = \Delta d \sin \alpha$$

where α is the angle moved through by the conductor in circular motion. Substituting in Equation (17-1), the *instantaneous emf* is

$$e = B\ell\,\frac{\Delta d'}{\Delta t}$$

$$= B\ell\,\frac{\Delta d \sin \alpha}{\Delta t}$$

$$e = B\ell v \sin \alpha \qquad \text{(17-3)}$$

Chap. 17   Alternating Current and Voltage

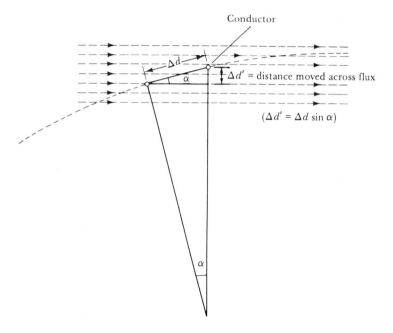

Conductor

$\Delta d'$ = distance moved across flux

$(\Delta d' = \Delta d \sin \alpha)$

FIGURE 17-3. A conductor moving through a distance Δd at an angle α with respect to a magnetic field, moves $\Delta d'$ across the field, where $\Delta d' = \Delta d \sin \alpha$. The emf generated in the conductor is directly proportional to $\sin \alpha$.

$B\ell v$ is the expression obtained in Equation (17-2) for the peak or maximum value (E_m) of the voltage generated. Therefore,

$$\boxed{e = E_m \sin \alpha} \qquad (17\text{-}4)$$

In Equation (17-4), α is usually termed the *phase angle* of the instantaneous voltage e. E_m is, of course, the peak value, but it is also referred to as the *amplitude* of the waveform.

EXAMPLE 17-1   In the hand-cranked generator illustrated in Figure 17-1, the length of each conductor within the magnetic field is 25 cm. The distance from each conductor to the axis of rotation is $r = 5$ cm and the flux density of the magnetic field is $B = 0.1$ T. Calculate the maximum output voltage from the generator when the conductors are rotated at 100 revolutions/minute.

SOLUTION

Distance traveled by each conductor in one revolution is the circumference of the circle traced out by rotation. Therefore,

$$\text{distance, } d = 2\pi r$$

$$= 2\pi \times 5 \times 10^{-2} \text{ m}$$

$$= 0.1\pi \text{ m}$$

$$\text{distance traveled per minute} = d \times 100 \text{ rpm}$$

$$= 0.1\pi \text{ m} \times 100$$

$$= 10\pi \text{ meters}$$

Therefore, conductor peripheral velocity, $v = \dfrac{10\pi}{60}$ meters/sec

$$\cong 0.524 \text{ m/s}$$

From Equation (17-2), *the peak emf generated in the loop is*

$$E_m = B\ell v$$

where $\ell = 2 \times$ (length of each conductor). Therefore

$$E_m \cong 0.1 \text{ T} \times 2 \times 25 \times 10^{-2} \times 0.524 \text{ m/s}$$

$$\cong \mathbf{26.2 \ mV}$$

EXAMPLE 17-2 For the generator described in Example 17-1, calculate the instantaneous output voltage at points 2, 3, 4, and 6 on the graph of e/t illustrated in Figure 17-2. Assume that the phase angles of the conductors at these points with respect to the flux direction are 45°, 90°, 135°, and 225°, respectively.

SOLUTION

From Equation (17-4), *the instantaneous emf generated is*

$$e = E_m \sin \alpha$$

At point 2,

$$e = 26.2 \text{ mV} \sin 45°$$

$$\cong \mathbf{18.5 \ mV}$$

At point 3,

$$e = 26.2 \text{ mV} \sin 90°$$

$$= \mathbf{26.2 \ mV}$$

At point 4,

$$e = 26.2 \text{ mV} \sin 135°$$

$$\cong \mathbf{18.5 \ mV}$$

At point 6,

$$e = 26.2 \text{ mV} \sin 225°$$

$$\cong -18.5 \text{ mV}$$

PRACTICE PROBLEMS

17-2.1 A single-loop generator as in Figure 17-1 is to produce a peak output of 100 mV. The conductor dimensions are $l = 6$ cm and $r = 2$ cm, and the magnetic flux density is 0.5 T. Determine the required rate of rotation.

17-2.2 For the generator in Problem 17-2.1, calculate the angle of the conductors with respect to the field flux for instantaneous output voltages of 66 mV, 70.7 mV, and 85 mV.

17-3

FREQUENCY, PHASE ANGLE, AND WAVELENGTH

FREQUENCY. Consider Figure 17-4, which once again illustrates the sine-wave output from an ac generator. It has been shown that when the conductors in the simple generator are rotated through one complete revolution, the output voltage goes through one complete cycle of sinusoidal change. The time taken for *one cycle* of change (from zero volts through $+E_m$, zero, and $-E_m$ back to zero again) is referred to as the *time period* of the waveform and is designated T (see Figure 17-4). If $T = 1$ s, it is said that the waveform has a *frequency* (f) of 1 *cycle per second* or 1 *hertz* (Hz).

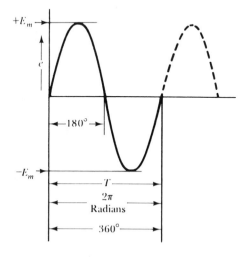

FIGURE 17-4. One cycle of a voltage sine wave. The instantaneous voltage e increases from zero to $+E_m$, then decreases through zero to $-E_m$, and finally returns to zero.

*A waveform has a frequency of **1 hertz*** (**Hz**) when it goes through one complete cycle of change in a period of 1 second.*

When two complete cycles of the waveform occur in one second, the frequency is two cycles per second, or 2 Hz. In this case, the time period of one cycle is $T = 0.5$ s, and

$$f = 1/0.5 \text{ s} = 2 \text{ Hz}$$

Similarly, if the waveform has a time period of 2 s, then its frequency is one cycle per 2 s, or

$$f = 1/2 \text{ s} = 0.5 \text{ Hz}$$

The frequency of the waveform is always found by inverting the time period.

$$\boxed{f = \frac{1}{T}}$$

(17-5)

where f is in hertz and T is in seconds.

The usual (North American) frequency for domestic and industrial ac supplies is 60 Hz. However, in electronics, the frequencies of ac waveforms can range from zero (i.e., dc) through kilohertz (kHz), megahertz (MHz), and gigahertz (GHz).

PHASE ANGLE. For one complete revolution, the conductors in the simple generator obviously rotate through 360°. If the angle of rotation of the conductors is measured in *radians* instead of degress, one complete revolution represents 2π radians. Since T is the time period in seconds (or time for one complete revolution), the angle through which the conductors move in 1 s is given by

$$\text{angular velocity} = \frac{360}{T} \text{ degrees/second}$$

or

$$\text{angular velocity} = \frac{2\pi}{T} \text{ radians/second}$$

and the phase angle at any instant t_1, t_2, t_3, etc., measured from $t = 0$ is

$$\text{angle} = \frac{2\pi}{T} \times t \text{ radians}$$

---

* Named for the German physicist Heinrich R. Hertz (1857–1894).

Chap. 17   Alternating Current and Voltage

Since $$f = \frac{1}{T}$$

the equation can be rewritten

$$\text{angle} = 2\pi f t \quad \text{radians}$$

or

$$\text{angle} = \omega t \quad \text{radians}$$

where ω is the angular velocity in rad/s (i.e., $\omega = 2\pi/T$).

Equation (17-4) can now be written in the form normally employed to represent ac waveforms:

$$\boxed{e = E_m \sin \omega t} \qquad \text{(17-6)}$$

EXAMPLE 17-3 An ac waveform with a frequency of 1.5 kHz has a peak value of 3.3 V. Calculate the instantaneous levels of voltage at $t_1 = 0.65$ μs and at $t_2 = 1.2$ ms.

SOLUTION

$$\omega = 2\pi f = 2\pi \times 1.5 \text{ kHz}$$
$$= 3\pi \times 10^3 \text{ rad/s}$$

Equation (17-6):

$$e = E_m \sin \omega t$$

At t_1,

$$e = 3.3 \text{ V} \sin \left[(3\pi \times 10^3 \times 0.65 \text{ μs})\text{rad} \right]$$
$$e_1 \cong \mathbf{20.2 \ mV}$$

At t_2,

$$e = 3.3 \text{ V} \sin \left[(3\pi \times 10^3 \times 1.2 \text{ ms})\text{rad} \right]$$
$$e_2 \cong \mathbf{-3.1 \ V}$$

In Figure 17-5, four different waveforms are illustrated. Waveforms *B*, *C*, and *D* obviously have smaller amplitudes than waveform *A*. Because waveform *B* goes through its cycle of change exactly at the same time as waveform *A*, the two are said to be *in phase* with each other. Waveforms *C* and *D*, however, are *out of phase* with waveforms *A* and *B*. Because waveforms *A* and *B* commence their cycles ahead of *C* by an angle θ (degrees or radians), *A* and *B* are said to *lead C* by θ.

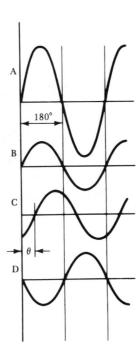

FIGURE 17-5. In-phase and out-of-phase sine waves. Waveform *B* is in phase with waveform *A*. Waveform *C lags* waveforms *A* and *B*.

Alternatively, waveform *C* may be said to *lag* waveforms *A* and *B* by θ. θ may also be referred to as the *phase difference* between the waveforms. Waveform *D* lags *A* and *B* by 180° and is said to be in *antiphase* to *A* and *B* because it is at its negative peak when *A* and *B* are at their positive peak levels, and vice versa.

WAVELENGTH. The *wavelength* of an ac waveform depends on its velocity. In the case of radio waves, the velocity is the speed of light, which is $c = 3 \times 10^8$ m/s. For voltage waves moving along widely spaced conductors, the transmission speed is also 3×10^8 m/s. When the conductors are close together, the velocity depends on the insulation employed. In Figure 17-6(a), e_1 represents the initial or zero level of a voltage waveform at the start of a long conductor. A fraction of a second later, in Figure 17-6(b), e_1 has traveled some distance along the conductor and the instantaneous level of the voltage at the generator terminals has changed to e_2. Both e_1 and e_2 are traveling along the conductor at a velocity of 3×10^8 m/s. In Figure 17-6(c) the voltage at the generator terminals has changed to e_3. By the time the generator terminal voltage has gone through one complete cycle [Figure 17-6(d)], e_1 has been moving along

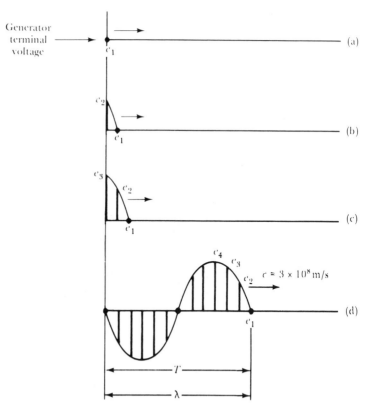

$c = 3 \times 10^8\,\text{m/s}$

(a)

(b)

(c)

(d)

FIGURE 17-6. An alternating voltage wave moves along widely spaced conductors at the speed of light, $c = 3 \times 10^8$ m/s. The time period of one cycle is $T = 1/f$, and the wavelength is $\lambda = c/f$.

the conductor for the time period T. Therefore, during one complete cycle, the distance travelled by e is,

$$\text{distance} = (3 \times 10^8 \text{ m/s} \times T \text{ seconds}) \qquad \text{meters}$$

This distance represents the *wavelength* λ of the alternating voltage. The equation for wavelength is,

$$\lambda = c \times T$$

or

$$\boxed{\lambda = \frac{c}{f}} \qquad \text{(17-7)}$$

where λ is wavelength in meters, c is velocity of light in m/s, and f is the frequency of the waveform in hertz.

EXAMPLE 17-4   Determine the wavelengths of 60 Hz and 1 MHz alternating waves. Also, calculate the frequency of a voltage that has a wavelength of 33 m.

SOLUTION

Equation (17-7):

$$\lambda = \frac{c}{f}$$

For 60 Hz,

$$\lambda = \frac{3 \times 10^8 \text{ m/s}}{60 \text{ Hz}}$$

$$= 5 \times 10^6 \text{ m}$$

For 1 MHz,

$$\lambda = \frac{3 \times 10^8 \text{ m/s}}{1 \text{ MHz}}$$

$$= 300 \text{ m}$$

From Equation (17-7)

$$f = \frac{c}{\lambda} = \frac{3 \times 10^8 \text{ m/s}}{33 \text{ m}}$$

$$= 9.09 \text{ MHz}$$

PRACTICE PROBLEMS

17-3.1   A voltage waveform has a peak value of 60 V and a frequency of 50 kHz. Calculate the instantaneous amplitudes at $t_1 = 3$ μs and $t_2 = 17$ μs.

17-3.2   Calculate the wavelength of the 740 kHz and 1200 kHz radio broadcast frequencies.

17-4

RESISTIVE LOAD WITH AC SUPPLY

CURRENT LEVEL.   Figure 17-7(a) shows a resistor R connected to an ac voltage source. Note the graphic symbol for the ac voltage source. Unlike the case of a dc circuit, a continuous current direction cannot be shown, because the current through R will change direction each time the polarity of e reverses. However, it is usually convenient to show an instantaneous current direction. The instantaneous current level can be found by Ohm's law:

Chap. 17   Alternating Current and Voltage

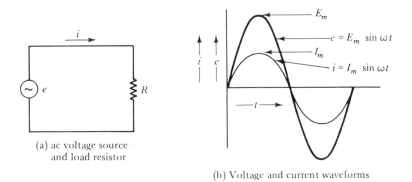

A.C. signal generator

A.C. voltages can be generated by electronic circuits, (as well as by the method described in Section 18-1). The laboratory type sine wave generator illustrated produces an output with amplitude adjustable from 0 V to 1 V, and frequencies ranging from 10 Hz to 100 kHz.

$$i = \frac{e}{R}$$

or,

$$i = \frac{E_m}{R} \sin \omega t \qquad (17\text{-}8)$$

The peak value of the current occurs at the peak value of the instantaneous voltage (i.e., when $e = E_m$). Therefore,

(a) ac voltage source and load resistor

(b) Voltage and current waveforms

FIGURE 17-7. The alternating current in a purely resistive circuit is in phase with the alternating voltage. As the voltage increases or decreases, the current also increases or decreases.

$$\text{maximum or peak current level, } I_m = \frac{E_m}{R}$$

and

$$\boxed{i = I_m \sin \omega t} \tag{17-9}$$

Since e is a sinusoidal quantity and i is directly proportional to e, the current in the circuit of Figure 17-7(a) is also a sinusoidal quantity. Furthermore, when e is zero, i must be zero, and when e is a maximum, i is at its peak value. This means that i is in phase with e, and the two waveforms can be plotted as shown in Figure 17-7(b).

POWER DISSIPATION. From Equation (3-5), the power dissipated in a resistor is

$$P = I^2 R$$

and the instantaneous power is

$$p = i^2 R$$
$$= (I_m \sin \omega t)^2 R$$

or

$$\boxed{p = I_m^2 R \sin^2 \omega t} \tag{17-10}$$

The maximum instantaneous power dissipation that occurs is

$$P_m = I_m^2 R$$

Therefore,

$$\boxed{p = P_m \sin^2 \omega t} \tag{17-11}$$

The instantaneous power dissipated in the load resistance at any instant can easily be calculated using the equations derived above. Note that when I_m is a negative quantity, Equation (17-10) gives

$$p = (-I_m)^2 R \sin^2 \omega t$$

which becomes

$$p = I_m^2 R \sin^2 \omega t$$

Chap. 17   Alternating Current and Voltage

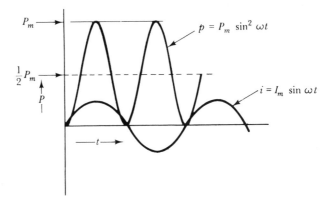

FIGURE 17-8. The instantaneous power dissipated in a purely resistive circuit is zero when the current is zero and increases to a peak as the current increases to a positive or negative peak level.

which is a positive quantity. Obviously, power is dissipated by a current flowing in either direction through a resistor. The fact that p is always a positive quantity means that all the power supplied is dissipated in the resistor R. If the instantaneous power were to become a negative quantity at any time, it would imply that the resistor was supplying power to the signal source, which is something a resistor cannot do. The fact that the instantaneous power is always a positive quantity also explains the shape of the graph of instantaneous power plotted in Figure 17-8. Peak power dissipation occurs both when i is a positive peak and when i is a negative peak. Zero power dissipation occurs when i is zero.

Examination of the waveform of instantaneous power dissipation in Figure 17-8 reveals that it is a perfect sine wave symmetrical about the level $\frac{1}{2}P_m$. Therefore, $\frac{1}{2}P_m$ is the average value of the power dissipated in the resistor; or, average power dissipated over one complete cycle is half the peak power:

$$\boxed{P = \tfrac{1}{2}P_m} \tag{17-12}$$

EXAMPLE 17-5 The circuit shown in Figure 17-7(a) has a 60 Hz supply voltage with a maximum value of 160 V. If $R = 10\ \Omega$, calculate the values of instantaneous current and power at phase angles of $\pi/4$, $\pi/2$, $5\pi/4$, and $3\pi/2$ radians.

SOLUTION

$$I_m = \frac{E_m}{R} = \frac{160\ \text{V}}{10\ \Omega}$$

$$= 16\ \text{A}$$

$$P_m = I_m^2 R = (16\ \text{A})^2 \times 10\ \Omega$$

$$= 2.56\ \text{kW}$$

Current calculations:

Equation (17-9):

$$i = I_m \sin \omega t$$

For $\omega t = \pi/4,$

$$i = 16 \text{ A} \sin \frac{\pi}{4} \quad \cong \textbf{11.3 A}$$

For $\omega t = \pi/2,$

$$i = 16 \text{ A} \sin \frac{\pi}{2} \quad = \textbf{16 A}$$

For $\omega t = 5\pi/4,$

$$i = 16 \text{ A} \sin \frac{5\pi}{4} \quad \cong \textbf{-11.3 A}$$

For $\omega t = 3\pi/2,$

$$i = 16 \text{ A} \sin \frac{3\pi}{2} \quad = \textbf{-16 A}$$

Power calculations:

Equation (17-11):

$$p = P_m \sin^2 \omega t$$

For $\omega t = \pi/4,$

$$p = 2.56 \text{ kW} \sin^2 \frac{\pi}{4} \quad = \textbf{1.28 kW}$$

For $\omega t = \pi/2,$

$$p = 2.56 \text{ kW} \sin^2 \frac{\pi}{2} \quad = \textbf{2.56 kW}$$

For $\omega t = 5\pi/4,$

$$p = 2.56 \text{ kW} \sin^2 \frac{5\pi}{4} \quad = \textbf{1.28 kW}$$

For $\omega t = 3\pi/2,$

$$p = 2.56 \text{ kW} \sin^2 \frac{3\pi}{2} \quad = \textbf{2.56 kW}$$

PRACTICE PROBLEM

17-4.1   An instantaneous peak power of 30 mW is dissipated in a 1.2 kΩ resistor with an ac voltage source. Calculate the instantaneous resistor voltages at phase angles of $\pi/3$ and $2\pi/3$ radians.

PEAK, AVERAGE, AND RMS VALUES OF SINE WAVES

PEAK AND PEAK-TO-PEAK VALUES. As already discussed, the peak value of a sinusoidal waveform is the maximum value: E_m for a voltage wave; I_m in the case of a current waveform. The *peak-to-peak* value is $2 E_m$ or $2 I_m$, as illustrated in Figure 17-9. For sinusoidal waveforms, there are direct relationships between peak, average, and rms values.

AVERAGE VALUE. The average value of a waveform can be determined by differential calculus, or by the more laborious process of taking the average value of a number of equally spaced instantaneous levels. The process is illustrated in Figure 17-9, where each 180° half (or half-cycle) of the current waveform is divided into nine 20° sections. The instantaneous values of the waveform at the center of each section is measured, and the average value calculated:

$$I_{av} = \frac{i_1 + i_2 + i_3 + \cdots + i_9}{9}$$

This process can be applied to waveforms of any shape; in fact, it may be the only way of determining the average value of some irregular wave-

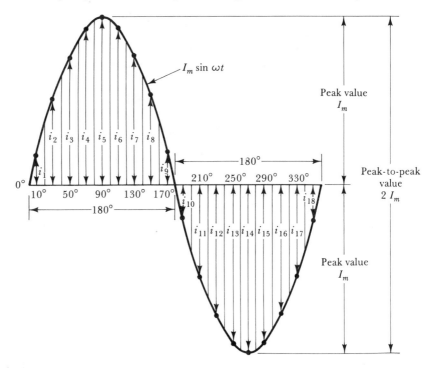

FIGURE 17-9. The *average value* of a waveform can be determined by averaging a number of equally spaced instantaneous amplitudes. The *root mean squared* (*rms*) *value* can also be calculated from a number of instantaneous amplitudes. Differential calculus affords simpler techniques.

forms. For the sine wave, however, it is not necessary to measure each instantaneous value, since they can be calculated from the maximum value and the sine of each angle. Thus, the average value of the *positive half-cycle* of the current wave is

$$I_{av} = \frac{I_m \sin 10° + I_m \sin 30° + I_m \sin 50° + \cdots + I_m \sin 170°}{9}$$

Substituting the appropriate values (or using calculus), it is found that

$$\boxed{I_{av} = \frac{2}{\pi} I_m \cong 0.637 \, I_m} \qquad (17\text{-}13)$$

Similarly, the average value of the *positive half-cycle* of a sinusoidal voltage wave is

$$\boxed{E_{av} = \frac{2}{\pi} E_m \cong 0.637 \, E_m} \qquad (17\text{-}14)$$

Returning to Figure 17-9, the average value of the second 180° (or *negative half-cycle*) of the waveform is found to be

$$I_{av} = -0.637 \, I_m$$

For the complete cycle,

$$I_{av} = 0.637 \, I_m - 0.637 \, I_m$$
$$= 0$$

Therefore, over the entire 360° sinusoidal waveform, the average value of the current (or voltage) is zero.

RMS VALUE. Taking each instantaneous current value in Figure 17-9 in turn, the instantaneous power dissipated in a resistor through which the current is flowing is obviously

$$p_1 = i_1^2 R, \; p_2 = i_2^2 R, \; \cdots, \; p_n = i_n^2 R$$

The average power dissipated in the resistor is then

$$P = \frac{i_1^2 R + i_2^2 R + \cdots + i_n^2 R}{n}$$

As explained in Section 17-4, the negative instantaneous current values also give a positive power dissipation in the load.

When a direct current is flowing through a resistor, the power dissipated is given by Equation (3-5):

$$P = I^2 R$$

To find the equivalent alternating current that dissipates the same amount of power as a direct current I flowing through the resistor, the dc and ac power dissipation expressions are equated:

$$I^2 R = \frac{i_1^2 R + i_2^2 R + \cdots + i_n^2 R}{n}$$

which gives,

$$I^2 = \frac{i_1^2 + i_2^2 + \cdots + i_n^2}{n}$$

It is seen that I^2 (the dc current squared) is equal to the average value of the instantaneous ac *current squared* or the ac *mean-squared value*. So the effective value (or dc equivalent value) of the alternating current is

$$I = \sqrt{\frac{i_1^2 + i_2^2 + \cdots + i_n^2}{n}}$$

Here, I is the *root of the mean-squared value,* or *rms value* of the alternating current. Thus, the rms value of an alternating current is the effective value, or *dc equivalent value,* as far as power dissipation is concerned. For this reason, the rms value is frequently termed the *effective value.*

Like the average value, the relationship between the rms value and the peak value of an ac waveform can be determined by calculus. However, it can also be worked out quite simply. From Equation (17-12), the ac power dissipation is

$$P = \tfrac{1}{2} P_m$$
$$= \tfrac{1}{2} I_m^2 R$$

and from Equation (3-5) for dc power dissipation,

$$P = I^2 R$$

Therefore,

$$I^2 R = \tfrac{1}{2} I_m^2 R$$
$$I^2 = \tfrac{1}{2} I_m^2$$
$$I = \sqrt{\tfrac{1}{2} I_m^2}$$

$$I = \frac{1}{\sqrt{2}} I_m = 0.707 \, I_m \qquad \text{(17-15)}$$

or

$$I = \frac{1}{1.414} I_m$$

Similarly, for an *alternating voltage,* the *effective value* is the rms value of the waveform, and where the waveform is sinusoidal,

$$E = \frac{1}{\sqrt{2}} E_m = 0.707 \, E_m \qquad \text{(17-16)}$$

or

$$E = \frac{1}{1.414} E_m$$

With alternating current and voltage the rms values are the *normally quoted* values. A domestic alternating voltage supply of 115 V, for example, has a peak value of

$$E_m = 1.414 \times 115 \text{ V} \cong 163 \text{ V}$$

Rms values are also normally used in all Ohm's law calculations involving current, voltage, and resistance.

*It is important to note that the relationships stated in Equations (17-13) through (17-16) apply **only to pure sine waves**. In the case of other waveforms, the peak, average, and rms quantities are related by other (different) factors.*

EXAMPLE 17-6 A 300 V sinusoidal ac supply is applied to a 50 Ω resistor. Determine the peak, rms, and average values of the current through the resistor. Also, calculate the power dissipated in the resistor.

SOLUTION

From Equation (17-16),

$$\text{peak voltage, } E_m = 1.414 \times E$$
$$= 1.414 \times 300 \text{ V}$$
$$\cong \mathbf{424 \text{ V}}$$

$$\text{peak current, } I_m = \frac{E_m}{R} = \frac{424 \text{ V}}{50 \text{ Ω}}$$
$$= \mathbf{8.48 \text{ A}}$$

$$\text{rms current, } I = \frac{E}{R} = \frac{300 \text{ V}}{50 \text{ Ω}}$$
$$= \mathbf{6 \text{ A}}$$

From Equation (17-13),

$$\text{average current for half-cycle, } I_{av} = 0.637\, I_m$$

$$= 0.637 \times 8.48 \text{ A}$$

$$\cong \mathbf{5.4\ A}$$

$$\text{average current for whole cycle} = 0$$

$$\text{power dissipation, } P = I^2 \times R$$

$$= (6\text{A})^2 \times 50\ \Omega$$

$$P = \mathbf{1.8\ kW}$$

or,
$$P = \frac{E^2}{R}$$

$$= \frac{(300\ \text{V})^2}{50\ \Omega}$$

$$= \mathbf{1.8\ kW}$$

PRACTICE PROBLEMS

17-5.1 The power dissipated in a 560 Ω resistor is measured as 250 mW. Calculate the peak, average, and rms values of the voltage applied to the resistor.

17-5.2 A 220 Ω resistor has an ac supply of 33 V. Determine the resistor power dissipation and peak current.

17-6

CATHODE RAY OSCILLOSCOPE

OSCILLOSCOPE CONTROLS. The *cathode ray oscilloscope* (CRO) is the basic instrument for the study of waveforms. It can be employed for measuring voltage, frequency, time interval, and phase difference. A knowledge of the use of the oscilloscope is essential for laboratory investigations involving alternating waveforms.

Figure 17-10 shows a line drawing of the front panel and controls of a representative oscilloscope. For discussion purposes, the controls and terminals are identified in groups which are numbered counterclockwise, starting from the top left hand corner. A discussion of the function of each item follows:

(1) *Screen and Graticule.* Waveforms under investigation are displayed on the screen. This is a flat end of a glass tube with its inside surface coated with fluorescent chemicals that glow when struck by a *beam* of electrons. The screen is protected by a (vertically and horizontally calibrated) flat piece of transparent plastic called a *graticule*. The graticule is used for measuring the amplitude (in vertical divisions) and the time period (in horizontal divisions) of a displayed waveform.

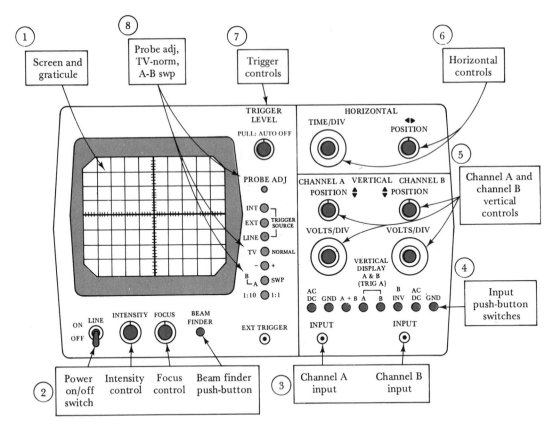

FIGURE 17-10. A cathode ray oscilloscope is the instrument normally used in the study of waveforms. The front panel and controls of a typical oscilloscope are shown. (Courtesy of Hewlett-Packard.)

(2) *ON/OFF, Intensity, Focus, Beam Finder.* The 115 V supply to the oscilloscope is switched on or off by the *LINE ON/OFF* switch. The *INTENSITY* control permits the brightness of the displayed waveform to be adjusted, and the *FOCUS* knob allows it to be focused to a fine line. The purpose of the push-button *BEAM FINDER* is to locate a display that has been shifted off the screen.

(3) *Channel A and Channel B Inputs.* A *dual-trace* oscilloscope can display two waveforms at the same time. Two coaxial-type input terminals are provided for connection of the signals. The inputs are identified as *CHANNEL A* and *CHANNEL B*. Additional labeling shows that the input resistance and input capacitance are 1 MΩ and 30 pF respectively.

(4) *Input Switches.* The *AC-DC* push-button switches immediately above each input terminal facilitate ac or dc connection of the voltage to be displayed. Sometimes it is necessary to display an alternating voltage and block a dc component of the input waveform. This is achieved by setting the *AC-DC* button to ac, where a coupling capacitor blocks the

dc and passes the ac quantity. The *GND* (*ground*) buttons alongside each ac-dc button disconnect the input voltages and grounds the input terminals. This permits each trace to be set to a convenient zero position on the screen. When an input voltage is displayed, its dc level may be measured with respect to the zero position.

The *VERTICAL DISPLAY A and B* push-buttons allow Channel *A* input (press button *A*), Channel *B* input (press *B*), or both (press both at the same time), to be displayed on the screen. The *B INV* button inverts the Channel *B* input, and the A + B button permits Channel *A* and Channel *B* inputs to be added together and displayed. With A + B and *B INV* buttons pressed, the displayed waveform is the difference of the two inputs.

(5) *Channel A and Channel B Vertical Controls.* For vertical adjustment of two displayed waveforms, there are separate *CHANNEL A* and *CHANNEL B* vertical controls. The vertical *POSITION* controls are used to move each waveform up or down the screen to set them in the best position for viewing. The *VOLTS/DIV* switch for each channel selects the vertical sensitivity of the display. When this control is set to 1 V, a signal having a peak-to-peak amplitude of 1 V would occupy one vertical division of the screen graticule. A signal that occupies four divisions at this setting would have a peak-to-peak amplitude of 4 V. Such a waveform is shown in the upper half of the screen in Figure 17-11. The *VOLTS/DIV* setting is correct only when the *vernier* knob at its center is in the CAL (calibrated) position. The vernier knob provides continuous volts/div adjustment, so that the display amplitude may be increased as desired.

(6) *Horizontal Controls.* The *TIME/DIV* switch selects the horizontal deflection sensitivity of the display in *ms/div* or *μs/div*. Refer to the two waveforms in Figure 17-11, and assume that the *TIME/DIV* selector is at 1 ms. One cycle of the upper waveform occupies four horizontal divisions, therefore its time period is $(4 \times 1 \text{ ms}) = 4$ ms. For the lower waveform, three complete cycles occupy 4 ms. Consequently, the time period is (4/3) ms, or 1.3 ms. Here again, the setting is correct only when the *vernier* knob at the center of the *TIME/DIV* control is in its CAL position. This vernier knob provides continuous time/division adjustment, so that a cycle of displayed waveform may be widened up to ten times the horizontal *TIME/DIV* setting. The *HORIZONTAL POS-ITION* knob performs a similar function to the *VERTICAL POSITION* control. The displayed waveforms may be moved horizontally across the screen as desired.

(7) *Trigger Controls.* In order to provide a stable waveform display, the display must be made to commence (at the left hand side of the screen) exactly when one of the input waveforms is at its zero position. Thus, the *time-base* circuits (which causes the electron beam to sweep horizontally across the screen) are said to be *triggered* at this instant.

The three *TRIGGER SOURCE* push-buttons permit the selection

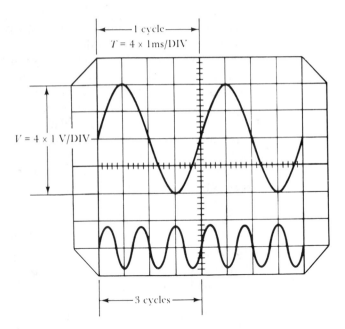

FIGURE 17-11. Two sine waves displayed on the screen of an oscilloscope. The voltage peak-to-peak amplitude is measured in terms of the *volts/(vertical division)*, and the time period of one cycle is measured according to the *time/ (horizontal division)*.

of *INT* (internal), *EXT* (external), or *LINE* as triggering sources for the time base. With *INT* selected, the time base is triggered from one of the input waveforms. This is the most commonly used triggering source. When both Channel *A* and Channel *B* inputs are displayed, the internal trigger source is Channel *A* in this particular oscilloscope. When the *EXT* button is depressed, the time base is triggered from an external source connected to the *EXT TRIGGER* terminal. Selection of *LINE* as the trigger source causes the time base to be triggered from the line or ac power frequency. The *1:10–1:1* push-button located immediately above the *EXT TRIGGER* input terminal is used only with external triggering. It is put in its 1:10 position when a large amplitude external triggering voltage must be potentially divided ten times before applying it to the triggering circuits.

Triggering may be *positive* (+) or *negative* (−) (selected by the −/+ push button). This simply means that the displayed waveform may be made to commence either when it is going positive or when it is going negative (see Figure 17-12).

The *TRIGGER LEVEL* knob continuously adjusts the instant in time at which the waveform display commences. The display may be made to commence exactly when the signal goes through its zero position, or at some time shortly before or after this instant. Usually the time base operates automatically to sweep the electron beam horizontally

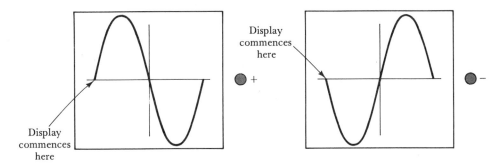

FIGURE 17-12. By use of the − /+ trigger control, a displayed waveform may be made to commence positive-going or negative-going.

across the screen, and in this case the time base is synchronized with the input signal. When the *TRIGGER LEVEL* control is set to *AUTO OFF,* the time base no longer operates automatically. Each horizontal sweep of the electron beam must then be triggered by an input waveform.

(8) *Probe Adj, TV-Norm and A-B SWP.* An internally generated 2 kHz, 0.5 V square-wave output is available from the *Probe Adj* (probe adjust) terminal. This is used as a known signal source when calibrating the oscilloscope *probes* (see below).

For most purposes the *TV-NORM* push-button is kept in its *NORMAL* position. It is set to the *TV* position only when television signals are to be investigated.

The *A-B SWP* button is usually left in the *SWP* (sweep) position, where it allows the time base to sweep the electron beam horizontally across the screen for waveform display. When the button is in the *A-B* position, the time base is disconnected. Now signals applied to *CHANNEL B* input terminal produce vertical deflection, while *CHANNEL A* inputs produce horizontal deflection. This allows two interdependent voltages to be plotted as a graph of *B* versus *A*.

MEASURING VOLTAGE AND TIME PERIOD. Figure 17-13 illustrates the method that should be used to achieve the most accurate measurement of voltage and time period. The appropriate *VERTICAL POSITION* control is used to shift the unwanted trace off the screen. The *VOLTS/ DIV* and *TIME/DIV* knobs are then adjusted to give the largest possible display of one cycle of the input waveform. Note that the *VOLTS/DIV* and *TIME/DIV* (center) vernier knobs must be in their CAL positions. From Figure 17-13(a), the time period of the waveform is (number of horizontal divisions) × (time/division). Figure 17-13(b) and (c) shows how the displayed waveform should be moved horizontally (to the calibrated vertical center line on the graticule) by the *HORIZONTAL POS-ITION* control, in order to accurately measure its peak-to-peak amplitude. The measured amplitude is (number of vertical divisions) × (volts/division).

PHASE DIFFERENCE MEASUREMENT. Figure 17-14 shows how the phase difference between two waveforms can be determined. From a measurement of the time period and the fact that one cycle represents 360°, the degrees/(horizontal division) can be calculated. The difference (in horizontal divisions) between commencement of each waveform is measured. This is then converted into degrees of phase shift using the calculated degrees/division.

EXAMPLE 17-7   Determine the frequency and amplitude of the waveforms illustrated in Figure 17-13. Also, determine the phase difference between the two waveforms shown in Figure 17-14.

SOLUTION

For Figure 17-13(a),

$$T = (8 \text{ horizontal divisions}) \times 5 \text{ ms/div}$$

$$= 40 \text{ ms}$$

$$f = 1/T = 1/40 \text{ ms}$$

$$\mathbf{= 25 \ Hz}$$

For Figure 17-13(b) and (c),

$$\text{peak amplitude} = (3.4 \text{ vertical divisions}) \times 2 \text{ V/div}$$

$$\mathbf{= 6.8 \ V}$$

$$\text{peak-to-peak amplitude} = 2 \times 6.8 \text{ V}$$

$$\mathbf{= 13.6 \ V}$$

For Figure 17-14,

$$T = 4 \text{ divisions}$$

$$= 360°$$

$$1 \text{ div} = 360°/4 = 90°$$

$$\text{phase difference} = (1.3 \text{ div}) \times 90°/\text{div}$$

$$\mathbf{= 117°}$$

OSCILLOSCOPE PROBES. Input signals are normally connected to an oscilloscope via coaxial cables with *probes* on their ends (see Figure 17-15). These are ordinarily just convenient to use insulated connecting clips. Each probe has two connections, as illustrated, one *input* and one *ground.* It is important to connect both ground terminals to the same (grounded) point in the circuit under investigation; otherwise, there will

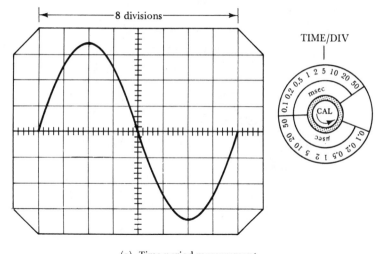

(a) Time period measurement

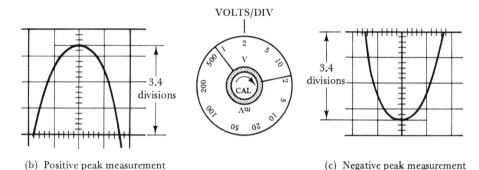

(b) Positive peak measurement

(c) Negative peak measurement

FIGURE 17-13. The time period of a waveform is measured by multiplying the number of horizontal divisions occupied by one cycle by the TIME/DIV selection. To determine the amplitude of a waveform, the number of vertical divisions is multiplied by the VOLTS/DIV setting.

be more than one grounded point, and the circuit may not function correctly. As illustrated in Figure 17-15, a coaxial cable consists of a central conductor with its insulation surrounded by another (braided circular) conductor. The circular conductor is grounded, and it acts as a *screen*, which helps to prevent unwanted radio frequency signals being picked up by the oscilloscope input.

Some probes, termed *attenuator probes*, have resistors inside them to increase the input resistance from the normal 1 MΩ input resistance of the oscilloscope to 10 MΩ. These probes also have the effect of reducing the voltage applied to the oscilloscope by a factor of 10. Thus, they are usually referred to as *10 : 1 probes*, and the ordinary (nonattenuator) probes are termed *1 : 1 probes*. Each attenuator probe has an adjustable capacitor. It is important that the probe capacitor be correctly adjusted

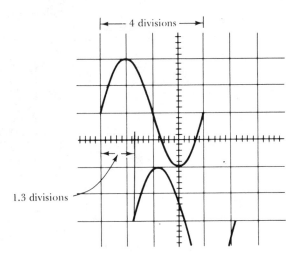

FIGURE 17-14. To estimate the phase difference between two waveforms, the number of horizontal divisions between the start of each wave is measured. Since one cycle represents 360°, the *degrees/horizontal division* can be determined, and the phase difference in divisions can be converted into degrees.

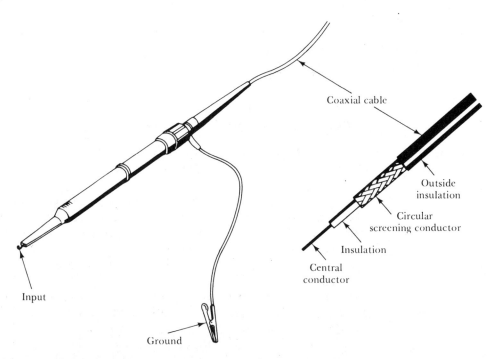

FIGURE 17-15. Oscilloscope probes are used to connect inputs to an oscilloscope. Coaxial cable is employed to exclude unwanted signals. Some probes contain adjustable capacitors to eliminate distortion in displayed waveforms.

when it is first connected for use with a particular oscilloscope. As already mentioned, an internally generated square wave is available from the oscilloscope *PROBE ADJ* terminal. The probe is connected to this terminal, and the probe capacitor is adjusted until a perfectly square waveform is displayed on the screen.

PRACTICE PROBLEMS

17-6.1 A 15 V sine wave occupies approximately 4.2 divisions peak-to-peak on an oscilloscope screen. What is the setting of the VOLTS/DIV control?

17-6.2 Two cycles of a sinusoidal ac waveform occupy 6.5 vertical divisions peak-to-peak and 8 horizontal divisions on an oscilloscope screen. If the VOLTS/DIV setting is 0.1 V, and the TIME/DIV control is at 50 μs, determine the rms value and frequency of the signal.

17-6.3 Two waveforms as in Problem 17-6.2 are to have a phase difference of 63°. Estimate the horizontal divisions difference between the two when displayed on the oscilloscope.

SUMMARY OF FORMULAS

Peak voltage generated in a conductor rotating in a magnetic field:
$$E_m = B\ell v$$

Frequency:
$$f = \frac{1}{T}$$

Wavelength:
$$\lambda = \frac{c}{f}$$

Instantaneous ac voltage:
$$e = E_m \sin \omega t$$

Instantaneous ac current:
$$i = I_m \sin \omega t$$

Instantaneous ac power:
$$p = p_m \sin^2 \omega t$$

Average ac power:
$$P = \tfrac{1}{2} P_m$$

Average (of half-cycle) ac current:
$$I_{av} = \frac{2}{\pi} I_m = 0.637 \, I_m$$

Average (of half-cycle) ac voltage:

$$E_{av} = \frac{2}{\pi} E_m = 0.637\ E_m$$

Rms ac current:

$$I = \frac{1}{\sqrt{2}} I_m = 0.707\ I_m$$

Rms ac voltage:

$$E = \frac{1}{\sqrt{2}} E_m = 0.707\ E_m$$

**REVIEW
QUESTIONS**

17-1 Draw sketches to illustrate, and explain, how an alternating current is generated in a conducting loop rotated in a magnetic field.

17-2 Derive an expression for the maximum voltage induced in a conducting loop rotated in a magnetic field. Also, derive an expression for the instantaneous value of the voltage.

17-3 Define ac current, ac voltage, sinusoidal waveform, sine wave, peak value, phase angle, and amplitude.

17-4 Sketch two sine waves that
a. Are in-phase.
b. Are in antiphase.
c. Have a phase difference of $\pi/2$ radians.

17-5 Derive equations for the instantaneous level of current through a resistor and for the instantaneous power dissipated when a sinusoidal ac voltage is applied to the resistor.

17-6 Draw a graph of instantaneous power versus time for the power dissipated in a resistor when a sinusoidal ac voltage is applied. Also, show that the average power dissipated is half the maximum instantaneous power.

17-7 Explain how the average value of a sinusoidal voltage may be calculated and write an expression for the average value in terms of the maximum value:
a. For a half-cycle of the waveform.
b. For a full cycle of the waveform.

17-8 Define the rms value of an alternating current and write an expression for the rms value in terms of the maximum value. Also, show that the rms value is the equivalent dc value, or effective value.

17-9 Referring to Figure 17-10, identify each control of the cathode ray oscilloscope and explain the function of each.

17-10 Using illustrations, explain how an oscilloscope is used to measure voltage, frequency, and phase difference.

17-1   A conducting loop rotated in a magnetic field has an axial length of 30 cm and a width of 8 cm. The magnetic field flux density is 0.25 T, and the loop is rotated at 140 rpm. Calculate the output frequency and the peak output voltage from the loop, (a) if the loop has one turn, (b) if the loop has 10 turns.

17-2   The generator described in Problem 17-1(a) is to be modified to produce a peak output of 100 mV. The alternatives are (a) increase the conductor length, (b) increase the speed of rotation, (c) increase the flux density. Determine the required new quantity in each case.

17-3   Calculate the instantaneous levels of output voltage from the generator described in Problem 17-1(b) at $\pi/4$, $3\pi/4$, and $5\pi/4$ radians from $e = 0$.

17-4   For the generator described in Problems 17-1 and 17-2, calculate the angles of the conducting loop at which the instantaneous output voltage is 5 mV, 10 mV, 12 mV, and 19 mV.

17-5   A 25 turn coil of insulated copper wire with an axial length of 5 cm and a width of 4 cm is rotated at 2400 rpm in a magnetic field with a flux density of 0.13 T. Calculate the peak output voltage.

17-6   Determine the angles of the rotating coil in Problem 17-5 at the instants that the output voltage is 50 mV, 100 mV, 200 mV, 400 mV, and 800 mV.

SECTION 17-3

17-7   For the generator described in Problem 17-1(b) calculate the instantaneous levels of output voltage at 5 ms, 10 ms, 25 ms, and 30 ms from the instant of zero output.

17-8   For the generator described in Problem 17-5, determine the times from zero output until the instantaneous output voltages are 70 mV, 140 mV, 280 mV, and 560 mV.

17-9   An ac waveform has a peak value of 9 V and a frequency of 150 kHz. Determine the instantaneous voltage levels at 1.1 μs, 5 μs, and 29 μs from the instant of zero voltage.

17-10   Calculate the frequencies and wavelengths of the outputs from the generators in (a) Example 17-1, (b) Problem 17-1, (c) Problem 17-5.

17-11   Calculate the frequency of an ac voltage with a wavelength of 300 km. Also, calculate the wavelengths of 120 kHz and 12 MHz ac waveforms.

17-12   Calculate the peak value of a 60 Hz waveform if its instantaneous level is measured as 110 V at 2 ms from the instant of zero output.

SECTION 17-4

17-13 A 120 Hz ac supply with a peak level of 100 V is applied to a 27 Ω resistor. Calculate the instantaneous levels of power dissipated at phase angles of $\pi/3$, $2\pi/3$, $5\pi/3$, and $3\pi/2$ radians.

17-14 The ac voltage in Problem 17-12 is applied to a 332 Ω resistor. Calculate the instantaneous power dissipations at 0.5 ms, 1 ms, 2 ms, and 4 ms from the instant of zero output. Also, determine the average power dissipated in the resistor.

17-15 A sinusoidal waveform with an rms value of 6 V and a frequency of 30 Hz is applied to a 120 Ω resistor. Determine the instantaneous power dissipated in the resistor at 5.5 ms from the instant of zero voltage. Also, calculate the peak instantaneous power and the average power dissipated in the resistor.

17-16 A resistor connected to the generator described in Problem 17-1(b) is to have a peak instantaneous power dissipation of 352 mW. Calculate the required resistance value. Also, determine the instantaneous power dissipation that occurs at 100 ms, 200 ms, and 400 ms from the instant of zero output voltage.

17-17 An instantaneous power of 410 μW is to be dissipated in a resistor connected to the generator described in Problem 17-5 at a time of 2.1 ms from zero output. Calculate the resistor value and determine the peak instantaneous power and the average power dissipated in the resistor.

17-18 A 400 Hz supply produces a peak power dissipation of 10 W in a 500 Ω resistor. Calculate the instantaneous power dissipated by the supply in a 3.3 kΩ resistor at angles of $\pi/4$, $\pi/2$, and $\pi/3$ radians from the instant of zero output.

SECTION 17-5

17-19 For Problem 17-13, calculate the rms and average levels of current in the resistor. Also, calculate the average power dissipation.

17-20 For Problem 17-14, determine the rms and average levels of current in the resistor and the average power dissipation.

17-21 A 297 V sine wave is applied to a 3.9 kΩ resistor. Calculate the peak, average, and rms levels of current flowing in the resistor. Also, determine the resistor power dissipation.

17-22 Calculate the average and rms output voltages from the generator in Problem 17-1(b). Also, determine the power dissipation in a 1 kΩ resistor connected to the output terminals.

17-23 Determine the rms and average output voltages from the generator in Problem 17-5. Also, calculate the value of a resistor which will dissipate 1 mW when connected to the generator output terminals.

17-24 A sine wave with a 166 V peak value dissipates 1 kW in a certain resistor. Calculate the resistor value and the average and rms levels of the current.

SECTION 17-6

17-25 Determine the peak-to-peak, average, and rms values, and the frequency of the waveform shown in Figure 17-16. Also, calculate the instantaneous amplitude of the wave at a phase angle of 120° and determine the amplitude at that angle from the display in Figure 17-16.

17-26 A voltage waveform with an amplitude of 1 V and a frequency of 25 Hz is to be displayed on an oscilloscope with the controls set as in Figure 17-16. Determine the number of vertical and horizontal divisions that would be occupied by the waveform.

17-27 (a) If the TIME/DIV control in Figure 17-16 is changed to 10 ms, how many cycles of the waveform shown will be displayed on the oscilloscope screen?

(b) How many cycles of a 500 Hz waveform could be displayed on an oscilloscope with the controls set as in Figure 17-16?

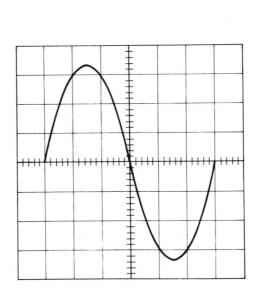
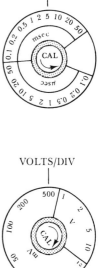

FIGURE 17-16.

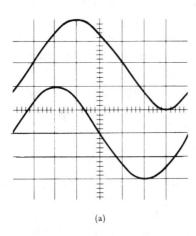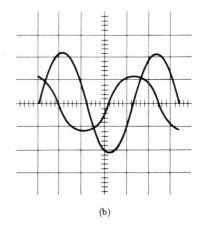

(a)   (b)

FIGURE 17-17.

17-28 Calculate the rms value of the largest amplitude sinusoidal wave-form that could be displayed on an oscilloscope, (a) with the controls set as in Figure 17-13, (b) as in Figure 17-16.

17-29 How many vertical divisions of an oscilloscope screen will a waveform with an rms value of 0.9 V occupy when the oscilloscope controls are as illustrated in Figure 17-16, (a) when a $1:1$ probe is used, (b) when a $10:1$ probe is employed.

17-30 Calculate the minimum frequency of a sine wave that can be displayed as one cycle on an oscilloscope with the *TIME/DIV* control set (a) as in Figure 17-13, (b) to 0.1 ms/div.

17-31 Determine the phase difference between the two waveforms shown in Figure 17-17(a), and between the two in Figure 17-17(b).

17-32 Determine the rms value of each waveform in Figure 17-17 if the oscilloscope controls are as in Figure 17-16.

17-33 Determine the frequency and wavelength of each waveform in Figure 17-17 if the oscilloscope controls are as in Figure 17-16.

17-34 For each of the waveforms illustrated in Figure 17-17, determine the frequency and rms value if the oscilloscope controls are set to 2 ms/div and 100 mV/div.

COMPUTER PROBLEM

17-35 Write a computer program to determine the instantaneous levels of a sine wave at several given time intervals from zero when the peak amplitude and frequency are specified.

ANSWERS TO PRACTICE PROBLEMS

17-2.1	796 rpm
17-2.2	41.3°, 45°, 58.2°
17-3.1	48.5 V, −48.5 V
17-3.2	405 m, 250 m
17-4.1	5.2 V, 5.2 V
17-5.1	16.73 V, 10.65 V, 11.83 V
17-5.2	4.95 W, 212 mA
17-6.1	10 V/div
17-6.2	230 mV, 5 kHz
17-6.3	0.7 div

18

PHASORS AND COMPLEX NUMBERS

Objectives

You will be able to:

Show how the instantaneous value of a sinusoidal waveform can be represented by a phasor.

Show how two sinusoidal waveforms can be represented by two phasors with a phase angle between them.

Determine the resultant of two sinusoidal waveforms by addition of instantaneous levels and by phasor addition.

Perform phasor addition and subtraction by resolving each phasor into horizontal and vertical components.

Express sinusoidal quantities in polar form and in rectangular form and convert from one form to another.

Solve problems involving addition, subtraction, multiplication, and division of complex quanitites.

Introduction

A sinusoidal alternating current or voltage can be graphically represented by a line of fixed length rotating about one of its ends. Such a line is termed a *phasor,* and the instantaneous value of a sinusoidal quantity can be determined from the length of the phasor and its angle with respect to the horizontal at the particular instant.

A phasor can be mathematically represented by its length and angle, or by its vertical and horizontal components. The vertical and horizontal components of a phasor can be written in the form of a *complex number.* Provided that certain rules are followed, phasors can be added, subtracted, multiplied, or divided. The same is true for complex numbers.

18-1

PHASOR REPRESENTATION OF ALTERNATING VOLTAGE

SCALARS AND VECTORS. Quantities that can be completely represented by a number of units are termed *scalar* quantities. Some examples of scalar quantities are temperature, volume, and resistance. Other quantities, like force and velocity, must have a direction indicated as well as the number of units stated if they are to be completely specified. Such quantities can be graphically represented by a line that has a length proportional to the number of units and that is drawn at an angle with respect to some reference direction. This line is termed a *vector*, and the quantity it represents is classified as a *vector quantity*.

PHASORS. The instantaneous levels of alternating current and voltage are vector quantities, but since the instantaneous levels are continuously changing, an ac waveform must be represented by a *rotating vector*, or *phasor*. A phasor is a vector that is rotating at a constant angular velocity.

The sinusoidal output voltage from the simple generator discussed in Section 17-1 can be represented by the *phasor diagram* shown in Figure 18-1. Here *OA* is a rotating vector (or phasor) with a constant angular velocity and a length that represents the peak output voltage E_m. The arrowhead identifies the end of the phasor that moves, and the other end is the axis of rotation. By convention, the direction of phasor rotation is taken as counterclockwise.

The instantaneous value of the generated voltage represented by the phasor obviously depends on the angle of the phasor with respect to the zero level. At time $t = 0$, the angle is $\theta = 0$ and the instantaneous voltage level is zero. At t_1, the angle is θ_1 and the instantaneous level is e_1, as illustrated in the figure, where

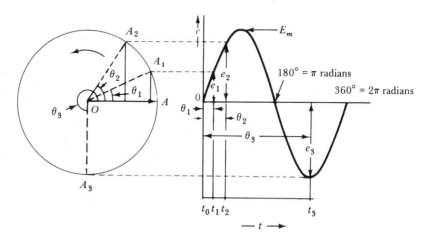

FIGURE 18-1. Phasor representation of a sine wave. Phasor *OA* rotates in a counter-clockwise direction, and its length equals the peak voltage E_m. At any angle θ, *OA sin θ* equals the instantaneous value of the sine wave.

$$e_1 = (OA) \sin \theta_1$$

$$= E_m \sin \theta_1$$

At t_2,

$$e_2 = E_m \sin \theta_2$$

and at t_3,

$$e_3 = E_m \sin \theta_3$$

At each of these instants, the voltage level is repesented by the *stopped phasor* (or vector) OA_1 at an angle θ_1, OA_2 at angle θ_2, and OA_3 at angle θ_3. Any sinusoidal ac voltage or current can be represented by a phasor, and instead of degrees the angles may be expressed in radians, as explained in Section 17-3.

18-2
ADDITION AND SUBTRACTION OF PHASORS

PHASOR ADDITION. In Figure 18-2 two sinusoidal waveforms A and B are shown together with the phasor representing each. It is seen that waveform B has a larger amplitude than waveform A. Therefore, phasor OB is longer than phasor OA. Also, waveform B starts to grow positively from its zero level $\phi°$ after the beginning of waveform A. Consequently, waveform B *lags* waveform A by $\phi°$, and the phasor OB is shown $\phi°$ behind OA. It can also be said that waveform A *leads* waveform B by $\phi°$.

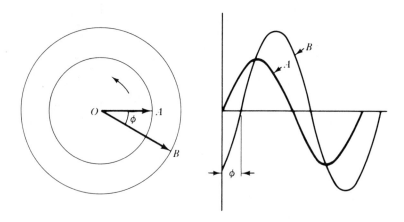

FIGURE 18-2. Two waveforms and their phasors. OB is larger than OA because waveform B has a higher peak value than waveform A. Waveform A leads waveform B by angle ϕ; consequently, phasor OA leads phasor OB by ϕ.

The instantaneous level of waveform A can be written

$$e_A = OA \sin \theta$$

and the instantaneous level of waveform B is

$$e_B = OB \sin(\theta - \phi)$$

Now consider Figure 18-3(a), in which the two waveforms are shown to be outputs from each of two series-connected generators. The resultant output of the two in series is the *phasor sum* of the two waveforms. Figure 18-3(b) shows how the instantaneous levels of A and B may be added to give the resultant waveform C. At t_1, $a_1 + b_1$ gives point 1 on waveform C. At t_2, $a_2 + b_2$ gives point 2 on waveform C, and so on. It is seen that waveform C lags waveform A by an angle $\alpha°$. Therefore, waveform C is represented by phasor OC lagging $\alpha°$ behind OA [see Figure 18-3(c)].

Referring to Figure 18-3(c), another method of obtaining the resultant of waveforms A and B is illustrated. A parallelogram is constructed, with AC being drawn equal to and parallel with OB, and BC equal to and parallel with OA. The resultant diagonal of the parallelogram OC represents the amplitude of waveform C, and the angle $\alpha°$ is the angle by which waveform C lags waveform A. Another approach to obtain the same result is shown in Figure 18-3(d). OA, identified as E_1, is first drawn horizontally. Then E_2 (which is OB) is drawn from the end of E_1 and at angle $-\phi$ with respect to E_1. The resultant E_3 (or OC) is found by drawing a line from O to the end of E_2. The angle of E_3 with respect to E_1 is again seen to be $-\alpha°$.

Instead of graphically adding the two phasors, the resultant may be obtained mathematically. Referring to Figure 18-4, each phasor is resolved into horizontal and vertical components (by trigonometry). The arithmetical sum of the horizontal components is then the horizontal component of the resultant. Similarly, the arithmetical sum of the vertical components is the vertical component of the resultant.

horizontal component of $E_1 = OA$

horizontal component of $E_2 = OD = E_2 \cos \phi$

total of horizontal components $= OE = OA + OD$

$$= E_1 + E_2 \cos \phi$$

vertical component of $E_1 = 0$

vertical component of $E_2 = OF = -E_2 \cos(90° - \phi) = -E_2 \sin \phi$

total of vertical components $= 0 + OF = -E_2 \sin \phi$

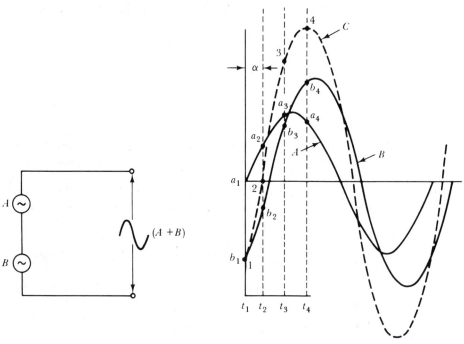

(a) Two sine wave
generators in series

(b) Addition of instantaneous
levels of two sine waves

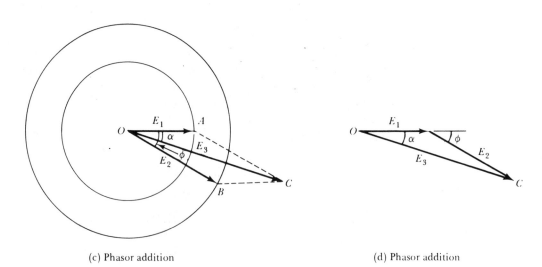

(c) Phasor addition

(d) Phasor addition

FIGURE 18-3.   To determine the resultant output from two series-connected ac
generators, the instantaneous levels of the waveforms can be added together.
Alternatively, phasors *OA* and *OB* representing each voltage can be added graphically to obtain the resultant *OC*.

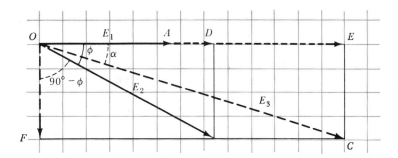

FIGURE 18-4. The resultant of two phasors can be determined by first resolving each into its horizontal and vertical components. The two horizontal quantities are then added together, and the two vertical quantities are added to give the horizontal and vertical components of the resultant.

The negative sign here indicates that the vertical component is measured down from the horizontal, rather than up.

$$\text{resultant } E_3 = OC = \sqrt{(OE)^2 + (OF)^2}$$

$$\text{angle } \alpha = \arctan\left(\frac{OF}{OE}\right)$$

PHASOR SUBTRACTION. Sometimes instead of phasor addition, *phasor subtraction* is required. An example of this occurs with the circuit shown in Figure 18-5(a), where

$$E_3 = E_1 - E_2$$

Figure 18-5(b) shows the process of phasor subtraction. E_2 is shown as $\theta°$ ahead of E_1. For subtraction, E_2 is reversed to become $-E_2$, as illus-

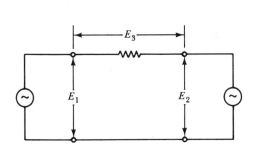

(a) E_3 is the phasor difference of E_1 and E_2

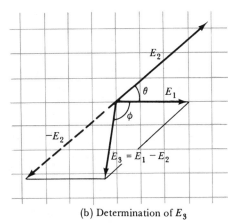

(b) Determination of E_3

FIGURE 18-5. The phasor difference of two phasors E_1 and E_2 can be determined by summing E_1 and $-E_2$.

trated. The resultant E_3 is then the phasor sum of E_1 and $-E_2$. Again, the resultant may be found mathematically, or by either of the two graphical methods described above.

Phasor addition and subtraction are not limited to two phasor quantities. The resultant of any number of quantities can be found, and some of these may require phasor addition while others may have to be subtracted.

EXAMPLE 18-1 Two sinusoidal quantities are given as $v_1 = 120 \sin \theta$ and $v_2 = 75 \sin (\theta - 30°)$. Determine the resultant obtained

a. When v_1 and v_2 are added,
b. When v_2 is subtracted from v_1.

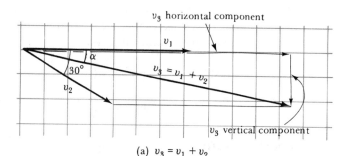

(a) $v_3 = v_1 + v_2$

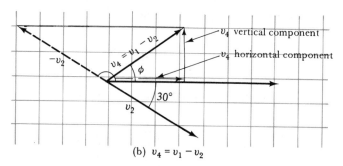

(b) $v_4 = v_1 - v_2$

FIGURE 18-6. Sum of two phasors, $v_3 = v_1 + v_2$, and difference of two phasors, $v_4 = v_1 - v_2$.

SOLUTION

a. $v_1 + v_2$ [*refer to* Figure 18-6(a)]:

$$horizontal\ component\ of\ v_1 = 120$$
$$horizontal\ component\ of\ v_2 = 75 \cos 30° \cong 65$$
$$horizontal\ component\ of\ v_3 = 120 + 65 = 185$$
$$vertical\ component\ of\ v_1 = 0$$
$$vertical\ component\ of\ v_2 = -75 \sin 30° = -37.5$$

$$\text{vertical component of } v_3 = 0 - 37.5 = -37.5$$

$$v_3 = \sqrt{185^2 + 37.5^2}$$

$$v_3 = 188.8$$

$$\text{angle } \alpha = \arctan \left[\frac{v_3 \text{ vertical component}}{v_3 \text{ horizontal component}} \right]$$

$$= \arctan \frac{-37.5}{185}$$

$$\alpha = -11.5°$$

$$\mathbf{\mathit{v_3} = 188.8 \sin (\theta - 11.5°)}$$

b. $v_1 - v_2$ [*refer to* Figure 18-6(b)]:

$$\text{horizontal component of } v_1 = 120$$

$$\text{horizontal component of } -v_2 = 75 \cos (180° - 30°) \cong -65$$

$$\text{horizontal component of } v_4 = 120 - 65 = 55$$

$$\text{vertical component of } v_1 = 0$$

$$\text{vertical component of } -v_2 = 75 \sin (180° - 30°) = 37.5$$

$$\text{vertical component of } v_4 = 0 + 37.5 = 37.5$$

$$v_4 = \sqrt{55^2 + 37.5^2}$$

$$v_4 = 66.6$$

$$\text{angle } \phi = \arctan \frac{37.5}{55}$$

$$\phi = 34.3°$$

$$\mathbf{\mathit{v_4} = 66.6 \sin (\theta + 34.3°)}$$

EXAMPLE 18-2   Graphically find the resultant of $v_1 + v_2 - v_3$, where $v_1 = 50 \sin \theta$, $v_2 = 30 \sin (\theta + 25°)$, and $v_3 = 25 \sin (\theta - 90°)$.

SOLUTION

1. The phasor diagram for v_1, v_2, and v_3 is drawn in Figure 18-7(a).
2. In Figure 18-7(b), v_1 is drawn to scale.
3. v_2 is drawn to scale from the end of v_1, and at an angle of $+25°$ with respect to v_1.
4. v_3 is drawn to scale from the end of v_2, and at an angle of $-90°$ with respect to v_1.
5. $-v_3$ is drawn opposite v_3, as illustrated.
6. v_4 is drawn by joining the origin of v_1 to the end of $-v_3$.
7. v_4 and ϕ are measured as: 86 and 26°:

$$\mathbf{\mathit{v_4} = 86 \sin (\theta + 26°)}$$

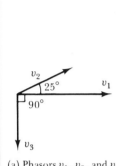

(a) Phasors v_1, v_2, and v_3

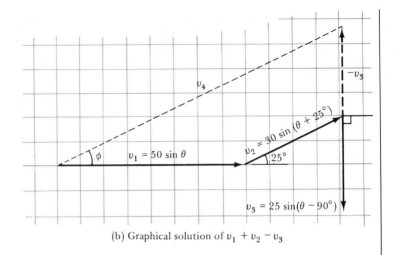

(b) Graphical solution of $v_1 + v_2 - v_3$

PRACTICE PROBLEMS

18-2.1 Two ac signal generators with identical frequencies have peak amplitudes of $E_1 = 100$ mV and $E_2 = 75$ mV. If E_2 lags E_1 by 25°, determine the resultant when the two outputs are connected in series.

18-2.2 For the quantities in Example 18-2, graphically determine $v_1 - v_2 + v_3$.

18-3

POLAR AND RECTANGULAR FORM, THE *J* OPERATOR

POLAR FORM. Another way of mathematically expressing a sinusoidal quantity, termed *polar form*, leaves out *sin* and substitutes $\angle$ for angle. Thus

$$e_1 = E_m \sin \theta \qquad \text{becomes } e_1 = E_m \underline{/\theta}$$

and
$$i_1 = I_m \sin(\theta + \alpha) \qquad \text{becomes } i_1 = I_m \underline{/\theta + \alpha}$$

Polar form is found to be very convenient for phasor quantities. The terms *modulus* and *argument* are sometimes used for the magnitude and angle, respectively, of a polar quantity. For e_1 above, the modulus is E_m, while θ is the argument.

Since there is a direct (0.707) relationship between the rms value and the peak value of a sinusoidal waveform, rms values are normally employed when using polar form. In fact, *with alternating voltages and*

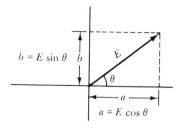

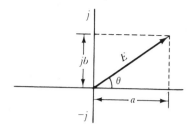

(a) Resolution of a phasor into vertical and horizontal components

(b) Use of the j operator

FIGURE 18-8. The j operator can be used to identify the vertical component of a phasor. Vertical component b is written as jb, while horizontal component a is simply written as a. Thus, phasor $E \underline{/\theta}$ is represented as $a + jb$.

currents, all quantities are assumed to be rms quantities unless otherwise indicated. Phasor addition and subtraction using rms quantities follows exactly the same graphical or mathematical procedure as when using peak quantities. The resultant obtained is, of course, an rms quantity. Rms values and peak values must not be mixed together in calculations. Where different types of quantities are to be combined, *all* must be converted into one type before phasor addition or subtraction.

As already explained, phasors must be resolved into horizontal and vertical components for phasor addition and subtraction. Thus, in Figure 18-8(a) the polar form, $e = E \underline{/\theta}$, becomes

$$\text{horizontal component, } a = E \cos \theta$$

$$\text{vertical component, } b = E \sin \theta$$

RECTANGULAR FORM, THE J OPERATOR. For convenience in writing vertical and horizontal components the j *operator* is employed, as illustrated in Figure 18-8(b). All positive vertical components are given the prefix j, while all negative vertical components are prefixed with $-j$. The quantity $E \underline{/\theta}$ can now be written

$$E \underline{/\theta} = a + jb$$

or

$$\boxed{E \underline{/\theta} = E \cos \theta + jE \sin \theta} \tag{18-1}$$

A phasor stated in this way is said to be expressed in *rectangular form*. Conversion from polar to rectangular form is simply a matter of resolving the phasor into its horizontal and vertical components. Conversion from rectangular to polar form is then the reverse of that process:

$$E = \sqrt{a^2 + b^2}$$

and
$$\theta = \arctan\left(\frac{b}{a}\right)$$

or

$$\boxed{E \underline{/\theta} = \sqrt{a^2+b^2} \bigg/ \arctan\left(\frac{b}{a}\right)} \qquad \text{(18-2)}$$

The j operator is not merely a prefix employed for convenience in identifying the vertical component of a phasor. Expressions such as Equation (18-1), using the j operator, are termed *complex numbers,* and the mathematics of complex numbers require that certain rules be followed. Consider phasor $E \underline{/0°}$ in Figure 18-9. When multiplied by j, the phasor is rotated counterclockwise by 90° or $\pi/2$ radians. Thus, it becomes

$$E \underline{/90°} = jE$$

When jE is multiplied by j, the phasor is

$$E \underline{/180°} = j^2 E = -E$$

Therefore,

$$\boxed{j^2 = -1} \qquad \text{(18-3)}$$

or
$$j = \sqrt{-1}$$

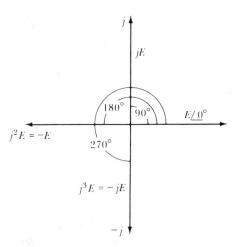

FIGURE 18-9. Multiplying $E \underline{/0}$ by j gives jE. Multiplying by j^2 gives $-E$. Multiplying by j^3 produces $-jE$.

Chap. 18   Phasors and Complex Numbers

The actual mathematical square root of +1 is 1, but the square root of −1 cannot be determined. For this reason j is frequently referred to as an *imaginary number*. For the complex number $a + jb$, a is sometimes termed the *real part*, while jb is called the *imaginary part*.

Returning to Figure 18-9, when $-E$ is once more multiplied by j, the phasor is rotated counterclockwise by a further 90°, giving

$$E \underline{/270°} = -jE$$

and when multiplied by a fourth j, the quantity is

$$E \underline{/0°} = -j^2E$$
$$= -(-1)E$$
$$= E$$

EXAMPLE 18-3   Convert the following quantities into rectangular form:

   a. $12\underline{/30°}$,   *b*. $270\underline{/1.7\pi}$,   *c*. $40\underline{/105°}$.

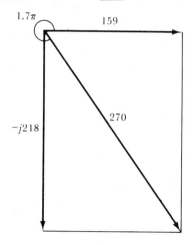

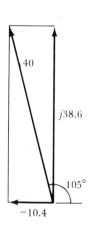

(a) $12\underline{/30°}$ (b) $270\underline{/1.7\pi}$ (c) $40\underline{/105°}$

FIGURE 18-10. Polar quantities are converted into rectangular quantities by resolution into their horizontal and vertical components. $E\underline{/\theta}$ is resolved into $E \cos \theta$ and $jE \sin \theta$.

SOLUTION

a. Equation (18-1):

$$E \underline{/\theta} = E \cos \theta + jE \sin \theta$$
$$12\underline{/30°} = 12 \cos 30° + j12 \sin 30°$$
$$= \mathbf{10.4 + j6} \quad [\textit{see} \text{ Figure 18-10(a)}]$$

b. $270\underline{/1.7\pi} = 270 \cos{(1.7\pi)} + j270 \sin{(1.7\pi)}$

      $= \mathbf{159 - j218}$ [*see* Figure 18-10(b)]

c. $40\underline{/105°} = 40 \cos{105°} + j40 \sin{105°}$

      $= \mathbf{-10.4 + j38.6}$ [*see* Figure 18-10(c)]

PRACTICE PROBLEMS

18-3.1   Taking $\theta = 0$, convert each of the quantities in Example 18-2 into rectangular form.

18-3.2   Express the following quantities in polar form:
$a = 3 + j4$, $b = 77 - j220$, $c = -53 - j47$.

18-4

MATHEMATICS OF COMPLEX QUANTITIES

ADDITION AND SUBTRACTION. When mathematically processing complex quantities in rectangular form, the usual rules of algebra apply as long as it is remembered that $j^2 = -1$. For addition and subtraction, the real parts of the numbers are added or subtracted, and the imaginary parts are added or subtracted. This is, of course, simply adding or subtracting the horizontal and vertical components of a phasor.

$$(a + jb) + (c + jd) = (a + c) + j(b + d) \qquad \text{(18-4)}$$

$$(a + jb) - (c + jd) = (a - c) + j(b - d) \qquad \text{(18-5)}$$

MULTIPLICATION. Multiplication of complex quantities in rectangular form follows the normal procedure for multiplication of algebraic expressions:

$(a + jb) \times (c + jd)$,

multiplying the top line by c,

multiplying the top line by jd,

$$\begin{array}{r} a + jb \\ \times (c + jd) \\ \hline ac + jbc \\ jad + j^2bd \\ \hline ac + jbc + jad - bd \end{array}$$

Therefore,

$$(a + jb) \times (c + jd) = (ac - bd) + j(bc + ad) \qquad \text{(18-6)}$$

DIVISION. Division of complex quantities in rectangular form requires the use of the *conjugate* of the denominator. The conjugate of a complex number is the same number with the sign changed on the j component. Thus, the conjugate of $(a + jb)$ is $(a - jb)$.

$$\frac{a + jb}{c + jd} = \frac{a + jb}{c + jd} \times \frac{c - jd}{c - jd}$$

Multiplying by $(c - jd)/(c - jd)$ is the same as multiplying by 1. Therefore, nothing has been changed

$$\frac{a + jb}{c + jd} \times \frac{c - jd}{c - jd} = \frac{ac + jbc - jad - j^2bd}{c^2 + jcd - jcd - j^2d^2}$$

$$= \frac{(ac + bd) + j(bc - ad)}{c^2 + d^2}$$

Therefore,

$$\boxed{\frac{a + jb}{c + jd} = \frac{ac + bd}{c^2 + d^2} + j\frac{bc - ad}{c^2 + d^2}} \qquad (18\text{-}7)$$

Another important consideration in processing j quantities is the reciprocal of a j quantity:

$$\frac{1}{jx} = \frac{1}{jx} \times \left(\frac{-j}{-j}\right)$$

$$= \frac{-j}{-j^2x} = \frac{-j}{-(-1)x} = \frac{-j}{x}$$

Therefore,

$$\frac{1}{jx} = -j\frac{1}{x}$$

or

$$\boxed{\frac{1}{j} = -j} \qquad (18\text{-}8)$$

EXAMPLE 18-4 Determine: $a.$ $(75 - j50) \times (25 + j5)$, $b.$ $(75 - j50) \div (25 + j5)$.

SOLUTION

$a.$

$$\begin{array}{r} 75 - j50 \\ \times\,(25 + j5) \\ \hline 1875 - j1250 \\ + j375 - j^2250 \\ \hline 1875 - j1250 + j375 + 250 \\ = 2125 - j875 \end{array}$$

b.

$$\frac{75 - j50}{25 + j5} \times \frac{25 - j5}{25 - j5} = \frac{1875 - j1250 - j375 + j^2250}{625 + j125 - j125 - j^225}$$

$$= \frac{1625 - j1625}{650}$$

$$= \mathbf{2.5 - j2.5}$$

POLAR FORM MATHEMATICS. Multiplication and division in polar form are much simpler than in rectangular form:

$$\boxed{E_1 \underline{/\theta_1} \times E_2 \underline{/\theta_2} = E_1 E_2 \underline{/\theta_1 + \theta_2}} \tag{18-9}$$

and

$$\boxed{\frac{E_1 \underline{/\theta_1}}{E_2 \underline{/\theta_2}} = \frac{E_1}{E_2} \underline{/\theta_1 - \theta_2}} \tag{18-10}$$

For addition and subtraction, quantities stated in polar form must first be converted to rectangular form.

EXAMPLE 18-5 Determine:

    a. $90\underline{/-33.7°} \times 25.5\underline{/11.3°}$.
    b. $90\underline{/-33.7°} \div 25.5\underline{/11.3°}$.

SOLUTION

    a. $90\underline{/-33.7°} \times 25.5\underline{/11.3°} = (90 \times 25.5)\underline{/-33.7° + 11.3°}$
                               $= \mathbf{2295\underline{/-22.4°}}$

    b. $90\underline{/-33.7°} \div 25.5\underline{/11.3°} = \frac{90}{25.5}\underline{/-33.7° - 11.3°}$
                               $= \mathbf{3.5\underline{/-45°}}$

Note that $90\underline{/-33.7°}$ and $25.5\underline{/11.3°}$ are the polar form of the quantities in Example 18-4, and that when the answers above are checked, they are found to be the polar form of the answers in Example 18-4.

PRACTICE PROBLEMS

18-4.1 Using the quantities in Problem 18-3.2, determine $a + b - c$.

18-4.2 Using the quantities in Problem 18-3.2, determine $a \times b/c$.

18-4.3 Solve $(100 + j85)/(37 - j29)$ in rectangular form. Then, convert each quantity to polar form and find the result in polar form.

                                                          Chap. 18   Phasors and Complex Numbers

SUMMARY OF FORMULAS

Polar to rectangular conversion:

$$E\underline{/\theta} = E \cos \theta + jE \sin \theta$$

Rectangular to polar conversion:

$$a + jb = \sqrt{a^2 + b^2}\underline{/\arctan\left(\frac{b}{a}\right)}$$

j^2 *and* $1/j$:

$$j^2 = -1$$

$$1/j = -j$$

Addition:

$$(a + jb) + (c + jd) = (a + c) + j(b + d)$$

Subtraction:

$$(a + jb) - (c + jd) = (a - c) + j(b - d)$$

Multiplication:

$$(a + jb) \times (c + jd) = (ac - bd) + j(bc + ad)$$

Division:

$$\text{for } \frac{a + jb}{c + jd} \text{ multiply by } \frac{c - jd}{c - jd}$$

Multiplication:

$$E_1\underline{/\theta_1} \times E_2\underline{/\theta_2} = E_1 E_2\underline{/\theta_1 + \theta_2}$$

Division:

$$\frac{E_1\underline{/\theta_1}}{E_2\underline{/\theta_2}} = \frac{E_1}{E_2}\underline{/\theta_1 - \theta_2}$$

REVIEW QUESTIONS

18-1 Sketch a sinusoidal waveform and show how its instantaneous value may be represented by a phasor.

18-2 Show how two sinusoidal waveforms that have a phase difference may be represented by phasors. Briefly explain.

18-3 Sketch the waveform from two series connected ac generators that have ac voltages which differ in amplitude and phase. Show how the resultant of the two waveforms may be obtained by adding the instantaneous voltage values. Also, show how the resultant may be obtained by phasor addition.

18-4 Sketch two phasors with different amplitudes and phase angles. Show how the resultant of the two can be obtained by resolving each into horizontal and vertical components.

18-5 Draw the diagram of a circuit that involves phasor subtraction. Sketch two phasors differing in amplitude and phase and show how one should be subtracted from the other.

18-6 Sketch a phasor diagram to show the relationship between polar and rectangular forms of phasor representation. Also, show the effect of multiplying $E \underline{/0}$ by j, j^2, and j^3. Explain briefly.

PROBLEMS

SECTION 18-2

18-1 Graphically determine $v_1 + v_2$, where $v_1 = 47 \sin \phi$ and $v_2 = 33 \sin(\phi + 20°)$.

18-2 For the quantities in Problem 18-1, graphically determine $v_1 - v_2$.

18-3 Graphically determine the resultant of $v_1 + v_2 + v_3 - v_4$, where $v_1 = 10 \sin \phi$, $v_2 = 15 \sin(\phi - 15°)$, $v_3 = 20 \sin(\phi + 10°)$, and $v_4 = 18 \sin(\phi + 25°)$.

18-4 For the quantities given in Problem 18-3, graphically determine $v_1 - v_2 - v_3 + v_4$.

18-5 Graphically determine $a - 2b$, where $a = 150\underline{/22°}$ and $b = 85\underline{/2.5}\ \pi$.

18-6 Graphically determine $b + c$, where $b = 85\underline{/2.5}\ \pi$ and $c = 64\underline{/72°}$.

18-7 Draw the phasor diagram for $i_1 + i_2 - i_3$, where $i_1 = 12\underline{/125°}$, $i_2 = 10\underline{/0°}$, and $i_3 = 15\underline{/86°}$.

18-8 For the quantities given in Problem 18-7, graphically determine $i_1 - i_2 + i_3$.

18-9 Graphically determine the resultant of $v_1 + v_2 - v_3$, where $v_1 = 45\underline{/30°}$, $v_2 = 27\underline{/21°}$, and $v_3 = 30\underline{/42°}$.

18-10 For the quantities given in Problem 18-9, graphically determine $v_1 - v_2 - v_3$.

18-11 For the quantities given in Problems 18-5 and 18-6, graphically determine $a + b - c$.

18-12 For the quantities given in Problems 18-5 and 18-6, graphically determine $2a + b - 2c$.

SECTION 18-3

18-13 Convert the quantities stated in Example 18-4 into polar form.

18-14 Convert the quantities stated in Problem 18-7 into rectangular form.

Chap. 18   Phasors and Complex Numbers

18-15 Convert the quantities stated in Problems 18-5 and 18-6 into rectangular form.

18-16 Convert the following quantities into polar form: $3 + j4$, $120 - j75$, $680 + j170$, $-35 + j84$.

18-17 Convert the quantities stated in Problem 18-9 into rectangular form.

18-18 Convert the following quantities into polar form: $-23 + j23$, $-333 - j270$, $80 + j17$, $-75 + j56$.

SECTION 18-4

18-19 Algebraically solve Problem 18-1.

18-20 Algebraically solve Problem 18-2.

18-21 For the quantities given in Problem 18-9, determine $(v_1 \times v_2)/v_3$.

18-22 Solve $\dfrac{(16 + j12) \times (22 - j18)}{14 + j25}$.

18-23 Algebraically solve Problem 18-11.

18-24 For the quantities given in Problems 18-5 and 18-6, determine $(a - 2b)/(b + c)$.

18-25 Solve $\dfrac{[(25 + j15) + (45 - j50)] \times (33 - j29)}{(62 + j70) - (32 + j100)}$.

18-26 For the quantitites given in Problem 18-1, determine $(v_1 \times v_2)/[2(v_1 + v_2)]$.

18-27 Solve $[(47\underline{/35°} - 22\underline{/60°}) \times 54\underline{/30°}]/(76 + j29)$.

18-28 Add $6.4\underline{/22°}$, $3.9\underline{/45°}$, and $4.1\underline{/68°}$.

18-29 Algebraically solve Problem 18-7 for $i_1 + i_2 - i_3$.

18-30 Solve $Z = (v_1 + v_2)/i_1$, where $v_1 = 115\underline{/0°}$, $v_2 = 115\underline{/-120°}$, and $i_1 = 22\underline{/73°}$.

18-31 Find $(a \times c)/b$, where $a = 29 - j73$, $b = 64 + j55$, and $c = 49 - j22$.

18-32 Algebraically solve Problem 18-9.

18-33 Algebraically solve Problem 18-8.

18-34 Algebraically solve Problem 18-4.

18-35 Algebraically solve Problem 18-12.

COMPUTER PROBLEMS

18-36 Write a computer program for converting several polar quantities into rectangular form.

18-37 Write a computer program for converting several rectangular quantities into polar form.

18-2.1 171 mV$\underline{/-10.7°}$

18-2.2 44 sin(0 − 59°)

18-3.1 $(50 + j0)$, $(27.2 + j12.7)$, $(0 − j25)$

18-3.2 5$\underline{/53.1°}$, 233$\underline{/-70.7°}$, 70.8$\underline{/222°}$

18-4.1 $133 − j169$

18-4.2 16.5$\underline{/-239°}$

18-4.3 $(0.559 + j2.74)$, 2.8$\underline{/78.4°}$

19

INDUCTANCE AND CAPACITANCE IN AC CIRCUITS

Objectives

You will be able to:

Sketch the waveforms of supply voltage and current in a purely inductive circuit. Explain the phase relationships between the waveforms.

Calculate the inductive reactance of a given inductor and the current that flows when the inductor is connected to a supply with a given voltage and frequency.

Sketch the waveforms of supply voltage and current in a purely capacitive circuit. Explain the phase relationships between the waveforms.

Calculate the capacitive reactance of a given capacitor and the current that flows when the capacitor is connected to a supply with a given voltage and frequency.

Sketch the waveforms of supply voltage, current, resistor voltage, and inductor voltage in a series *RL* circuit. Explain the waveforms.

Sketch the waveforms of supply voltage, current, resistor voltage, and capacitor voltage in a series *RC* circuit. Explain the waveforms.

Calculate all current levels, voltage levels, and phase relationships in series *RC* and series *RL* circuits when each circuit is connected to a supply with a given voltage and frequency.

Calculate all current levels, voltage levels, and phase relationships in series *RLC* circuits and parallel *RLC* circuits when each circuit is connected to a supply with a given voltage and frequency.

Sketch phasor diagrams, impedance diagrams, and admittance diagrams where appropriate for series *RL* circuits, series *RC* circuits, series *RLC* circuits, and parallel *RLC* circuits.

Sketch and explain the operation of *RC* and *LC* low-pass and high-pass filters.

Calculate the output voltage obtained from low-pass and high-pass filters when a given input is applied.

Sketch typical frequency response graphs for low-pass and high-pass filters.

Introduction

Alternating current flow in an inductor depends on the applied voltage and on the *inductive reactance* of the inductor. The inductive reactance is proportional to the inductance value and the frequency of the alternating supply voltage. Similarly, the ac current flow in a capacitor is dependent on the supply voltage and the *capacitive reactance*. The capacitance value and the supply frequency determine the capacitive reactance.

When an alternating voltage is applied to a pure inductance, the current that flows through the inductance lags its terminal voltage by 90°. Conversely, the alternating current through a capacitor leads the capacitor terminal voltage by 90°. For *RL* and *RC* series circuits, the phase angle of current with respect to supply voltage is less than 90°.

A series *RLC* circuit may behave as a resistive-inductive circuit or as a resistive-capacitive circuit, depending on the component values and supply frequency. The same can be said of a parallel *RLC* circuit.

19-1
ALTERNATING CURRENT AND VOLTAGE IN AN INDUCTIVE CIRCUIT

In Section 17-4 it was explained that when an alternating voltage is applied to a purely resistive circuit, the resultant current through the resistor is in phase with the applied voltage. This is illustrated once again in Figure 19-1. Because of the presence of a *counter-emf* in an inductive circuit, the current and voltage phase relationships are considerably different from the case of the resistive circuit.

Figure 19-2 shows an inductance L with an ac voltage applied to it. For the waveforms illustrated, it is assumed that the resistance of the coil is very much smaller than the inductance. Thus, the circuit is assumed to be purely inductive. Since the counter-emf is proportional to the rate of

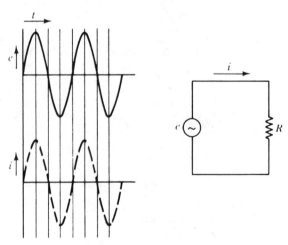

FIGURE 19-1. In a purely resistive circuit with an ac voltage source, the circuit current is in-phase with the applied voltage.

Chap. 19   Inductance and Capacitance in AC Circuits

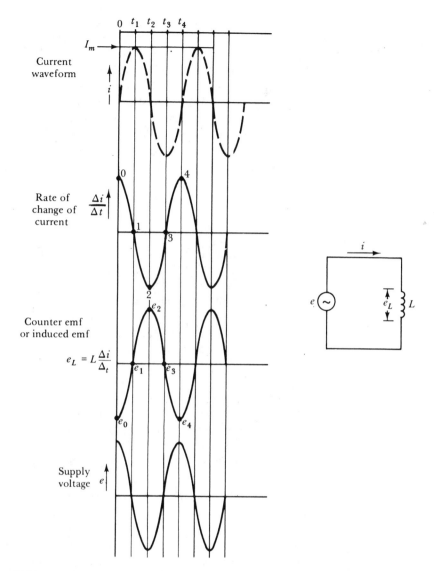

FIGURE 19-2. For a purely inductive circuit, $\Delta i/\Delta t$ is a maximum negative quantity when i crosses zero in a negative-going direction. Counter-emf e_L opposes Δi and is proportional to $\Delta i/\Delta t$. So, e_L is at its positive maximum when $\Delta i/\Delta t$ is at negative maximum. The supply voltage e is in anti-phase to e_L.

change of current through the inductance, it is appropriate to start with the current waveform.

As shown in the illustration, the current waveform is sinusoidal with a zero level at $t = 0$, positive peak at t_1, zero again at t_2, and so on. At t_1, the current has stopped increasing positively, and has not yet started to decrease. Therefore, the rate of change of current at t_1 is zero. This gives point 1 on the waveform representing *rate of change of cur-*

rent. At t_2, the current is decreasing at its maximum rate, which means it has a maximum negative rate of change. Thus, point 2 on the $\Delta i/\Delta t$ waveform is a peak negative value. At t_3, the current has a zero rate of change once again, and the $\Delta i/\Delta t$ value is therefore plotted at point 3. Finally, at t_4, the current has a maximum positive rate of change, giving point 4 on the waveform of current rate of change. Similarly, at $t = 0$ the current has a maximum positive rate of change, and this gives point 0 on the $\Delta i/\Delta t$ waveform.

Now recall from Equation (14-3) that the counter-emf in an inductor is

$$e_L = L\,\frac{\Delta i}{\Delta t}$$

Therefore, the counter-emf is directly proportional to the rate of change of current. However, because the counter-emf opposes the change of current, it has a maximum positive value when the rate of change of current is at its maximum negative value, and vice versa. The outcome of this is that the waveform of counter-emf is in *antiphase* with the waveform of rate of change of current. In Figure 19-2, $e_L = e_1 = 0$ when $\Delta i/\Delta t$ is zero at t_1. At t_2, $\Delta i/\Delta t$ is a peak negative quantity, and $e_L = e_2$, which is a peak positive quantity. Similarly, at t_4, $\Delta i/\Delta t$ has a peak positive value, and $e_L = e_4$, which is the peak negative level of e_L. The counter-emf is always in antiphase to the applied emf or ac supply voltage. Therefore the final waveform in Figure 19-2 shows the supply voltage waveform in antiphase to that of the counter-emf.

Putting the waveforms of supply voltage and inductor current together, as in Figure 19-3(a), it is seen that

*the current in a purely inductive circuit **lags** the voltage by 90° ($\pi/2$ radians).*

Alternatively, it may be stated that the supply voltage leads the current by 90°. The phasor diagram of voltage and current in an inductive circuit [Figure 19-3(b)] uses the voltage as a reference and shows the current phasor 90° behind the voltage. Where the inductor terminal voltage is

$$e = E_m \sin \omega t$$

the inductor current is represented as

$$\boxed{i = I_m \sin\left(\omega t - \frac{\pi}{2}\right)} \qquad \text{(19-1)}$$

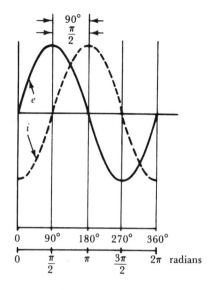

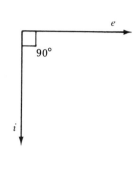

(a) Current and voltage waveforms
in a purely inductive circuit

(b) Phasor diagram of e and i in
a purely inductive circuit

FIGURE 19-3. In a purely inductive circuit, the current lags the voltage by 90°
($\pi/2$ radians).

19-2

**INDUCTIVE
REACTANCE
AND
SUSCEPTANCE**

INDUCTIVE REACTANCE. Referring once again to the current wave-form in Figure 19-2, it is seen that the current grows from zero to its maximum value I_m in one fourth of a cycle. The time period of one cycle is

$$T = \frac{1}{f}$$

Therefore, the time taken for the current to grow from zero to I_m is

$$t = \frac{1}{4f}$$

The *average* rate of change of current during time t is

$$\frac{\Delta i}{\Delta t} = \frac{I_m}{1/(4f)} = 4fI_m$$

From Equation (14-3), $e_L = L(\Delta i/\Delta t)$. Substituting for $\Delta i/\Delta t$, the average counter-emf induced is

$$e_L = 4fLI_m$$

And since the counter-emf is equal to the supply voltage in a purely inductive circuit, the average supply voltage is

$$E_{av} = 4fLI_m$$

Actually there is always a slight difference between counter-emf and supply voltage; otherwise, no current would flow and no counter-emf would be generated. From Equation (17-14),

$$E_{av} = \frac{2}{\pi} E_m$$

Therefore,

$$\frac{2}{\pi} E_m = 4fLI_m$$

or

$$\frac{E_m}{I_m} = \frac{4\pi fL}{2}$$

$$\frac{E_m}{I_m} = 2\pi fL$$

Substituting rms quantities for maximum quantities,

$$\frac{1.414 \ E}{1.414 \ I} = 2\pi fL$$

or, *for an inductor:*

$$\boxed{\frac{E}{I} = 2\pi fL} \qquad \qquad \textbf{(19-2)}$$

where E and I are rms quantities.

From Ohm's law $(E \ volts)/(I \ amps)$ gives resistance in ohms. However, in a pure inductance there is zero resistance, but $2\pi fL$ does represent an *opposition* to current flow. The name given to this quantity is *inductive reactance*, and like resistance, inductive reactance is measured in ohms. The symbol X_L is employed for inductive reactance.

Since $2\pi f = \omega$, the angular velocity in radians per second, Equation (19-2) can be rewritten as

$$\frac{E}{I} = X_L = \omega L$$

or,

$$\boxed{\textit{inductive reactance}, X_L = \omega L = 2\pi fL} \qquad (19\text{-}3)$$

When f is in hertz and L is in henrys, X_L is given in ohms. Note that X_L is directly proportional to frequency and inductance.

Substituting for I_m in Equation (19-1), the expression for the instantaneous current in an inductive circuit becomes

$$\boxed{i = \frac{E_m}{\omega L} \sin\left(\omega t - \frac{\pi}{2}\right)} \qquad (19\text{-}4)$$

INDUCTIVE SUSCEPTANCE. The reciprocal of resistance R is conductance G, which is the measure of a resistive circuit's ability to pass current. Similarly, inductive reactance X_L has its reciprocal in *inductive susceptance*, for which the symbol is B_L. Inductive susceptance is a measure of a purely inductive circuit's ability to pass current, and like conductance its unit is the siemens, S.

$$\boxed{\textit{inductive susceptance}, B_L = \frac{1}{X_L}} \qquad (19\text{-}5)$$

When X_L is in ohms, B_L is in siemens.

EXAMPLE 19-1 A 500 mH inductor is supplied from a 115 V source with a frequency of 60 Hz. Calculate the inductive reactance and the current that flows in the circuit.

SOLUTION

Equation (19-3),

$$\textit{inductive reactance}, X_L = 2\pi fL$$
$$= 2\pi \times 60 \text{ Hz} \times 500 \text{ mH}$$
$$\cong \textbf{188.5 } \mathbf{\Omega}$$
$$\textit{current}, I = \frac{E}{X_L} = \frac{115 \text{ V}}{188.5 \text{ }\Omega}$$
$$= \textbf{610 mA}$$

EXAMPLE 19-2 An inductor supplied with 50 V ac with a frequency of 10 kHz passes a current of 7.96 mA. Determine the inductance of the inductor.

SOLUTION

$$X_L = \frac{E}{I} = \frac{50 \text{ V}}{7.96 \text{ mA}}$$

$$= 6.28 \text{ k}\Omega$$

From Equation (19-3),

$$L = \frac{X_L}{2\pi f} = \frac{6.28 \text{ k}\Omega}{2\pi \times 10 \text{ kHz}}$$

$$\cong \textbf{100 mH}$$

EFFECT OF MUTUAL INDUCTANCE. By substituting Equation (14-7) for mutual inductance instead of Equation (14-3) for self inductance in the derivation of Equation (19-2), an expression is obtained for the alternating voltage induced in the secondary of two mutually coupled coils (see Section 14-4). When an alternating current flows in the primary,

$$\frac{E_2}{I_1} = 2\pi f M$$

or

$$\boxed{E_2 = \omega M I_1} \tag{19-6}$$

where E_2 is the rms voltage induced in the secondary winding, M is the mutual inductance, and I_1 is the rms current in the primary.

EXAMPLE 19-3 When a 100 mA alternating current with a frequency of 1 kHz flows in the primary of two coupled coils, the secondary voltage is 1 V. Calculate the mutual inductance between the two coils.

SOLUTION

From Equation (19-6),

$$M = \frac{E_2}{\omega I_1}$$

$$= \frac{1 \text{ V}}{2\pi \times 1 \text{ kHz} \times 100 \text{ mA}}$$

$$= \textbf{1.59 mH}$$

19-2.1 An inductor passes a current of 25 mA when connected to a 40 mV, 200 Hz supply. Calculate the inductance.

19-2.2 Determine the current that flows in a 50 μH inductor supplied from a 35 V, 1 MHz source.

19-2.3 A secondary coil is wound on the inductor in Problem 19-2.2. The mutual inductance between the two coils is 3.3 μH. Determine the emf induced in the secondary winding.

19-3

ALTERNATING CURRENT AND VOLTAGE IN A CAPACITIVE CIRCUIT

In the circuit shown in Figure 19-4, it is assumed that C is a pure capacitance with a dielectric having infinite resistance, and that the connecting wires have zero resistance. The first waveform illustrated is that of the instantaneous level of supply voltage e. This is also the voltage that appears at the capacitor terminals. The waveform representing rate of change of capacitor voltage is drawn by considering the supply voltage waveform.

At t_1, the capacitor voltage has stopped increasing positively and has not yet started to decrease. Consequently, the rate of change of voltage at t_1 is zero. This gives point 1 on the waveform of *rate of change* of voltage. At t_2, the voltage is decreasing at its maximum rate; thus, it has a maximum negative rate of change, and this is plotted at point 2 on the $\Delta v / \Delta t$ waveform. At t_3, the voltage has zero rate of change again, giving point 3 on the $\Delta v / \Delta t$ wave. The voltage is increasing positively at its maximum rate of change at t_4; therefore, point 4 is plotted as the positive peak of the $\Delta v / \Delta t$ waveform. Similarly, point 0 represents the maximum positive $\Delta v / \Delta t$ that occurs at $t = 0$.

From Equation (15-10),

$$I = \frac{CE}{t}$$

While a capacitor is being charged through Δv volts for a time Δt seconds, the instantaneous current is

$$i = C \frac{\Delta v}{\Delta t}$$

It is seen that the instantaneous current in a purely capacitive circuit is directly proportional to the rate of change of capacitor voltage. Now consider the $\Delta v / \Delta t$ waveform once again in Figure 19-4. The value of $\Delta v / \Delta t$ is zero at time t_1. Therefore, the instantaneous current i_1 is plotted as zero on the current waveform. At time t_2, the instantaneous

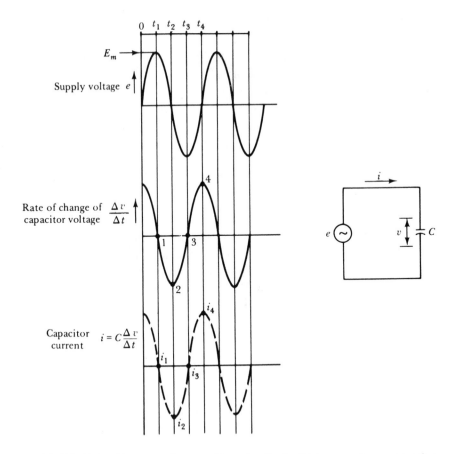

FIGURE 19-4. For a purely capacitive circuit, $\Delta v/\Delta t$ is a maximum negative quantity when supply voltage e crosses zero in a negative-going direction. Because i is proportional to $C\Delta v/\Delta t$, the capacitor current is in-phase with $\Delta v/\Delta t$.

value of $\Delta v/\Delta t$ is maximum negative rate of change; consequently, the instantaneous current at t_2 is a maximum negative quantity i_2, as illustrated. At t_3, $\Delta v/\Delta t$ is zero once again, giving i_3 equal to zero. When the rate of change of voltage is a maximum positive quantity at t_4, the instantaneous current is also a positive maximum, plotted as i_4 on the current waveform.

In Figure 19-5(a) the waveforms of supply voltage and capacitor current are shown together. The waveforms, and the phasor diagram in Figure 19-5(b), show that

*the current in a purely capacitive circuit **leads** the voltage by 90°*
($\pi/2$ radians).

When the capacitor terminal voltage is represented as,

$$e = E_m \sin \omega t$$

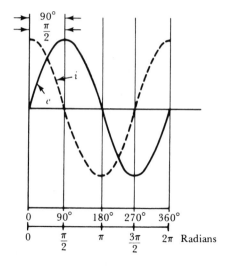

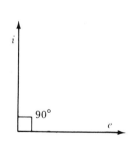

(a) Current and voltage waveform
in a purely capacitive circuit

(b) Phasor diagram of e and i in a
purely capacitive circuit

FIGURE 19-5. In a purely capacitive circuit, the current leads the voltage by 90°
($\pi/2$ radians).

the capacitor current is

$$i = I_m \sin\left(\omega t + \frac{\pi}{2}\right)$$
(19-7)

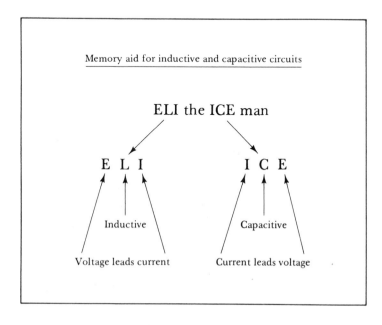

19-4

CAPACITIVE REACTANCE AND SUSCEPTANCE

CAPACITIVE REACTANCE. As already explained for the inductive circuit, the capacitor current grows from zero to its maximum level I_m in a time of $t = 1/(4f)$. From Equation (17-13),

$$I_{av} = \frac{2}{\pi} I_m$$

and

$$Q = CE_m = I_{av} \times t$$

Therefore,

$$CE_m = \frac{2}{\pi} I_m \times \frac{1}{4f}$$

$$= I_m \times \frac{1}{2\pi f}$$

for which

$$\frac{E_m}{I_m} = \frac{1}{2\pi f C}$$

Substituting rms quantities for maximum voltage and current,

$$\frac{1.414\ E}{1.414\ I} = \frac{1}{2\pi f C}$$

or, *for a capacitor:*

$$\boxed{\frac{E}{I} = \frac{1}{2\pi f C}} \tag{19-8}$$

where E and I are rms quantities.

As in the case of the pure inductor, a pure capacitor has no resistive component, but $1/(2\pi f C)$ represents an opposition to current flow. Termed *capacitive reactance,* this quantity is measured in ohms and is given the symbol X_C:

$$\boxed{\textit{capacitive reactance, } X_C = \frac{1}{\omega C} = \frac{1}{2\pi f C}} \tag{19-9}$$

When f is in hertz and C in farads, X_C is given in ohms. Note that X_C is inversely proportional to frequency and capacitance.

Substituting for I_m in Equation (19-7), the instantaneous current level in a capacitive circuit is given by

$$\boxed{i = E_m\ (\omega C)\sin\left(\omega t + \frac{\pi}{2}\right)} \tag{19-10}$$

CAPACITIVE SUSCEPTANCE. The reciprocal of capacitive reactance X_C is *capacitive susceptance B_C*, which is a measure of a purely capacitive circuit's ability to pass current:

$$\boxed{\textit{capacitive susceptance, } B_C = \frac{1}{X_C}} \qquad \text{(19-11)}$$

where X_C is in ohms and B_C is in siemens.

EXAMPLE 19-4   A 50 μF capacitor is supplied from a 115 V source with a frequency of 60 Hz. Determine the capacitive reactance and calculate the current that flows in the circuit.

SOLUTION

Equation (19-9),

$$\textit{capacitive reactance, } X_C = \frac{1}{2\pi f C}$$

$$= \frac{1}{2\pi \times 60 \text{ Hz} \times 50 \text{ μF}}$$

$$\cong 53.1 \; \Omega$$

$$\textit{current, } I = \frac{E}{X_C} = \frac{115 \text{ V}}{53.1 \; \Omega}$$

$$\cong 2.2 \text{ A}$$

EXAMPLE 19-5   A capacitor passes a current of 12.6 mA when supplied with 20 V ac with a frequency of 1 kHz. Determine the capacitance of the capacitor.

SOLUTION

$$X_C = \frac{E}{I} = \frac{20 \text{ V}}{12.6 \text{ mA}}$$

$$\cong 1.59 \text{ k}\Omega$$

From Equation (19-9),

$$C = \frac{1}{2\pi f X_C}$$

$$= \frac{1}{2\pi \times 1 \text{ kHz} \times 1.59 \text{ k}\Omega}$$

$$= 0.1 \text{ μF}$$

19-4.1 A 1000 pF capacitor has a 1.5 V terminal voltage when passing a 6.6 mA current. Determine the frequency of the supply.

19-4.2 The supply current in the circuit of Problem 19-4.1 changes to 6.93 mA when an oscilloscope is connected across the 1000 pF capacitor. Calculate the input capacitance of the oscilloscope.

19-5
SERIES *RL* CIRCUITS

CURRENT AND VOLTAGE WAVEFORMS. A series circuit consisting of inductance L and resistance R is shown in Figure 19-6(a), while the waveforms and phasor diagram for the circuit are shown in Figure 19-6(b) and (c), respectively. Referring to the circuit diagram, it is seen that (as for all series circuits) the current I is common to both R and L. The circuit waveforms are therefore drawn starting with the waveform of current [Figure 19-6(b)].

The voltage V_R across the resistance R is always in phase with the current through the resistance. Thus, the waveform of V_R in Figure 19-6(b) is drawn in phase with the current waveform. In the case of the inductor, the voltage V_L leads the current through the inductor by 90° (see Section 19-1). The waveform of V_L is therefore drawn in Figure 19-6(b) 90° ahead of the waveform of I. The applied voltage E is obviously the resultant of the two component voltages V_R and V_L, and its waveform is obtained simply by summing the instantaneous levels of V_R and V_L. It is seen that E leads I by an angle $\phi°$ which is less than 90°.

PHASOR DIAGRAM. The phasor diagram for the circuit is drawn by once again starting with current, since the current is the common quantity in a series circuit. A horizontal line is drawn to scale representing current I [see Figure 19-6(c)]. Since V_R is in phase with I, another horizontal line is drawn alongside I to represent V_R. V_L is 90° ahead of I; therefore, the phasor for V_L is drawn vertically at an angle of 90° with respect to I. The phasor addition of V_L and V_R gives a resultant that represents the applied voltage E. Once again, it is seen that the applied voltage leads the circuit current by an angle of $\phi°$, which is less than the 90° angle that would exist between E and I in a pure inductive circuit. Note that the phasor representing the counter-emf in the inductor is $-V_L$, as shown by the broken line in Figure 19-6(c).

CIRCUIT EQUATIONS. As explained in Chapter 18, E can be expressed in rectangular or polar form. Referring to Figure 19-6(c),

$$\boxed{E = V_R + jV_L} \qquad (19\text{-}12)$$

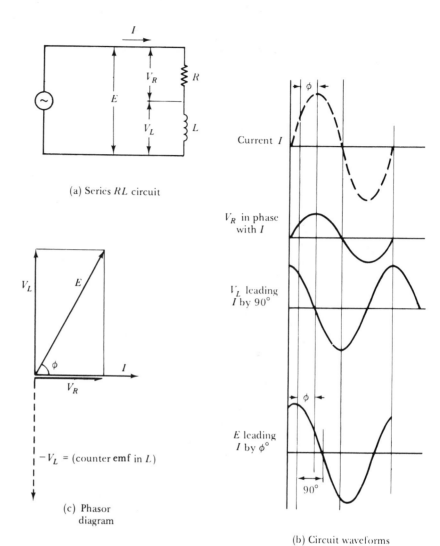

(a) Series *RL* circuit

(c) Phasor diagram

$-V_L = (\text{counter emf in } L)$

Current *I*

V_R in phase with *I*

V_L leading *I* by 90°

E leading *I* by φ°

90°

(b) Circuit waveforms

FIGURE 19-6. In a series *RL* circuit, the current lags the inductor voltage (V_L) by 90° and lags the supply voltage (E) by an angle less than 90°.

or from Equation (18-2),

$$E = \sqrt{V_R^2 + V_L^2} \Big/ \arctan\left(\frac{V_L}{V_R}\right) \qquad \text{(19-13)}$$

Dividing Equation (19-12) by *I* gives

$$\frac{E}{I} = \frac{V_R}{I} + j\frac{V_L}{I} \qquad \text{(19-14)}$$

V_R/I is, of course, the voltage across the resistance R divided by the current through the resistance. Therefore,

$$\frac{V_R}{I} = R$$

Also, V_L/I is the inductor voltage divided by the current through the inductor. Thus,

$$\frac{V_L}{I} = X_L$$

In Equation (19-14) E/I obviously represents opposition to current flow; however, it is neither resistance nor inductive reactance since it has both as component parts. In this case E/I is termed *impedance* and is given the symbol Z. Thus, Equation (19-14) can be restated:

$$\boxed{impedance,\ Z = R + jX_L} \tag{19-15}$$

When R and X_L are in ohms, the units of Z are also ohms.

The numerical value or modulus of Z (i.e., not including the angle) is written $|Z|$, and

$$\boxed{|Z| = \sqrt{R^2 + X_L^2}} \tag{19-16}$$

Also, the angle of Z is

$$\boxed{\phi = \arctan \frac{X_L}{R}} \tag{19-17}$$

Therefore,

$$\boxed{Z = \sqrt{R^2 + X_L^2}\ \big/\!\arctan \frac{X_L}{R}} \tag{19-18}$$

Equation (19-18) represents conversion from the rectangular form $Z = R + jX$ to polar form $Z\underline{/\phi}$. For conversion from polar to rectangular form, Equation (18-1) is rewritten using impedances, $Z\underline{/\phi} = Z \cos \phi + jZ \sin \phi$.

IMPEDANCE DIAGRAM. Z, R, and X_L can be represented on a vector diagram. They are not phasor quantities because they have fixed values; that is, unlike ac voltage and current, Z, R and X_L do *not* have continuously changing instantaneous values. To distinguish it from a phasor

Chap. 19   Inductance and Capacitance in AC Circuits

diagram, the vector diagram of impedance, reactance, and resistance is usually drawn in triangular form, and is referred to as an *impedance diagram*. Figure 19-7 shows the impedance diagram for the series *RL* circuit of Figure 19-6(a). A horizontal line is first drawn to represent the resistive component *R*. The +*j* component (i.e., X_L) is then drawn at +90° with respect to *R*, as illustrated. The *Z* component is the hypotenuse of the triangle, and the angle ϕ is the phase angle of the impedance *Z* with respect to resistance *R*.

ADMITTANCE.   As with resistance and reactance, the reciprocal of impedance is sometimes employed. The *admittance*, symbol *Y*, is the reciprocal of impedance *Z*, and its unit is the siemens, *S*.

$$admittance, \ Y = \frac{1}{Z}$$

(19-19)

When Z is in ohms, Y is in siemens.

PRACTICAL INDUCTORS.   All inductors are coils that have some winding resistance, and as already explained, they can be represented by a pure inductance in series with the winding resistance. Therefore, the diagrams in Figures 19-6 and 19-7 could apply to the case of an inductor connected directly to an ac supply. Where an external resistance is connected in series with the inductor, the coil resistance should also be shown as a series component in the equivalent circuit.

SERIES *RL* CIRCUIT ANALYSIS PROCEDURE

1. *Calculate the inductive reactance, $X_L = 2\pi fL$.*
2. *If there is more than one resistive component, calculate the total resistance, $R = R_1 + \cdots$.*
3. *Calculate the impedance, $|Z| = \sqrt{R^2 + X_L^2}$.*

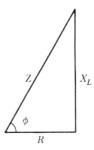

FIGURE 19-7.   Impedance diagram (or impedance triangle) for a series *RL* circuit. Inductive reactance vector X_L is always drawn up from the resistance vector at an angle of 90°. The impedance vector *Z* is the resultant of *R* and X_L.

4. Calculate the phase angle, φ = arctan (X_L/R).

5. Calculate the current, $I = E/Z$.

6. Determine the resistive voltage, $V_R = IR$ or $V_R = E$ cos φ.

7. Determine the inductive voltage, $V_L = IX_L$ or $V_L = E$ sin φ.

EXAMPLE 19-6 A series circuit consists of $R = 20$ Ω, $L = 20$ mH, and an ac supply of 60 V with $f = 100$ Hz. Calculate the current, the voltage across R, the voltage across L, and the phase angle of current with respect to the supply voltage.

SOLUTION

Equation (19-3), $X_L = 2\pi f L$

$$= 2\pi \times 100 \text{ Hz} \times 20 \text{ mH}$$

$$\cong 12.57 \text{ Ω}$$

Equation (19-16),

$$|Z| = \sqrt{R^2 + X_L^2}$$

$$= \sqrt{20^2 + 12.57^2}$$

$$\cong 23.6 \text{ Ω}$$

Equation (19-17),

$$\phi = \arctan \frac{X_L}{R}$$

$$= \arctan \frac{12.57 \text{ Ω}}{20 \text{ Ω}}$$

$$= \mathbf{32.1°}$$

$$|I| = \frac{E}{|Z|} = \frac{60 \text{ V}}{23.6 \text{ Ω}}$$

$$\cong \mathbf{2.54 \text{ A}}$$

$$V_R = E \cos \phi$$

$$= 60 \text{ V} \cos 32.1°$$

$$= \mathbf{50.8 \text{ V}} = I \times R$$

$$V_L = E \sin \phi$$

$$= 60 \text{ V} \sin 32.1°$$

$$= \mathbf{31.9 \text{ V}} = I \times X_L$$

Chap. 19   Inductance and Capacitance in AC Circuits

EXAMPLE 19-7   A 64 mH inductor with a winding resistance of $R_W = 700 \, \Omega$ is connected in series with a resistor $R_1 = 3.3$ kΩ. A 10 V ac supply with a frequency of 5 kHz is connected to the series circuit. Calculate the circuit current and the terminal voltage of the inductor.

SOLUTION

Equation (19-3),

$$X_L = 2\pi f L$$

$$= 2\pi \times 5 \text{ kHz} \times 64 \text{ mH}$$

$$\cong 2 \text{ k}\Omega$$

$$\text{total resistance} = R_1 + R_W$$

$$= 3.3 \text{ k}\Omega + 700 \, \Omega$$

$$= 4 \text{ k}\Omega$$

Equation (19-16),

$$Z = \sqrt{R^2 + X_L^2}$$

$$= \sqrt{(4 \text{ k}\Omega)^2 + (2 \text{ k}\Omega)^2}$$

$$= 4.47 \text{ k}\Omega$$

Equation (19-17),

$$\phi = \arctan \frac{X_L}{R}$$

$$= \arctan \frac{2 \text{ k}\Omega}{4 \text{ k}\Omega}$$

$$= 26.6°$$

$$I = \frac{E}{Z} = \frac{10 \text{ V}}{4.47 \text{ k}\Omega}$$

$$\cong \textbf{2.24 mA}$$

$$V_R = I \times R$$

$$= 2.24 \text{ mA} \times 4 \text{ k}\Omega$$

$$= 8.96 \text{ V}$$

The phasor diagram in Figure 19-8(a) *shows the total resistive component of the input voltage* (V_R) *in phase with the current. The inductive component* V_L *leads* V_R *by 90°.*

$$V_L = I \times X_L$$

$$= 2.24 \text{ mA} \times 2 \text{ k}\Omega$$

$$= 4.48 \text{ V}$$

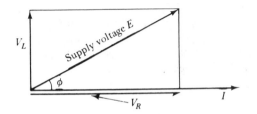

(a) Phasor diagram showing relationship between supply voltage, current, and the inductive and resistive voltage components

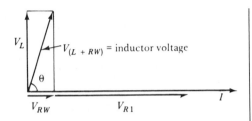

(b) Phasor components of inductor voltage

FIGURE 19-8. Phasor diagrams for Example 19-7.

The voltage drop across the resistive component of the inductor is

$$V_{RW} = I \times R_W$$
$$= 2.24 \text{ mA} \times 700 \text{ } \Omega$$
$$= 1.57 \text{ V}$$

In Figure 19-8(b) *it is seen that V_R consists of two components V_{RW} and V_{R1}, and that the voltage across the inductor is the resultant of V_L and V_{RW}. Therefore,*

$$\text{inductor voltage, } V_{(L+RW)} = \sqrt{V_{RW}^2 + V_L^2}$$
$$= \sqrt{1.57^2 + 4.48^2}$$
$$= \mathbf{4.75 \text{ V}}$$

The inductor voltage leads the current by an angle θ [see Figure 19-8(b)]*, where*

$$\theta = \arctan \frac{V_L}{V_{RW}}$$
$$= \arctan \frac{4.48 \text{ V}}{1.57 \text{ V}}$$
$$\cong \mathbf{70.7°}$$

PRACTICE PROBLEMS

19-5.1 A series circuit has $R_1 = 120$ Ω, $L = 30$ mH, and the coil resistance of the inductor is $R_W = 40$ Ω. If the circuit is connected to a 24 V, 400 Hz supply, calculate the circuit current and its phase angle with respect to the supply.

19-5.2 An inductor in series with a 2.7 kΩ resistor is connected to a 100 mV, 250 kHz supply. The resistor voltage is measured as 40.5 mV. Calculate the inductance.

Chap. 19   Inductance and Capacitance in AC Circuits

19-6

SERIES
RC CIRCUITS

CURRENT AND VOLTAGE WAVEFORMS. Figure 19-9(a) shows a capacitor C connected in series with a resistor R, with an ac supply voltage. The circuit waveforms are shown in Figure 19-9(b), and the phasor diagram for the circuit is drawn in Figure 19-9(c).

Starting with the waveform of current I, which is common to both R and C, the waveform of voltage V_R across R is drawn in phase with I. The current through the capacitor leads the capacitor terminal voltage by 90° (see Section 19-3); therefore, the V_C waveform is drawn 90° behind the current wave. The applied voltage E is the vector sum of V_R and V_C, and it is seen that I leads E by an angle ϕ, which is less than 90°.

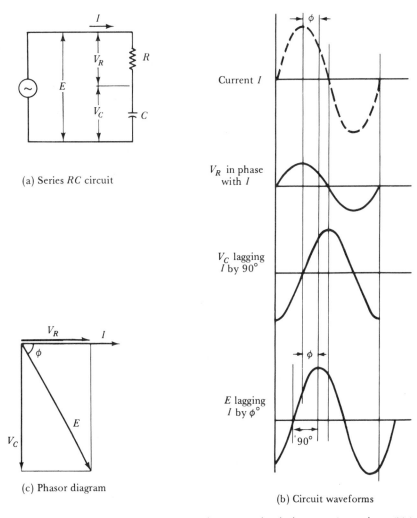

(a) Series RC circuit

(c) Phasor diagram

(b) Circuit waveforms

FIGURE 19-9. In a series RC circuit, the current leads the capacitor voltage (V_C) by 90° and leads the supply voltage (E) by an angle less than 90°.

PHASOR DIAGRAM. For the phasor diagram in Figure 19-9(c), a horizontal line is drawn to represent I, and the phasor for the voltage V_R is drawn in phase with I. The capacitor voltage phasor V_C is drawn 90° behind the current phasor. The supply voltage E is the resultant of V_R and V_C, and it lags I by the angle ϕ, as illustrated.

CIRCUIT EQUATIONS. From Chapter 18 and Figure 19-9(c), the rectangular form expression for E in the series resistance capacitance circuit is

$$E = V_R - jV_C \tag{19-20}$$

and from Equation (18-2),

$$E = \sqrt{V_R^2 + V_C^2} \; \underline{/\arctan \dfrac{V_C}{V_R}} \tag{19-21}$$

Dividing Equation (19-20) through by I:

$$\frac{E}{I} = \frac{V_R}{I} - j\frac{V_C}{I}$$

or

$$impedance, \; Z = R - jX_C \tag{19-22}$$

where R is resistance in ohms, X_C is capacitive reactance in ohms and Z is impedance in ohms. Also,

$$|Z| = \sqrt{R^2 + X_C^2} \tag{19-23}$$

and

$$\phi = \arctan \frac{X_C}{R} \tag{19-24}$$

giving

$$Z = \sqrt{R^2 + X_C^2} \; \underline{/\arctan \dfrac{X_C}{R}} \tag{19-25}$$

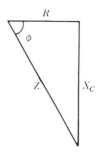

FIGURE 19-10. Impedance diagram (or impedance triangle) for a series *RC* circuit. Capacitive reactance vector X_C is drawn down from the resistance vector at an angle of $-90°$. The impedance vector Z is the resultant of R and X_C.

Equation (19-25) converts from rectangular to polar form. As stated in Section 19-5, conversion from polar to rectangular form is performed using Equation (18-1) rewritten for impedances, $Z\underline{/\phi} = Z \cos \phi + jZ \sin \phi$.

IMPEDANCE DIAGRAM. The impedance diagram for a series resistance-capacitance circuit is shown in Figure 19-10. As in the case of the *RL* circuit, a horizontal line is first measured off to represent R, then the $-j$ component X_C is drawn at $-90°$ with respect to R. Z is the hypotenuse of the impedance triangle, and the phase angle of Z with respect to R is ϕ.

SERIES *RC* CIRCUIT ANALYSIS PROCEDURE

1. *Calculate the capacitive reactance, $X_C = 1/(2\pi f C)$.*
2. *If there is more than one resistive component, calculate the total resistance, $R = R_1 + \cdots$.*
3. *Calculate the impedance, $|Z| = \sqrt{R^2 + X_C^2}$.*
4. *Calculate the phase angle, $\phi = \arctan(X_C/R)$.*
5. *Calculate the current, $I = E/Z$.*
6. *Determine the resistive voltage, $V_R = IR$ or $V_R = E \cos \phi$.*
7. *Determine the capacitive voltage, $V_C = IX_C$ or $V_C = E \sin \phi$.*

EXAMPLE 19-8   A series circuit consists of $R = 47 \, \Omega$, $C = 10 \, \mu F$, and an ac supply of 100 V with $f = 300$ Hz. Calculate the current, the voltage across R, the voltage across C, and the phase angle of the current with respect to the supply voltage.

SOLUTION

Equation (19-9), $X_C = \dfrac{1}{2\pi f C}$

$$= \frac{1}{2\pi \times 300 \text{ Hz} \times 10 \ \mu F}$$

$$\cong 53.1 \ \Omega$$

Equation (19-23),

$$|Z| = \sqrt{R^2 + X_C^2}$$
$$= \sqrt{47^2 + 53.1^2}$$
$$= 70.9 \ \Omega$$

Equation (19-24),

$$\phi = \arctan \frac{X_C}{R}$$
$$= \arctan \frac{53.1 \ \Omega}{47 \ \Omega}$$
$$\cong \mathbf{48.5°}$$
$$|I| = \frac{E}{|Z|} = \frac{100 \ \text{V}}{70.9 \ \Omega}$$
$$\boldsymbol{I} \cong \mathbf{1.41 \ A}$$
$$V_R = E \cos \phi$$
$$= 100 \ \text{V} \cos 48.5°$$
$$\boldsymbol{V_R} \cong \mathbf{66.3 \ V} = \boldsymbol{I \times R}$$
$$V_C = E \sin \phi$$
$$= 100 \ \text{V} \sin 48.5°$$
$$\boldsymbol{V_C} \cong \mathbf{74.9 \ V} = \boldsymbol{I \times X_C}$$

PRACTICE PROBLEMS

19-6.1 A series circuit with $R = 470 \ \Omega$ and $C = 0.1 \ \mu\text{F}$ is connected to a 50 V, 1.8 kHz supply. Calculate V_C, V_R, and the phase angle of the current with respect to the supply.

19-6.2 A 33 V signal is applied to a 10 kΩ resistor in series with a 4000 pF capacitor. Calculate the capacitor voltage when the signal frequency is 10 kHz, and when it is 55 kHz.

19-7

SERIES *RLC* CIRCUITS

When resistance, inductance, and capacitance are connected in series, the circuit behaves either as a series *RL* circuit or as a series *RC* circuit, depending on which of the two reactances (X_L or X_C) is the larger. *The special case where X_L and X_C are equal is considered in Chapter 23.*

PHASOR DIAGRAM. In the *RLC* series circuit shown in Figure 19-11(a), the current is once again common to all series components. The phasor diagram is drawn as before, by starting with the horizontal current phasor [see Figure 19-11(b)]. The resistor voltage V_R is in phase with the current; therefore, the phasor of V_R is drawn in phase with the *I* phasor, as shown. Since the current lags the voltage V_L across the inductor, the inductor voltage phasor is drawn 90° ahead of the current phasor. Similarly, the capacitor voltage phasor V_C is drawn 90° behind the current phasor, because the current leads the voltage across the capacitor by 90°.

Figure 19-11(b) shows a phasor diagram for a circuit where the inductive reactance is greater than the capacitive reactance. This means that V_L is greater than V_C, and the circuit behaves as an *RL* series circuit. The two reactive phasors (V_L and V_C) must be added vectorially to find

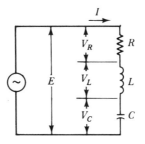

(a) Series *RLC* circuit

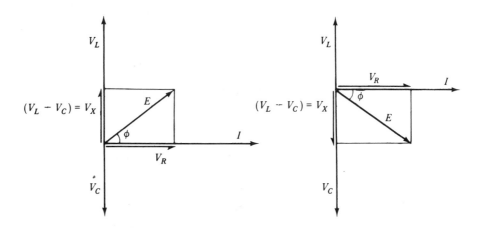

(b)   Phasor diagram for series
RLC circuit in which $X_L > X_C$

(c)   Phasor diagram for a series
RLC circuit in which $X_C > X_L$

FIGURE 19-11.   Series *RLC* circuit and phasor diagrams. Where $X_L > X_C$, the current lags the supply voltage. With $X_C > X_L$, the current leads the supply voltage.

the resultant reactive phasor V_X. Then the resultant of V_X and V_R is the supply voltage E, and ϕ is its phase angle with respect to the current. In Figure 19-11(c), the phasor diagram illustrates the case where X_C is greater than X_L, giving a capacitor voltage V_C which is larger than the inductor voltage V_L. In this case V_X is capacitive, E lags I by the phase angle ϕ, and the circuit behaves as a RC series circuit.

CIRCUIT EQUATIONS. The rectangular expression for the voltage in a series RLC circuit is

$$\boxed{E = V_R + j\,(V_L - V_C)} \qquad (19\text{-}26)$$

or

$$\boxed{E = V_R + j\,V_X} \qquad (19\text{-}27)$$

where
$$V_X = V_L - V_C$$

Converting from rectangular form, the expression for the voltage in polar form is

$$\boxed{E = \sqrt{V_R^2 + V_X^2}\,\Big/\,\arctan \frac{V_X}{V_R}} \qquad (19\text{-}28)$$

Recall from Equation (18-1) that to convert from polar form to rectangular form:

$$E\underline{/\theta} = E \cos\theta + jE \sin\theta$$

Dividing Equation (19-26) through by I,

$$\frac{E}{I} = \frac{V_R}{I} + j\left(\frac{V_L}{I} - \frac{V_C}{I}\right)$$

which gives

$$\boxed{impedance,\ Z = R + j\,(X_L - X_C)} \qquad (19\text{-}29)$$

Also,

$$\boxed{|Z| = \sqrt{R^2 + X_{eq}^2}} \qquad (19\text{-}30)$$

where

$$X_{eq} = X_L - X_C$$

and

$$\boxed{\phi = \arctan \frac{X_{eq}}{R}} \qquad \text{(19-31)}$$

giving

$$\boxed{Z = \sqrt{R^2 + X_{eq}^2} \, \bigg/ \, \arctan \frac{X_{eq}}{R}} \qquad \text{(19-32)}$$

Impedance conversion from polar to rectangular uses Equation (18-1) rewritten for impedances:

$$Z\underline{/\phi} = Z \cos \phi + jZ \sin \phi$$

IMPEDANCE DIAGRAMS. The impedance diagrams shown in Figure 19-12(a) and (b) relate to the phasor diagrams in Figure 19-11(b) and (c) respectively, and are self-explanatory.

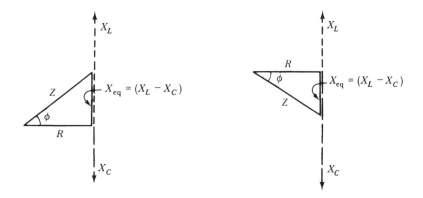

(a) Impedance diagram
corresponding to
Figure 19–11(b)

(b) Impedance diagram
corresponding to
Figure 19–11(c)

FIGURE 19-12. Impedance diagrams for a series *RLC* circuit. If the circuit is largely inductive, the impedance triangle is that of an *RL* circuit. When the circuit is largely capacitive, an *RC* impedance triangle results.

SERIES *RLC* CIRCUIT ANALYSIS PROCEDURE

1. *Calculate the inductive reactance, $X_L = 2\pi f L$.*
2. *Calculate the capacitive reactance, $X_C = 1/(2\pi f C)$.*
3. *Determine the equivalent reactance $X_{eq} = X_L - X_C$.*
4. *If there is more than one resistive component, calculate the total resistance $R = R_1 + \cdots$.*
5. *Calculate the impedance $|Z| = \sqrt{R^2 + X_{eq}^2}$.*
6. *Calculate the phase angle $\phi = \arctan(X_{eq}/R)$.*
7. *Calculate the current $I = E/Z$.*
8. *Determine the resistive voltage $V_R = IR$ or $V_R = E \cos\phi$.*
9. *Determine the reactive voltage $V_x = IX_{eq}$ or $V_x = E \sin\phi$.*
10. *Determine the inductive voltage $V_L = IX_L$.*
11. *Determine the capacitive voltage $V_C = IX_C$.*

EXAMPLE 19-9   A series *RLC* circuit has $R = 33\ \Omega$, $L = 50$ mH, and $C = 10\ \mu$F. The supply voltage is 75 V with a frequency of 200 Hz. Calculate I, V_R, V_L, V_C, and the phase angle of the current with respect to the supply voltage.

SOLUTION

Equation (19-3), $X_L = 2\pi f L = 2\pi \times 200\ \text{Hz} \times 50\ \text{mH}$

$$= 62.8\ \Omega$$

Equation (19-9), $X_C = \dfrac{1}{2\pi f C} = \dfrac{1}{2\pi \times 200\ \text{Hz} \times 10\ \mu\text{F}}$

$$= 79.6\ \Omega$$

Equation (19-29):

$$Z = R + j(X_L - X_C)$$

$$= 33\ \Omega + j(62.8 - 79.6)\ \Omega$$

$$= 33\ \Omega - j16.8\ \Omega$$

$$X_{eq} = -16.8\ \Omega \qquad (\textit{because } X_C > X_L,\ X_{eq} \textit{ is capacitive})$$

$$Z = \sqrt{R^2 + X_{eq}^2} \Big/ \arctan \frac{X_{eq}}{R}$$

$$= \sqrt{33^2 + 16.8^2} \Big/ \arctan \frac{-16.8\ \Omega}{33\ \Omega}$$

$$\cong 37\ \Omega\underline{/-27^\circ}$$

$$I = \frac{E}{Z} = \frac{75\ \text{V}}{37\ \Omega\underline{/-27^\circ}}$$

$$= 2.03\ \text{A}\underline{/27^\circ}$$

Since X_{eq} is a capacitive reactance, the circuit current I leads the supply voltage E.

$$V_R = I \times R = 2.03 \text{ A} \times 33 \text{ } \Omega$$

$$\cong \textbf{67 V}$$

$$V_L = I \times X_L = 2.03 \text{ A} \times 62.8 \text{ } \Omega$$

$$= \textbf{127 V}$$

$$V_C = I \times X_C = 2.03 \text{ A} \times 79.6 \text{ } \Omega$$

$$= \textbf{162 V}$$

PRACTICE PROBLEMS

19-7.1 For the circuit in Example 19-9, recalculate the quantities when the supply frequency is 300 Hz.

19-7.2 A series circuit consists of a 10 mH inductor, 0.1 μF capacitor, and 1 kΩ resistor. If the resistor voltage is measured as 1 V with a frequency of 10 kHz, calculate the supply voltage and its phase angle with respect to the current.

19-8
PARALLEL *RLC* CIRCUITS

PHASOR DIAGRAMS. When resistance, inductance, and capacitance are connected in parallel as in Figure 19-13(a), the supply voltage E is common to all components. Since voltage E is common to all components, the phasor diagram is commenced by first drawing the phasor for E horizontally [see Figure 19-13(b)]. The rest of the phasor diagram consists of current phasors, with i_R being drawn in phase with E. The inductive current i_L lags the applied voltage; therefore, the i_L phasor is drawn 90° behind E. Similarly, the capacitive current leads the applied voltage, so its phasor is drawn 90° ahead of E. The phasors of i_L and i_C are in antiphase, so, as illustrated in Figure 19-13(b), their resultant is simply

$$i_X = i_C - i_L$$

The supply current, *I,* is the resultant of the phasor sum of i_X and i_R, and the phase angle of I with respect to E is ϕ, as shown.

In the phasor diagram illustrated, i_L is greater than i_C, and the complete circuit behaves as a resistive–inductive parallel combination. Obviously, it is also possible for i_C to be greater than i_L, giving a circuit that performs as a resistor and capacitor in parallel. *The special case where i_L and i_C are equal is considered in Chapter 23.*

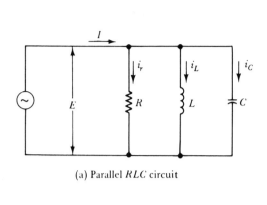

(a) Parallel RLC circuit

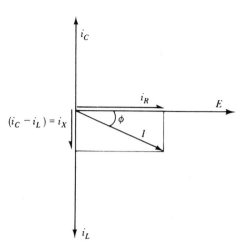

(b) Phasor diagram for a parallel RLC
circuit in which $i_L > i_C$

FIGURE 19-13. Parallel RLC circuit and phasor diagrams. If $i_L > i_C$, the supply current lags the supply voltage. With $i_C > i_L$, the supply current leads the supply voltage.

CIRCUIT EQUATIONS. Consideration of the circuit and phasor diagram in Figure 19-13 reveals that the equation for the total current is written

$$I = i_R + j(i_C - i_L) \qquad (19\text{-}33)$$

or

$$I = i_R + ji_X \qquad (19\text{-}34)$$

where

$$i_X = i_C - i_L$$

The expression for the current in polar form is

$$I = \sqrt{i_R^2 + i_X^2} \Big/ \arctan \frac{i_X}{i_R} \qquad (19\text{-}35)$$

Dividing Equation (19-33) through by E:

$$\frac{I}{E} = \frac{i_R}{E} + j\left(\frac{i_C}{E} - \frac{i_L}{E}\right)$$

Chap. 19   Inductance and Capacitance in AC Circuits

which gives

$$\boxed{\frac{1}{Z} = \frac{1}{R} + j\left(\frac{1}{X_C} - \frac{1}{X_L}\right)} \qquad \text{(19-36)}$$

or

$$\boxed{admittance, \ Y = G + j(B_C - B_L)} \qquad \text{(19-37)}$$

where Y is admittance, G is conductance, B_L is inductive susceptance, and B_C is capacitive susceptance, all of which are measured in siemens, S. Also,

$$Y = G + jB_{eq}$$

where

$$B_{eq} = B_C - B_L$$

and

$$\boxed{|Y| = \sqrt{G^2 + B_{eq}^2}} \qquad \text{(19-38)}$$

and

$$\boxed{\phi = \arctan \frac{B_{eq}}{G}} \qquad \text{(19-39)}$$

giving

$$\boxed{Y = \sqrt{G^2 + B_{eq}^2} \ \Big/ \arctan \frac{B_{eq}}{G}} \qquad \text{(19-40)}$$

Equation (19-40) converts from the rectangular $G + jB$ to polar $Y\underline{/\phi}$. Here again Equation (18-1) can be rewritten for conversion of admittance from polar to retangular form, $Y\underline{/\phi} = Y \cos \phi + jY \sin \phi$.

ADMITTANCE DIAGRAM. It is impossible to draw an impedance diagram for a parallel *RLC* circuit; instead, an *admittance diagram* is prepared, as shown in Figure 19-14. A horizontal line is first drawn to scale to represent the conductance G. The capacitive susceptance vector B_C is next drawn vertically at an angle of 90° with respect to the conductance vector. Then the inductive susceptance B_L vector is drawn at an angle of −90° with respect to G. The resultant susceptance vector B_{eq} is found by

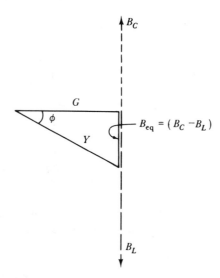

FIGURE 19-14. Admittance diagram for a parallel *RLC* circuit. Capacitive susceptance vector B_C is drawn up from the conductance vector G at an angle of 90°. Inductive susceptance vector B_L is drawn down from G at an angle of −90°. The admittance Y is the resultant of G and $(B_C − B_L)$.

vectorially summing B_C and B_L. Finally, the resultant of B_{eq} and G is found, giving the admittance Y.

PARALLEL *RLC* CIRCUIT ANALYSIS PROCEDURE

1. *Calculate the inductive reactance, $X_L = 2\pi f L$.*
2. *Calculate the capacitive reactance, $X_C = 1/(2\pi f C)$.*
3. *Determine the currents through each component:*

$$i_R = E/R, \qquad i_L = E/X_L, \qquad i_C = E/X_C.$$

4. *Calculate the reactive circuit, $i_X = i_C − i_L$.*
5. *Calculate the supply current, $I = \sqrt{i_R^2 + i_X^2}$.*
6. *Calculate the phase angle, $\phi = \arctan(i_X/i_R)$.*

Alternatively:

1. *Calculate the inductive reactance, $X_L = 2\pi f L$.*
2. *Calculate the capacitive reactance, $X_C = 1/(2\pi f C)$.*
3. *Determine the conductance and susceptances:*

$$G = 1/R, \qquad B_L = 1/X_L, \qquad B_C = 1/X_C.$$

4. *Determine the currents through each component:*

$$i_R = EG, \qquad i_L = EB_L, \qquad i_C = EB_C.$$

5. *Calculate the reactive current, $i_X = i_C − i_L$.*

Chap. 19   Inductance and Capacitance in AC Circuits

6. *Calculate the supply current, $I = \sqrt{i_R^2 + i_X^2}$.*

7. *Calculate the phase angle, $\phi = \arctan(i_X/i_R)$.*

EXAMPLE 19-10 A parallel *RLC* circuit as in Figure 19-15 has $R = 100\ \Omega$, $L = 20$ mH, and $C = 10\ \mu$F. The supply voltage is 35 V with a frequency of 500 Hz. Calculate i_R, i_L, i_C, and the supply current I. Also, determine the phase angle of I with respect to the supply voltage.

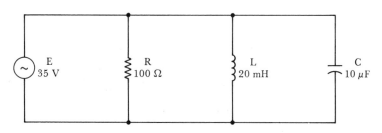

FIGURE 19-15. Parallel *RLC* circuit for Example 19-10.

SOLUTION

$$X_L = 2\pi f L = 2\pi \times 500\ \text{Hz} \times 20\ \text{mH}$$
$$= 62.8\ \Omega$$

$$X_C = \frac{1}{2\pi f C} = \frac{1}{2\pi \times 500\ \text{Hz} \times 10\ \mu\text{F}}$$
$$= 31.8\ \Omega$$

$$i_R = \frac{E}{R} = \frac{35\ \text{V}}{100\ \Omega}$$
$$= \textbf{350 mA}$$

$$i_L = \frac{E}{X_L} = \frac{35\ \text{V}}{62.8\ \Omega}$$
$$= \textbf{557 mA}$$

$$i_C = \frac{E}{X_C} = \frac{35\ \text{V}}{31.8\ \Omega}$$
$$= \textbf{1.1 A}$$

$$i_X = i_C - i_L$$
$$= 1.1\ \text{A} - 557\ \text{mA}$$
$$i_X = 543\ \text{mA}$$

Equation (19-34),

$$I = i_R + ji_X$$
$$= 350\ \text{mA} + j543\ \text{mA}$$

Equation (19-35),

$$I = \sqrt{i_R^2 + i_X^2} \Big/ \arctan \frac{i_X}{i_R}$$

$$= \sqrt{0.35^2 + 0.543^2} \Big/ \arctan \frac{0.543 \text{ A}}{0.35 \text{ A}}$$

$$= \textbf{646 mA} \underline{/\textbf{57.2°}}$$

EXAMPLE 19-11   Draw a phasor diagram and an admittance diagram for the circuit in Example 19-10, and determine the impedance of the total circuit.

SOLUTION

Phasor diagram: Figure 19-16(a) is drawn using the current levels, angles, etc., obtained in Example 19-10.

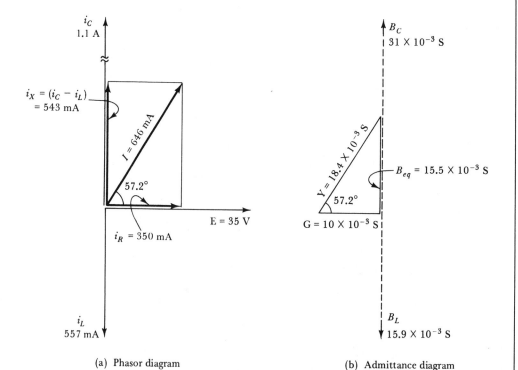

(a) Phasor diagram

(b) Admittance diagram

FIGURE 19-16. Phasor diagram and admittance diagram for Example 19-11.

Admittance diagram:

$$G = \frac{1}{R} = \frac{1}{100 \ \Omega}$$

$$= \textbf{10 mS}$$

$$B_L = \frac{1}{X_L} = \frac{1}{62.8 \ \Omega}$$

$$= \textbf{15.9 mS}$$

$$B_C = \frac{1}{X_C} = \frac{1}{31.8 \ \Omega}$$

$$= \textbf{31.4 mS}$$

$$B_{eq} = B_C - B_L$$

$$= (31.4 \times 10^{-3}) - (15.9 \times 10^{-3})$$

$$= \textbf{15.5 mS}$$

Equation (19-40),

$$Y = \sqrt{G^2 + B_{eq}^2} \ \underline{/\arctan \frac{B_{eq}}{G}}$$

$$= \sqrt{(10 \times 10^{-3})^2 + (15.5 \times 10^{-3})^2} \ \underline{/\arctan \frac{15.5 \times 10^{-3}}{10 \times 10^{-3}}}$$

$$= 18.4 \ \text{mS} \underline{/57.2°}$$

Equation (19-19),

$$Z = \frac{1}{Y} = \frac{1}{18.4 \times 10^{-3} \underline{/57.2°}}$$

$$= \textbf{54.3} \ \boldsymbol{\Omega} \underline{/\textbf{--57.2°}}$$

Figure 19-16(b) is now drawn using B, G, and Y values obtained above.

Example 19-11 shows that the parallel *RLC* circuit with the component values and frequency as stated has an impedance of

$$Z = 54.3 \ \Omega \underline{/-57.2°}$$

Using Equation (18-1), the impedance can be resolved from the polar form into rectangular form:

$$Z = 54.3 \cos(-57.2°) + j54.3 \sin(-57.2°)$$

$$= 29.4 \ \Omega - j45.6 \ \Omega$$

This value of Z would also be obtained with a resistance of $R = 29.4 \ \Omega$, connected *in series* with a capacitive reactance of $X_C = 45.6 \ \Omega$. Thus, it is seen this particular parallel RLC circuit has an equivalent series RC circuit. Depending on the component values, the equivalent series circuit of a parallel RLC circuit could be an RL circuit.

PRACTICE PROBLEMS

19-8.1 For the circuit in Example 19-10, recalculate the quantities when the supply frequency is 200 Hz.

19-8.2 Draw phasor and admittance diagrams for Problem 19-8.1 and determine the total circuit impedance.

19-9

LOW-PASS FILTERS

CR **LOW-PASS FILTERS.** Sometimes a direct voltage has an unwanted alternating voltage component, which must be reduced or removed. The circuit employed to do this is known as a *low-pass filter*. Similarly, an unwanted high frequency that is superimposed on a low-frequency alternating voltage can be attenuated by the use of a low-pass filter. The function of a low-pass filter is to pass low frequencies, and reduce or *attenuate* high frequencies.

The circuit of Figure 19-17(a) shows a simple *CR* low-pass filter. The input is a direct voltage with an ac waveform superimposed, and the output is a direct voltage and an attenuated alternating quantity. Apart from a voltage drop across R due to any direct-current flow, the presence of C and R has no significant effect upon the dc voltage. However, the capacitor offers a definite impedance $[X_C = 1/(\omega C)]$ to alternating input voltages, and this impedance decreases as the input frequency increases. Thus, as illustrated in Figure 19-17(b), the ac input is potentially divided, or attenuated, by the resistance and the capacitive impedance. The higher the ac frequency, the lower the capacitive impedance, and consequently the greater the attenuation.

The filter attenuation is easily calculated for a given input frequency. Input current:

$$|I| = \frac{V_i}{\sqrt{R^2 + X_C^2}}$$

and ac output voltage is

$$|V_0| = IX_C$$

$$V_0 = \frac{V_i X_C}{\sqrt{R^2 + X_C^2}}$$

Chap. 19   Inductance and Capacitance in AC Circuits

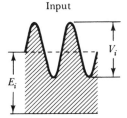

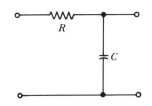

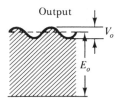

(a) *CR* filter low-pass filter circuit

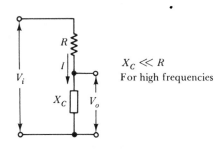

$X_C \ll R$
For high frequencies

(b) ac inputs are potentially
divided across R and X_C

FIGURE 19-17. A *CR* low-pass filter passes dc and low-frequency signals and attenuates high-frequency inputs. Capacitive impedance X_C is high at low frequencies and low at high frequencies. The potential divider effect of X_C and R produces attenuation.

The filter attenuation is

$$\boxed{\frac{V_0}{V_i} = \frac{X_C}{\sqrt{R^2 + X_C^2}}} \qquad \text{(19-41)}$$

EXAMPLE 19-12   The low-pass filter in Figure 19-17(a) has $R = 330\ \Omega$ and $C = 1\ \mu\text{F}$. The ac input voltage is $V_i = 20\ \text{V}$, and the input frequency is 1 kHz. Calculate the level of the alternating output voltage.

SOLUTION

$$X_C = \frac{1}{2\pi f C} = \frac{1}{2\pi \times 1\ \text{kHz} \times 1\ \mu\text{F}}$$

$$X_C \cong 159\ \Omega$$

From Equation (19-41),

$$V_0 = \frac{20 \text{ V} \times 159 \text{ } \Omega}{\sqrt{330^2 + 159^2}}$$

$$V_0 = 8.68 \text{ V}$$

FREQUENCY RESPONSE.   The performance of a filter may be illustrated by plotting its *frequency response graph*. The input (signal) voltage is maintained at a constant level while its frequency is varied. The output voltage levels are measured at several convenient signal frequencies. These output levels are then plotted versus frequency to a logarithmic base.

A typical frequency response graph for a low-pass filter is illustrated in Figure 19-18. It is seen that the output voltage remains equal to the input for all signal frequencies up to approximately 300 Hz. For greater signal frequencies, the output level is progressively reduced. The shape of the frequency response graph clearly shows that the filter passes low signal frequencies (up to 300 Hz in this case), and attenuates higher frequencies. Hence the name, *low-pass filter*.

Returning again to Figure 19-18, it is seen that at $f = 500$ Hz , the output is $V_0 = 0.707 \text{ } V_i$, or $(1/\sqrt{2})V_i$. This frequency is known as the *half-power point*. For a given load resistor (R_L) connected to the out-

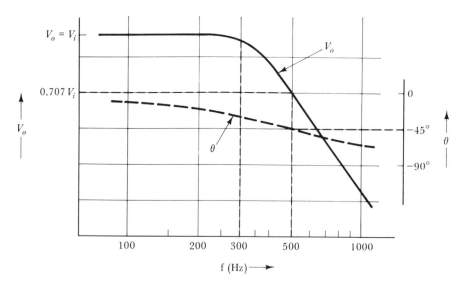

FIGURE 19-18.   Amplitude versus frequency and phase versus frequency graphs for a low-pass filter. Output amplitude V_0 and phase-shift θ are plotted for a constant level of V_i. At low frequencies, $V_0 \approx V_i$ and $\theta \approx 0$. At the frequency at which $X_C = R$, the output is 0.707 V_i and there is a phase lag of 45°.

put of the filter, the output power at low frequencies is $P_0 = V_i^2/R_L$. At the half-power frequency, the output power becomes $P_0 = (V_i/\sqrt{2})^2/R_L$, or $P_0 = V_i^2/(2R_L)$. This is half the normal output power. The half-power point is usally taken as the *cut-off frequency* of the filter.

PHASE-SHIFT. As well as voltage attenuation, a low-pass filter produces a phase-shift between input and output voltages. Like the output voltage amplitude, the phase-shift may be plotted versus frequency (see Figure 19-18). At low frequencies, the impedance of the RC circuit is largely capacitive, and the circuit current *leads* the input voltage by 90° (refer to Figure 19-9). However, because $V_0 = IX_C$, the output (capacitor) voltage *lags* the circuit current by 90°. This means that there is no phase shift between V_0 and V_i. At higher signal frequency, the circuit impedance becomes increasingly more resistive, and the angle by which I leads V_i gets progressively smaller. But V_0 (across C) still lags I by 90°. Thus, as the signal frequency increases, V_0 increasingly *lags* V_i. At $X_C = R$, V_0 lags V_i by $\theta = 45°$. The phase lag continues to increase with increasing signal frequency but levels off at a maximum of 90° at frequencies where $X_C \ll R$.

LC **LOW-PASS FILTER.** Figure 19-19(a) shows the circuit of a low-pass filter that uses a series-connected inductance L in place of the resistance R. The input is shown as a low-frequency sine wave with a low-amplitude high-frequency ac component superimposed. The inductive reactance of L is much higher for the high-frequency input than for the low-frequency input, and the capacitive reactance is much lower for the high frequency than for the low frequency. The combined effect is that both ac quantities are attenuated by X_L and X_C [see Figure 19-19(b)], with the high-frequency component suffering the greatest attenuation. For the kind of input waveform shown in Figure 19-19(a), the unwanted high-frequency component is sometimes referred to as *noise*, while the wanted voltage is termed the *signal*. The *signal-to-noise* (S/N) *ratio* is an important quantity. At the output of a filter, the signal-to-noise ratio should be much greater than at the input. The type of waveform shown in Figure 19-17(a) could also be filtered by the circuit in Figure 19-19(a).

Assuming that the resistive component of the inductance is much smaller than the inductive reactance, the input current to the LC filter is

$$|I| = \frac{V_i}{X_L - X_C}$$

and the output voltage is

$$|V_0| = IX_C$$

$$V_0 = \frac{V_i X_C}{X_L - X_C}$$

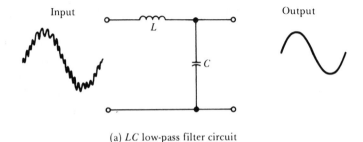

Input   Output

(a) *LC* low-pass filter circuit

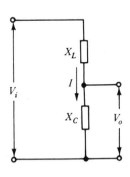

$X_C \ll X_L$
For high frequencies

(b)   High-frequency
ac inputs are potentially
divided across X_L and X_C

FIGURE 19-19. An *LC* low-pass filter passes dc and low-frequency signals and attenuates high-frequency inputs. $X_C > X_L$ at low frequencies, and $X_L > X_C$ at high frequencies. The potential divider effect produces attenuation.

The filter attenuation is

$$\boxed{\frac{V_0}{V_i} = \frac{X_C}{X_L - X_C}}$$ **(19-42)**

EXAMPLE 19-13   A low-pass filter has $L = 20$ mH and $C = 0.12$ μF. The input signal amplitude is 2 V peak-to-peak, and its frequency is 5 kHz. An unwanted noise input is also present with an amplitude of 0.2 V peak-to-peak and a frequency of 50 kHz. Calculate the signal-to-noise ratio at the input and at the output.

SOLUTION

The input signal-to-noise ratio is

$$\frac{S_i}{N_i} = \frac{2 \text{ V}}{0.2 \text{ V}} = 10$$

At 5 kHz,

$$X_L = 2\pi fL = 2\pi \times 5 \text{ kHz} \times 20 \text{ mH}$$
$$= 628 \text{ }\Omega$$

$$X_C = \frac{1}{2\pi fC} = \frac{1}{2\pi \times 5 \text{ kHz} \times 0.12 \text{ }\mu\text{F}} = 265 \text{ }\Omega$$

From Equation (19-42) *the signal output is*

$$S_o = \frac{S_i \times X_C}{X_L - X_C} = \frac{2 \text{ V} \times 265 \text{ }\Omega}{628 \text{ }\Omega - 265 \text{ }\Omega}$$
$$S_o = 1.46 \text{ V peak-to-peak}$$

At 50 kHz,

$$X_L = 2\pi fL = 2\pi \times 50 \text{ kHz} \times 20 \text{ mH}$$
$$= 6.28 \text{ k}\Omega$$

$$X_C = \frac{1}{2\pi fC} = \frac{1}{2\pi \times 50 \text{ kHz} \times 0.12 \text{ }\mu\text{F}}$$
$$= 26.5 \text{ }\Omega$$

Using Equation (19-42) *the noise output is*

$$N_o = \frac{N_i \times X_C}{X_L - X_C}$$
$$= \frac{0.2 \text{ V} \times 26.5 \text{ }\Omega}{6.28 \text{ k}\Omega - 26.5 \text{ }\Omega}$$
$$N_o \cong 0.85 \text{ mV peak to peak}$$

The output signal-to-noise ratio is

$$\frac{S_o}{N_o} = \frac{1.46 \text{ V}}{0.85 \text{ mV}} \cong \mathbf{1700}$$

In Example 19-13, the input signal amplitude is just 10 times the unwanted noise voltage amplitude. The output signal is seen to be 1700 times the output noise.

PRACTICE PROBLEMS

19-9.1 For the filter in Example 19-12, determine the output voltage when the signal frequency is 150 Hz. Also, calculate the frequency at which $V_o = 0.707 \text{ } V_i$.

19-9.2 An *LC* *l*ow-pass filter uses a 300 μH inductor. Calculate the required capacitor if a 200 kHz signal is to be attenuated by a factor of 100.

19-10
HIGH-PASS FILTERS

CR HIGH-PASS FILTER. The function of a *high-pass filter* is the converse of that of a low-pass filter. A high-pass filter passes high-frequency input voltages, and blocks low frequencies and dc quantities. Usually, direct voltages are completely blocked, while low-frequency ac components are attenuated. Figure 19-20(a) shows the circuit of a *CR* high-pass filter, with a combined dc and ac input. The presence of the series-connected capacitor prevents the direct voltage input from passing to the output. The ac input is potentially divided across X_C and R [see Figure 19-20(b)], so the ac output is an attenuated version of the input. Where a large enough capacitor is employed, the ac attenuation may be negligible.

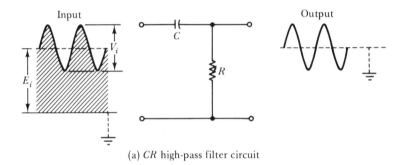

(a) *CR* high-pass filter circuit

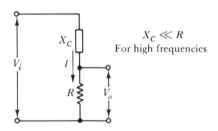

(b) ac inputs are only slightly affected by potential division across X_C and R

FIGURE 19-20. An *RC* high-pass filter passes high-frequency signals, and attenuates dc and low-frequency inputs. Capacitive impedance X_C is high at low frequencies and low at high frequencies. Attenuation is the result of the potential divider effect of X_C and R.

The actual attenuation of the input signal for the circuit in Figure 19-20(a) can be calculated in a similar way as for the low-pass filter:

$$|I| = \frac{V_i}{\sqrt{R^2 + X_C^2}}$$

$$V_o = IR$$

$$= \frac{V_i R}{\sqrt{R^2 + X_C^2}}$$

Therefore,

$$\frac{V_o}{V_i} = \frac{R}{\sqrt{R^2 + X_C^2}} \qquad \text{(19-43)}$$

FREQUENCY RESPONSE. The frequency response graph for a high-pass filter is obtained in the same way as for the low-pass filter. The input signal frequency is varied while its voltage level is maintained constant. The output voltage is measured at several convenient frequencies, and the measured levels are plotted versus frequency to a logarithmic base.

A typical frequency response graph for a high-pass filter is shown in Figure 19-21. In this particular case, all signals with frequencies higher than approximately 500 Hz are passed without attenuation. As the frequency is reduced below 500 Hz, the output voltage is progressively

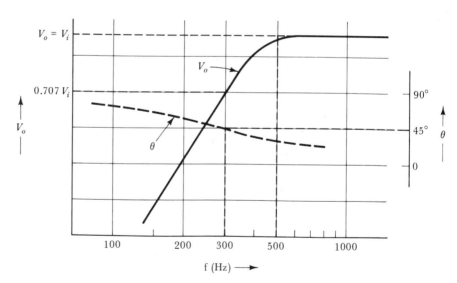

FIGURE 19-21. Amplitude versus frequency and phase versus frequency graphs for a high-pass filter. Output amplitude V_o and phase-shift θ are plotted for a constant level of V_i. At low frequencies, $V_o \ll V_i$, and there is a 90° phase-lead. At the frequency at which $X_C = R$, $V_o = 0.707\ V_i$, and $\theta = 45°$.

reduced. Clearly, the filter passes high frequencies and attenuates lower frequencies, and it is appropriately named.

As already discussed for the low-pass filter, the frequency at which $V_o = 0.707$ V_i (300 Hz in Figure 19-21), is known as the half-power point and is taken as the filter cut-off frequency. This is the frequency at which $X_C = R$.

EXAMPLE 19-14   A high-pass CR filter has a resistance of $R = 2$ kΩ. The lowest input frequency to be passed is 300 Hz. Calculate the value of a suitable capacitor. Also, determine the attenuation of the filter for a 60 Hz frequency.

SOLUTION

For $V_o = 0.707$ V_i, $X_C = R$

or,
$$C = \frac{1}{2\pi f R} = \frac{1}{2 \times \pi \times 300\,\text{Hz} \times 2\,\text{k}\Omega}$$

$$\approx 0.27\ \mu\text{F}$$

For $f = 60$ Hz,

$$X_C = \frac{1}{2\pi f C} = \frac{1}{2 \times \pi \times 60\,\text{Hz} \times 0.27\,\mu\text{F}}$$

$$= 9.8\ \text{k}\Omega$$

From Equation (19-43), the attenuation at 60 Hz is,

$$\frac{V_o}{V_i} = \frac{2\ \text{k}\Omega}{\sqrt{(2\ \text{k}\Omega)^2 + (9.8\ \text{k}\Omega)^2}}$$

$$\approx 0.2$$

PHASE-SHIFT.   As in a low-pass filter, there is a phase-shift between input and output voltages in a high-pass filter. Phase-shift θ is plotted versus frequency in Figure 19-21. At low frequencies, the RC circuit impedance is largely capacitive, and the circuit current leads the input voltage by 90° (refer to Figure 19-9). Because $V_o = IR$, the output leads the input by $\theta = 90°$. At higher signal frequency, X_C becomes progressively smaller, and as the circuit becomes more resistive, the angle θ is progressively reduced. At $X_C = R$, V_o leads V_i by 45°. The phase lead continues to decrease with increasing signal frequency, approximating zero at frequencies where $X_C \ll R$.

LC **HIGH-PASS FILTER.**   The *LC* high-pass filter circuit in Figure 19-22(a) substitutes an inductance L in place of the resistance in the CR circuit.

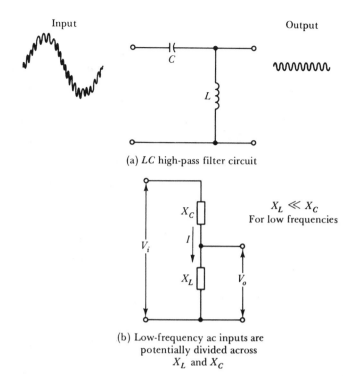

(a) *LC* high-pass filter circuit

$X_L \ll X_C$
For low frequencies

(b) Low-frequency ac inputs are
potentially divided across
X_L and X_C

FIGURE 19-22. An *LC* high-pass filter passes high-frequency signals and attenuates dc and low-frequency inputs. $X_C > X_L$ at low frequencies, and $X_L > X_C$ at high frequencies. The potential divider effect produces attenuation.

At low frequencies the inductance offers a low impedance and the capacitance offers a high impedance. Consequently, X_L is much smaller than X_C [see Figure 19-22(b)]. Therefore, low-frequency inputs are attenuated. With high frequency, however, the inductance has a high impedance and the capacitor impedance is low. Consequently, high-frequency input voltages experience very little attenuation. The input waveform shown in Figure 19-22(a) is a high-frequency wanted signal combined with an unwanted low-frequency component. If the filter does its job, the low frequency is severely attenuated, while the high-frequency signal is passed to the output.

As before, the attenuation of the filter is easily calculated:

$$\text{input current, } |I| = \frac{V_i}{X_C - X_L}$$

$$V_o = IX_L$$

$$= \frac{V_i X_L}{X_C - X_L}$$

$$\boxed{\textbf{\textit{attenuation}}, \; \frac{V_o}{V_i} = \frac{X_L}{X_C - X_L}} \qquad \textbf{(19-44)}$$

EXAMPLE 19-15 The high-pass filter circuit shown in Figure 19-22(a) has $C = 1\ \mu F$ and $L = 0.47$ mH. A 5 kHz input signal is applied, with a 60 Hz unwanted noise frequency combined with the signal. If the signal amplitude is 2 V peak-to-peak and the 60 Hz frequency is 10 V peak-to-peak at the input, calculate the output amplitudes of each.

SOLUTION

At 5 kHz,

$$X_L = 2\pi fL = 2\pi \times 5\ \text{kHz} \times 0.47\ \text{mH}$$

$$= 14.8\ \Omega$$

$$X_C = \frac{1}{2\pi fC} = \frac{1}{2\pi \times 5\ \text{kHz} \times 1\mu F}$$

$$= 31.8\ \Omega$$

From Equation (19-44):
5 kHz *signal output*:

$$S_o = \frac{S_i X_L}{X_C - X_L}$$

$$= \frac{2\ \text{V} \times 14.8\ \Omega}{31.8\ \Omega - 14.8\ \Omega}$$

$$= \mathbf{1.74\ V}$$

At 60 Hz,

$$X_L = 2\pi fL = 2\pi \times 60\ \text{Hz} \times 0.47\ \text{mH}$$

$$= 0.177\ \Omega$$

$$X_C = \frac{1}{2\pi fC} = \frac{1}{2\pi \times 60\ \text{Hz} \times 1\ \mu F}$$

$$= 2.65\ \text{k}\Omega$$

From Equation (19-44):
60 Hz *noise output:*

$$N_o = \frac{N_i X_L}{X_C - X_L}$$

$$= \frac{10\ \text{V} \times 0.177\ \Omega}{2.65\ \text{k}\Omega - 0.177\ \Omega}$$

$$= \mathbf{668\ \mu V}$$

PRACTICE PROBLEMS

19-10.1 The filter in Example 19-14 has a 250 mV, 2 kHz input signal. A 100 mV, 30 Hz noise signal is also present at the input. Calculate the signal-to-noise ratio at the input and at the output.

19-10.2 A high-pass filter circuit with $C = 1000$ pF and $L = 400$ μH, has a 50 mV, 150 kHz signal, and a 1 V, 120 Hz noise input. Determine the signal-to-noise ratio at the output.

SUMMARY OF FORMULAS

Inductive reactance:

$$X_L = \omega L = 2\pi f L$$

Inductive susceptance:

$$B_L = \frac{1}{X_L}$$

Capacitive reactance:

$$X_C = \frac{1}{\omega C} = \frac{1}{2\pi f C}$$

Capacitive susceptance:

$$B_C = \frac{1}{X_C}$$

For LR series circuit:

$$E = \sqrt{V_R^2 + V_L^2} \bigg/ \arctan \frac{V_L}{V_R}$$

$$Z = \sqrt{R^2 + X_L^2} \bigg/ \arctan \frac{X_L}{R}$$

For CR series circuit:

$$E = \sqrt{V_R^2 + V_C^2} \bigg/ \arctan \frac{V_C}{V_R}$$

$$Z = \sqrt{R^2 + X_C^2} \bigg/ \arctan \frac{X_C}{R}$$

For RLC series circuit:

$$E = \sqrt{V_R^2 + V_X^2}\,\underline{/\arctan \dfrac{V_X}{V_R}}$$

$$Z = \sqrt{R^2 + X_{eq}^2}\,\underline{/\arctan \dfrac{X_{eq}}{R}}$$

For RLC parallel circuit:

$$I = \sqrt{i_R^2 + i_X^2}\,\underline{/\arctan \dfrac{i_X}{i_R}}$$

Admittance:

$$Y = \dfrac{1}{Z}$$

$$Y = G + j(B_C - B_L)$$

$$Y = \sqrt{G^2 + B_{eq}^2}\,\underline{/\arctan \dfrac{B_{eq}}{G}}$$

Attenuation of low-pass CR filter:

$$\dfrac{V_o}{V_i} = \dfrac{X_C}{\sqrt{R^2 + X_C^2}}$$

Attenuation of low-pass LC filter:

$$\dfrac{V_o}{V_i} = \dfrac{X_C}{X_L - X_C}$$

Attenuation of high-pass CR filters:

$$\dfrac{V_o}{V_i} = \dfrac{R}{\sqrt{R^2 + X_C^2}}$$

Attenuation of high-pass LC filters:

$$\dfrac{V_o}{V_i} = \dfrac{X_L}{X_C - X_L}$$

REVIEW QUESTIONS

19-1 Define inductive reactance, inductive susceptance, capacitive reactance, capacitive susceptance, impedance, admittance, impedance diagram, and admittance diagram.

19-2 Define low-pass filter, high-pass filter, noise, signal-to-noise ratio, and coupling capacitor.

19-3   For an inductance with an alternating voltage supply, sketch the waveforms of current, rate of change of current, induced voltage, and supply voltage. Show the phase relationships between each of the waveforms, and carefully explain the origin of the relationships.

19-4   Sketch the phasor diagram and the waveforms of terminal voltage and current for an inductance with an ac supply. Also, write equations for instantaneous current. Explain briefly.

19-5   Using the waveforms sketched for Review Question 19-3, develop equations for $\Delta i/\Delta t$, e_L, E_{av}, and X_L. Define X_L and B_L.

19-6   For a capacitor with an alternating voltage supply, sketch the waveforms of supply voltage, rate of change of capacitor voltage, and instantaneous capacitor current. Show the phase relationships between each waveform and carefully explain the origin of the relationships.

19-7   Sketch the phasor diagram and the waveforms of terminal voltage and current for a capacitor with an ac supply. Also, write equations for instantaneous supply voltage and instantaneous current. Explain briefly.

19-8   Using the waveforms sketched for Review Question 19-6, develop an equation for X_C. Define X_C and B_C.

19-9   For a series RL circuit with an ac supply, sketch typical waveforms of current, resistor voltage, inductor voltage, and supply voltage. Explain the phase relationships between the waveforms.

19-10   Sketch a typical phasor diagram for an RL series circuit, and briefly explain. Also, write the equations for supply voltage and for the circuit impedance in rectangular and polar form.

19-11   For a series RC circuit with an ac supply, sketch typical waveforms of current, resistor voltage, capacitor voltage, and supply voltage.

19-12   Sketch a typical phasor diagram for an RC circuit and explain briefly. Also, write the equations for supply voltage and for circuit impedance in rectangular and polar form.

19-13   For a series RLC circuit, sketch typical phasor diagrams for the case of:
a. $X_C > X_L$.
b. $X_L > X_C$.
Explain briefly.

19-14   For a series RLC circuit sketch typical impedance diagrams for:
a. $X_C > X_L$.
b. $X_L > X_C$.
Explain briefly. Also, write equations for supply voltage and for circuit impedance in polar and rectangular form.

19-15 For a parallel *RLC* circuit, derive rectangular form expressions for the circuit admittance. Also, write an equation for the circuit impedance.

19-16 Sketch typical phasor and admittance diagrams for a parallel *RLC* circuit. Explain briefly.

19-17 Sketch the circuits of low-pass filters:
a. Using a capacitor and resistor.
b. Using a capacitor and inductor.
Explain briefly how each circuit operates.

19-18 Sketch the circuits of high-pass filters.
a. Using a capacitor and resistor.
b. Using a capacitor and inductor.
Explain briefly how each circuit operates.

19-19 Derive equations for the attenuation of a *CR* high-pass filter and for an *LC* high-pass filter. Also, derive the equations for *CR* and *LC* low-pass filters.

19-20 Sketch typical frequency response graphs for low-pass and high-pass filters. Briefly explain.

PROBLEMS

SECTION 19-2

19-1 Calculate the current that flows in a 300 mH inductor connected to a 100 V, 250 Hz supply. Sketch the voltage and current waveforms and draw the phasor diagram for the circuit.

19-2 A 50 mH inductor passes a current of 398 mA when connected to a 400 Hz supply. Calculate the supply voltage.

19-3 An inductor with an inductance of 100 μH passes a current of 10 mA when its terminal voltage is 6.3 V. Calculate the frequency of the ac supply.

19-4 A 29 mA current flows in an inductor which is connected to a 15 V, 1 kHz ac supply. Determine the inductance value.

SECTION 19-4

19-5 A 100 V, 250 Hz supply is applied to a 30 μF capacitor. Calculate the capacitive reactance and the current that flows. Sketch the voltage and current waveforms and draw the phasor diagram for the circuit.

19-6 A 4.5 μF capacitor passes a current of 34 mA when connected to a 100 Hz supply. Calculate the supply voltage.

19-7 A 100 μF capacitor passes a current of 0.25 mA when its terminal voltage is 4 V. Determine the supply frequency.

19-8 A current of 270 mA flows in a capacitor which is connected to a 25 V, 300 Hz ac supply. Determine the capacitance value.

SECTION 19-5

19-9   A series RL circuit with $R = 68\ \Omega$ and $L = 10$ mH is connected to a 33 V supply with a frequency of 1 kHz. Determine the circuit current, the resistor voltage, the inductor voltage, and the phase angle of the current with respect to the supply voltage. Sketch a phasor diagram and an impedance diagram for the circuit.

19-10   In the circuit described in Problem 19-9, the phase angle of the current with respect to the supply voltage is to be altered to 35°. Calculate a suitable new resistance value to effect this change and determine the new current level.

19-11   A 25 V, 10 kHz supply is applied to a 10 mH inductor in series with a 1.2 kΩ resistor. The inductor has a winding resistance of 220 Ω. Calculate the circuit current and the inductor terminal voltage. Also, draw a phasor diagram and an impedance diagram for the circuit.

19-12   In the circuit described in Problem 19-11, the circuit current is to be changed to 10 mA by adjusting the supply frequency. Calculate the appropriate frequency and determine the new level of inductor terminal voltage.

SECTION 19-6

19-13   A series RC circuit with $R = 120\ \Omega$ and $C = 3.3\ \mu$F is connected to a 12 V supply with a frequency of 1 kHz. Determine the circuit current, the resistor voltage, the capacitor voltage, and the phase angle of the current with respect to the supply voltage. Sketch a phasor diagram and an impedance diagram for the circuit.

19-14   In the circuit described in Problem 19-13, the circuit current is to be changed to 50 mA by adjusting the capacitor. Determine the required new capacitance and the new capacitor terminal voltage.

19-15   A 35 V, 500 Hz supply applied to a capacitor in series with a 220 Ω resistor produces a circuit current of 66 mA. Calculate the capacitance of the capacitor.

19-16   In the circuit described in Problem 19-15, the phase angle of the current is to be altered to $-35°$ by adjusting the supply frequency. Determine the appropriate frequency and the new current level in the circuit.

SECTION 19-7

19-17   A 20 mH inductor is connected in series with a 2 μF capacitor and a 200 Ω resistor. The supply voltage is 15 V with a frequency of 600 Hz. Calculate the circuit current and its phase angle with respect to the supply voltage. Also, calculate the terminal voltage of all components.

19-18 For the circuit described in Problem 19-17, accurately sketch phasor and impedance diagrams.

19-19 The resistor in Problem 19-17 is to be adjusted to give an inductor terminal voltage of 7.5 V. Calculate the appropriate resistor value.

19-20 Recalculate all of the quantities in Problem 19-17 for a supply frequency of 900 Hz.

19-21 A coil with an inductance of 390 mH is connected in series with a 620 Ω resistor and a 6 μF capacitor. The supply is 45 V. Determine the frequency at which $X_C = 300\ \Omega$, and at this frequency find the circuit current and $(V_L + V_R)$. Also, draw the phasor and impedance diagrams.

19-22 Recalculate the quantities in Problem 19-21 for the case of $X_L = 300\ \Omega$.

SECTION 19-8

19-23 A 220 mH inductor is connected in parallel with an 8 μF capacitor and a 330 Ω resistor. The supply is 10 V with a frequency of 100 Hz. Calculate i_R, i_L, i_C, and the supply current. Also, determine the phase angle of the current with respect to the supply voltage.

19-24 The supply voltage to the circuit in Problem 19-23 is to be altered to give $i_C = 61$ mA. Calculate the required new level of supply voltage, and recalculate the total supply current and its phase angle with respect to the voltage.

19-25 For the circuit described in Problem 19-23, accurately sketch phasor and impedance diagrams.

19-26 The three components in Problem 19-17 are reconnected in parallel with the supply. Calculate the total supply current and its phase angle with respect to the voltage.

19-27 The three components in Problem 19-21 are connected in parallel and supplied from a 120 V, 60 Hz source. Determine the supply current and $(i_R + i_L)$. Also, sketch the phasor and admittance diagrams.

19-28 A 200 mH inductor is connected in parallel with a 1.5 kΩ resistor, a capacitor, and a 33 V, 400 Hz supply. If the supply current is 40 mA, determine the capacitance of the capacitor.

SECTION 19-9

19-29   A low-pass CR filter has $C = 100$ μF and $R = 100$ Ω. Determine the attenuation of the circuit, (a) for a 10 Hz input frequency, (b) for a 250 Hz frequency.

19-30   For the CR filter in Problem 19-29, calculate the frequency at which the attenuation is 0.707. Also, determine the phase shift from input to output at that frequency.

19-31   A low-pass LC filter has $L = 50$ mH and $C = 0.2$ μF. The input signal is 1 V peak-to-peak with a frequency of 3 kHz. An unwanted noise with an amplitude of 0.3 V and a frequency of 25 kHz is also present at input. Determine the output signal and noise amplitudes and the signal-to-noise ratio.

SECTION 19-10

19-32   A high-pass filter has a 12 kHz input signal with an unwanted 100 Hz noise voltage. The signal amplitude is 5 V, and the noise has a 100 mV amplitude. If $L = 0.1$ mH and $C = 0.1$ μF, calculate the signal-to-noise ratio at the filter output.

19-33   A CR high-pass filter is to pass all signals above a frequency of 200 Hz. A maximum attenuation of 0.707 is acceptable for passed signals. If R is 5 kΩ, calculate a suitable capacitance value. Also, determine the signal phase shift at 200 Hz.

19-34   A high-pass filter has a 1 V, 25 kHz input signal with an unwanted 0.1 V, 120 Hz noise voltage. A 0.03 μF capacitor and a 50 μH inductor are employed. Calculate the signal-to-noise ratio at the filter output.

COMPUTER PROBLEMS

19-35   Write a computer program to analyze a series RLC circuit to determine circuit current, component voltages, and the phase angle of the current with respect to supply voltage.

19-36   Write a computer program to analyze a series LC low-pass filter with a wanted signal and an unwanted noise. The following quantities are to be determined: signal phase shift, noise attenuation, S/N ratio at input, and S/N ratio at output.

ANSWERS TO PRACTICE PROBLEMS

19-2.1	1.27 mH
19-2.2	111 mA
19-2.3	2.3 V
19-4.1	700 kHz

20

SERIES AND PARALLEL AC CIRCUITS

Introduction

AC series impedance circuits can be analyzed in a similar way to dc series resistance circuits, as long as the phase angle of each impedance is taken into consideration. For addition impedances must be in rectangular form, and for division or multiplication it is convenient to convert to polar form. The ac voltage divider can be treated in a similar way to the dc voltage divider as long as the phase angles are correctly considered. Parallel impedances should be converted into admittances and added together; then the result may be inverted to determine the equivalent total impedance. In dealing with series-parallel impedance combinations, each parallel group of impedances should be resolved into a single equivalent impedance, so that the entire configuration reduces to a simple series impedance circuit. For every simple series impedance circuit there is an equivalent parallel impedance circuit, and vice versa. The equations for conversion from one to the other are sometimes very useful.

SERIES-CONNECTED IMPEDANCES

Impedances in series in an ac circuit can be treated similarly to resistances in series in a dc circuit. The equivalent total impedance is determined, and using the applied voltage, the current through the circuit can be calculated. However, series impedances cannot be added directly, but must be resolved into rectangular form; then the resistive and reactive components are added separately. Figure 20-1 illustrates the process. The three series impedances in Figure 20-1(a) are each converted to their rectangular form, Figure 20-1(b). The total values of R and X are calculated separately and then converted back to polar form, to give the equivalent circuit of Figure 20-1(c).

ANALYSIS PROCEDURE FOR A SERIES IMPEDANCE CIRCUIT

1. *Convert each impedance into rectangular form.*
2. *Sum the individual resistive components to determine the total equivalent resistance.*
3. *Sum the individual reactive components to determine the total equivalent reactance.*
4. *Convert the equivalent resistance and reactance values into the polar form equivalent impedance.*

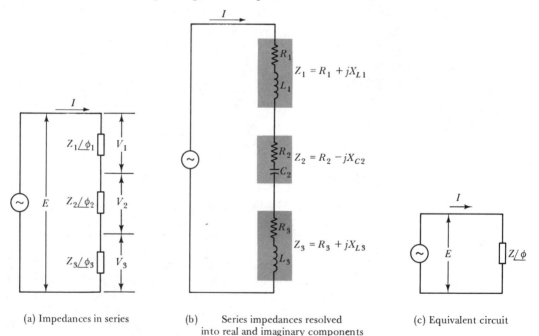

(a) Impedances in series   (b)   Series impedances resolved into real and imaginary components   (c) Equivalent circuit

FIGURE 20-1. To analyze a series impedance circuit, the individual impedances are first converted into rectangular form. All the resistive components are summed, and all the reactive components are summed, to give the equivalent circuit impedance.

5. *Divide the equivalent impedance into the applied voltage to determine the supply current.*

EXAMPLE 20-1   The three impedances shown in Figure 20-1 are $Z_1 = 70.7 \ \Omega\underline{/45°}$, $Z_2 = 92.4 \ \Omega\underline{/330°}$, $Z_3 = 67 \ \Omega\underline{/60°}$, and the supply voltage is $100 \ V$. Analyze the circuit to determine the current taken from the supply. Also, draw a phasor diagram of supply voltage and current.

SOLUTION

Converting to rectangular form:
From Equation (18-1):

$$Z_1 = 70.7 \cos 45° + j70.7 \sin 45°$$
$$= (50 + j50) \ \Omega$$
$$Z_2 = 92.4 \cos 330° + j92.4 \sin 330°$$
$$= (80 - j46.2) \ \Omega$$
$$Z_3 = 67 \cos 60° + j67 \sin 60°$$
$$= (33.5 + j58) \ \Omega$$

Adding impedances in series:

$$Z = Z_1 + Z_2 + Z_3$$
$$= (50 + 80 + 33.5) + j(50 - 46.2 + 58)$$
$$= (163.5 + j61.8) \ \Omega$$

Converting to polar form:
From Equation (19-18):

$$Z = \sqrt{163.5^2 + 61.8^2} \ \underline{/\arctan \frac{61.8}{163.5}}$$
$$= 174.8 \ \Omega\underline{/20.7°}$$

Determining the current:

$$I = \frac{E}{Z}$$
$$= \frac{100 \ V}{174.8 \ \Omega\underline{/20.7°}}$$
$$= \mathbf{572 \ mA}\underline{\mathbf{/-20.7°}}$$

The current is 572 mA, *lagging the supply voltage by 20.7° (see* Figure 20-2).

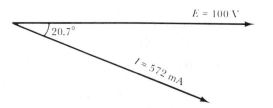

FIGURE 20-2. Phasor diagram of supply voltage and circuit current for the circuit in Figure 20-1.

PRACTICE PROBLEMS

20-1.1 A 0.1 μF capacitor is connected in series with a 10 mH inductor which has a coil resistance of 400 Ω. Calculate the circuit current when the supply is a 1 V, 3.18 kHz source.

20-1.2 A fourth impedance, $Z_4 = 26 \; \Omega \underline{/30°}$, is connected in series with the three impedances in Example 20-1. Calculate the new current level.

20-2
AC VOLTAGE DIVIDER

The voltage divider rule as applied to series resistors in a dc circuit can also be applied to series impedances in an ac circuit. As always in an ac circuit, care must be taken to ensure that phase angles are taken into consideration. Referring again to Figure 20-1(a), it is seen that the voltage across Z_3 is

$$V_3 = IZ_3$$

Also, as already shown, the current through Z_1, Z_2, and Z_3 is

$$I = \frac{E}{Z_1 + Z_2 + Z_3}$$

Therefore,

$$\boxed{V_3 = \frac{EZ_3}{Z_1 + Z_2 + Z_3}} \qquad \text{(20-1)}$$

Similarly,

$$V_2 = \frac{EZ_2}{Z_1 + Z_2 + Z_3}$$

and

$$V_1 = \frac{EZ_1}{Z_1 + Z_2 + Z_3}$$

EXAMPLE 20-2 For the series impedance circuit described in Example 20-1 in which $Z_1 = 70.7\ \Omega\underline{/45°}$, $Z_2 = 92.4\ \Omega\underline{/330°}$, $Z_3 = 67\ \Omega\underline{/60°}$, and the supply is 100 V, use the voltage divider equation to determine the values of V_1, V_2, and V_3. Also, draw a complete phasor diagram for the circuit voltages.

SOLUTION

From Example 20-1:

$$Z_1 + Z_2 + Z_3 = 174.8\ \Omega\underline{/20.7°}$$

From Equation (20-1):

$$V_3 = \frac{100\ \text{V}(67\ \Omega\underline{/60°})}{174.8\ \Omega\underline{/20.7°}}$$

$$= \frac{100\ \text{V} \times 67\ \Omega}{174.8\ \Omega}\ \underline{/(60° - 20.7°)}$$

$$\mathbf{= 38.3\ V\underline{/39.3°}}$$

$$V_2 = \frac{100\ \text{V}(92.4\ \Omega\underline{/330°})}{174.8\ \Omega\underline{/20.7°}}$$

$$\mathbf{= 52.9\underline{/309.3°}}$$

$$V_1 = \frac{100\ \text{V}(70.7\ \Omega\underline{/45°})}{174.8\ \Omega\underline{/20.7°}}$$

$$\mathbf{= 40.4\ V\underline{/24.3°}}$$

The voltage phasor diagram is as shown in Figure 20-3. *Note that the phasor sum of V_1, V_2, and V_3 gives the supply voltage E.*

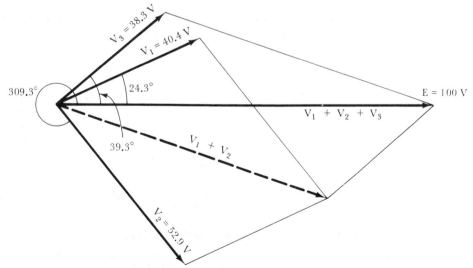

FIGURE 20-3. Phasor diagram of supply voltage and individual impedance voltages for the circuit in Figure 20-1.

20-2.1 Use the voltage divider equation to calculate the voltage across each of the two components in Problem 20-1.1

20-2.2 Calculate the voltages across each of the components in the circuit described in Problem 20-1.2

20-3

IMPEDANCES IN PARALLEL

As discussed in Section 19-8, impedances in parallel must be converted into admittances in order to determine the total circuit admittance. This can then be inverted to give the total impedance of the circuit. Once the total impedance is known, the current taken from the supply can be determined simply by dividing the impedance into the supply voltage. Alternatively, the supply current can be found by first calculating the individual branch currents in rectangular form, and then adding the real and imaginary components of the branch currents.

The analysis procedure for a parallel impedance circuit is illustrated in Figure 20-4. Each impedance is first converted into the equivalent admittance [Figures 20-4(a) and (b)]. The admittances are then resolved into conductance and susceptance [see Figure 20-4(c)]. The total parallel admittance is next calculated, converted into polar form, and then inverted to determine the equivalent circuit impedance [Figures 20-4(d), (e), and (f)].

ANALYSIS PROCEDURE FOR PARALLEL IMPEDANCE CIRCUITS

1. *Operating in polar form, each individual impedance is converted into an admittance simply by using Equation (19-19):*

$$Y = \frac{1}{Z}$$

2. *The individual admittances are resolved into rectangular form using Equation (19-37):*

$$Y = G + jB_{eq}$$

where $\quad B_{eq} = B_C - B_L$

or $\quad \boxed{Y = Y \cos \theta + jY \sin \theta} \qquad (20\text{-}2)$

3. *The total parallel admittance is found as*

$$\boxed{Y = Y_1 + Y_2 + Y_3} \qquad (20\text{-}3)$$

where the admittances are in rectangular form.

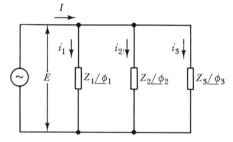
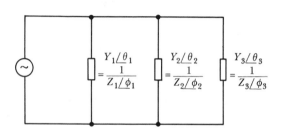

(a) Parallel impedance circuit

(b) Impedances converted to admittances

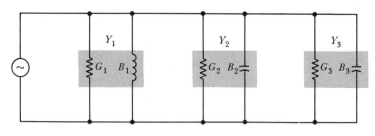

(c) Admittances resolved into conductances
and susceptances

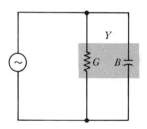

(d) Conductances are
summed and
susceptances
are summed

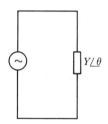

(e) Conductance and
susceptance
converted into
admittance

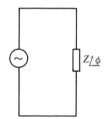

(f) Admittance
converted into
impedance

FIGURE 20-4. To analyze a parallel impedance circuit, the individual impedances are first converted into admittances which are then resolved into conductances and susceptances. All the conductances are summed, and all the susceptances are summed, to give the equivalent circuit admittance.

4. The total admittance is converted back to polar form, using Equation (19-40):

$$Y = \sqrt{G^2 + B_{eq}^2} \; \bigg/ \; \arctan \frac{B_{eq}}{G}$$

5. *The polar-form admittance is inverted to determine the total circuit impedance:*

$$Z = \frac{1}{Y}$$

EXAMPLE 20-3 Three impedances in parallel consist of $Z_1 = 1606 \; \Omega \underline{/51°}$, $Z_2 = 977 \; \Omega \underline{/-33°}$, and $Z_3 = 953 \; \Omega \underline{/-19°}$. If the supply voltage is 33 V, calculate the total circuit impedance and the current taken from the supply.

SOLUTION

Refer to Figure 20-4.
Determining individual admittances:
From Equation (19-19):

$$Y_1 = \frac{1}{Z_1} = \frac{1}{1606 \; \Omega \underline{/51°}}$$

$$= 622.7 \; \mu S \underline{/-51°}$$

$$Y_2 = \frac{1}{977 \; \Omega \underline{/-33°}}$$

$$= 1.02 \; mS \underline{/33°}$$

$$Y_3 = \frac{1}{953 \; \Omega \underline{/-19°}}$$

$$= 1.05 \; mS \underline{/19°}$$

Converting to rectangular form:
From Equation (20-2):

$$Y_1 = Y_1 \cos \phi_1 + jY_1 \sin \phi_1$$

$$= 622.7 \; \mu S[\cos(-51°) + j \sin(-51°)]$$

$$= 392 \; \mu S - j484 \; \mu S$$

$$Y_2 = 1.02 \; mS(\cos 33° + j \sin 33°)$$

$$= 855 \; \mu S + j556 \; \mu S$$

$$Y_3 = 1.05 \; mS(\cos 19° + j \sin 19°)$$

$$= 993 \; \mu S + j342 \; \mu S$$

Adding admittances:

$$Y = Y_1 + Y_2 + Y_3$$

$$= (392 + 855 + 993)\mu S + j \, (-484 + 556 + 342)\mu S$$

$$= 2.24 \; mS + j0.414 \; mS$$

Converting back to polar form:
From Equation (19-40):

$$Y = \sqrt{G^2 + B_{eq}^2} \bigg/ \arctan \frac{B_{eq}}{G}$$

$$= \sqrt{(2.24 \text{ mS})^2 + (0.414 \text{ mS})^2} \bigg/ \arctan \frac{0.414 \text{ mS}}{2.24 \text{ mS}}$$

$$= 2.28 \text{ mS} \big/ 10.5°$$

Total impedance:

$$Z = \frac{1}{Y} = \frac{1}{2.28 \text{ mS} \big/ 10.5°}$$

$$= \mathbf{439 \ \Omega \big/ {-10.5°}}$$

Supply current:

$$I = \frac{33 \text{ V}}{439 \ \Omega \big/ {-10.5°}}$$

$$= \mathbf{75.2 \text{ mA} \big/ 10.5°}$$

EXAMPLE 20-4   For the parallel impedance circuit in Example 20-3 determine the individual branch currents and the total supply current.

SOLUTION

Refer to Figure 20-4
Branch currents:

$$i_1 = \frac{E}{Z_1} = \frac{33 \text{ V}}{1606 \ \Omega \big/ 51°}$$

$$= \mathbf{20.5 \text{ mA} \big/ {-51°}}$$

$$i_2 = \frac{33 \text{ V}}{977 \ \Omega \big/ {-33°}}$$

$$= \mathbf{33.8 \text{ mA} \big/ 33°}$$

$$i_3 = \frac{33 \text{ V}}{953 \ \Omega \big/ {-19°}}$$

$$= \mathbf{34.6 \text{ mA} \big/ 19°}$$

Converting to rectangular form:

$$i_1 = i_1 \cos \phi + ji_1 \sin \phi$$

$$= 20.5 \text{ mA}[\cos(-51°) + j \sin(-51°)]$$

$$= 12.9 \text{ mA} - j15.9 \text{ mA}$$

$$i_2 = 33.8 \text{ mA}(\cos 33° + j \sin 33°)$$

$$= 28.3 \text{ mA} + j18.4 \text{ mA}$$

$$i_3 = 34.6 \text{ mA}(\cos 19° + j \sin 19°)$$

$$= 32.7 \text{ mA} + j11.3 \text{ mA}$$

Adding the currents:

$$I = i_1 + i_2 + i_3$$

$$= (12.9 + 28.3 + 32.7) \text{ mA} + j \, (-15.9 + 18.4 + 11.3) \text{ mA}$$

$$I = 73.9 \text{ mA} + j13.8 \text{ mA}$$

Converting back to polar form:

$$I = \sqrt{(73.9 \text{ mA})^2 + (13.8 \text{ mA})^2} \Big/ \arctan \frac{13.8 \text{ mA}}{73.9 \text{ mA}}$$

$$\boldsymbol{I = 75.2 \text{ mA} \underline{/10.6°}}$$

PRACTICE PROBLEMS

20-3.1 A parallel impedance circuit has $Z_1 = 5.6$ k$\Omega\underline{/30°}$, $Z_2 = 2.2$ k$\Omega\underline{/-20°}$, $Z_3 = 3.3$ k$\Omega\underline{/45°}$, and a supply voltage of 12 V. Determine the total circuit admittance and the supply current.

20-3.2 For the circuit in Problem 20-3.1, calculate the branch currents and the total supply current.

20-4
AC CURRENT DIVIDER

The current divider rule can be applied to ac impedances in parallel as well as to parallel resistors in a dc circuit. Referring to Figure 20-5, Equations (6-8) and (6-9) for a *two*-branch parallel circuit (as developed in Section 6-4) are rewritten:

$$i_1 = I \, \frac{Z_2}{Z_1 + Z_2} \tag{20-4}$$

and

$$i_2 = I \, \frac{Z_1}{Z_1 + Z_2} \tag{20-5}$$

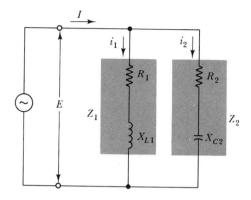

FIGURE 20-5. Alternating current divides between two parallel-connected impedances in a similar way to direct current dividing between two parallel-connected resistors.

In applying these equations, it is necessary to convert the denominator impedances into rectangular form to add them together. Once Z_1 and Z_2 are added, the denominator should be returned to polar form so that it may be easily divided into the numerator.

EXAMPLE 20-5   The two impedances in the circuit shown in Figure 20-5 are $Z_1 = (1 \text{ k}\Omega + j2.7 \text{ k}\Omega)$ and $Z_2 = (790 \ \Omega - j1.6 \text{ k}\Omega)$. The total current taken from the supply is 15 mA. Using the current divider rule, calculate the two branch currents.

SOLUTION

$$Z_1 = \sqrt{(1 \text{ k}\Omega)^2 + (2.7 \text{ k}\Omega)^2} \ \Big/ \arctan \frac{2.7 \text{ k}\Omega}{1 \text{ k}\Omega}$$

$$= 2.88 \text{ k}\Omega \underline{/69.7°}$$

$$Z_2 = \sqrt{(790 \ \Omega)^2 + (1.6 \text{ k}\Omega)^2} \ \Big/ \arctan \frac{-1.6 \text{ k}\Omega}{790 \ \Omega}$$

$$= 1.78 \text{ k}\Omega \underline{/-63.7°}$$

$$Z_1 + Z_2 = (1 \text{ k}\Omega + j2.7 \text{ k}\Omega) + (790 \ \Omega - j1.6 \text{ k}\Omega)$$

$$= 1.79 \text{ k}\Omega + j1.1 \text{ k}\Omega$$

$$= \sqrt{(1.79 \text{ k}\Omega)^2 + (1.1 \text{ k}\Omega)^2} \ \Big/ \arctan \frac{1.1 \text{ k}\Omega}{1.79 \text{ k}\Omega}$$

$$= 2.1 \text{ k}\Omega \underline{/31.6°}$$

Equation (20-4):

$$i_1 = I \frac{Z_2}{Z_1 + Z_2}$$

$$= (15 \text{ mA}) \frac{1.78 \text{ k}\Omega \underline{/-63.7°}}{2.1 \text{ k}\Omega \underline{/31.6°}}$$

$$i_1 = 12.7 \text{ mA} \underline{/-95.3°}$$

Equation (20-5):

$$i_2 = I \frac{Z_1}{Z_1 + Z_2}$$

$$= (15 \text{ mA}) \frac{2.88 \text{ k}\Omega \underline{/69.7°}}{2.1 \text{ k}\Omega \underline{/31.6°}}$$

$$= 20.6 \text{ mA} \underline{/38.1°}$$

When a phasor diagram is drawn for i_1 and i_2, it is seen that their resultant is $I = 15$ mA (i.e., the supply current).

Equations (20-4) and (20-5) apply only to a *two*-branch parallel circuit. For a multibranch current divider circuit, Equations (6-10) and (6-11) can be rewritten in the form of impedances and admittances. Rewriting Equations (6-10):

$$\boxed{I_n = I \left[\frac{Z}{Z_n} \right]} \tag{20-6}$$

In Equation (20-6), I_n is the current in branch n, I is the supply current, Z is the impedance of the complete circuit, and Z_n is the impedance of branch n.

Substituting $Z = 1/Y$ and $Z_n = 1/Y_n$ into Equation (20-6) gives the current divider equation in the form of admittances:

$$I_n = I \left[\frac{Y_n}{Y} \right]$$

or,

$$\boxed{I_n = I \left[\frac{Y_n}{Y_1 + Y_2 + Y_3 + \cdots} \right]} \tag{20-7}$$

Equation (20-7) is comparable to voltage divider Equation (20-1).

EXAMPLE 20-6 The three impedances in Figure 20-6 are: $Z_1 = 1606\ \Omega\underline{/51°}$, $Z_2 = 977\ \Omega\underline{/-33°}$, and $Z_3 = 953\ \Omega\ \underline{/-19°}$. (These are the same component values as used in Examples 20-3 and 20-4.) The supply current is $I = 75.2\ \text{mA}\underline{/10.5°}$. Determine I_1, I_2, and I_3.

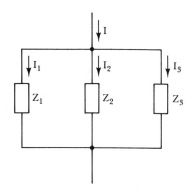

FIGURE 20-6. Alternating current division in a multi-impedance parallel circuit. The current through any branch is $I_n = I(Z/Z_n)$, where Z is the complete circuit impedance.

SOLUTION

From Example 20-3:

$$Y_1 = \frac{1}{Z_1} = 622.7\ \mu\text{S}\underline{/-51°}$$

$$= 392\ \mu\text{S} - j484\ \mu\text{S}$$

$$Y_2 = \frac{1}{Z_2} = 1.02\ \text{mS}\underline{/33°}$$

$$= 855\ \mu\text{S} + j556\ \mu\text{S}$$

$$Y_3 = \frac{1}{Z_3} = 1.05\ \text{mS}\underline{/19°}$$

$$= 993\ \mu\text{S} + j343\ \mu\text{S}$$

$$Y = Y_1 + Y_2 + Y_3$$

$$= 2.24\ \text{mS} + j0.414\ \text{mS}$$

$$= 2.28\ \text{mS}\underline{/10.5°}$$

Equation (20-7): $I_1 = I\left[\dfrac{Y_1}{Y}\right]$

$$= 75.2\ \text{mA}\underline{/10.5°}\left[\frac{622.7\ \mu\text{S}\underline{/-51°}}{2.28\ \text{mS}\underline{/10.5°}}\right]$$

$$= \mathbf{20.5\ mA\underline{/-51°}}$$

$$I_2 = I\left[\frac{Y_2}{Y}\right]$$

$$= 75.2 \text{ mA}\underline{/10.5°}\left[\frac{1.02 \text{ mS}\underline{/33°}}{2.28 \text{ mS}\underline{/10.5°}}\right]$$

$$= \mathbf{33.6 \text{ mA}\underline{/33°}}$$

$$I_3 = I\left[\frac{Y_3}{Y}\right]$$

$$= 75.2 \text{ mA}\underline{/10.5°}\left[\frac{1.05 \text{ mS}\underline{/19°}}{2.28 \text{ mS}\underline{/10.5°}}\right]$$

$$= \mathbf{34.6 \text{ mA}\underline{/19°}}$$

These results are similar to those obtained in Example 20-4.

PRACTICE PROBLEMS

20-4.1 Use the current divider equation to determine the branch currents in a parallel circuit consisting of $Z_1 = 470 \text{ } \Omega\underline{/39°}$ and $Z_2 = 390 \text{ } \Omega\underline{/-75°}$. The supply current is 51.7 mA$\underline{/26.2°}$.

20-4.2 If an additional parallel impedance $Z_3 = 560 \text{ } \Omega\underline{/15°}$ is included in the circuit of Problem 20-4.1, determine the new level of supply current.

20-5

SERIES-PARALLEL IMPEDANCES

ANALYSIS PROCEDURE FOR SERIES-PARALLEL IMPEDANCE CIRCUITS

1. *Resolve all series-connected impedances into a single equivalent impedance.*

2. *Resolve all parallel-connected impedances into a single equivalent impedance.*

3. *Repeat procedures 1 and 2 until a single equivalent impedance is determined for the whole circuit.*

4. *Calculate the supply current.*

5. *Use the current divider rule to determine the individual branch currents.*

Figure 20-7(a) shows a series-parallel impedance circuit, and Figures 20-7(b) and (c) show the stages in analyzing the circuit. Impedances Z_3 and Z_4 are added together (in rectangular form) to give a single equivalent impedance. Then converting Z_2 and $(Z_3 + Z_4)$ into admittances, the equivalent impedance of Z_2 in parallel with $(Z_3 + Z_4)$ is found. At this stage the circuit has become two impedances in series, and these may be

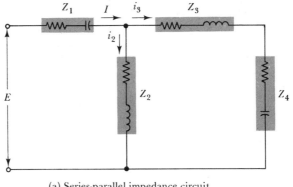

(a) Series-parallel impedance circuit

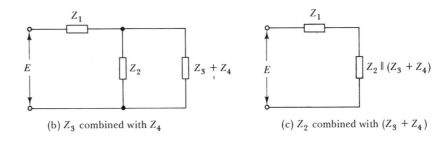

(b) Z_3 combined with Z_4 (c) Z_2 combined with $(Z_3 + Z_4)$

FIGURE 20-7. To analyze a series-parallel impedance circuit, all series-connected impedances are resolved into a single impedance, and all parallel-connected impedances are resolved into a single impedance.

added in rectangular form to give the equivalent impedance for the whole circuit.

EXAMPLE 20-7 For the circuit shown in Figure 20-7(a): $Z_1 = (560 \ \Omega - j620 \ \Omega)$, $Z_2 = (330 \ \Omega + j470 \ \Omega)$, $Z_3 = (390 + j270 \ \Omega)$, and $Z_4 = (220 \ \Omega - j220 \ \Omega)$. If the supply voltage is 30 V, determine the current through Z_2.

SOLUTION

$$Z_3 + Z_4 = (390 \ \Omega + j270 \ \Omega) + (220 \ \Omega - j220 \ \Omega)$$

$$= 610 \ \Omega + j50 \ \Omega$$

$$= \sqrt{610^2 + 50^2} \ \Big/ \arctan \frac{50 \ \Omega}{610 \ \Omega}$$

$$= 612 \ \Omega \underline{/4.7°}$$

$$Y_{3,4} = \frac{1}{Z_3 + Z_4} = \frac{1}{612 \ \Omega \underline{/4.7°}}$$

$$= 1.63 \ \text{mS} \underline{/-4.7°}$$

$$Y_{3,4} = 1.63 \text{ mS}[\cos(-4.7°) + j \sin(-4.7°)]$$
$$= 1.62 \text{ mS} - j0.134 \text{ mS}$$
$$Z_2 = \sqrt{330^2 + 470^2} \left/ \arctan \frac{470}{330}\right.$$
$$= 574 \text{ } \Omega \underline{/54.9°}$$
$$Y_2 = \frac{1}{Z_2} = \frac{1}{574 \text{ } \Omega \underline{/54.9°}}$$
$$= 1.74 \text{ mS} \underline{/-54.9°}$$
$$= 1.74 \text{ mS}[\cos(-54.9°) + j \sin(-54.9°)]$$
$$= 1 \text{ mS} - j1.42 \text{ mS}$$
$$Y_{3,4} + Y_2 = (1.62 \text{ mS} - j0.134 \text{ mS}) + (1 \text{ mS} - j1.42 \text{ mS})$$
$$= 2.62 \text{ mS} - j1.55 \text{ mS}$$
$$= \sqrt{(2.62 \text{ mS})^2 + (1.55 \text{ mS})^2} \left/ \arctan \frac{-1.55 \text{ mS}}{2.62 \text{ mS}}\right.$$
$$= 3.04 \text{ mS} \underline{/-30.6°}$$
$$Z_2 \| (Z_3 + Z_4) = \frac{1}{Y_{3,4} + Y_2}$$
$$= \frac{1}{3.04 \text{ mS} \underline{/-30.6°}}$$
$$= 329 \text{ } \Omega \underline{/30.6°}$$
$$= 329 \text{ } \Omega[\cos(30.6°) + j \sin(30.6°)]$$
$$= 283 \text{ } \Omega + j167 \text{ } \Omega$$
$$Z_T = Z_1 + [Z_2 \| (Z_3 + Z_4)]$$
$$= (560 \text{ } \Omega - j620 \text{ } \Omega) + (283 \text{ } \Omega + j167 \text{ } \Omega)$$
$$= 843 \text{ } \Omega - j453 \text{ } \Omega$$
$$= \sqrt{843^2 + 453^2} \left/ \arctan \frac{-453 \text{ } \Omega}{843 \text{ } \Omega}\right.$$
$$= 957 \text{ } \Omega \underline{/-28.3°}$$
$$I = \frac{E}{Z_T} = \frac{30 \text{ V}}{957 \text{ } \Omega \underline{/-28.3°}}$$
$$= 31.3 \text{ mA} \underline{/28.3°}$$

The total current I flows through Z_1 and splits up into i_2 and i_3, as indicated in Figure 20-7(a).

Using the current divider rule:

$$i_2 = I \frac{Z_3 + Z_4}{Z_2 + (Z_3 + Z_4)}$$

$$Z_2 + (Z_3 + Z_4) = (330 \ \Omega + j470 \ \Omega) + (610 \ \Omega + j50 \ \Omega)$$

$$= 940 + j520$$

$$= \sqrt{940^2 + 520^2} \ \Big/ \arctan \frac{520}{940}$$

$$= 1074 \ \Omega \underline{/29°}$$

Therefore,

$$i_2 = 31.3 \ \text{mA}\underline{/28.3°} \left[\frac{612 \ \Omega\underline{/4.7°}}{1074 \ \Omega\underline{/29°}} \right]$$

$$= \textbf{17.8 mA}\underline{\textbf{/4°}}$$

PRACTICE PROBLEMS

20-5.1 An impedance network consists of Z_1 in series with parallel-connected Z_2 and Z_3, and in series with Z_4. The impedances are $Z_1 = 1.46 \ \text{k}\Omega\underline{/68.3°}$, $Z_2 = 200 \ \Omega\underline{/53.1°}$, $Z_3 = 707 \ \Omega\underline{/-45°}$, and $Z_4 = 566 \ \Omega\underline{/-58°}$. Determine the circuit impedance.

20-5.2 If a supply of 18 V is applied to the circuit in Problem 20-5.1, determine the current through Z_2.

20-6

SERIES AND PARALLEL EQUIVALENT CIRCUITS

Sometimes it is convenient to replace a series *RL* circuit with a parallel *RL* circuit which offers exactly the same impedance characteristics, and vice versa. In this case each circuit is the *equivalent circuit* of the other, and if each were contained in a sealed box with only their terminals showing, it might be difficult to distinguish between them.

Figure 20-8 shows a series *RL* circuit consisting of R_s and X_s, and its equivalent parallel circuit, R_p and X_p. From Figure 20-8(a),

$$Z_s = R_s + jX_s$$

and from Figure 20-8(b),

$$Y_p = \frac{1}{R_p} - j\frac{1}{X_p}$$

Also,

$$Z_s = Z_p$$

or

$$\frac{1}{Z_s} = Y_p$$

Therefore,

$$\frac{1}{R_s + jX_s} = \frac{1}{R_p} - j\frac{1}{X_p}$$

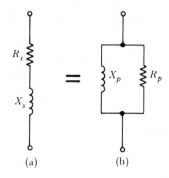

(a)   (b)

FIGURE 20-8. An *LR* series circuit may be replaced with an equivalent *LR* parallel circuit which has a similar terminal impedance. The circuit terminal resistances differ.

$$\frac{1}{R_s + jX_s} \times \frac{R_s - jX_s}{R_s - jX_s} = \frac{1}{R_p} - j\frac{1}{X_p}$$

$$\frac{R_s - jX_s}{R_s^2 + X_s^2} = \frac{1}{R_p} - j\frac{1}{X_p}$$

Equating the resistive terms,

$$\frac{R_s}{R_s^2 + X_s^2} = \frac{1}{R_p}$$

or

$$\boxed{R_p = \frac{R_s^2 + X_s^2}{R_s}}$$ (20-8)

Equating the reactive terms,

$$\frac{X_s}{R_s^2 + X_s^2} = \frac{1}{X_p}$$

or

$$\boxed{X_p = \frac{R_s^2 + X_s^2}{X_s}}$$ (20-9)

Equations (20-8) and (20-9) also apply to the series *CR* circuit and its parallel equivalent shown in Figure 20-9. These equations enable the component values of the parallel *CR* or *LR* equivalent circuit to be calculated for any given series *CR* or *LR* circuit.

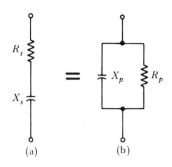

FIGURE 20-9. A *CR* series circuit may be replaced with an equivalent *CR* parallel circuit which has a similar terminal impedance. Here again, the circuit terminal resistances differ.

To derive equations for calculating the series *CR* or *LR* equivalent circuit for a given parallel circuit, Y_p is written in terms of a conductance and susceptance. Z_s is then equated to the reciprocal of Y_p.

For the circuits in Figure 20-9,

$$Z_s = R_s - jX_s$$

and

$$Y_p = G_p + jB_p$$

$$Z_s = \frac{1}{Y_p}$$

Therefore,

$$R_s - jX_s = \frac{1}{G_p + jB_p}$$

$$R_s - jX_s = \frac{1}{G_p + jB_p} \times \frac{G_p - jB_p}{G_p - jB_p}$$

$$R_s - jX_s = \frac{G_p - jB_p}{G_p^2 + B_p^2}$$

Equating the resistive terms and the reactive terms

$$\boxed{R_s = \frac{G_p}{G_p^2 + B_p^2}} \tag{20-10}$$

and

$$\boxed{X_s = \frac{B_p}{G_p^2 + B_p^2}} \tag{20-11}$$

Since it may be more convenient to use R_p and X_p rather than G_p and B_p, the equations can be rewritten:

Equation (20-10):

$$R_s = \frac{1/R_p}{1/R_p^2 + 1/X_p^2}$$

$$= \frac{1/R_p}{1/R_p^2 + 1/X_p^2} \times \frac{R_p^2 X_p^2}{R_p^2 X_p^2}$$

or

$$\boxed{R_s = \frac{R_p X_p^2}{X_p^2 + R_p^2}} \qquad \text{(20-12)}$$

and from Equation (20-11),

$$\boxed{X_s = \frac{X_p R_p^2}{X_p^2 + R_p^2}} \qquad \text{(20-13)}$$

EXAMPLE 20-8   At a frequency of 1 kHz, an unknown impedance behaves as a 0.01 μF capacitor in series with a resistance of 10 kΩ. A dc measurement at the terminals of the impedance gives a resistance of 35.3 kΩ. Determine the actual components and how they are connected.

SOLUTION

$$X_s = \frac{1}{2\pi f C} = \frac{1}{2\pi \times 1 \text{ kHz} \times 0.01 \text{ μF}}$$

$$= 15.9 \text{ kΩ}$$

$$R_s = 10 \text{ kΩ}$$

Equation (20-8):

$$R_p = \frac{R_s^2 + X_s^2}{R_s} = \frac{(10 \text{ kΩ})^2 + (15.9 \text{ kΩ})^2}{10 \text{ kΩ}}$$

$$= 35.3 \text{ kΩ}$$

Equation (20-9):

$$X_p = \frac{R_s^2 + X_s^2}{X_s} = \frac{(10 \text{ kΩ})^2 + (15.9 \text{ kΩ})^2}{15.9 \text{ kΩ}}$$

$$= 22.2 \text{ kΩ}$$

$$C_p = \frac{1}{2\pi f X_p} = \frac{1}{2\pi \times 1 \text{ kHz} \times 22.2 \text{ kΩ}}$$

$$= 0.007 \text{ μF}$$

If the actual impedance consisted of a series-connected resistor and capacitor, the dc resistance measurement would give a value greater than R_s; that is, it would be $R_s + $(the capacitor dielectric resistance). Since the measured resistance corresponds with the value of R_p, the actual impedance consists of a 35.3 kΩ resistor connected in parallel with a 0.007 μF capacitor.

PRACTICE PROBLEM

20-6.1 The components in Figure 20-5 are $Z_1 = 1\ k\Omega + j2.7\ k\Omega$ and $Z_2 = 790\ \Omega - j1.6\ k\Omega$. Resolve each impedance into its parallel equivalent circuit.

SUMMARY OF FORMULAS

AC voltage divider:

$$V_3 = \frac{EZ_3}{Z_1 + Z_2 + Z_3}$$

Resolving polar form admittances into rectangular form:

$$Y = Y \cos \theta + jY \sin \theta$$

AC current divider:

$$i_1 = I\frac{Z_2}{Z_1 + Z_2}, \qquad i_2 = I\frac{Z_1}{Z_1 + Z_2}$$

$$I_n = I\left(\frac{Z}{Z_n}\right) = I\left(\frac{Y_n}{Y_1 + Y_2 + Y_3 + \cdots}\right)$$

Series-parallel equivalent circuits:

$$R_p = \frac{R_s^2 + X_s^2}{R_s}$$

$$X_p = \frac{R_s^2 + X_s^2}{X_s}$$

$$R_s = \frac{G_p}{G_p^2 + B_p^2} = \frac{R_p X_p^2}{X_p^2 + R_p^2}$$

$$X_s = \frac{B_p}{G_p^2 + B_p^2} = \frac{X_p R_p^2}{X_p^2 + R_p^2}$$

PROBLEMS

SECTION 20-1

20-1 A coil with a resistance of 100 Ω and an inductance of 300 mH is connected in series with a 10 μF capacitor, another inductor of 150 mH, and a resistance of 180 Ω. The supply voltage is 115 V with a frequency of 60 Hz. Calculate the current taken from the supply.

20-2 Two impedances, $Z_1 = 120 \ \Omega \underline{/33°}$ and $Z_2 = 56 \ \Omega \underline{/60°}$, are connected in series to a 15 V ac supply. Calculate the circuit current and the terminal voltage of each impedance.

20-3 For the circuit described in Problem 20-1, calculate the terminal voltages of the capacitor and inductor. Also, draw a phasor diagram showing the supply voltage and current and the voltages across the capacitor and inductor.

20-4 A 3 V, 12 kHz voltage is applied to a series circuit consisting of $R = 4.7 \ k\Omega$, $L = 56$ mH, and $C = 0.001 \ \mu$F. The inductor has a coil resistance of 800 Ω. Calculate the terminal voltage of each component and the phase angle of each voltage with respect to the supply.

SECTION 20-2

20-5 An impedance of $(3.3 \ k\Omega + j3.9 \ k\Omega)$ is connected in series with an impedance of $(6.8 \ k\Omega - j5.6 \ k\Omega)$ to a 55 V supply. Use the voltage divider rule to determine the voltage across each impedance.

20-6 Use the voltage divider rule to determine the voltage across each component in Problem 20-1.

SECTION 20-3

20-7 Three impedances connected in parallel are: $Z_1 = 5.6 \ k\Omega - j3.3 \ k\Omega$, $Z_2 = 1.8 \ k\Omega + j2.2 \ k\Omega$, $Z_3 = 8.2 \ k\Omega - j6.2 \ k\Omega$. Determine the total circuit impedance and the current taken from a 19 V supply.

20-8 If the impedances in Problem 20-4 are connected in parallel to the supply, calculate the total circuit impedance and the supply current.

SECTION 20-4

20-9 The two impedances described in Problem 20-5 are connected in parallel to a supply. If the total current taken from the supply is 22 mA, use the current divider rule to calculate the current through each impedance.

20-10 The two impedances in Problem 20-2 are connected in parallel to an ac supply. If the current through Z_1 is 83 mA$\underline{/-33°}$, use the current divider rule to determine the current through Z_2.

20-11 For the circuit described in Problem 20-7, use the calculated supply current and the appropriate current divider equation to determine each of the branch currents.

20-12 A 0.1 μF capacitor in series with a 470 Ω resistor is connected in parallel with a 75 mH inductor which has a 330 Ω coil re-

        Chap. 20   Series and Parallel AC Circuits

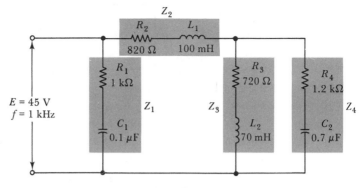

FIGURE 20-10.

sistance. A 7.5 V, 1 kHz supply is applied to the parallel circuit. Calculate the total supply current, then use the current divider rule to determine the branch currents.

SECTION 20-5

20-13 Calculate the equivalent impedance of Z_3 and Z_4 in parallel in the circuit shown in Figure 20-10.

20-14 Calculate the total equivalent impedance of the circuit shown in Figure 20-10.

20-15 For the circuit shown in Figure 20-10, determine the voltage drops acorss L_2 and C_2.

20-16 In the circuit shown in Figure 20-10, L_1 is changed to 200 mH, and C_1 is altered to 0.3 μF. Calculate the new supply current.

20-17 Determine the total equivalent impedance for the circuit shown in Figure 20-11.

20-18 Calculate the current flowing in R_3 and L_2 in the circuit in Figure 20-11.

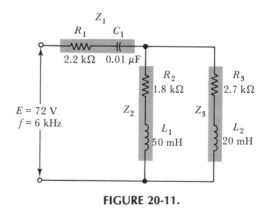

FIGURE 20-11.

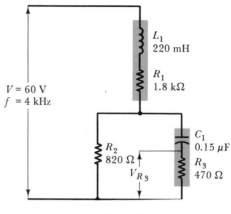

$V = 60$ V
$f = 4$ kHz

L_1
220 mH

R_1
1.8 kΩ

C_1
0.15 μF

R_2
820 Ω

V_{R3}

R_3
470 Ω

FIGURE 20-12.

20-19 Calculate the current flowing in R_2 and L_1 in the circuit in Figure 20-11.

20-20 A 6.8 kΩ resistor is connected in parallel with Z_1 in Figure 20-11. Calculate the new supply current.

20-21 Determine the total impedance of the circuit in Figure 20-12.

20-22 Determine V_{R3} in the circuit in Figure 20-12.

20-23 Calculate the impedance of $Z_2 \| Z_3$ in the circuit in Figure 20-13.

20-24 Calculate the total impedance of the circuit in Figure 20-13.

20-25 For the circuit in Figure 20-13, calculate the voltages across R_3 and C_2.

20-26 An 8.2 kΩ resistor is connected in parallel with Z_1 in Figure 20-11. Calculate the new level of supply current.

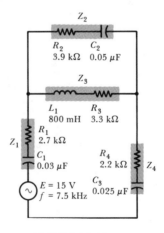

Z_2

R_2 C_2
3.9 kΩ   0.05 μF

Z_3

L_1 R_3
800 mH   3.3 kΩ

R_1
2.7 kΩ

Z_1

C_1
0.03 μF

R_4
2.2 kΩ

Z_4

$E = 15$ V
$f = 7.5$ kHz

C_3
0.025 μF

FIGURE 20-13.

Chap. 20   Series and Parallel AC Circuits

20-27 Derive the equations for converting a series-connected inductance and resistance into a parallel-connected inductance and resistance.

20-28 Derive the equations for converting a parallel-connected capacitance and resistance into a series-connected capacitance and resistance.

20-29 An unknown impedance measured at a frequency of 5 kHz behaves as a 500 mH inductor in series with a 10 kΩ resistor. A dc measurement gives a terminal resistance of 10 kΩ. Determine the parallel equivalent circuit of the impedance and decide the actual component values.

20-30 Calculate i_1, i_2, and I in Figure 20-14.

20-31 Replace R_1 and L_1 in Figure 20-14 with their parallel equivalent circuits, then once again calculate all branch currents.

20-32 In the circuit in Figure 20-10, calculate the parallel equivalent impedances for Z_1, Z_2, and Z_4.

20-33 In the circuit in Figure 20-10, replace Z_1, Z_2, and Z_4 with their parallel equivalent impedances, then calculate the total equivalent impedance of the circuit.

20-34 An unknown impedance behaves as a 2000 pF capacitor in series with a 1.5 kΩ resistor when measured at a frequency of 100 kHz. The dc resistance between the terminals is found to be greater than 10 MΩ. Calculate the parallel equivalent circuit of the impedance and determine which of the two circuits correctly represents the impedance.

COMPUTER PROBLEMS

20-35 Write a computer program to analyze a series circuit consisting of components R_1, R_2, L_1, and C_1. The program should accept any given component values, supply voltage, and frequency, and should print out the circuit impedance, the supply current, and the component voltages.

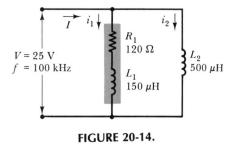

FIGURE 20-14.

20-36 Write a computer program to determine the equivalent impedance of the circuit configuration shown in Figure 20-11 for any given set of component values and supply frequency.

21

POWER IN
AC CIRCUITS

Objectives   You will be able to:

Sketch and explain the waveforms of current, voltage, and power for a pure resistance connected in an ac circuit.

Solve problems involving current voltage and power dissipation in a purely resistive ac circuit.

Sketch and explain the waveforms of current, voltage, and power for a pure inductance in an ac circuit.

Sketch and explain the waveforms of current, voltage, and power for a pure capacitance in an ac circuit.

Sketch and explain the waveforms of current, voltage, and power for series *LR* and *CR* ac circuits.

Calculate the true power and reactive power in purely resistive, purely inductive, and purely capacitive ac circuits.

Calculate the apparent power, true power, and reactive power in series *LR* and *CR* circuits.

Define power factor and discuss power factor correction.

Sketch power triangles for series *LR* and *CR* circuits and calculate power factor correction capacitor values for inductive loads.

Calculate power level changes in decibels.

Introduction

In an alternating-current circuit, power is dissipated in a resistor, but not in a pure inductor or capacitor. Because the current in an *LR* circuit lags the supply voltage by an angle ϕ, the amount of useful power supplied to the circuit is proportional to cos ϕ. Similarly, in a *CR* circuit, the useful power is proportional to the angle by which the current leads the voltage. The presence of the reactive components causes *reactive power* to be supplied to the circuit. This gives an *apparent power,* which is greater than the *true power* utilized in the circuit. The ratio of true power to apparent power is termed the *power factor* of the circuit.

The unit of true power is the *watt* (W), the unit of apparent power is the *volt-amp* (VA), and the reactive power unit is the *volt-amp-reactive* (var). These quantities all refer to ac circuits, which might have supply frequencies ranging from 25 Hz to 400 Hz. For audio and radio-frequency power measurements, the unit used is the *decibel* (dB).

21-1
POWER DISSIPATED IN A RESISTANCE

In Section 17-4 it was shown that when an alternating current flows in a resistance, the power dissipated can be calculated as follows. Equation (17-12):

$$P = \tfrac{1}{2} P_m \quad \text{watt}$$

where

$$P_m = E_m I_m$$

Therefore,

$$P = \tfrac{1}{2} E_m I_m$$

$$= \frac{E_m}{\sqrt{2}} \times \frac{I_m}{\sqrt{2}}$$

or

$$\boxed{P = EI} \tag{21-1}$$

where E and I are rms values. This is exactly the same as for Equation (3-3), which was derived for a dc circuit. As long as rms quantities are used for voltage and current, the power calculations *for a resistor* in an ac circuit are exactly the same as those for a dc circuit. Thus, from Equations (3-4) and (3-5),

$$\boxed{P = \frac{E^2}{R}} \tag{21-2}$$

and

$$\boxed{P = I^2 R} \tag{21-3}$$

where E and I are again rms values.

Figure 17-8 shows that the instantaneous power dissipated in a resistance alternates between zero and a peak level as the current rises from zero to its positive and negative peak values. If the resistance were the tungsten filament of a lamp, and if the frequency of the alternating current were 0.1 Hz, the filament could clearly be seen to go bright and dim at a frequency of 0.2 Hz. The lamp would be bright when the current is $+I_m$, dim when the current is zero, and bright again when the current is $-I_m$. The waveform of power in Figure 17-8 shows that two positive peaks of power occur during each cycle of current. Since the normal domestic and industrial power frequency is 60 Hz (in North America), any fluctuations in the brightness of an electric lamp are much too fast for the human eye to detect.

EXAMPLE 21-1 A 100 W electric lamp is supplied from a 115 V, 60 Hz source. Calculate:

 a. The level of current that flows.
 b. The resistance of the filament.
 c. The peak instantaneous power dissipated in the lamp filament.

SOLUTION

a. Equation (21-1):

$$P = EI$$

Therefore, $$I = \frac{P}{E}$$

$$I = \frac{100 \text{ W}}{115 \text{ V}} \cong 870 \text{ mA}$$

b. Equation (21-2):

$$P = \frac{E^2}{R}$$

or $$R = \frac{E^2}{P}$$

$$R = \frac{(115 \text{ V})^2}{100 \text{ W}} = 132 \ \Omega$$

c. Equation (17-12):

$$P = \tfrac{1}{2} P_m$$

giving $$P_m = 2 \times P$$

$$P_m = 2 \times 100 \text{ W} = 200 \text{ W}$$

PRACTICE PROBLEM

21-1.1 Three series-connected resistors, $R_1 = 500 \ \Omega$, $R_2 = 680 \ \Omega$, and $R_3 = 820 \ \Omega$, are supplied from an ac source with a peak value of 39.6 V. Calculate the power dissipation in each resistor and the peak power supplied to the circuit.

21-2

POWER IN AN INDUCTANCE

When an alternating voltage is applied to a pure inductance, the current lags the applied voltage by 90°. This was discussed in Section 19-1. The waveforms of current and voltage for an inductance are reproduced in Figure 21-1. Since the instantaneous power supplied to any component is calculated as

$$p = e \times i$$

the waveform of power can easily be derived from the current and voltage waves. At time t_1 on Figure 21-1,

$$i = 0 \quad \text{and} \quad e = E_m$$

Therefore,

$$p = i \times E_m$$
$$= 0 \times E_m$$
$$p = 0$$

At t_2,

$$i = I_m \sin 45°$$
$$i = \frac{I_m}{\sqrt{2}} = I \quad (rms \ value)$$

and

$$e = E_m \sin 135°$$
$$e = \frac{E_m}{\sqrt{2}} = E \quad (rms \ value)$$

Therefore,

$$p_m = IE$$

At t_3,

$$i = I_m \quad \text{and} \quad e = 0$$
$$= I_m \times 0$$
$$p = 0$$

At t_4,

$$i = \frac{I_m}{\sqrt{2}} \quad \text{and} \quad e = \frac{-E_m}{\sqrt{2}}$$
$$p_m = \left[\frac{I_m}{\sqrt{2}}\right] \times \left[\frac{-E_m}{\sqrt{2}}\right]$$

giving

$$p_m = -IE$$

At this point (i.e., when $p = -IE$) the power supplied is a negative quantity, which means that the inductance is not absorbing power but *supplying power*. Continuing the process of calculating and plotting the

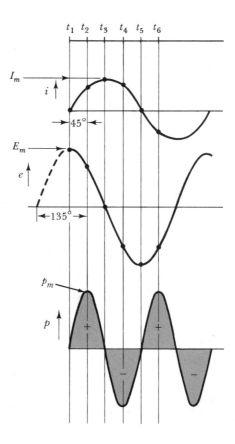

FIGURE 21-1. The waveform of power supplied from an ac source to a pure inductance can be derived from the voltage and current waveforms (using instantaneous levels, $p = ei$). The average power absorbed by the inductance is zero.

instantaneous power levels in Figure 21-1, it is seen that the waveform of power supplied has a frequency which is twice that of the voltage and current. Also, since the *negative* half-cycles of power supplied are equal to the positive half-cycles, the average power supplied to the inductance is zero.

In Chapter 14 it was shown that energy can be stored in an inductor. Therefore, referring to Figure 21-1 again, it can be said that the energy supplied to the inductor is stored during the time that energy input is a positive quantity, and that the stored energy is returned to the source of supply when the energy input is negative.

PRACTICE PROBLEM

21-2.1 For the waveforms illustrated in Figure 21-1, $I_m = 15$ mA and $E_m = 20$ V. Determine the instantaneous power dissipations at points halfway between t_1 and t_2 and halfway between t_3 and t_4.

21-3
POWER IN A CAPACITANCE

In the case of a pure capacitance supplied with an alternating voltage, the current leads the voltage by 90° (see Section 19-3). Figure 21-2 shows the current and voltage waveforms for a capacitance, and the waveform of power supplied, as derived from the current and voltage waves.

At t_1,

$$p = i \times e$$
$$= 0 \times (-E_m)$$
$$= 0$$

At t_2,

$$i = I_m \sin 45°$$
$$i = \frac{I_m}{\sqrt{2}} = I \quad \text{(rms value)}$$

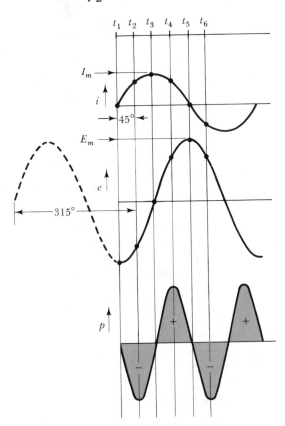

FIGURE 21-2. The waveform of power supplied from an ac source to a pure capacitance can be derived from the voltage and current waveforms (using instantaneous levels, $p = ei$). The average power absorbed by the capacitance is zero.

Chap. 21   Power in AC Circuits

and

$$e = E_m \sin 315°$$

$$e = \frac{-E_m}{\sqrt{2}} = -E \quad \text{(rms value)}$$

Therefore, $\quad p_m = -IE$

Continuing the process, exactly as was done for the inductive circuit, it is seen that the waveform of power supplied to a capacitor has a frequency which is twice that of the voltage and current. It is also seen that the power supplied to the capacitor is alternatively positive and negative, meaning that the capacitor stores the power supplied to it then returns the stored power to the source of supply. The result of this is that the average power supplied to the capacitance is zero.

21-4
TRUE POWER AND REACTIVE POWER

The power supplied to a resistance is sometimes referred to as *resistive power*. Similarly, the power supplied to a reactance (inductive or capacitive) can be termed *reactive power*. Also, because the power supplied to a resistance is actually dissipated in the form of heat, while that supplied to a reactance averages out to zero, the *resistive power* is known as *true power*.

It is very important to distinguish between true power (symbol P) and reactive power (symbol Q). Consequently, true power is always measured in *watts* (W), while the reactive power unit is given the name *volt-amp-reactive* (var). For a pure inductance or capacitance, the reactive power is calculated from

$$\boxed{Q_L = E_L I_L \quad \text{or} \quad Q_C = E_C I_C} \tag{21-4}$$

where Q is in var, and E_L, I_L, E_C, and I_C are the device voltage and current levels.

Reactive power can also be calculated as

$$\boxed{Q_L = I_L^2 X_L \quad \text{or} \quad Q_C = I_C^2 X_C} \tag{21-5}$$

and

$$\boxed{Q_L = \frac{E_L^2}{X_L} \quad \text{or} \quad Q_C = \frac{E_C^2}{X_C}} \tag{21-6}$$

EXAMPLE 21-2 Calculate the power supplied when a 120 V, 60 Hz source is connected to:

   a. A 60 Ω resistor.
   b. A 50 mH inductor.
   c. A 33 μF capacitor.

SOLUTION

a.

$$P = \frac{E^2}{R} = \frac{(120 \text{ V})^2}{60 \text{ }\Omega}$$

$$= \textbf{240 W} \quad (\textbf{\textit{true power}})$$

b.

$$X_L = 2\pi f L$$

$$= 2 \times \pi \times 60 \text{ Hz} \times 50 \text{ mH}$$

$$X_L \cong 18.8 \text{ }\Omega$$

From Equation (21-6):

$$Q_L = \frac{E_L^2}{X_L} = \frac{(120 \text{ V})^2}{18.8 \text{ }\Omega}$$

$$= \textbf{766 var} \quad (\textbf{\textit{reactive power}})$$

c.

$$X_C = \frac{1}{2\pi f c}$$

$$= \frac{1}{2 \times \pi \times 60 \text{ Hz} \times 33 \text{ }\mu\text{F}}$$

$$\cong 80.4 \text{ }\Omega$$

From Equation (21-6),

$$Q_C = \frac{E_C^2}{X_C} = \frac{(120 \text{ V})^2}{80.4 \text{ }\Omega}$$

$$= \textbf{179 var} \quad (\textbf{\textit{reactive power}})$$

PRACTICE PROBLEM

21-4.1 A 0.1 μF capacitor, 100 mH inductor, and a 1 kΩ resistor connected in parallel are supplied from a 400 Hz ac source. The power dissipated in the resistor is measured as 100 mW. Calculate the reactive power supplied to the capacitor and the reactive power supplied to the inductor.

21-5

POWER IN RL AND RC CIRCUITS

RL CIRCUIT POWER. In a series circuit consisting of inductance and resistance, the current lags the voltage by an angle ϕ which is less than 90° (see Section 19-5). Figure 21-3 shows typical voltage and current waveforms for an *RL* circuit, along with the waveform of power in the circuit. The power waveform is derived in the usual way by multiplying together instantaneous values of voltage and current. During the period from t_1 to t_2, both i and e are positive quantities; therefore, power p remains positive throughout that time. From t_2 to t_3, i is positive while e is a negative quantity; consequently, the product of i and e is negative, and the power waveform is below the zero line. After t_3, the current and

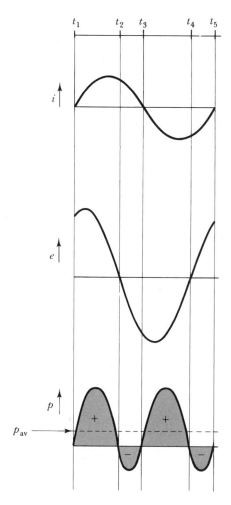

FIGURE 21-3. The power waveform for an *LR* circuit can be derived from the voltage and current waveforms. Some power is dissipated in the resistance, so the positive and negative portions of the power waveform are unequal.

voltage are both negative until t_4. Since $(-e) \times (-i)$ is a positive quantity, from t_3 to t_4 the power is positive again.

The power waveform clearly shows that more positive power than negative power is supplied to the circuit. This is to be expected, of course, since power is dissipated in the resistance, while the average power supplied to the inductance remains at zero. The average power supplied to the circuit can be represented by the broken line shown on the power waveform.

Now consider the phasor diagram for a series RL circuit, as reproduced in Figure 21-4. The current I is shown lagging the applied voltage E by the angle ϕ. The voltage across the resistance is E_R and it is in phase with I, so E_R also lags E by angle ϕ. The voltages across the inductance is E_L, and it leads the current by an angle of 90°. The phasor sum of E_R and E_L gives the supply voltage E, as illustrated.

The true power dissipated in the RL circuit is, of course, the power dissipated in the resistor. Therefore, true power,

$$P = E_R \times I$$

and from Figure 21-4,

$$E_R = E \cos \phi$$

giving

$$P = (E \cos \phi) \times I$$

or

$$\boxed{P = EI \cos \phi} \tag{21-7}$$

The reactive power is the product of the inductive voltage and current, E_L and I_L. From Figure 21-4,

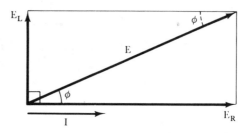

FIGURE 21-4. Phasor diagram for an LR circuit. Because power is dissipated only in the resistive component, the circuit power is $P = I \times$ (*voltage in phase with I*), or $P = EI \cos \theta$.

$$E_L = E \sin \phi$$

and $$I_L = I$$

Therefore, $$Q = (E \sin \phi) \times I$$

or

$$\boxed{Q = EI \sin \phi} \tag{21-8}$$

If the supply voltage E and the measured current I in an RL circuit were multiplied together, the product would be neither the true power nor the reactive power. However, it would give a quantity that might appear to be the power supplied to the circuit. The term *apparent power* (symbol S) is applied to this quantity, and units of apparent power are *volts-amps* (VA):

$$\boxed{S = EI} \tag{21-9}$$

where S is the apparent power and E and I are rms values of supply voltage and current.

EXAMPLE 21-3   In a series RL circuit supplied with 50 V, the current is measured as 100 mA with a phase angle of 25°. Calculate the apparent power, reactive power, and true power supplied to the circuit.

SOLUTION

Equation (21-9):

$$S = EI \qquad \text{volt-amp}$$
$$= 50 \text{ V} \times 100 \text{ mA}$$
$$\mathbf{= 5 \ VA}$$

Equation (21-8):

$$Q = EI \sin \phi \text{ var}$$
$$= 50 \text{ V} \times 100 \text{ mA} \times (\sin 25°)$$
$$\cong \mathbf{2.1 \ var}$$

Equation (21-7):

$$P = EI \cos \phi \text{ watts}$$
$$= 50 \text{ V} \times 100 \text{ mA} \times (\cos 25°)$$
$$\cong \mathbf{4.5 \ W}$$

RC CIRCUIT POWER. When the current and voltage waveforms in a series *CR* circuit are employed to derive the waveform of power supplied (Figure 21-5), it is seen that a similar result is obtained as in the case of the *LR* circuit. The average power supplied to the circuit is a positive quantity, and this represents the power dissipated in the resistance. The phasor diagram for the *CR* circuit, as shown in Figure 21-6, gives the same equations for true power, reactive power, and apparent power as those derived for the *LR* current.

The power supplied to *parallel CR* and *LR* circuits can also be resolved into true power, reactive power, and apparent power components. Using the supply voltage and current, and the current-voltage phase angle, Equations (21-7), (21-8), and (21-9) are just as valid for

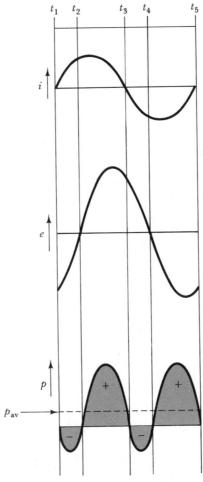

FIGURE 21-5. The power waveform for a *CR* circuit can be derived from the voltage and current waveforms. Some power is dissipated in the resistance, so the positive and negative portions of the power waveform are unequal.

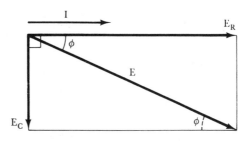

FIGURE 21-6. Phasor diagram for a *CR* circuit. Because power is dissipated only in the resistive component, the circuit power is *P = I × (voltage in phase with I)*, or *P = EI cos* φ.

parallel circuits as they are for series circuits. However, in the next section it is shown that capacitors connected in parallel with a load can be employed in correcting problems that arise because of the phase difference between supply current and voltage.

EXAMPLE 21-4   A series circuit consisting of $R = 1.2$ kΩ and $C = 0.1$ μF is supplied with 45 V at a frequency of 1 kHz. Determine the apparent power, true power, and reactive power in the circuit.

SOLUTION

$$X_C = \frac{1}{2\pi fc}$$

$$= \frac{1}{2 \times \pi \times 1 \text{ kHz} \times 0.1 \text{ μF}}$$

$$= 1.59 \text{ kΩ}$$

$$|Z| = \sqrt{R^2 + X_C^2}$$

$$= \sqrt{(1.2 \text{ kΩ})^2 + (1.59 \text{ kΩ})^2}$$

$$= 1.99 \text{ kΩ}$$

$$\phi = \arctan \frac{X_C}{R}$$

$$= \arctan \left(\frac{1.59 \text{ kΩ}}{1.2 \text{ kΩ}} \right)$$

$$\cong 53°$$

$$|I| = \frac{E}{|Z|} = \frac{45 \text{ V}}{1.99 \text{ kΩ}}$$

$$= 22.6 \text{ mA}$$

apparent power, $S = EI = 45 \text{ V} \times 22.6 \text{ mA}$

$$\cong 1 \text{ VA}$$

---

true power, $P = EI \cos \phi$

$= 45 \text{ V} \times 22.6 \text{ mA} \times (\cos 53°)$

$= \mathbf{0.61 \text{ W}}$

reactive power, $Q = EI \sin \phi$

$= 45 \text{ V} \times 22.6 \text{ mA} \times (\sin 53°)$

$= \mathbf{0.81 \text{ var}}$

Note that since the true power is the power dissipated in the resistor, it can also be calculated as I^2R, Equation (21-3):

$$P = I^2 R$$

$$= (22.6 \text{ mA})^2 \times 1.2 \text{ k}\Omega$$

$$= \mathbf{0.61 \text{ W}}$$

Also note that the reactive power can be calculated as I^2X_C, Equation (21-5):

$$Q = I^2 X_C$$

$$= (22.6 \text{ mA})^2 \times 1.59 \text{ k}\Omega$$

$$= \mathbf{0.81 \text{ var}}$$

PRACTICE PROBLEMS

21-5.1   A series LR circuit with a 3.3 V, 250 Hz supply has $R = 1.8 \text{ k}\Omega$. The circuit current is measured as 1.5 mA lagging the input voltage by 35.1°. Determine the inductance value.

21-5.2   Calculate the apparent power, reactive power, and true power supplied to the circuit in Problem 21-5.1.

21-5.3   A series CR circuit with a 24 V, 400 Hz supply has $R = 1.2 \text{ k}\Omega$ and $C = 0.2 \text{ }\mu\text{F}$. Calculate the apparent power, reactive power, and true power in the circuit.

21-6
POWER FACTOR

POWER FACTOR EQUATION.   Ideally, all the supply voltage and current should be converted into true power in a load. When this is not the case, a certain kind of inefficiency occurs. The ratio of *true power* to *apparent power* is termed the *power factor* of the load,

$$\text{power factor} = \frac{\text{true power}}{\text{apparent power}} = \frac{P}{EI}$$

As explained in the previous section, the true power in a load is calculated from Equation (21-7) as

$$P = EI \cos \phi$$

where E and I are the supply voltage and current and ϕ is the phase angle between them. Equation (21-7) can be rewritten to give

$$\cos \phi = \frac{P}{EI} = \frac{\text{true power}}{\text{apparent power}}$$

Therefore,

$$\boxed{\text{power factor} = \cos \phi = \frac{P}{EI}} \qquad \text{(21-10)}$$

If the power factor was 1, the phase angle ϕ would be zero and all of $(E \times I)$ from the supply would be dissipated as true power in the load. Where ϕ is greater than zero, $\cos \phi$ is less than 1 and only a portion of supply $(E \times I)$ is converted into useful power. Thus, it is seen that the power factor always has a maximum value of 1, and is normally less than 1.

The power factor can be expressed as a ratio or as a percentage, and is also usually defined as *leading* or *lagging*. A 60% *lagging power factor* implies an inductive load in which the supply current lags the voltage by an angle with a cosine of 0.6 (i.e., by approximately 53°). A *90% leading power factor* would indicate a capacitive load in which the current leads the voltage by an angle with a cosine of 0.9, or approximately 26°.

POWER FACTOR CORRECTION. Most industrial loads consist of electric motors. Thus they are inductive and have lagging power factors. Consider the situation illustrated in Figure 21-7(a), and assume that the supply voltage and current are measured as $E = 120$ V and $I = 100$ A, with the current lagging the voltage by an angle of 33.5°. Using the appropriate equations, the apparent power, true power, and reactive power are calculated as:

apparent power,
$$S = EI$$
$$= 120 \text{ V} \times 100 \text{ A}$$
$$= \textbf{12 kVA}$$

true power,
$$P = EI \cos \phi$$
$$= 12 \text{ kVA} \times (\cos 33.5°)$$
$$= \textbf{10 kW}$$

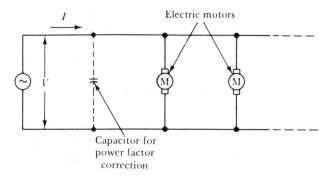

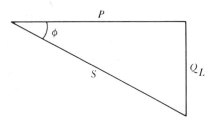

(a) Generator with an inductive load

(b) Power triangle for an inductive load

FIGURE 21-7. Because electric motors are inductive, the supply current lags the supply voltage; they are said to have a *lagging power factor*. A power triangle can be drawn to show the relationship between *true power P, apparent power S,* and *reactive power Q_L*.

inductive reactive power, $Q_L = EI \sin \phi$

$$= 12 \text{ kVA} \times (\sin 33.5°)$$

$$\mathbf{= 6.6 \text{ kvar}}$$

A *power triangle* can be drawn for the circuit, as illustrated in Figure 21-7(b). The true power P is represented by a horizontal vector, and the vector for apparent power S is drawn lagging the true power by the phase angle ϕ. The reactive power vector Q_L completes the triangle.

If the power factor were 1 (i.e., $\phi = 0$), then to supply a power of 10 kW to the load would require a current of

$$I = \frac{P}{E} = \frac{10 \text{ kW}}{120 \text{ V}}$$

$$I \cong 83 \text{ A}$$

This shows that if the power factor could be increased to unity (ϕ reduced to zero), the generators supplying the load would have to produce a current of only 83 A instead of 100 A, while still supplying the required amount of true power. Also, the conducting cables could be selected to carry 83 A instead of 100 A, and consequently they would be less expensive. Thus it is always best to have a power factor as near unity as possible, and in fact the power factor of an inductive load can be adjusted toward unity. The process is known as *power factor correction*.

Power factor correction for an inductive load consists simply of connecting capacitance in parallel with the load. Suppose that the 100 A load discussed above had been capacitive instead of inductive. The current would then lead the supply voltage by the phase angle, and the power triangle would be drawn as illustrated in Figure 21-8(a). The *apparent power* vector S is shown leading the *true power* vector P by the angle ϕ, and the *capacitive reactive power* vector Q_C is drawn vertically *up* from the true power vector, as shown.

When capacitance is connected in parallel with an inductive load, the power triangle has a *capacitive reactive power* component as well as *inductive reactive power*. The diagram in Figure 21-8(b) illustrates the situation. The capacitive reactive power is represented by the vector Q_C, drawn vertically *up* while Q_L drawn *down* represents the inductive reactive power. The net reactive power is the difference between Q_L and Q_C, and in the diagram it is so small that it gives a very small phase angle ϕ, which results in a near-unity power factor. Unity power factor would, of course, be achieved when Q_C and Q_L are equal. *This is to be avoided because it would create a state of resonance (see Chapter 23).*

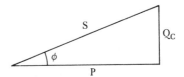

(a) Power triangle for a capacitive load

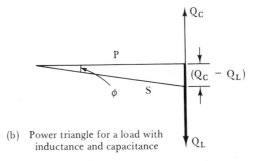

(b) Power triangle for a load with inductance and capacitance

FIGURE 21-8. A capacitive load has a *leading power factor*; its reactive power vector Q_C is drawn vertically up from the vector for the true power. Capacitance can be used to partially correct a lagging power factor.

EXAMPLE 21-5 The current taken from a 115 V, 60 Hz supply is measured as 20 A with a lagging power factor of 75%. Calculate the apparent power, true power, and reactive power. Also, determine the amount of capacitance that must be connected in parallel with the load to correct the power factors to 95% lagging.

SOLUTION

$$power\ factor = 75\%$$

or
$$\cos \phi = 0.75$$

$$\phi = \arccos 0.75$$

$$= 41.4°$$

$$apparent\ power,\ S = EI$$

$$= 115\ V \times 20\ A$$

$$\mathbf{= 2.3\ kVA}$$

$$true\ power,\ P = EI \cos \phi$$

$$= 2300\ VA \times 0.75$$

$$\mathbf{= 1.725\ kW}$$

$$reactive\ power,\ Q_L = EI \sin \phi$$

$$= 2300\ VA \times (\sin 41.4°)$$

$$\mathbf{= 1.52\ kvar}$$

For a 95% power factor,

$$\cos \phi = 0.95$$

and
$$\phi = \arccos 0.95$$

$$= 18.2°$$

true power remains,

$$P = 1.725\ kW$$

and
$$P = EI \cos \phi$$

new value of

$$S = EI = \frac{P}{\cos \phi}$$

$$EI = \frac{1.725\ kW}{0.95}$$

$$= 1.82\ kVA$$

and new value of

$$Q = EI \sin \phi$$
$$= 1.82 \text{ kVA} \times (\sin 18.2°)$$
$$= 568 \text{ var}$$

and
$$Q = Q_L - Q_C$$

giving
$$Q_C = Q_L - Q$$
$$= 1.52 \text{ kvar} - 568 \text{ var}$$
$$= 952 \text{ var}$$

From Equation (21.6):

$$\frac{E^2}{X_C} = Q_C$$

or,
$$X_C = \frac{(115 \text{ V})^2}{952 \text{ var}}$$
$$= 13.9 \ \Omega$$

$$X_C = \frac{1}{2\pi fC}$$

$$C = \frac{1}{2\pi f X_C}$$
$$= \frac{1}{2 \times \pi \times 60 \text{ Hz} \times 13.9 \ \Omega}$$
$$= 191 \ \mu\text{F}$$

PRACTICE PROBLEMS

21-6.1 An industrial load which uses a 120 V, 60 Hz supply has a 1000 μF power factor correction capacitor. The measured power is 23 kW with a power factor of 0.93. Calculate the power factor with the capacitor removed.

21-6.2 For Problem 21-6.1, determine the supply current with the capacitor connected and without the capacitor connected.

21-7
POWER MEASURE-MENT IN DECIBELS

The *change* in output from an amplifier, signal generator, or other electronics equipment is measured in *decibels* (dB). The decibel is one-tenth of the unit known as the *Bel**, which is inconveniently large. When the power output changes from P_1 to P_2, the change A_p (which may

---

* Named in honor of the Scottish-born inventor of the telephone, Alexander Graham Bell (1847–1922).

be a gain or an attenuation) is expressed as the log of the ratio of P_2 to P_1:

$$A_P = \log_{10}(P_2/P_1) \; Bel$$

or,

$$\boxed{A_P = 10 \log_{10}(P_2/P_1) \; \text{dB}} \tag{21-11}$$

If P_1 and P_2 are dissipated in a load resistor, R_L, the measured output voltages or currents can be employed to calculate the power change:

$$P_1 = V_1^2/R_L \quad \text{and} \quad P_2 = V_2^2/R_L$$

or

$$P_1 = I_1^2 R_L \quad \text{and} \quad P_2 = I_2^2 R_L$$

Substituting the voltages into Equation (21-11),

$$A_P = 10 \log_{10}\left[\frac{V_2^2/R_L}{V_1^2/R_L}\right] \text{dB}$$

$$= 10 \log_{10}(V_2^2/V_1^2) \; \text{dB}$$

or

$$\boxed{A_P = 20 \log_{10}(V_2/V_1) \; \text{dB}} \tag{21-12}$$

Substituting the currents into Equation (21-11),

$$\boxed{A_P = 20 \log_{10}(I_2/I_1) \; \text{dB}} \tag{21-13}$$

When power levels P_1 and P_2 are directly measured, the output power change in decibels is calculated by the use of Equation (21-11). When voltage or current measurements are made, Equation (21-12) or (21-13) is employed.

EXAMPLE 21-6   Two output voltage levels from an amplifier are measured as $V_1 = 1$ V and $V_2 = 2$ V. Calculate the output power change. Also, determine the new level for V_2 which will give a power output change of -3 dB when $V_1 = 1$ V.

SOLUTION

Equation (21-12):

$$A_P = 20 \log_{10}(V_2/V_1)$$

$$= 20 \log_{10}(2 \text{ V}/1 \text{ V})$$

$$\cong 6 \text{ dB}$$

When P changes by −3 dB,

$$-3 \text{ dB} = 20 \log_{10}(V_2/V_1)$$

$$\text{antilog}(-3 \text{ dB}/20) = V_2/V_1$$

$$= V_2/1 \text{ V}$$

$$V_2 = 1 \text{ V} \times \text{antilog}(-3 \text{ dB}/20)$$

$$\cong \textbf{0.707 V}$$

Equations (21-11), (21-12), and (21-13) can be used to determine the *power gain* of one or more electronic circuits. In this case *the equations are correct only when the load resistance connected to the circuit output is equal to the circuit input resistance.* However, it is common practice in the electronics industry to apply these equations even where the input and load resistances are not equal!

A major advantage of measuring power gains in decibels is illustrated in Figure 21-9. If the power gain of amplifier 1 is X dB, and that of amplifier 2 is Y dB, the overall gain is $(X + Y)$ dB. This is more convenient than calculating each stage gain as a ratio of output power to input power, and then multiplying the two together to determine the overall gain.

When the absolute output power of a device is to be measured (rather than a change in output power), the reference level employed is 1 mW dissipated in a resistance of 600 Ω. The unit of absolute power level for this situation is known as the *decibel-milliwatt* (dBm). A signal

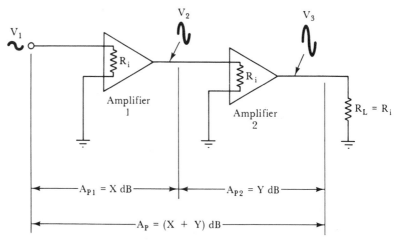

FIGURE 21-9. The power gains in decibels (dB) in each of several stages of amplification may be added together to determine the overall power gain.

generator that produces a output of 100 mW into a 600 Ω load has an absolute output power of: $10 \log_{10} (100 \text{ mW}/1 \text{ mW}) = 20$ dBm.

PRACTICE PROBLEM

21-7.1 The signal voltage levels in the two-stage amplifier illustrated in Figure 21-9 are measured as $V_1 = 10$ mV, $V_2 = 22.4$ mV, $V_3 = 126$ mV. Calculate the power gain of each stage and the overall power gain in decibels.

SUMMARY OF FORMULAS

True power:

$$P = EI \qquad \text{watt}$$

$$P = \frac{E^2}{R} \qquad \text{watt}$$

$$P = I^2 R \qquad \text{watt}$$

Reactive power:

$$Q_L = E_L I_L \quad \text{or} \quad Q_C = E_c I_c \qquad \text{var}$$

$$Q_L = I_L^2 X_L \quad \text{or} \quad Q_C = I_c^2 X_c \qquad \text{var}$$

$$Q_L = \frac{E_L^2}{X_L} \quad \text{or} \quad Q_C = \frac{E_c^2}{X_c} \qquad \text{var}$$

True power:

$$P = EI \cos \phi \qquad \text{watt}$$

Reactive power:

$$Q = EI \sin \phi \qquad \text{var}$$

Apparent power:

$$S = EI \text{ volt-amp}$$

Power factor:

$$\cos \phi = \frac{P}{EI}$$

Power change:

$$A_P = 10 \log_{10} (P_2/P_1) \text{ dB}$$

$$A_P = 20 \log_{10} (V_2/V_1) \text{ dB}$$

$$A_P = 20 \log_{10} (I_2/I_1) \text{ dB}$$

REVIEW QUESTIONS

21-1 Sketch typical waveforms of current and voltage for a pure resistance connected in an alternating current circuit. From the voltage and current waveforms, derive the waveform of instantaneous power dissipation. Explain briefly.

21-2 Sketch typical waveforms of current and voltage for a pure inductance with an ac supply. Derive the waveform for instantaneous power. Briefly explain.

21-3 Sketch typical waveforms of current and voltage for a pure capacitance with an ac supply. Derive the waveform for instantaneous power. Briefly explain.

21-4 Sketch typical current and voltage waveforms for a series *RL* circuit with an ac supply. Derive the waveform for instantaneous power. Briefly explain.

21-5 Sketch typical current and voltage waveforms for a series *RC* circuit with an ac supply. Derive the waveform for instantaneous power. Briefly explain.

21-6 Define reactive power, apparent power, and true power. Write equations for each.

21-7 Define power factor, write an equation for it, and discuss its importance.

21-8 Sketch typical power triangles for a series *LR* circuit and for a series *CR* circuit. Show the effect that a parallel capacitor can have on the power triangle for a series *LR* circuit.

21-9 Explain power measurement in decibels, and write the equation for calculating the power change when measurement is made of: (1) output power levels, (2) output voltages, and (3) output currents. Explain the unit known as the decibel-milliwatt.

PROBLEMS

SECTION 21-1

21-1 A 2 kW heating element is supplied from a 220 V ac source. Determine (a) the resistance of the element, (b) the current level, (c) the peak power dissipated in the element.

21-2 Determine the resistance of a 60 W lamp that is supplied from a 50 V ac source. Also, calculate the current and the peak power dissipated in the lamp.

SECTION 21-2

21-3 Calculate the peak instantaneous power dissipated in a 300 mH pure inductance which is connected to a 115 V, 60 Hz source.

21-4 A 56 mH pure inductance has a 60 mA, 400 Hz current. Calculate the peak instantaneous power dissipated in the inductor.

SECTION 21-3

21-5 Determine the peak instantaneous power dissipated in a 0.1 μF pure capacitance which is connected to a 115 V, 60 Hz source.

21-6 A 100 pF pure capacitance has a supply of 0.9 V at a frequency of 1 MHz. Calculate the maximum instantaneous power dissipated in the capacitor.

SECTION 21-4

21-7 Calculate the power (true and reactive) supplied by a 240 V, 100 Hz source when it is connected to (a) a 200 mH inductor, (b) a 100 Ω resistor, (c) a 100 μF capacitor.

21-8 Three components, $L = 400$ μH, $C = 300$ pF, and $R = 1.8$ kΩ, are connected in parallel to a 3 V, 200 kHz supply. Calculate the power (true and reactive) supplied to each component.

SECTION 21-5

21-9 Calculate the apparent power, true power, and reactive power dissipated in a series LR circuit when the supply voltage is 75 V, the current is 1 A, and the phase angle is 30° lagging. Also, draw a power triangle for the circuit.

21-10 Determine the apparent power, true power, and reactive power in the circuit described in Problem 19-9.

21-11 A 200 mH inductor is connected in series with a 600 Ω resistor. The ac supply voltage is 15 V with a frequency of 1 kHz. Determine the apparent power, true power, and reactive power in the circuit. Draw the power triangle.

21-12 The supply frequency for the circuit in Problem 21-11 is altered until the true power is measured as 50 mW. Calculate the new frequency and determine the new apparent power and reactive power.

21-13 A 50 V, 400 Hz supply is applied to a load consisting of a 25 μF capacitor in series with a 4.7 Ω resistor. Determine the apparent power, true power, and reactive power in the circuit. Draw the power triangle.

21-14 The supply frequency for the circuit in Problem 21-13 is altered until the true power is measured as 30 W. Calculate the new frequency and determine the new apparent power and reactive power.

21-15 Determine the apparent power, true power, and reactive power for the circuit described in Problem 19-13.

21-16 Determine the apparent power, true power, and reactive power for the circuit described in Problem 19-23.

SECTION 21-6

21-17 The current taken by a certain load connected to a 220 V, 100 Hz supply is measured as 4 A with a power factor of 69%

lagging. Determine the apparent power, true power, and reactive power. Draw the power triangle for the circuit.

21-18 For the load in Problem 21-17, calculate the parallel capacitance required to correct the power factor to 97% lagging. Draw the new power triangle for the circuit.

21-19 An inductive circuit with a 250 μF parallel-connected capacitor has a power factor of 93% lagging. The supply voltage is 120 V with a frequency of 60 Hz, and the supply current is 15 A. Determine the apparent power, true power, and reactive power.

21-20 For the circuit in Problem 21-19, determine the apparent power, true power, and reactive power when the capacitor is disconnected.

21-21 A 24 V, 400 Hz supply is connected to a 4 kW load with a 65% lagging power factor. Calculate the current that must be carried by the conductors. Also, calculate the new level of conductor current when the power factor is corrected to 85% lagging and determine the capacitance required for power factor correction.

21-22 For Problem 21-21, calculate the capacitor required to reduce the conductor current to 185 A.

21-23 A 120 mH inductor is connected in series with resistance of 70 Ω to a 50 V, 400 Hz supply. Determine the power dissipated in the circuit. Also, calculate the load current when a 1 μF capacitor is connected in parallel with the load.

21-24 For Problem 21-23, calculate the capacitor required to reduce the power factor to 0.87. Determine the new level of load current.

21-25 A 230 V, 120 Hz supply is connected to a 12 kW load with an 80% lagging power factor via conductors that can carry a maximum current of 58 A. Calculate the actual current level in the conductors if the load remains unaltered. Determine how the load should be changed to reduce the conductor current to 58 A. Draw the power triangle for the circuit.

21-26 For Problem 21-25, determine the conductor current that flows when a 75 μF power factor correction capacitor is employed.

SECTION 21-7

21-27 Two output power levels from a signal generator are measured as $P_1 = 5$ mW and $P_2 = 80$ mW. Determine the power change in dB from P_1 to P_2. Also, calculate the absolute output power levels in dBm if $R_L = 600\ \Omega$.

21-28 The following input and output voltages are measured in a three-stage electronic amplifier: $V_{i1} = 1.5$ mV, $V_{o1} = 45$ mV, $V_{o2} = 600$ mV, $V_{o3} = 7$ V. Calculate the power gain of each stage and the overall power gain in decibels.

21-29 An amplifier with 600 Ω input and load resistances has a signal input of 7 mV and an output of 250 mV. Calculate the power gain and determine the new voltage gain required to give an absolute output power of 30 dBm with a 7 mV input.

21-30 An electronic amplifier with an input resistance of 1 MΩ has an input current of 200 nA. The output current into a 1 MΩ load is measured as 10 μA. Calculate the power gain in dB and determine the new level of output current when the power gain is reduced by 3 dB.

COMPUTER PROBLEM

21-31 Write a computer program to solve the type of problem presented in Example 21-5. The program should accept any given supply voltage, frequency, load current, power factor, and corrected power factor. The apparent power, true power, reactive power, and required power factor correction capacitance should all be printed out.

ANSWERS TO PRACTICE PROBLEMS	**21-1.1**	98 mW, 133 mW, 161 mW, 784 mW
	21-2.1	106 mW, −106 mW
	21-4.1	25.1 mvar, 398 mvar
	21-5.1	809 mH
	21-5.2	4.95 mVA, 2.85 mvar, 4.05 mW
	21-5.3	247 mVA, 212 mvar, 127 mW
	21-6.1	0.845
	21-6.2	206 A, 227 A
	21-7.1	7 dB, 15 dB, 22 dB

22 AC NETWORK ANALYSIS

Objectives   You will be able to:

State Kirchhoff's voltage and current laws as applied to impedance networks.

Analyze complex ac networks by use of loop equations.

State the superposition theorem as applied to ac networks. Apply the superposition theorem to ac network analysis.

Solve complex ac networks by the use of nodal analysis.

State Thévenin's theorem as applied to ac network analysis. Apply Thévenin's theorem to ac network analysis.

State Norton's theorem as applied to ac network analysis. Apply Norton's theorem to ac network analysis.

State the maximum power transfer theorem as applied to ac networks. Apply the maximum power transfer theorem to determine optimum loads for ac sources.

Apply delta-wye transformations to simplify the analysis of ac networks.

Introduction

The analysis techniques and theorems employed to solve dc network problems in Chapters 8 and 9 can also be applied to ac networks. However, all impedances, voltages, and currents must be treated as phasor quantities. Consequently, there are a lot more calculations involved in analyzing an ac impedance network than for a similar dc resistance circuit. As in the case of purely resistive circuits, the calculations can be greatly simplified by application of the various network theorems.

22-1

AC SOURCES AND KIRCHHOFF'S LAWS FOR AC CIRCUITS

In analyzing an ac network, it is necessary to know the phase angle of each source as well as the amplitude of the voltage or current generated at each source. Also, although alternating voltages are continuously reversing polarity, + and − terminals must be identified at each generator. This is necessary because if a generator is connected in reverse, the phase angle of the generator output is altered by 180°.

Figure 22-1(a) shows a circuit that has two voltage sources: $E_1 = 6$ V $\underline{/0°}$ and $E_2 = 12$ V $\underline{/-20°}$. The waveforms in the illustration show the phase relationship between these two sources. Because the phase angle of E_2 is $-20°$ and that of E_1 is $0°$, the zero level of E_2 occurs 20° after the zero level of E_1, as illustrated. Now assume that the generator which produces E_2 is reconnected in reverse, as shown in Figure 22-1(b). Then E_2 still has its zero level 20° after E_1, but instead of increasing positively from zero, its output increases in the *negative* direction, as illustrated. Thus, the reversal of the output terminals has added a further 180° to the phase of E_2 with respect to E_1. It is for this reason that the terminals of all ac sources in a circuit diagram must have a polarity identification.

Kirchhoff's Voltage Law can be stated for ac circuits as follows:

Kirchhoff's Voltage Law (ac circuits)

In any closed electric circuit, the phasor sum of the voltage drops must equal the phasor sum of the applied voltages.

The statement differs from that made for dc networks (see Section 5-2) only in that it refers to the *phasor sum* of the voltages instead of the *algebraic sum*. The same remarks apply to the currents referred to in Kirchhoff's Current Law (Section 6-1 for dc), which for ac circuits is stated thus:

Kirchhoff's Current Law (ac circuits)

The phasor sum of the currents entering a point in an electric circuit must equal the phasor sum of the currents leaving that point.

It is obvious that in solving ac networks, the phase angles of all impedances, and of all voltages and currents, must be carefully considered. Where two quantities are to be multiplied or divided, they should be stated in polar form. Where they are to be added or subtracted, they must be converted into rectangular form.

22-2

AC CIRCUIT LOOP EQUATIONS

The procedure for analysis of ac networks by loop currents is exactly the same as that for dc networks, with the exception that impedances are being dealt with instead of resistances. Loop currents are drawn first, usually in a clockwise direction, as for dc circuits. When the analysis is complete, those branch currents that come out as positive quantities are (instantaneously) in the same direction as that selected for the loop

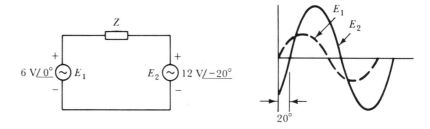

(a) Two signal sources 20° out of phase

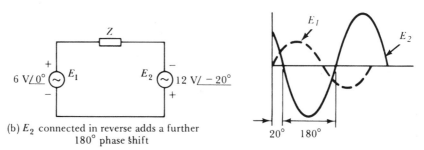

(b) E_2 connected in reverse adds a further 180° phase shift

FIGURE 22-1. All ac sources must have a terminal polarity identification to indicate whether the output is initially increasing in a positive direction or in a negative direction. This is necessary to define the phase relationships between two or more sources.

currents. The branch currents that have negative signs have an additional 180° phase shift in relation to the loop currents.

PROCEDURE FOR AC NETWORK ANALYSIS BY LOOP EQUATIONS

1. *Convert all current sources to voltage sources.*
2. *Draw all loop currents in clockwise direction and identify them by number.*
3. *Identify all impedance voltage drops as + to − in the direction of the loop current.*
4. *Identify all voltage sources with their correct polarity.*
5. *Write the equations for the voltage drops around each loop.*
6. *Solve the equations to find an equation for the required branch current.*
7. *Reduce the equations to the simplest possible form before substituting the impedance quantities into each equation.*

Because of the necessity of converting from polar to rectangular form, and vice versa, during calculations, there is much more work involved in analyzing an ac network than for a similar dc network. Consequently, the possibility of calculation errors is increased many times. To minimize the calculations, the equations for the unknown quantities should be reduced to their very simplest state. The following example illustrates the approach that should be taken.

EXAMPLE 22-1 Using loop equations, analyze the impedance network shown in Figure 22-2 to derive an equation for the current through Z_3.

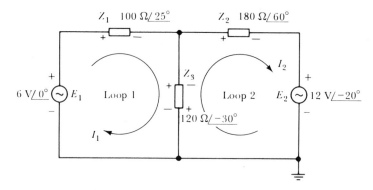

FIGURE 22-2. In the loop equation (or mesh equation) method of network analysis, the voltage drops around each circuit loop are equated to zero. For loop 1, $0 = I_1 Z_1 + I_1 Z_3 - I_2 Z_3 - E_1$.

SOLUTION

Loop currents I_1 and I_2 are drawn clockwise, as shown on the circuit diagram, and voltage drops are identified as $+$ to $-$ in the direction of the loop current. For loop 1:

$$E_1 = I_1(Z_1 + Z_3) - I_2 Z_3 \qquad (1)$$

For loop 2:

$$0 \text{ V} = I_2(Z_2 + Z_3) - I_1 Z_3 + E_2$$

Therefore,

$$-E_2 = I_2(Z_2 + Z_3) - I_1 Z_3 \qquad (2)$$

Equation (1) $\times \dfrac{(Z_2 + Z_3)}{Z_3}$:

$$\frac{E_1(Z_2 + Z_3)}{Z_3} = \frac{I_1(Z_1 + Z_3)(Z_2 + Z_3)}{Z_3} - I_2(Z_2 + Z_3) \qquad (3)$$

Adding Equation (2) to Equation (3) to eliminate I_2 gives

$$\frac{E_1(Z_2 + Z_3)}{Z_3} - E_2 = I_1 \left[\frac{(Z_1 + Z_3)(Z_2 + Z_3)}{Z_3} - Z_3 \right] \qquad (4)$$

Multiplying through by Z_3:

$$E_1(Z_2 + Z_3) - E_2Z_3 = I_1[(Z_1 + Z_3)(Z_2 + Z_3) - Z_3^2]$$

and
$$I_1 = \frac{E_1(Z_2 + Z_3) - E_2Z_3}{(Z_1 + Z_3)(Z_2 + Z_3) - Z_3^2}$$

$$= \frac{E_1Z_2 + E_1Z_3 - E_2Z_3}{Z_1Z_2 + Z_1Z_3 + Z_2Z_3 + Z_3^2 - Z_3^2}$$

$$= \frac{E_1Z_2 + E_1Z_3 - E_2Z_3}{Z_1Z_2 + Z_1Z_3 + Z_2Z_3} \tag{5}$$

Equation (2) $\times \dfrac{Z_1 + Z_3}{Z_3}$:

$$-\frac{E_2(Z_1 + Z_3)}{Z_3} = \frac{I_2(Z_2 + Z_3)(Z_1 + Z_3)}{Z_3} - I_1(Z_1 + Z_3) \tag{6}$$

Adding Equation (1) *to* Equation (6) *to eliminate I_1 gives*

$$E_1 - \frac{E_2(Z_1 + Z_3)}{Z_3} = I_2\left[\frac{(Z_2 + Z_3)(Z_1 + Z_3)}{Z_3} - Z_3\right]$$

Multiplying through by Z_3:

$$E_1Z_3 - E_2(Z_1 + Z_3) = I_2[(Z_2 + Z_3)(Z_1 + Z_3) - Z_3^2]$$

and
$$I_2 = \frac{E_1Z_3 - E_2(Z_1 + Z_3)}{(Z_2 + Z_3)(Z_1 + Z_3) - Z_3^2}$$

$$= \frac{E_1Z_3 - E_2Z_1 - E_2Z_3}{Z_1Z_2 + Z_2Z_3 + Z_1Z_3 + Z_3^2 - Z_3^2}$$

$$= \frac{E_1Z_3 - E_2Z_1 - E_2Z_3}{Z_1Z_2 + Z_1Z_3 + Z_2Z_3} \tag{7}$$

current through Z_3,

$$I_3 = I_1 - I_2$$

$$= \frac{E_1Z_2 + E_1Z_3 - E_2Z_3 - E_1Z_3 + E_2Z_1 + E_2Z_3}{Z_1Z_2 + Z_1Z_3 + Z_2Z_3}$$

$$= \frac{E_1Z_2 + E_2Z_1}{Z_1Z_2 + Z_1Z_3 + Z_2Z_3} \tag{8}$$

The equation above for the current through Z_3 is stated in its simplest possible form. The component values for the equation can now be calculated separately, and then used for determination of I_3.

EXAMPLE 22-2 Substitute the component values into the equation derived in Example 22-1 to calculate the current through Z_3.

SOLUTION

$$Z_1 = 100 \ \Omega \underline{/25°} = 90.6 \ \Omega + j42.3 \ \Omega$$

$$Z_2 = 180 \ \Omega \underline{/60°} = 90 \ \Omega + j156 \ \Omega$$

$$Z_3 = 120 \ \Omega \underline{/-30°} = 104 \ \Omega - j60 \ \Omega$$

$$E_1 = 6 \ V \underline{/0°} = 6 \ V + j0 \ V$$

$$E_2 = 12 \ V \underline{/-20°} = 11.3 \ V - j4.1 \ V$$

$$Z_1 Z_2 = 18 \ k\Omega \underline{/85°} = 1.57 \ k\Omega + j17.9 \ k\Omega$$

$$Z_1 Z_3 = 12 \ k\Omega \underline{/-5°} = 11.95 \ k\Omega - j1.05 \ k\Omega$$

$$Z_2 Z_3 = 21.6 \ k\Omega \underline{/30°} = 18.7 \ k\Omega + j10.8 \ k\Omega$$

$$Z_1 Z_2 + Z_1 Z_3 + Z_2 Z_3 = 32.22 \ k\Omega + j27.65 \ k\Omega$$

$$= 42.46 \ k\Omega \underline{/40.6°}$$

$$E_1 Z_2 = 1080 \underline{/60°} = 540 + j935$$

$$E_2 Z_1 = 1200 \underline{/5°} = 1195 + j105$$

$$E_1 Z_2 + E_2 Z_1 = 1735 + j1040$$

$$= 2023 \underline{/30.9°}$$

Substituting into Equation (8):

$$I_3 = \frac{2023 \underline{/30.9°}}{42.46 \times 10^3 \underline{/40.6°}}$$

$$= \mathbf{47.6 \ mA \underline{/-9.7°}}$$

Although the circuit in Figure 22-2 is not an extremely complex network, the analysis of it (in Examples 22-1 and 22-2) is quite lengthy. In following sections it is shown that the analysis can be simplified by application of one of the circuit theorems. In general, the solution of more complex networks requires use of the circuit theorems.

PRACTICE PROBLEMS

22-2.1 For the circuit in Figure 22-2, use loop equations to determine I_2 when $E_2 = 0$.

22-2.2 For the circuit in Figure 22-2, use loop equations to determine I_1 when $E_1 = 0$.

22-3

SUPER-POSITION THEOREM APPLIED TO AC NETWORKS

The superposition theorem allows a complex network involving several voltage and/or current sources to be reduced to several simpler networks each of which has only one voltage or current source. For application to ac networks, the *superposition theorem* is stated:

Superposition Theorem (ac circuits)

In a network containing more than one source of voltage or current, the current through any branch is the phasor sum of the currents produced by each source acting independently.

PROCEDURE FOR DETERMINING AN AC NETWORK BRANCH CURRENT BY THE SUPERPOSITION THEOREM:

1. *Select one source, and replace all other sources with their internal impedances.*

2. *Determine the amplitude and phase angle of the current that flows through the desired branch as a result of the single source acting alone.*

3. *Repeat steps 1 and 2 using each source in turn until the branch current components have been calculated for all sources.*

4. *Determine the phasor sum of the current components to obtain the actual branch current.*

EXAMPLE 22-3 Using the superposition theorem, analyze the impedance network shown in Figure 22-2 to determine the current through Z_3.

SOLUTION

Selecting source E_1, and replacing source E_2 with its internal impedance, gives the circuit shown in Figure 22-3.

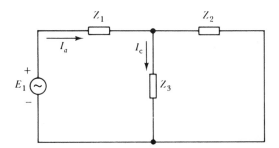

FIGURE 22-3. Application of the superposition theorem to the circuit in Figure 22-2 to determine the current through Z_3. Voltage source E_2 is first replaced by its internal impedance, then current I_c is determined.

From Figure 22-3:

$$I_a = \frac{E_1}{Z_1 + Z_2 \| Z_3}$$

$$= \frac{E_1}{Z_1 + \dfrac{Z_2 Z_3}{Z_2 + Z_3}}$$

By the current divider rule:

$$I_c = I_a \times \frac{Z_2}{Z_2 + Z_3}$$

Therefore,

$$I_c = \frac{E_1}{Z_1 + \dfrac{Z_2 Z_3}{Z_2 + Z_3}} \times \frac{Z_2}{Z_2 + Z_3}$$

$$= \frac{E_1 Z_2}{Z_1 Z_2 + Z_1 Z_3 + Z_2 Z_3} \qquad (1)$$

Selecting source E_2, and replacing source E_1 with its internal impedance, gives the circuit shown in Figure 22-4.

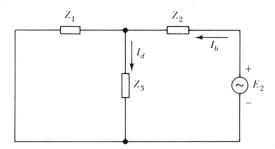

FIGURE 22-4. Continued application of the superposition theorem to Figure 22-2 to determine the current through Z_3. Voltage source E_1 is replaced by its internal impedance, and current I_d is determined. Then $I_3 = I_C + I_d$.

From Figure 22-4:

$$I_b = \frac{E_2}{Z_2 + \dfrac{Z_1 Z_3}{Z_1 + Z_3}}$$

and

$$I_d = I_b \times \frac{Z_1}{Z_1 + Z_3}$$

Therefore,

$$I_d = \frac{E_2}{Z_2 + \dfrac{Z_1 Z_3}{Z_1 + Z_3}} \times \frac{Z_1}{Z_1 + Z_3}$$

$$= \frac{E_2 Z_1}{Z_2 Z_1 + Z_2 Z_3 + Z_1 Z_3} \qquad (2)$$

$$I_3 = I_c + I_d$$

$$I_3 = \frac{E_1 Z_2 + E_2 Z_1}{Z_1 Z_2 + Z_1 Z_3 + Z_2 Z_3} \tag{3}$$

Equation (3) *is exactly the same equation obtained for* I_3 *in* Example 22-1. *Therefore, substituting the values of impedances and voltages into the equation, exactly as in* Example 22-2, *gives*

$$I_3 = 47.6 \text{ mA} \underline{/-9.7°}$$

When Examples 22-1 and 22-3 are compared, it is seen that the equation for the unknown branch current was arrived at much more quickly by the use of the superposition theorem than when the loop currents method was used. However, in the case of a more complex circuit, the superposition theorem may not simplify the derivation of the equations.

PRACTICE PROBLEMS

22-3.1 A network consists of two ac sources (E_1 and E_2 with source impedances Z_1 and Z_2) connected in parallel with a load Z_L. Use the superposition theorem to develop an equation for the current from source E_2.

22-3.2 For Problem 22-3.1, the source voltages and impedances are: $E_1 = 100$ mV$\underline{/0°}$, $E_2 = 70$ mV$\underline{/-38°}$, $Z_1 = 600$ $\Omega\underline{/0°}$, $Z_2 = 200$ $\Omega\underline{/90°}$, $Z_L = 283$ $\Omega\underline{/-45°}$. Calculate the current from source E_2.

22-4

NODAL ANALYSIS FOR AC CIRCUITS

The procedure for nodal analysis of an impedance network is exactly the same as that for a dc resistance network, with the exception that once again the phase angle of all quantities must be taken into consideration.

PROCEDURE FOR NODAL ANALYSIS OF AC NETWORKS:

1. *Convert all voltage sources to current sources and redraw the circuit.*
2. *Identify all nodes and choose a reference node.*
3. *Write the equations for the currents flowing into and out of each node, with the exception of the reference node.*
4. *Solve the equations to determine the node voltage and the required branch currents.*

EXAMPLE 22-4 Using nodal analysis on the impedance network shown in Figure 22-2, determine the current through Z_3.

SOLUTION

Converting the voltage sources to current sources gives the circuit shown in Figures 22-5(a) *and* (b).

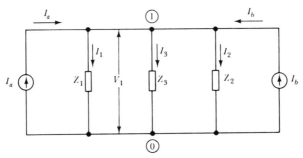

(a) Voltage sources replaced by current sources

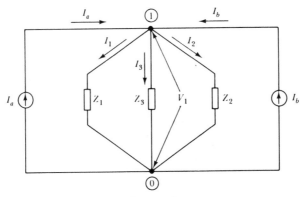

(b) Circuit redrawn to show currents entering and leaving node 1

FIGURE 22-5. For nodal analysis of a network, the voltage sources are first replaced by current sources. The voltage at each node is identified with respect to ground (node zero). Then, circuit equations are written to determine the branch currents and node voltages.

The current sources are:

$$I_a = \frac{E_1}{Z_1} = \frac{6\text{ V}\underline{/0°}}{100\ \Omega\underline{/25°}}$$

$$= 60\text{ mA}\underline{/-25°} = 54.4\text{ mA} - j25.4\text{ mA}$$

$$I_b = \frac{E_2}{Z_2} = \frac{12\text{ V}\underline{/-20°}}{180\ \Omega\underline{/60°}}$$

$$= 66.7\text{ mA}\underline{/-80°} = 11.58\text{ mA} - j65.69\text{ mA}$$

The current equation at node 1 *in* Figure 22-5(b) *is:*

$$I_a + I_b = I_1 + I_2 + I_3 \tag{1}$$

$$= \frac{V_1}{Z_1} + \frac{V_1}{Z_2} + \frac{V_1}{Z_3}$$

$$= V_1 \left(\frac{1}{Z_1} + \frac{1}{Z_2} + \frac{1}{Z_3} \right)$$

Therefore, $\quad V_1 = \dfrac{I_a + I_b}{\dfrac{1}{Z_1} + \dfrac{1}{Z_2} + \dfrac{1}{Z_3}}$ $\tag{2}$

and $\quad I_3 = \dfrac{V_1}{Z_3}$

$$I_3 = \frac{I_a + I_b}{Z_3 \left(\dfrac{1}{Z_1} + \dfrac{1}{Z_2} + \dfrac{1}{Z_3} \right)} \tag{3}$$

$$Z_1 = 100 \ \Omega \underline{/25°}$$

$$\frac{1}{Z_1} = 0.01 \ S \underline{/-25°} = (0.009 = j0.0042) \ S$$

$$Z_2 = 180 \ \Omega \underline{/60°}$$

$$\frac{1}{Z_2} = 0.0056 \ S \underline{/-60°} = (0.0028 - j0.0048) \ S$$

$$Z_3 = 120 \ \Omega \underline{/-30°}$$

$$\frac{1}{Z_3} = 0.0083 \ S \underline{/30°} = (0.0072 + j0.0042) \ S$$

$$\frac{1}{Z_1} + \frac{1}{Z_2} + \frac{1}{Z_3} = (0.019 - j0.0048) \ S$$

$$= 0.0196 \ S \underline{/-14.2°}$$

and $\quad I_a + I_b = 65.98 \ mA - j91.09 \ mA$

$$= 112.5 \ mA \underline{/-54.1°}$$

Substituting into Equation (3):

$$I_3 = \frac{112.5 \ mA \underline{/-54.1°}}{120 \ \Omega \underline{/-30°} \times 0.0196 \ S \underline{/-14.2°}}$$

$$= \mathbf{47.8 \ mA \underline{/-9.9°}}$$

Note that because of approximations in the impedance calculations, I_3 is slightly different from that determined in Example 22-2. Also note that by substitutions for I_a and I_b into Equation (3), an equation for I_3 can be derived exactly the same as that obtained in Examples 22-1 and 22-3.

The nodal analysis method of finding the current through the unknown impedance in Figure 22-2 is seen to be simpler than the use of either loop currents or the superposition theorem. Again, this method may not be the simplest for more complex circuits.

PRACTICE PROBLEM

22-4.1 Use nodal analysis to develop an equation for the current from source E_1 in the circuit in Problem 22-3.1

22-5

THÉVENIN'S THEOREM APPLIED TO AC CIRCUITS

Thévenin's theorem allows any single impedance in a network to be isolated. The rest of the network is replaced by a single impedance and a single voltage source. In this way the entire network is reduced to one voltage source in series with two impedances. Calculation of the current level through the desired impedance is then quite simple.

For application to ac networks, Thévenin's theorem is stated:

Thévenin's Theorem (ac circuits)

Any two-terminal network containing impedances and voltage and/or current sources may be replaced by a single voltage source in series with a single impedance. The emf of the voltage source is the open-circuit emf at the network terminals, and the series impedance is the impedance between the network terminals when all sources are replaced by their internal impedances.

PROCEDURE FOR THÉVENIZING AN AC NETWORK:

1. Calculate the open-circuit terminal voltage of the network.

2. Redraw the network with each voltage source replaced by a short-circuit in series with its internal impedance, and each current source replaced by an open-circuit in parallel with its internal impedance.

3. Calculate the impedance of the redrawn network as seen from the output terminals.

To apply Thévenin's theorem to the problem of finding the current through impedance Z_3 in the circuit of Figure 22-2, Z_3 is first removed as shown in Figure 22-6(a). The open-circuit terminal voltage (V_{th}) is then calculated. Next, the circuit is redrawn with the voltage sources replaced by their internal impedances [Figure 22-6(b)], and the impedance *looking into* the output terminals is calculated. Finally, as illustrated in Figure 22-6(c), the open-circuit output voltage is shown in series with the internal impedance and Z_3, and the output current is calculated.

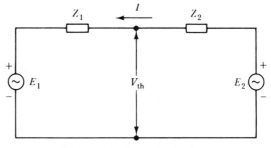

(a) Open-circuit terminal voltage

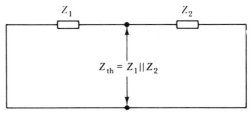

(b) Internal impedance

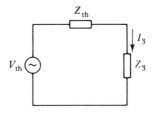

(c) Thévenin equivalent circuit

FIGURE 22-6. To thévenize an ac network, the open-circuit terminal voltage of the network is first determined. Then, with all sources replaced by their impedances, the network internal impedance is calculated.

EXAMPLE 22-5 Use Thévenin's theorem to determine the current through Z_3 in the circuit shown in Figure 22-2.

SOLUTION

Referring to Figure 22-6(a):

$$I = \frac{E_2 - E_1}{Z_1 + Z_2}$$

and
$$V_{th} = E_1 + IZ_1$$
$$= E_1 + \frac{Z_1(E_2 - E_1)}{Z_1 + Z_2}$$

$$= \frac{E_1(Z_1 + Z_2) + Z_1(E_2 - E_1)}{Z_1 + Z_2}$$

$$= \frac{E_1 Z_2 + E_2 Z_1}{Z_1 + Z_2} \qquad (1)$$

From Figure 22-6(b):

$$Z_{th} = Z_1 \| Z_2$$

$$= \frac{Z_1 Z_2}{Z_1 + Z_2} \qquad (2)$$

and from Figure 22-6(c):

$$I_3 = \frac{V_{th}}{Z_3 + Z_{th}}$$

$$= \frac{V_{th}}{Z_3 + \dfrac{Z_1 Z_2}{Z_1 + Z_2}}$$

$$= \frac{V_{th}(Z_1 + Z_2)}{Z_3(Z_1 + Z_2) + Z_1 Z_2}$$

$$= \frac{V_{th}(Z_1 + Z_2)}{Z_1 Z_2 + Z_1 Z_3 + Z_2 Z_3} \qquad (3)$$

Substituting Equation (1) *into* Equation (3):

$$I_3 = \frac{Z_1 + Z_2}{Z_1 Z_2 + Z_1 Z_3 + Z_2 Z_3} \times \frac{E_1 Z_2 + E_2 Z_1}{Z_1 + Z_2}$$

$$= \frac{E_1 Z_2 + E_2 Z_1}{Z_1 Z_2 + Z_2 Z_3 + Z_2 Z_3} \qquad (4)$$

Once again, this is the same equation for the current through Z_3 as derived in Examples 22-1 *and* 22-3. *Therefore,*

$$I_3 = 47.6 \text{ mA} \underline{/-9.7°}$$

Comparing Example 22-5 to Example 22-3, it is seen that the solution of this particular problem by Thévenin's theorem is approximately as simple (or as complex) as the use of the superposition theorem.

PRACTICE PROBLEM

22-5.1   A network consists of an ac source (E_1 with source impedance Z_1) connected in parallel with an impedance Z_2 and a load Z_L.

Chap. 22   AC Network Analysis

Use the Thévenin's theorem to develop an equation for the load current.

22-6

NORTON'S THEOREM APPLIED TO AC CIRCUITS

By application of Norton's theorem, any impedance in a network can be isolated, and the rest of the network replaced by a single current source in parallel with a single impedance.

For ac network applications, Norton's theorem is stated:

Norton's Theorem (ac circuits) **Any two-terminal network containing impedances and voltage sources, and/or current sources, may be replaced by a single current source in parallel with a single impedance. The output from the current source is the short-circuit current at the network terminals, and the parallel impedance is the impedance between the network terminals when all sources are replaced by their internal impedances.**

PROCEDURE FOR NORTONIZING AN AC NETWORK:

1. *Calculate the short-circuit current at the network terminals.*

2. *Redraw the network with each voltage source replaced by a short-circuit in series with its internal impedance, and each current source replaced by an open-circuit in parallel with its internal impedance.*

3. *Calculate the impedance of the redrawn networks as seen from the output terminals.*

EXAMPLE 22-6 Using Norton's theorem, determine the current through Z_3 in the circuit shown in Figure 22-2.

SOLUTION

From Figure 22-7(a):

$$I_1 = \frac{E_1}{Z_1}$$

and
$$I_2 = \frac{E_2}{Z_2}$$

$$I_{sc} = I_1 + I_2$$

$$= \frac{E_1}{Z_1} + \frac{E_2}{Z_2} \qquad (1)$$

From Figure 22-7(b):

$$Z_0 = Z_1 \| Z_2$$

$$= \frac{Z_1 Z_2}{Z_1 + Z_2} \qquad (2)$$

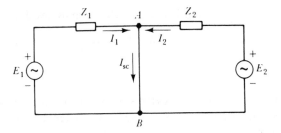

(a) Short-circuit output current

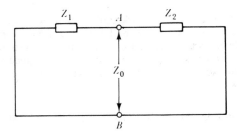

(b) Internal impedance

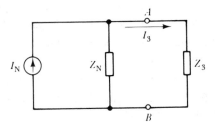

(c) Norton equivalent circuit

FIGURE 22-7. To nortonize an ac network, the short-circuit output current is first determined. Then, with all sources replaced by their impedances, the network internal impedance is calculated.

In Figure 22-7(c):

$$I_N = I_{sc} = \frac{E_1}{Z_1} + \frac{E_2}{Z_2}$$

and

$$Z_N = Z_0 = Z_1 \| Z_2$$

Using the current divider rule:

$$I_3 = \frac{I_N \times Z_N}{Z_3 + Z_N}$$

Substituting for Z_N:

$$I_3 = \frac{I_N \times \dfrac{Z_1 Z_2}{Z_1 + Z_2}}{Z_3 + \dfrac{Z_1 Z_2}{Z_1 + Z_2}}$$

$$= \frac{I_N Z_1 Z_2}{Z_1 Z_3 + Z_2 Z_3 + Z_1 Z_2} \tag{3}$$

Substituting for I_N:

$$I_3 = \frac{Z_1 Z_2 \left(\dfrac{E_1}{Z_1} + \dfrac{E_2}{Z_2} \right)}{Z_1 Z_3 + Z_2 Z_3 + Z_1 Z_2}$$

$$= \frac{E_1 Z_2 + E_2 Z_1}{Z_1 Z_3 + Z_2 Z_3 + Z_1 Z_2} \tag{4}$$

Once again this is the same equation for I_3 *as derived in other examples, and it gives*

$$I_3 = 47.6 \text{ mA} \underline{/-9.7°}$$

The use of Norton's theorem to determine the current through impedance Z_3 in Figure 22-2 is seen to simplify the solution of the problem to the same degree as Thévenin's theorem.

PRACTICE PROBLEM

22-6.1 Apply Norton's theorem to the circuit described in Problem 22-5.1 to determine an equation for the current through Z_2.

22-7

MAXIMUM POWER TRANSFER THEOREM APPLIED TO AC CIRCUITS

When considering purely resistive circuits, it was found that maximum power is transferred from a voltage source (or current source) when the load resistance equals the source resistance (see Section 9-5). To understand how the maximum power transfer theorem applies to impedance networks, consider the circuit shown in Figure 22-8. The source impedance is

$$Z = R_1 + jX_1$$

and the load impedance is

$$Z_L = R_2 - jX_2$$

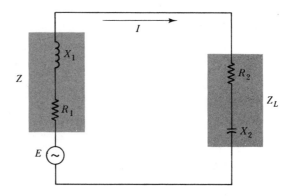

FIGURE 22-8. Maximum power transfer from an ac source to a load occurs when the load impedance is the conjugate of the network output impedance. That is, when $X_2 = -X_1$ and $R_2 = R_1$.

The output current is

$$I = \frac{E}{Z + Z_L}$$

$$= \frac{E}{(R_1 + jX_1) + (R_2 - jX_2)}$$

or $\qquad |I| = \dfrac{E}{\sqrt{(R_1 + R_2)^2 + (X_1 - X_2)^2}}$

The power dissipation in the load impedance occurs in the resistive portion of the load. Thus,

$$P_0 = I^2 R_2$$

or $\qquad \boxed{P_0 = \dfrac{E^2 R_2}{(R_1 + R_2)^2 + (X_1 - X_2)^2}} \qquad$ **(22-1)**

Maximum power output occurs when the denominator in Equation (22-1) has its minimum value (i.e., when $|X_1| = |X_2|$) which gives

$$\boxed{P_{0(max)} = \dfrac{E^2 R_2}{(R_1 + R_2)^2}} \qquad \textbf{(22-2)}$$

When this condition is obtained, the circuit behaves as if there were no reactive components, and therefore requires $R_1 = R_2$ for maximum power output.

It is seen that maximum power output is derived from an ac source when the resistive components of the source impedance and load impedance are equal, and when the reactive components of the source impedance and load impedance are equal in magnitude but opposite in sign. Thus, for maximum power transfer, an *LR* source must have a *CR* load in which the capacitive reactance has the same magnitude as the inductive reactance. Similarly, a *CR* source must have an *LR* load, with $|X_L| = |X_C|$. Another way of stating this requirement is that the load impedance must be the *conjugate* of the source impedance.

For application to ac impedance networks and ac sources, the maximum power transfer theorem is stated as follows:

Maximum Power Transfer Theorem (ac circuits) | **Maximum output power is obtained from a network or source when the load impedance is the conjugate of the output impedance of the network or source, as seen from the terminals of the load.**

EXAMPLE 22-7 A voltage source has an equivalent circuit consisting of $R_s = 100\ \Omega$ in series with $L = 20\ \mu\text{H}$. Calculate the optimum load for maximum output power at a frequency of 500 kHz.

SOLUTION

Reactive component of source impedance:

$$X_1 = 2\pi f L$$
$$= 2\pi \times 500\ \text{kHz} \times 20\ \mu\text{H}$$
$$X_1 = 62.83\ \Omega$$

Reactive component of load impedance:

$$X_2 = -X_1$$
$$= -62.83\ \Omega\ (capacitive)$$

Therefore, $\qquad \dfrac{1}{2\pi f C} = 2\pi f L$

and $\qquad\qquad C = \dfrac{1}{(2\pi f)^2 L}$

$$= \dfrac{1}{(2\pi \times 500\ \text{kHz})^2 \times 20\ \mu\text{H}}$$
$$= \textbf{0.005}\ \boldsymbol{\mu}\textbf{F}$$

Resistive component of load impedance:

$$\boldsymbol{R_L = R_s = 100\ \Omega}$$

EXAMPLE 22-8 For the voltage source described in Example 22-7, a parallel *CR* circuit is to be used as a load. Determine the required component values.

SOLUTION

From Example 22-7, *a parallel CR equivalent circuit is required for the series connected R* = 100 Ω *and C* = 0.005 μF.
  Series circuit is

$$R - jX = 100 \ \Omega - j62.83 \ \Omega$$

From Equation (20-8),

$$R_p = \frac{R_s^2 + X_s^2}{R_s}$$

$$= \frac{100^2 + 62.83^2}{100}$$

$$= \textbf{139.5 } \Omega$$

From Equation (20-9),

$$X_p = \frac{R_s^2 + X_s^2}{X_s}$$

$$= \frac{100^2 + 62.83^2}{62.83}$$

$$X_p = 222 \ \Omega$$

$$= \frac{1}{2\pi f C_p}$$

Therefore,
$$C_p = \frac{1}{2\pi f X_p}$$

$$= \frac{1}{2\pi \times 500 \text{ kHz} \times 222 \ \Omega}$$

$$= \textbf{0.0014 μF}$$

PRACTICE PROBLEM

22-7.1 For Problem 22-5.1, the source voltage and impedances are $E_1 = 500$ mV, $Z_1 = 600 \ \Omega\underline{/0°}$, and $Z_2 = 1.2 \text{ k}\Omega\underline{/60°}$. Determine an optimum load for maximum power output.

22-8

DELTA–WYE TRANSFORMA-TIONS FOR AC NETWORKS

Conversion between *delta* (Δ) and *wye* (Y) impedance networks can be performed using the equations derived in Section 8-5 for delta–wye conversion of resistance networks. Figure 22-9 shows Δ and Y imped-ance networks, and the conversion equations for the networks as illus-trated are listed below. As always, the impedances must be treated as vector quantities.

Equations (8-5), (8-6), and (8-7) are rewritten in the form of impedances. Converting from Δ to Y:

$$Z_a = \frac{Z_{ab}Z_{ac}}{Z_{ab} + Z_{ac} + Z_{bc}}$$

(22-3)

$$Z_b = \frac{Z_{ab}Z_{bc}}{Z_{ab} + Z_{ac} + Z_{bc}}$$

(22-4)

$$Z_c = \frac{Z_{ac}Z_{bc}}{Z_{ab} + Z_{ac} + Z_{bc}}$$

(22-5)

Equations (8-8), (8-9), and (8-10) are rewritten in the form of imped-ances. Converting from Y to Δ:

$$Z_{ab} = \frac{Z_a Z_b + Z_a Z_c + Z_b Z_c}{Z_c}$$

(22-6)

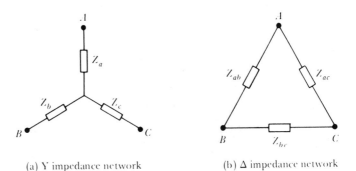

(a) Y impedance network　　　　(b) Δ impedance network

FIGURE 22-9. Wye and delta impedance networks identified for convenience in converting between types. In the wye network, each impedance is named for the terminal it is connected to—Z_a, Z_b, or Z_c. For the delta network, each impedance is named for the two terminals it is connected across—Z_{ab}, Z_{bc}, or Z_{ac}.

$$Z_{ac} = \frac{Z_a Z_b + Z_a Z_c + Z_b Z_c}{Z_b}$$

(22-7)

$$Z_{bc} = \frac{Z_a Z_b + Z_a Z_c + Z_b Z_c}{Z_a}$$

(22-8)

A typical application of Δ–Y conversion is illustrated in Figure 22-10. In the circuit shown in Figure 22-10(a), the current through impedance Z_3 is to be determined. To simplify the calculations, the delta-connected impedances Z_4, Z_5, and Z_6 are converted to the Y network Z_a, Z_b, and Z_c shown in Figure 22-10(b). Note that the value of Z_3 is unaffected by the conversion. Figure 22-10(c) shows that the network has been very much simplified by the Δ to Y conversion.

The procedure for Δ–Y and Y–Δ transformations of impedance networks is exactly the same as that listed for resistance networks in Section 8-5. As always, impedances must be in rectangular form for addition and subtraction, and in polar form for multiplication and division.

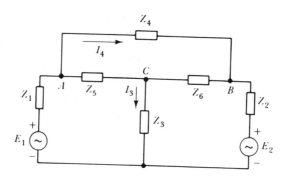

(a) Impedance network with Δ connected impedances Z_4, Z_5 and Z_6

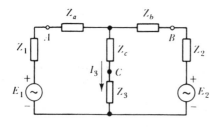

(b) Δ connected impedances replaced by equivalent Y network

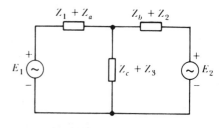

(c) Simplified impedance network

FIGURE 22-10. Use of delta–wye transformation to simplify a complex network. Delta network, Z_4, Z_5, Z_6 is replaced by its equivalent wye network Z_a, Z_b, Z_c.

EXAMPLE 22-9 Convert the delta-connected impedances Z_4, Z_5, and Z_6, in Figure 22-10(a), into the wye network Z_a, Z_b, and Z_c, shown in Figure 22-10(b). The impedance values are $Z_4 = 100 \, \Omega \underline{/30°}$, $Z_5 = 95 \, \Omega \underline{/40°}$, and $Z_6 = 60 \, \Omega \underline{/20°}$.

SOLUTION

$$Z_{ab} = Z_4 = 100 \, \Omega \underline{/30°} = 86.6 \, \Omega + j50 \, \Omega$$

$$Z_{ac} = Z_5 = 95 \, \Omega \underline{/40°} = 72.8 \, \Omega + j61.1 \, \Omega$$

$$Z_{bc} = Z_6 = 60 \, \Omega \underline{/20°} = 56.4 \, \Omega + j20.5 \, \Omega$$

$$Z_{ab} + Z_{ac} + Z_{bc} = 215.8 \, \Omega + j131.6 \, \Omega$$

$$= 253 \, \Omega \underline{/31.4°}$$

From Equation (22-3):

$$Z_a = \frac{100 \, \Omega \underline{/30°} \times 95 \, \Omega \underline{/40°}}{253 \, \Omega \underline{/31.4°}}$$

$$= \mathbf{37.5 \, \Omega \underline{/38.6°}}$$

From Equation (22-4):

$$Z_b = \frac{100 \, \Omega \underline{/30°} \times 60 \, \Omega \underline{/20°}}{253 \, \Omega \underline{/31.4°}}$$

$$= \mathbf{23.7 \, \Omega \underline{/18.6°}}$$

From Equation (22-5):

$$Z_c = \frac{95 \, \Omega \underline{/40°} \times 60 \, \Omega \underline{/20°}}{253 \, \Omega \underline{/31.4°}}$$

$$= \mathbf{22.5 \, \Omega \underline{/28.6°}}$$

PRACTICE PROBLEM

22-8.1 Convert the impedances in Figure 22-2 into the equivalent delta-network impedances.

SUMMARY OF FORMULAS

Power output from an ac source to an impedance:

$$P_o = \frac{E^2 R_2}{(R_1 + R_2)^2 + (X_1 - X_2)^2}$$

$$P_{o(\text{max})} = \frac{E^2 R_2}{(R_1 + R_2)^2}$$

Δ-*to-Y network conversion:*

$$Z_a = \frac{Z_{ab}Z_{ac}}{Z_{ab}+Z_{ac}+Z_{bc}}$$

$$Z_b = \frac{Z_{ab}Z_{bc}}{Z_{ab}+Z_{ac}+Z_{bc}}$$

$$Z_c = \frac{Z_{ac}Z_{bc}}{Z_{ab}+Z_{ac}+Z_{bc}}$$

*Y-to-*Δ *network conversion:*

$$Z_{ab} = \frac{Z_aZ_b+Z_aZ_c+Z_bZ_c}{Z_c}$$

$$Z_{ac} = \frac{Z_aZ_b+Z_aZ_c+Z_bZ_c}{Z_b}$$

$$Z_{bc} = \frac{Z_aZ_b+Z_aZ_c+Z_bZ_c}{Z_a}$$

REVIEW QUESTIONS

22-1 State Kirchhoff's voltage law and current law as applied to ac impedance networks.

22-2 Explain why ac voltage and current sources must have their terminals identified as + and −.

22-3 State the superposition theorem as applied to ac impedance networks.

22-4 State Thévenin's theorem as applied to ac impedance networks.

22-5 State Norton's theorem as applied to ac impedance networks.

22-6 State the maximum power transfer theorem as applied to ac sources. Sketch the circuit of an ac source and load, and derive the appropriate equation to show the truth of the theorem.

PROBLEMS

SECTION 22-2

22-1 Use loop equations to solve Problem 20-18.

22-2 Using loop equations, derive an equation for the current through Z_4 in the circuit shown in Figure 22-11.

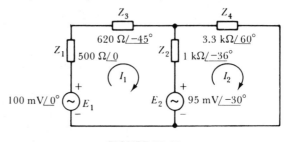

FIGURE 22-11.

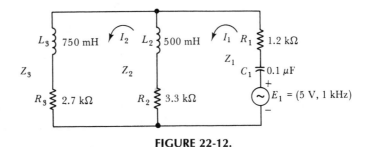

FIGURE 22-12.

22-3 Substitute the component values into the equation derived for Problem 22-2 to determine the current through Z_4.

22-4 Using loop equations, derive an equation for the voltage across R_2 in the circuit shown in Figure 22-12.

22-5 Substitute the component values into the equation derived for Problem 22-4 to determine the voltage across R_2.

22-6 Use the loop equation method to derive an equation for the current supplied by voltage source E_1 in Figure 22-13.

22-7 Substitute the component values into the equation derived for Problem 22-6 to determine the current supplied by E_1.

22-8 Use loop equations to solve Problem 20-19.

SECTION 22-3

22-9 Repeat Problem 22-2 using the superposition theorem.

22-10 Use the superposition theorem to derive an equation for the current through Z_3 in Figure 22-11.

22-11 Substitute the component values into the equation derived for Problem 22-10 to determine the current through Z_3.

22-12 Use the superposition theorem to derive an equation for the current through Z_2 in Figure 22-13.

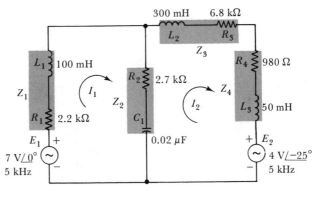

FIGURE 22-13.

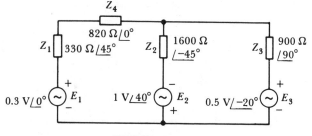

FIGURE 22-14.

22-13 Substitute the component values into the equation derived for Problem 22-12 to determine the current through Z_2.

22-14 Apply the superposition theorem to derive an equation for the current through Z_4 in the circuit shown in Figure 22-14.

22-15 Substitute the component values into the equation derived for Problem 22-14 to determine the current through Z_4.

22-16 Use the superposition theorem to derive an equation for the voltage drop across Z_4 in Figure 22-15.

22-17 Substitute the component values into the equation derived for Problem 22-16 to determine the voltage across Z_4.

22-18 Repeat Problem 22-6 using the superposition theorem.

SECTION 22-4

22-19 Use nodal analysis to solve Problem 20-18.

22-20 Use nodal analysis to solve Problem 22-16.

22-21 Use nodal analysis to solve Problem 22-2.

22-22 Use nodal analysis to solve Problem 22-4.

22-23 Use nodal analysis to solve Problem 22-6.

22-24 Use nodal analysis to derive an equation for the current through Z_3 in the circuit in Figure 22-16.

22-25 Substitute the component values into the equation derived for Problem 22-24 to determine the current through Z_3.

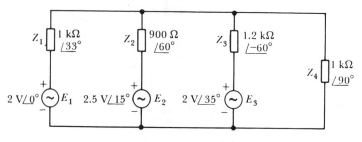

FIGURE 22-15.

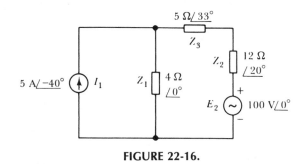

FIGURE 22-16.

SECTION 22-5

22-26 Use Thévenin's theorem to derive an equation for the current through Z_5 in Figure 22-17.

22-27 Substitute the component values into the equation derived for Problem 22-26 to determine the current through Z_5.

22-28 Use Thévenin's theorem to derive an equation for the current through R_3 in Figure 22-18.

22-29 Substitute the component values into the equation derived for Problem 22-28 to determine the current through R_3.

SECTION 22-6

22-30 Apply Norton's theorem to solve Problem 22-16.

22-31 Apply Norton's theorem to solve Problem 22-24.

22-32 Apply Norton's theorem to solve Problem 22-26.

22-33 Apply Norton's theorem to solve Problem 22-28.

22-34 Use the Norton's theorem to derive an equation for the current through R_2 in Figure 22-18.

22-35 Substitute the component values into the equation derived for Problem 22-34 to determine the current through R_2.

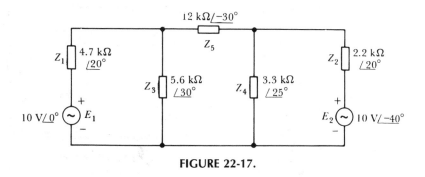

FIGURE 22-17.

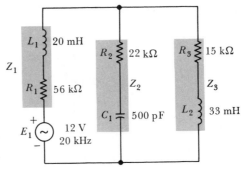

FIGURE 22-18.

SECTION 22-7

22-36 A voltage source has an equivalent circuit consisting of $R = 500 \ \Omega$ in series with $C = 0.01 \ \mu F$. Calculate the required component values for a series LR circuit that will draw maximum power from the source when the signal frequency is 1 MHz.

22-37 For Problem 22-36, calculate the required component values for a parallel LR circuit that will draw maximum power from the source.

22-38 For the circuit in Figure 22-15, calculate a suitable impedance that may be substituted in place of Z_4 to give maximum power dissipation.

22-39 For the circuit in Figure 22-18, calculate suitable series-connected components that may be substituted in place of R_3 and L_2 for maximum power dissipation.

SECTION 22-8

22-40 Apply delta-wye transformation and nodal analysis to derive an equation for the current through Z_4 in Figure 22-19.

22-41 Substitute the component values into the equation derived for Problem 22-40 to determine the current through Z_4.

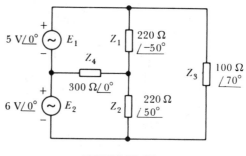

FIGURE 22-19.

Chap. 22   AC Network Analysis

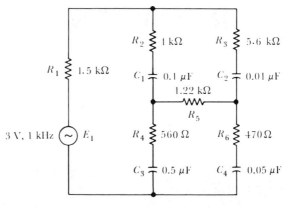

FIGURE 22-20.

22-42 Apply delta-wye transformation to derive an equation for the current supplied by the voltage source in Figure 22-20.

22-43 Substitute the component values into the equation derived for Problem 22-42 to determine the source current.

COMPUTER PROBLEMS

22-44 Write a computer program to convert delta-connected impedances into their equivalent wye-connected components.

22-45 Write a computer program to convert wye-connected impedances into their equivalent delta-connected components.

ANSWERS TO PRACTICE PROBLEMS

22-2.1 $17 \text{ mA} \underline{/-70.6°}$

22-2.2 $-34 \text{ mA} \underline{/-90.6°}$

22-3.1 $(E_2Z_1 + E_2Z_L - E_1Z_L)/(Z_1Z_2 + Z_2Z_L + Z_1Z_L)$

22-3.2 $181 \text{ μA} \underline{/-72.8°}$

22-4.1 $\dfrac{E_1}{Z_1} - \dfrac{1}{Z_1}\left[\dfrac{\dfrac{E_1}{Z_1} + \dfrac{E_2}{Z_2}}{\dfrac{1}{Z_1} + \dfrac{1}{Z_2} + \dfrac{1}{Z_L}}\right]$

22-5.1 $\dfrac{E_1Z_2}{Z_2Z_1 + Z_2Z_L + Z_1Z_L}$

22-6.1 $\dfrac{E_1Z_L}{Z_1Z_2 + Z_2Z_L + Z_1Z_L}$

22-7.1 $429 \text{ Ω} - j\,148 \text{ Ω}$

22-8.1 $236 \text{ Ω} \underline{/-19.3°},\ 354 \text{ Ω} \underline{/70.7°},\ 425 \text{ Ω} \underline{/15.7°}$

23

RESONANCE

Objectives

You will be able to:

Write the equations for the impedance of series and parallel *LCR* circuits and explain what occurs at the resonance frequency.

Draw graphs for inductive reactance, capacitive reactance, and total impedance versus frequency for a series *LCR* circuit.

Analyze series and parallel *LCR* circuits to determine the resonance frequency. Also, determine the component voltages and current levels at various frequencies.

Sketch phasor diagrams and component waveforms for series and parallel *LCR* circuits at resonance.

For series and parallel resonance circuits, define and calculate the circuit *Q* factor, half-power points, and bandwidth.

For coupled coils tuned to resonance, sketch the graphs of primary current and secondary current versus frequency. Calculate the critical value of the coefficient of coupling for the coils and determine the levels of current in each coil.

Sketch the circuits and frequency response graphs of band-stop and band-pass filters using series resonance and using parallel resonance. Calculate the frequency bands of band-pass and band-stop filters.

Introduction

When a series-connected *LCR* circuit has the frequency of its alternating supply voltage varied, it is found that $X_L = X_C$ at a particular frequency. The two reactances cancel, and the result is that the series impedance of the circuit has a minimum value of $Z = R$. The circuit is said to be in a state of *resonance,* and the frequency at which this occurs is termed the *resonance frequency.* Since the impedance is a minimum, the series current is a maximum, and there is a rise in the voltage developed across *L* and *C*.

A parallel *LC* circuit can also be in a state of resonance, and again this occurs when $X_L = X_C$. In this case, the circuit impedance has a maximum value, and the circuit current is a minimum, but there is a rise in the L and C current levels. A circuit can be tuned to resonate over a range of frequencies by making L or C adjustable.

23-1

SERIES RESONANCE

FREQUENCY EFFECT ON CIRCUIT IMPEDANCE. The series LCR circuit shown in Figure 23-1(a) has an impedance of

$$\boxed{Z = R + j(X_L - X_C)} \qquad (23\text{-}1)$$

Because $X_L = 2\pi f L$ and $X_C = 1/(2\pi f C)$, the actual value of the impedance obviously depends upon the frequency of the alternating supply. Figure 23-1(b) shows the values of jX_L and $-jX_C$ plotted versus supply frequency. It is seen that the inductive reactance increases linearly from zero with increase in frequency, because X_L is directly proportional to f. However, the capacitive reactance is inversely proportional to the frequency, and at low frequencies X_C is infinitely large. As f increases, X_C rapidly decreases from infinity, but never becomes zero. The resulting graph of X_C versus frequency is curved, as illustrated.

The total reactance is $(X_L - X_C)$, and the graph of this quantity is shown by the dashed line in Figure 23-1(b). At low frequencies, it is clear that X_L is much smaller than X_C, and thus the total reactance is largely capacitive. At the higher frequencies, where X_L is much larger than X_C, the total reactance becomes largely inductive. At one particular frequency, identified as f_r on the graph, X_L and X_C are numerically equal. Consequently, the impedance of the series LCR circuit becomes

$$Z = R + j(0)$$

or
$$Z = R$$

The frequency at which this phenomenon occurs is known as the *resonance frequency* (f_r), and at this frequency the circuit is said to be in a state of *electrical resonance*.

The resonance frequency for a series LCR circuit is defined as the frequency at which $X_L = X_C$.

Referring again to Figure 23-1(b), it is seen that at frequencies above and below resonance, the reactance $(X_L - X_C)$ has a large value (inductive or capacitive). Therefore, when the circuit impedance $(|Z| = \sqrt{R^2 + X^2})$ is plotted versus frequency [Figure 23-2(a)], it is found that the impedance has a high value above and below the resonance frequency, and at resonance it dips to its minimum value of $Z = R$. Thus,

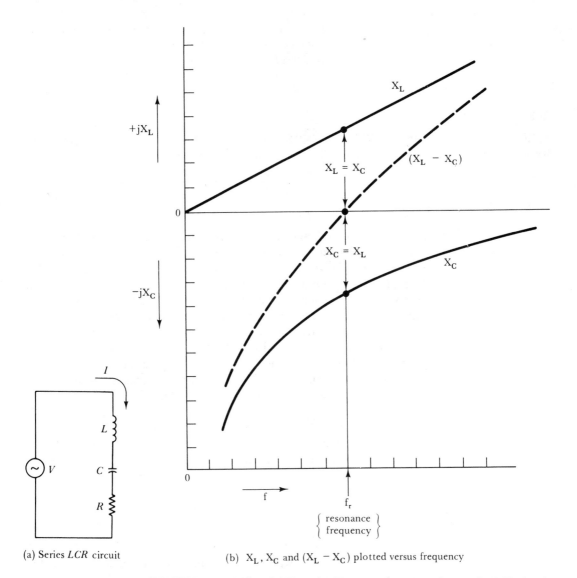

(a) Series *LCR* circuit

(b) X_L, X_C and $(X_L - X_C)$ plotted versus frequency

FIGURE 23-1. A plot of $+jX_L$ and $-jX_C$ versus frequency for a series *LCR* circuit shows that at a particular frequency, f_r, $X_L=X_C$, and consequently $(X_L-X_C) = 0$ at f_r. The frequency f_r is known as the *resonance frequency* for the circuit.

A series LCR circuit has a minimum impedance at the resonance frequency.

CURRENT AT RESONANCE. The current in the series *LCR* circuit is determined from

$$I = \frac{V}{R + j\,(X_L - X_C)}$$

(23-2)

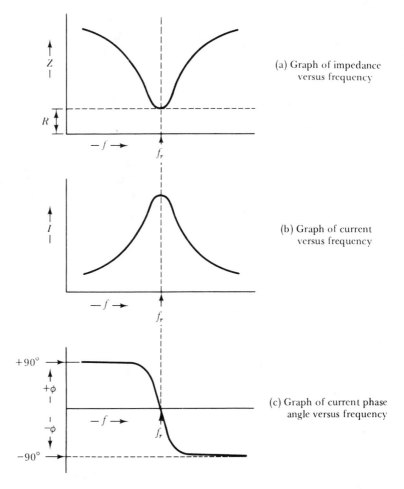

(a) Graph of impedance versus frequency

(b) Graph of current versus frequency

(c) Graph of current phase angle versus frequency

FIGURE 23-2. Graph of impedance, current, and phase angle versus frequency for a series *LCR* circuit. Because $Z = R + j\,(X_L - X_C)$, the impedance dips to R at f_r, and the current peaks. The current phase angle equals zero at f_r, is leading below f_r, and is lagging above f_r.

or
$$|I| = \frac{V}{\sqrt{R^2 + (X_L - X_C)^2}}$$

where $|I|$ is the numerical value of the current, without reference to its phase angle.

Thus, at resonance, when $X_L = X_C$, the current equation becomes

$$I = \frac{V}{R}$$

A typical graph of current versus frequency for the series *LCR* circuit is shown in Figure 23-2(b). The current is a minimum at frequencies above and below resonance, but peaks sharply at resonance as the impedance falls to a minimum. This I versus f graph is sometimes referred to as the *frequency response curve* of the circuit.

As already noted from Figure 23-1(b), the impedance of the series *LCR* circuit is largely capacitive at frequencies well below resonance. This means that the circuit current leads the applied voltage by a phase angle of approximately 90°. Conversely, because the impedance is largely inductive at frequencies much greater than the resonance frequency, the phase angle of the current above resonance is approximately −90°. The graph of phase angle versus frequency for the series *LCR* circuit [Figure 23-2(c)] shows the 90° leading phase angle at low frequencies, changing to 0° at the resonance frequency, and moving to a 90° lagging phase angle above f_r.

RESONANCE FREQUENCY. The frequency at which resonance occurs is easily calculated by equating X_L and X_C:

$$X_L = 2\pi f_r L$$

$$X_C = \frac{1}{2\pi f_r C}$$

Therefore,
$$2\pi f_r L = \frac{1}{2\pi f_r C}$$

$$f_r^2 = \frac{1}{(2\pi)^2 LC}$$

and

$$\boxed{f_r = \frac{1}{2\pi\sqrt{LC}}} \tag{23-3}$$

When L and C are in henrys and farads, respectively, Equation (23-3) gives f_r in hertz.

EXAMPLE 23-1 If the series *LCR* circuit shown in Figure 23-1(a) has $L = 85$ μH, $C = 298$ pF, $R = 100$ Ω, and $V = 10$ V, determine the resonance frequency. Also, calculate the circuit currents at $0.25\,f_r$, $0.5\,f_r$, $0.8\,f_r$, f_r, $1.25\,f_r$, $2\,f_r$, and $4\,f_r$. Plot a graph of I versus f to a logarithmic base.

SOLUTION

Equation (23-3):

$$f_r = \frac{1}{2\pi\sqrt{LC}} = \frac{1}{2\pi\sqrt{85\ \mu H \times 298\ pF}}$$

$$f_r = 1\ \text{MHz}$$

Equation (23-2):

$$I = \frac{V}{R + j(X_L - X_C)}$$

or

$$|I| = \frac{V}{\sqrt{R^2 + (X_L - X_C)^2}}$$

The values of the various quantities are tabulated below for each frequency, and the graph of current versus frequency is plotted in Figure 23-3.

f	X_L (Ω)	X_C (Ω)	$(X_L - X_C)$ (Ω)	$|z|$ (Ω)	I (mA)
250 kHz	134	2136	$-j2$ k	$\cong 2$ k	5
500 kHz	267	1068	$-j801$	$\cong 807$	12.4
800 kHz	427	668	$-j241$	$\cong 261$	38.3
1 MHz	534	534	0	100	100
1.25 MHz	668	427	$+j241$	$\cong 261$	38.3
2 MHz	1068	267	$+j801$	$\cong 807$	12.4
4 MHz	2136	134	$+j2$ k	$\cong 2$ k	5

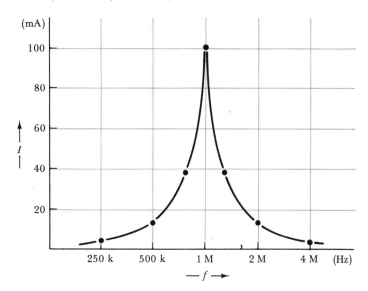

FIGURE 23-3. Plot of current versus frequency (to a logarithmic base) for a series *LCR* circuit. The current peaks sharply at the resonance frequency.

The graph of current versus frequency in Figure 23-3 clearly shows that the current reaches a maximum value at the resonance frequency, and that it falls off sharply at frequencies above and below resonance.

Note from the table that when the frequency is doubled from 250 kHz to 500 kHz, the current goes from 5 mA to 12.4 mA. Also note that above resonance the current goes from 12.4 mA to 5 mA when the frequency is doubled from 2 MHz to 4 MHz. The same sort of corresponding current ratios are evident for other frequency changes above and below the resonance frequency. It is seen that the ratio of I to f is *not* linear but logarithmic, and this is why a logarithmic frequency scale is employed on the graph.

RESONANCE RISE IN VOLTAGE. The voltage across the resistance of the series LCR circuit is $V_R = I \times R$, and at resonance this is $V_R =$ (the supply voltage), V in Figure 23-1(a).

The voltages across the capacitor and the inductor are

$$\boxed{V_C = IX_C} \qquad \text{(23-4)}$$

and

$$\boxed{V_L = IX_L} \qquad \text{(23-5)}$$

When the capacitor voltage and inductor voltage values are calculated for various frequencies, and plotted versus a logarithmic frequency scale, it is found that their *frequency response curves* are similar in shape to that of the I versus f graph. It is also found that

V_C and V_L at resonance can be many times greater than the supply voltage.

This effect is termed the *resonant rise in voltage*.

EXAMPLE 23-2   For the series LCR circuit referred to in Example 23-1, determine the values of capacitor voltage and inductor voltage at each frequency. Also plot V_R, V_C, and V_L versus frequency.

SOLUTION

$$V_C = IX_C,\ V_L = IX_L,\ \text{and}\ V_R = IR$$

The values of the various quantities are tabulated below for each frequency, and the graphs are plotted in Figure 23-4. Note from the graph that below resonance $V_C \gg V_L$ and above resonance $V_L \gg V_C$.

f	IR (V)	IX_L (V)	IX_C (V)
250 kHz	0.5	0.67	10.7
500 kHz	1.24	3.3	13.2
800 kHz	3.83	16.4	25.5
1 MHz	10	53.4	53.4
1.25 MHz	3.83	25.5	16.4
2 MHz	1.24	13.2	3.3
4 MHz	0.5	10.7	0.67

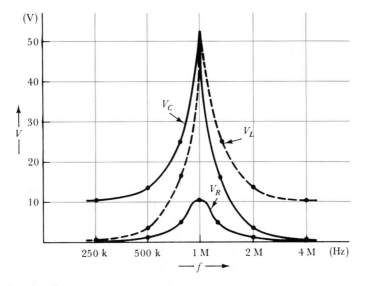

FIGURE 23-4. Graphs of capacitor voltage V_C, inductor voltage V_L, and resistor voltage V_R versus frequency (to a logarithmic base) for a series LCR circuit. At the resonance frequency, V_C and V_L become much greater than the supply voltage.

Example 23-2 demonstrates that at resonance the voltages across the capacitor and inductor become much greater than the supply voltage. The voltage across the resistance (V_R) is, of course, in phase with the circuit current, whereas V_L leads the current by 90°, and V_C lags the current by 90°. This is illustrated in the phasor diagram in Figure 23-5(a). From the phasor diagram it is clear that V_L and V_C cancel at resonance, and the supply voltage is equal to the voltage across the resistance.

ENERGY TRANSFER BETWEEN L AND C. When the waveforms of V_L, V_R, and V_C are plotted for the resonance frequency [see Figure 23-5(b)], it is evident that as V_L grows positively, V_C is becoming more negative, and

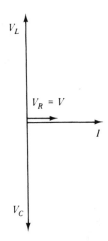

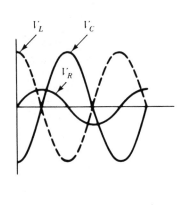

(a)　　Phasor diagram for a
　　　series *LCR* circuit at resonance

(b)Waveforms of V_L, V_C and V_R at resonance

FIGURE 23-5. Voltage phasor diagram and waveforms for a series *LCR* circuit at resonance. V_C and V_L cancel each other, V_R equals the supply voltage, and the current is in-phase with the supply voltage.

vice versa. The positive peak of V_L occurs at the same instant as the negative peak of V_C, and the positive peak of V_C corresponds in time with the negative peak of V_L. Note also that the positive and negative peaks of voltage across the resistance coincide with the instant when V_L and V_C are zero. The information to be derived from these waveforms is that there is energy stored in the resonant circuit. The energy is continuously being transferred from the inductor to the capacitor and back again at the resonance frequency. The resonance frequency is the only frequency at which this transfer of energy can take place for the particular values of capacitor and inductor involved. Note, however, that *there can be no energy stored and no transfer of energy between the reactive components unless there is an energy input at the resonance frequency.*

PRACTICE PROBLEMS

23-1.1 Determine the new resonance frequency for the series *LCR* circuit in Example 23-1 when an oscilloscope with an input capacitance of 40 pF is connected in parallel with the capacitor.

23-1.2 Calculate the new levels of I, V_L and V_C at 0.5 f_r and 0.8 f_r for the circuit in Problem 23-1.1.

23-2

TUNING FOR RESONANCE

The equation for calculating the resonance frequency has already been derived as Equation (23-3):

$$f_r = \frac{1}{2\pi\sqrt{LC}}$$

Chap. 23　Resonance

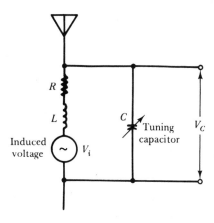

FIGURE 23-6. Resonance circuit with a tuning capacitor. Capacitor C can be adjusted to give $X_C = X_L$ at any desired frequency. Thus, the resonance frequency is selected.

When L and C have fixed values, the frequency of the supply must be made equal to f_r to achieve resonance. Suppose, instead, that f had a fixed value; then either L or C (or perhaps both) could be adjusted until the circuit resonates at the supply frequency. In fact, this is exactly what is done in a radio receiver. Figure 23-6 shows a circuit in which V_s represents a small voltage induced from the atmosphere into the inductance L. Resistance R represents the resistive component of L, and the capacitor C is variable. The electromagnetic signals sent into the atmosphere by different radio stations have different frequencies. Thus, by adjusting capacitor C the circuit may be made to resonate at any one of a wide range of radio frequencies. Adjusting the value of C (or L) is termed *tuning* the circuit, and C is referred to as a *tuning capacitor*. The resonance frequency causes a *resonance rise in voltage* across the tuning capacitor. The signal developed across the capacitor can then be passed on to the other circuits of the radio receiver.

EXAMPLE 23-3   The circuit shown in Figure 23-6 has $L = 100$ μH and is to be tuned to resonate at frequencies ranging from 500 kHz to 1 MHz. Calculate the range of adjustment of the tuning capacitor.

SOLUTION

Equation (23-3),

$$f_r = \frac{1}{2\pi\sqrt{LC}}$$

or

$$f_r^2 = \frac{1}{4\pi^2 LC}$$

which gives,

$$C = \frac{1}{4\pi^2 f_r^2 L}$$

At $f_r = 500$ kHz:

$$C = \frac{1}{4\pi^2 \times (500 \text{ kHz})^2 \times 100 \text{ } \mu\text{H}}$$

$$\cong 1000 \text{ pF}$$

At $f_r = 1$ MHz:

$$C = \frac{1}{4\pi^2 \times (1 \text{ MHz})^2 \times 100 \text{ } \mu\text{H}}$$

$$\cong 253 \text{ pF}$$

Capacitor range is from **253 pF** *to* **1000 pF.**

PRACTICE PROBLEM

23-2.1   A series *LCR* circuit has $R_L = 5.5 \text{ } \Omega$, $L = 56 \text{ } \mu\text{H}$, C adjustable from 300 pF to 1500 pF, and stray capacitance of 30 pF in parallel with C. Determine the range of resonance frequencies for the circuit.

23-3

Q FACTOR OF A SERIES RESONANT CIRCUIT

Reconsider the equations for I, V_L, and V_C at resonance:

$$V_L = IX_L$$

and

$$I = \frac{V}{R}$$

Therefore,

$$V_L = \frac{V}{R} X_L$$

or,

$$\boxed{\frac{V_L}{V} = \frac{X_L}{R}} \qquad \text{(23-6)}$$

Similarly,

$$\boxed{\frac{V_C}{V} = \frac{X_C}{R}} \qquad \text{(23-7)}$$

The ratio: (capacitor voltage, or inductor voltage, at resonance)/(supply voltage) is a measure of the *quality* of a resonance circuit. This is termed the *Q factor* of the circuit, and it is also known as the *voltage magnification factor*.

From Equations (23-6) and (23-7), the equations for Q factor are:

$$Q = \frac{X_L}{R}$$

or

$$\boxed{Q = \frac{\omega L}{R}} \tag{23-8}$$

and

$$Q = \frac{X_C}{R}$$

giving

$$\boxed{Q = \frac{1}{\omega C R}} \tag{23-9}$$

Since the coil resistance is often the only resistance in a series resonance circuit, the Q is sometimes referred to as the *Q factor of the coil*. Rewriting Equation (23-8),

$$Q = \frac{2\pi f_r L}{R}$$

and substituting for f_r from Equation (23-3),

$$Q = \frac{2\pi L \left(\dfrac{1}{2\pi \sqrt{LC}} \right)}{R}$$

which reduces to

$$\boxed{Q = \frac{1}{R} \sqrt{\frac{L}{C}}} \tag{23-10}$$

It is seen that the Q factor of a series resonance circuit may be increased either by reducing R or by increasing the L/C ratio.

The Q factor can also be defined in terms of the ratio of the reactive power to the power dissipated in the circuit resistance. Using this definition, the equations for Q come out exactly as derived above.

EXAMPLE 23-4   A series *LCR* circuit which resonates at $f_r \cong 500$ kHz has $L = 100$ μH, $R = 25$ Ω, and $C = 1000$ pF. Determine the Q factor of the circuit. Also, determine the new value of C required for resonance at 500 kHz when the value of L is doubled, and calculate the new Q factor.

SOLUTION

Equation (23-10):

$$Q = \frac{1}{R} \sqrt{\frac{L}{C}}$$

$$Q_1 = \frac{1}{25 \ \Omega} \sqrt{\frac{100 \ \mu H}{1000 \ pF}}$$

$$\cong 12.6$$

When L is doubled:

Equation (23-3):

$$f_r = \frac{1}{2\pi \sqrt{LC}}$$

which gives, $$C = \frac{1}{4\pi^2 f_r^2 L}$$

$$= \frac{1}{4\pi^2 \times (500 \ kHz)^2 \times 200 \ \mu H}$$

$$\cong 500 \ pF$$

$$Q_2 = \frac{1}{25 \ \Omega} \sqrt{\frac{200 \ \mu H}{500 \ pF}}$$

$$\cong 25$$

PRACTICE PROBLEM

23-3.1   Calculate the Q factor of the circuit in Problem 23-2.1 at maximum and minimum resonance frequencies.

23-4

BANDWIDTH OF A SERIES RESONANT CIRCUIT

Consider the current versus frequency response graph once again, as reproduced in Figure 23-7. It is clear from the graph that the current reaches a maximum level (I_m) at resonance. It is also clear that at frequencies close to resonance, the current level is only a little below its maximum value. Thus, the resonant circuit is said to select a *band* of

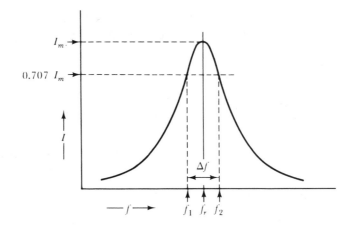

FIGURE 23-7. The lowest and highest frequencies for maximum current in a series *LCR* circuit are defined as those frequencies (f_1 and f_2) at which the current is 0.707 of its peak level; ($f_2 - f_1$) is known as the *bandwidth* of the circuit.

frequencies rather than just one frequency. The lowest and highest frequencies of the band are identified as f_1 and f_2, respectively, in Figure 23-7, and the *bandwidth* is

$$\boxed{\Delta f = f_2 - f_1} \tag{23-11}$$

The frequencies f_1 and f_2 are arbitrarily defined as those frequencies at which the power delivered to the circuit is half the power delivered at resonance.

For this reason, f_1 and f_2 are termed the *half-power points*. The power levels at these frequencies can also be shown to be 3 dB below the power level at resonance (see Section 21-7). The power delivered to the circuit at resonance is

$$P = I_m^2 R$$

Thus, for half-power,

$$\frac{1}{2} P = \frac{I_m^2 R}{2}$$

which gives

$$\boxed{\frac{1}{2} P = \left(\frac{I_m}{\sqrt{2}}\right)^2 R} \tag{23-12}$$

Equation (23-12) shows that at the half-power points, the current in the series LCR circuit is

$$I = \frac{I_m}{\sqrt{2}} = 0.707 \, I_m$$

From Equation (23-2), the current in the circuit is

$$|I| = \frac{V}{\sqrt{R^2 + (X_L - X_C)^2}}$$

and this can be rewritten as

$$|I| = \frac{V}{R\sqrt{1 + \left(\dfrac{X_L - X_C}{R}\right)^2}}$$

or

$$|I| = \frac{I_m}{\sqrt{1 + \left(\dfrac{X_L - X_C}{R}\right)^2}}$$

and from Equation (23-12) at the half-power points,

$$I = \frac{I_m}{\sqrt{2}}$$

Therefore, at the half-power points,

and

$$\frac{X_L - X_C}{R} = 1$$

or

$$\boxed{X_L - X_C = R} \qquad \text{(23-13)}$$

But at f_r, $X_L - X_C = 0$. Therefore, when the frequency increases from f_r to f_2, X_L must increase by $R/2$ and X_C must decrease by $R/2$, to fulfill Equation (23-13). So,

$$2\pi f_2 L - 2\pi f_r L = \tfrac{1}{2} R$$

which gives

$$\boxed{f_2 - f_r = \frac{R}{4\pi L}} \qquad \text{(23-14)}$$

Similarly, when the frequency decreases from f_r to f_1, X_C must increase by $R/2$ and X_L must decrease by $R/2$. Consequently,

$$2\pi f_r L - 2\pi f_1 L = \tfrac{1}{2}R$$

and

$$\boxed{f_r - f_1 = \frac{R}{4\pi L}} \qquad \text{(23-15)}$$

Adding Equation (23-14) to Equation (23-15):

$$f_2 - f_1 = \frac{R}{2\pi L}$$

or the circuit bandwidth is

$$\boxed{\Delta f = \frac{R}{2\pi L}} \qquad \text{(23-16)}$$

Multiplying the right-hand side of Equation (23-16) by f_r / f_r gives

$$\Delta f = \frac{R f_r}{2\pi f_r L}$$

or

$$\boxed{\Delta f = \frac{f_r}{Q}} \qquad \text{(23-17)}$$

The ability of a resonant circuit to select one particular frequency and to discriminate against other frequencies is termed the *selectivity* of the circuit. The circuits with the narrowest bandwidths obviously have the greatest selectivity. Equation (23-17) clearly shows that the greater the Q of a circuit, the better its selectivity.

EXAMPLE 23-5   For the circuit described in Example 23-4, calculate the half-power frequencies and the bandwidths:

a. When $L = 100$ μH.
b. When the value of L is doubled.

SOLUTION

a. for L = 100 μH:

From Equation (23-14):

$$f_2 = \frac{R}{4\pi L} + f_r$$

$$= \frac{25\ \Omega}{4\pi \times 100\ \mu\text{H}} + 500\ \text{kHz}$$

$$\cong \mathbf{520\ kHz}$$

From Equation (23-15):

$$f_1 = f_r - \frac{R}{4\pi L}$$

$$= 500\ \text{kHz} - \frac{25\ \Omega}{4\pi \times 100\ \mu\text{H}}$$

$$\cong \mathbf{480\ kHz}$$

$$\Delta f = f_2 - f_1$$

$$= \mathbf{40\ kHz}$$

Alternatively, from Equation (23-17):

$$\Delta f = \frac{f_r}{Q}$$

From Example 23-4, $Q = 12.6$ *when* $L = 100\ \mu\text{H}$. *Therefore,*

$$\Delta f = \frac{500\ \text{kHz}}{12.6}$$

$$\cong \mathbf{40\ kHz}$$

b. *for* $L = 200\ \mu\text{H}$,

$$f_2 = \frac{R}{4\pi L} + f_r$$

$$= \frac{25\ \Omega}{4\pi \times 200\ \mu\text{H}} + 500\ \text{kHz}$$

$$\cong \mathbf{510\ kHz}$$

$$f_1 = f_r - \frac{R}{4\pi L}$$

$$\cong \mathbf{490\ kHz}$$

$$\Delta f = \mathbf{20\ kHz}$$

Alternatively, from Example 23-4, $Q = 25$ *when* $L = 200\ \mu\text{H}$:

$$\Delta f = \frac{f_r}{Q} = \frac{500\ \text{kHz}}{25}$$

$$= \mathbf{20\ kHz}$$

23-4.1 Calculate the bandwidth of the circuit in Problem 23-2.1.

23-5

PARALLEL RESONANCE

IDEAL PARALLEL RESONANT CIRCUIT. Consider the parallel *LCR* circuit shown in Figure 23-8(a). The admittance of the circuit is

$$Y = \frac{1}{R_P} - j\frac{1}{X_L} + j\frac{1}{X_C}$$

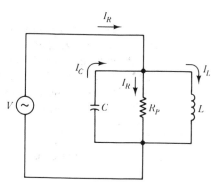

(a) Parallel resonance circuit

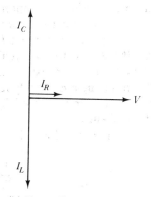

(b) Phasor diagram for parallel resonance circuit

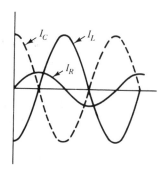

(c) Waveforms of I_C, I_L, and I_R at resonance

FIGURE 23-8. Parallel resonant circuit, current phasor diagram, and waveforms at resonance. I_C and I_L cancel each other. The total supply current is I_R, and this is in-phase with the supply voltage.

If the supply frequency is adjusted until X_L and X_C are equal, the admittance becomes

$$Y = \frac{1}{R_P}$$

and the circuit impedance is

$$Z = R_P$$

Consequently, the current taken from the supply source is

$$I = \frac{V}{R_P}$$

The current through R_P is in phase with the supply voltage, the current through L lags the supply voltage by 90°, and the capacitor current leads the supply voltage by 90°. This is illustrated by the phasor diagram in Figure 23-8(b), and by the current waveforms in Figure 23-8(c). When X_L and X_C are equal, the inductive and capacitive currents are equal and opposite, as illustrated in the phasor diagram. Thus, the total current supplied by the voltage source is I_R. I_C and I_L are the result of the energy stored in the circuit being continuously transferred from the inductor to the capacitor, and back again.

PRACTICAL PARALLEL RESONANT CIRCUIT. Since a practical inductor is not purely reactive but has a winding resistance, it must be represented as a resistance R_L in series with an inductance L. Some capacitors must also be shown as having a resistive component; however, in general, capacitors can be assumed to be purely reactive. The circuit shown in Figure 23-9(a) is that of a practical parallel LC circuit, and the phasor diagram in Figure 23-9(b) represents the circuit conditions at resonance. It is seen that the presence of the resistive component in the inductive branch makes the I_L phase angle ϕ slightly less than 90°. Resonance is achieved when the vertical component of I_L [I_L' in Figure 23-9(b)] is equal to the capacitor current I_C.

The admittance of the parallel circuit in Figure 23-9(a) is

$$Y = \frac{1}{R_L + jX_L} + j\frac{1}{X_C}$$

Multiplying

$$\frac{1}{R_L + jX_L} \quad \text{by} \quad \frac{R_L - jX_L}{R_L - jX_L}$$

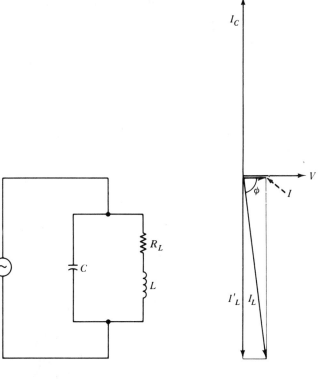

(a) Parallel LC circuit

(b) Phasor diagram for
parallel LC circuit

FIGURE 23-9. In a practical parallel resonance circuit, the presence of a resistive component in the inductance gives an inductor current phase angle less
than 90° at resonance.

gives

$$Y = \frac{R_L}{R_L^2 + X_L^2} - j\frac{X_L}{R_L^2 + X_L^2} + j\frac{1}{X_C} \qquad \textbf{(23-18)}$$

At resonance:

$$\frac{1}{X_C} = \frac{X_L}{R_L^2 + X_L^2}$$

or

$$X_C = \frac{R_L^2 + X_L^2}{X_L} \qquad \textbf{(23-19)}$$

which, when substituted into Equation (23-18), gives the admittance at resonance as

$$Y = \frac{R_L}{R_L^2 + X_L^2}$$

Thus,
$$Z = \frac{R_L^2 + X_L^2}{R_L}$$

From Equation (23-19):

$$X_L X_C = R_L^2 + X_L^2$$

Therefore,
$$Z = \frac{X_L X_c}{R_L}$$

$$= \frac{2\pi f_r L}{2\pi f_r C R_L}$$

or at resonance

$$Z = \frac{L}{CR_L} \qquad\qquad \textbf{(23-20)}$$

If the impedance of the parallel LC circuit [as represented by the inverse of Equation (23-18)] is plotted to a logarithmic frequency base, it is found that the impedance peaks at the resonance frequency [see Figure 23-10(a)]. *Thus,*

a parallel LC circuit has a maximum impedance at the resonance frequency,

while, as already discussed, a series resonant circuit has a *minimum* impedance at the resonance frequency. As a consequence of the peak in the impedance value of a parallel resonant circuit, there is a dip in the current taken from the supply at the resonance frequency, as illustrated

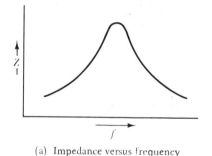

(a) Impedance versus frequency

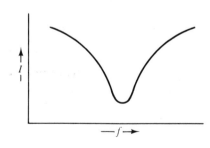

(b) Current versus frequency

FIGURE 23-10. Graph of impedance and current versus frequency for a parallel *LCR* circuit. Because X_L and X_C cancel each other out at f_r, the impedance peaks to R, and the current dips.

in Figure 23-10(b). Once again, this is the opposite of the case with series resonance.

PRACTICE PROBLEM

23-5.1 A parallel LC circuit has $R_L = 5.5 \, \Omega$, $L = 68 \, \mu H$, C adjustable from 200 pF to 1200 pF, and stray capacitance of 30 pF in parallel with C. Determine the maximum and minimum circuit impedance at resonance.

23-6

Q FACTOR FOR PARALLEL LC CIRCUITS

Referring again to Figure 23-9(b) it is seen that the total current drawn from the supply at resonance is I, which is in phase with supply voltage V. Also, I is very much less than I_C and I_L. Thus, in the parallel resonant LC circuit, there is a *current magnification,* which is analogous to the voltage magnification that occurs in a series resonant circuit. The Q factor, *or current magnification factor* of the parallel resonant circuit, can be determined as the ratio of I_L (or I_C) to the supply current I.

From Figure 23-9(b), the supply current is

$$I = I_L \cos \phi = I_L \frac{R_L}{X_L}$$

and

$$Q = \frac{I_L}{I} = \frac{X_L}{R_L}$$

or,

$$\boxed{Q = \frac{\omega L}{R_L}}$$

This equation is exactly the same as the Q factor equation for a series resonant circuit [Equation (23-8)]; that is, the Q is once again the Q factor of the inductance.

23-7

RESONANCE FREQUENCY FOR PARALLEL LC CIRCUITS

Returning to Equation (23-19), which gives the value of X_C at resonance:

$$X_C = \frac{R_L^2 + X_L^2}{X_L}$$

When $Q > 10$,

$$X_L^2 \gg R_L^2$$

and therefore, $\qquad\qquad X_C \cong X_L$

This gives the resonance frequency for a parallel LC circuit for $Q > 10$ as:

$$f_r = \frac{1}{2\pi\sqrt{LC}}$$

This is also the expression for the resonance frequency of a series LCR circuit [Equation (23-3)].

Equation (23-3) does not apply in parallel LC circuits where Q is less than 10. Where $Q < 10$, the value of f_r for a parallel LC circuit can be shown to be:

$$f_r = \frac{1}{2\pi\sqrt{LC}}\sqrt{1 - \frac{CR_L^2}{L}} \qquad\qquad \textbf{(23-21)}$$

The bandwidth of a parallel resonant circuit is determined in exactly the same way as that for a series resonant circuit [Equation (23-17)]:

$$\Delta f = \frac{f_r}{Q}$$

EXAMPLE 23-6　A parallel LC circuit has an inductance $L = 100$ μH, which has a coil resistance of 12 Ω. The capacitor C is adjustable over the range 200 pF to 300 pF. Determine the maximum and minimum resonance frequencies for the circuit. Also, calculate the Q factor and bandwidth of the circuit at the two resonance frequency extremes:

SOLUTION

Equation (23-3):

$$f_r = \frac{1}{2\pi\sqrt{LC}}$$

When $C = 200$ pF,

$$f_r = \frac{1}{2\pi\sqrt{100\ \mu H \times 200\ pF}}$$

$$f_{r(max)} = \textbf{1.13 MHz}$$

When $C = 300$ pF,

$$f_r = \frac{1}{2\pi\sqrt{100 \ \mu H \times 300 \ pF}}$$

$$f_{r(min)} = \textbf{919 kHz}$$

Equation (23-8):

$$Q = \frac{\omega L}{R_L}$$

At $f = 1.13$ MHz,

$$Q = \frac{2\pi \times 1.13 \ MHz \times 100 \ \mu H}{12 \ \Omega}$$

$$= \textbf{59.2}$$

At $f = 919$ kHz,

$$Q = \frac{2\pi \times 919 \ kHz \times 100 \ \mu H}{12 \ \Omega}$$

$$= \textbf{48}$$

Equation (23-16):

$$\Delta f = \frac{R_L}{2\pi L} = \frac{12 \ \Omega}{2\pi \times 100 \ \mu H}$$

$$= \textbf{19.1 kHz}$$

PRACTICE PROBLEMS

23-7.1 For the circuit in Problem 23-5.1, determine the maximum and minimum resonance frequencies, the circuit bandwidth, and the Q factors at each resonance frequency.

23-7.2 The highest resonance frequency for the circuit in Problem 23-5.1 is found to be 10 kHz below the calculated frequency. The effect could be due to stray inductance in series with L, or to additional stray capacitance in parallel with C. Calculate the stray quantities for each case.

23-8
RESISTANCE DAMPING OF PARALLEL *LC* CIRCUITS

Figure 23-11 shows a parallel *LC* circuit with a resistor R_P connected in parallel with the capacitor and inductance. R_L is the resistive component of the inductance. Since the impedance of the parallel *LC* circuit is a maximum at resonance, the presence of R_P reduces the total impedance of the circuit. As will be shown, it also reduces the Q factor of the circuit.

When R_P is not present, the supply current is determined as:

$$I = I_L \times \frac{R_L}{X_L} \qquad (see \text{ Section 23-6})$$

$$\cong \frac{V}{X_L} \times \frac{R_L}{X_L}$$

$$I \cong \frac{VR_L}{X_L^2}$$

The current through R_P is in phase with I and is

$$I_R = \frac{V}{R_P}$$

Therefore, the total supply current is

$$I_T = I + I_R$$

$$I_T = \frac{VR_L}{X_L^2} + \frac{V}{R_P}$$

The Q factor for the parallel resonance *LCR* circuit is determined as

$$Q = \frac{I_L}{I_T}$$

$$= \frac{V/X_L}{\dfrac{VR_L}{X_L^2} + \dfrac{V}{R_P}} = \frac{1}{\dfrac{R_L}{X_L} + \dfrac{X_L}{R_p}}$$

giving

$$Q = \frac{1}{\dfrac{R_L}{\omega L} + \dfrac{\omega L}{R_P}}$$

When

$$\frac{R_L}{\omega L} \ll \frac{\omega L}{R_P}$$

$$\boxed{Q = \frac{R_P}{\omega L}} \qquad (23\text{-}22)$$

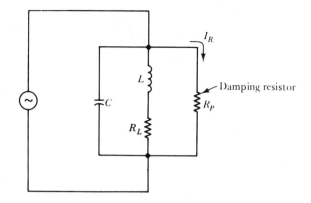

FIGURE 23-11. A *damping resistor* may be connected into a parallel *LCR* circuit to reduce the circuit *Q* factor, and thus minimize unwanted resonance effects.

Note that this equation for the *Q* of a parallel *LCR* circuit is inverted by comparison with the equation for the *Q* of the inductance ($Q = \omega L / R_L$). It is seen that when R_P is made smaller, the circuit *Q* is reduced. The effect is known as *damping,* and R_P is referred to as a *damping resistor.* Damping may be employed where resonance of *LC* circuits is undesirable, for example to avoid picking up radio frequencies in audio equipment. The *Q* factor for the inductance is given as $\omega L / R_L$. Thus, to obtain Equation (23-22), $\omega L / R_L \gg R_P / \omega L$; that is, the *Q* factor of the inductor must be at least 10 times the *Q* factor of the parallel *LCR* circuit.

PRACTICE PROBLEM

23-8.1 The circuit in Problem 23-5.1 is to be damped to have a maximum *Q* factor of 6. Determine the required damping resistance value.

23-9
TUNED COUPLED COILS

COUPLED COILS. In Chapter 14 it was shown that mutual inductance exists between adjacent coils, and that when a current flows in one coil, a terminal voltage may be induced in a mutually coupled coil. Figure 23-12(a) shows a typical arrangement of two mutually coupled coils wound on a nonmagnetic core. The coefficient of coupling between the coils (see Section 14-4) can be adjusted simply by altering the spacing between them. The circuit diagram for coupled coils is shown in Figure 23-12(b). Note the dots at one end of each coil. These are employed to show the phase relationship between input and output voltages. The dots simply identify in-phase ends of each coil. Thus, when an input voltage applied to the left-hand coil has the instantaneous polarity

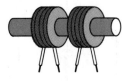
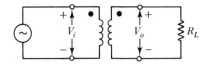

(a) Coupled coils   (b) Circuit diagram for two coupled coils

FIGURE 23-12. Coupled coils and their circuit diagram. The dot notation on the circuit diagram denotes the relative polarity of the coil input and output voltages.

shown (positive at the dotted end of the coil), the output voltage from the right-hand coil is also positive at the dotted end of the coil.

The equivalent circuit for the coupled coils is shown in Figure 23-13. R_1 represents the resistive component of the *primary* winding (the left-hand winding in Figure 23-13), and L_1 is the inductive component. In the *secondary* winding R_2 and L_2 are the total resistive and inductive components, respectively (R_2 includes the load resistance R_L). When the current I_1 flows in the primary winding, a voltage $\omega M I_1$ is induced in the secondary winding [see Equation (19-6)], causing current I_2 to flow. The flow of I_2 also induces a voltage $\omega M I_2$ in the primary winding. This induced voltage must be shown as having a polarity that opposes the primary current. Two voltage generators are included in the equivalent circuit to represent the voltages induced in each winding.

COUPLED IMPEDANCE. The equation for the voltage drops around the primary circuit may be written

$$V = I_1R_1 + jI_1X_1 + \omega M I_2 \qquad \text{(23-23)}$$

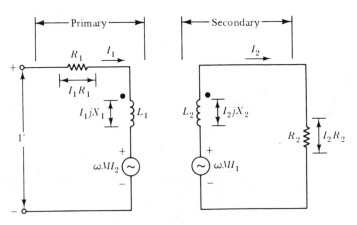

FIGURE 23-13. Equivalent circuit for two coupled coils. Each coil has resistive and inductive components (R and L), and each has a voltage ($\omega M I$) induced in it by the current flow in the other coil.

and the equation for the voltage drops around the secondary circuit is

$$\omega M I_1 = I_2 R_2 + j I_2 X_2$$

giving

$$\boxed{I_2 = \frac{\omega M I_1}{R_2 + j X_2}} \tag{23-24}$$

Substituting for I_2 in Equation (23-23),

$$V = I_1 R_1 + j I_1 X_1 + \frac{(\omega M)^2 I_1}{R_2 + j X_2}$$

from which the primary impedance is

$$Z_1 = \frac{V}{I_1}$$

or,

$$\boxed{Z_1 = R_1 + j X_1 + \frac{(\omega M)^2}{R_2 + j X_2}} \tag{23-25}$$

The primary circuit impedance is seen to consist of a *self-impedance* Z and a *coupled impedance* Z_2' (coupled from the secondary into the primary), where

$$\text{self-impedance } Z = R_1 + j X_1$$

and

$$\boxed{\textit{coupled impedance } Z_2' = \frac{(\omega M)^2}{R_2 + j X_2}} \tag{23-26}$$

The coupled impedance can be resolved into resistive and reactive components by multiplying it by the conjugate of the denominator (see Section 18-4):

$$
\begin{aligned}
Z_2' &= \frac{(\omega M)^2}{R_2 + j X_2} \times \frac{R_2 - j X_2}{R_2 - j X_2} \\
&= \frac{(\omega M)^2 R_2 - j (\omega M)^2 X_2}{R_2^2 + X_2^2}
\end{aligned}
$$

$$\boxed{Z_2' = \frac{(\omega M)^2 R_2}{R_2^2 + X_2^2} - j \frac{(\omega M)^2 X_2}{R_2^2 + X_2^2}} \tag{23-27}$$

It is seen that the coupled impedance is made up of a coupled resistance R_2' and a coupled reactance X_2':

$$\text{coupled resistance, } R_2' = \frac{(\omega M)^2 R_2}{R_2^2 + X_2^2} \qquad \text{(23-28)}$$

and

$$\text{coupled reactance, } X_2' = -j\,\frac{(\omega M)^2 X_2}{R_2^2 + X_2^2} \qquad \text{(23-29)}$$

COEFFICIENT OF COUPLING. Now consider the coupled circuits shown in Figure 23-14(a), in which the primary and secondary circuits include capacitors and are tuned to resonate at the same frequency. At the resonance frequency, the series reactance in each circuit is zero. Thus, when X_2 is zero, only the resistive component of the impedance is coupled from the secondary into the primary. Also, because of the presence of X_2 in the denominator of the expression for the coupled resistance [Equation (23-28)] the coupled resistance is a maximum when X_2 becomes zero (i.e., at the resonance frequency). This (maximum) coupled resistance adds to the coil resistance of the primary, and thus has the effect of reducing the primary current at resonance.

Figure 23-14(b) shows the equivalent circuit of the coupled circuits at resonance. R_1 and R_2 are the primary and secondary coil resistances, respectively, and R_2' is the secondary resistance coupled into the primary, where at resonance (with $X_2 = 0$) Equation 23-28 becomes,

$$R_2' = \frac{(\omega M)^2}{R_2} \qquad \text{(23-30)}$$

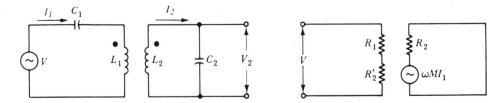

(a) Coupled coils tuned to resonate

(b) Equivalent circuit at resonance

FIGURE 23-14. Coupled coils may be tuned to resonate at the same frequency by including appropriate capacitors in the primary and secondary circuits. At resonance, X_{C1} and X_{C2} cancel X_{L1} and X_{L2} respectively, and maximum primary and secondary currents flow.

Viewed from the input terminals, the primary circuit (operating alone) is a series resonant circuit, with a maximum impedance at resonance:

$$Z = R_1$$

The secondary circuit behaves as a series resonant circuit with a supply voltage of $\omega M I_1$. The coupled resistance R_2' causes the maximum (resonance) impedance of the primary circuit to become

$$\boxed{Z_1 = R_1 + R_2'} \tag{23-31}$$

It is clear that the effect of R_2' is to increase Z, and consequently reduce the current flowing through the primary winding. The degree to which the primary current is affected by the secondary depends upon how tightly the coils are coupled, that is, upon the coefficient of coupling k.

The frequency response curves of I_1 and I_2 for *loosely coupled coils* are shown in Figure 23-15(a). The shape of the I_1 response curve is largely unaffected by the coupled impedance, and the I_2 curve is seen to be very much smaller than I_1 but with the same shape as the I_1 curve.

With *tight coupling* [Figure 23-15(b)], the coupled resistance has a considerable effect at the resonance frequency. The total primary impedance is increased, and the primary current is significantly reduced. Also, the primary response curve is distorted by the occurrence of two *humps,* one above and one below the resonance frequency. These humps result from the fact that the secondary impedance is capacitive below the resonance frequency and inductive above f_r. When coupled into the primary, the coupled impedances are inductive below f_r and capacitive above f_r. These result in minimum primary impedances, and therefore maximum currents, at two frequencies above and below resonance. Be-

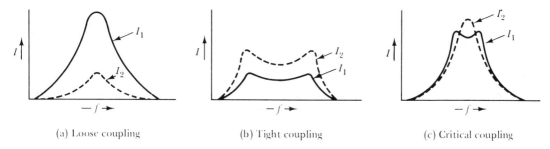

(a) Loose coupling   (b) Tight coupling   (c) Critical coupling

FIGURE 23-15. The current versus frequency response of tuned coupled coils depends upon the degree of coupling. Loose coupling gives low secondary current. Tight coupling produces distortion. Critical coupling affords maximum power transfer from primary to secondary.

cause of the tight coupling, the response curve of the secondary current is simply a slightly amplified version of the primary.

With *critical coupling* [Figure 23-15(c)], maximum power transfer occurs from the primary to the secondary, and thus the secondary current is greater than in either of the other two cases. The condition for maximum power transfer is that the source impedance and load impedance be equal (see Section 22-7). The two humps still occur in the primary response curve but are absent in the secondary. For maximum power transfer and critical coupling:

$$R_1 = R_2'$$

$$R_1 = \frac{(\omega M)^2}{R_2}$$

or

$$\boxed{M^2 = \frac{R_1 R_2}{\omega^2}} \qquad (23\text{-}32)$$

and from Equation (14-13),

$$M^2 = k^2 L_1 L_2$$

Therefore,

$$k^2 L_1 L_2 = \frac{R_1 R_2}{\omega^2}$$

$$k^2 = \frac{R_1}{\omega L_1} \times \frac{R_2}{\omega L_2} = \frac{1}{Q_1} \times \frac{1}{Q_2}$$

or

$$\boxed{\textit{for critical coupling, } k = \frac{1}{\sqrt{Q_1 Q_2}}} \qquad (23\text{-}33)$$

EXAMPLE 23-7 A 100 µH coil with a resistance of $R_1 = 20\ \Omega$ is connected in series with a capacitor and a supply voltage of 2.5 V rms. The supply frequency is 1 MHz and the capacitor is tuned for resonance at this frequency. A second coil with $L_2 = 50$ µH and $R_2 = 8\ \Omega$ is coupled to the first coil, and is also tuned for resonance at 1 MHz. Calculate the critical value of the coefficient of coupling, and determine the levels of the currents in each coil when critically coupled.

SOLUTION

$$Q_1 = \frac{\omega L_1}{R_1} = \frac{2\pi \times 1 \text{ MHz} \times 100 \text{ }\mu\text{H}}{20 \text{ }\Omega}$$

$$= 31.4$$

$$Q_2 = \frac{\omega L_2}{R_2} = \frac{2\pi \times 1 \text{ MHz} \times 50 \text{ }\mu\text{H}}{8 \text{ }\Omega}$$

$$= 39.3$$

Equation (23-33):

$$\text{critical } k = \frac{1}{\sqrt{Q_1 Q_2}}$$

$$= \frac{1}{\sqrt{31.4 \times 39.3}}$$

$$= \mathbf{0.028}$$

When critically coupled:

$$R_1 = R_2'$$

$$I_1 = \frac{V}{R_1 + R_2'}$$

$$= \frac{V}{2R_1} = \frac{2.5 \text{ V}}{2 \times 20 \text{ }\Omega}$$

$$= \mathbf{62.5 \text{ mA}}$$

From Equation (23-32):

$$(\omega M)^2 = R_1 R_2$$

$$\omega M = \sqrt{R_1 R_2}$$

$$= \sqrt{20 \text{ }\Omega \times 8 \text{ }\Omega}$$

$$= 12.65$$

secondary voltage, $e_2 = \omega M I_1$

and secondary current, $I_2 = \frac{e_2}{R_2}$

$$= \frac{\omega M I_1}{R_2} = \frac{12.65 \times 62.5 \text{ mA}}{8 \text{ }\Omega}$$

$$= \mathbf{98.8 \text{ mA}}$$

23-9.1 A primary coil with $R_1 = 12 \; \Omega$ and $L_1 = 68 \; \mu H$ is coupled to a secondary coil which has $R_2 = 12 \; \Omega$ and $L_2 = 75 \; \mu H$. The coils are in resonance at a supply frequency of 800 kHz and the coefficient of coupling between the coils is $k = 0.04$. Calculate the resistance that must be connected in series with the secondary coil to obtain critical coupling.

23-10

RESONANCE FILTERS

In Sections 19-9 and 19-10 it was shown that RC and LC circuits can be employed to *filter out* unwanted ac frequencies. These circuits were designated *low-pass filters* and *high-pass filters*. Resonance circuits can also be employed very effectively as filters. Because resonance occurs over a narrow band of frequencies, resonance filters are normally either *band-pass filters* or *band-stop filters*. This means that they will either pass or stop a band of frequencies.

BAND-PASS FILTERS. The circuit of a resonance band-pass filter is shown in Figure 23-16(a), and its equivalent circuit is drawn in Figure 23-16(b). The impedance of the series LC circuit is

$$Z = R_L + j\,(\omega L - 1/\omega C)$$

where R_L is the resistance of the inductor.

As shown in Figure 23-16(b), the input voltage V_i is potentially divided across Z and R_1 to give an output voltage of

$$V_o = \frac{V_i R_1}{R_1 + Z}$$

At frequencies far above and below resonance, Z is much greater than R_1, and consequently the output voltage V_o is very much smaller than V_i. However, at resonance $\omega L = 1/\omega C$ and the impedance of the series LC circuit becomes

$$Z = R_L$$

and the output voltage is

$$V_o = \frac{V_i R_1}{R_1 + R_L}$$

If R_L is very much smaller than R_1, very little of the input voltage is lost across R_L and the output voltage at resonance is almost equal to the input voltage.

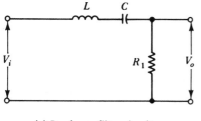

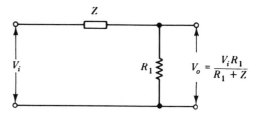

(a) Band-pass filter circuit

(b) Equivalent circuit

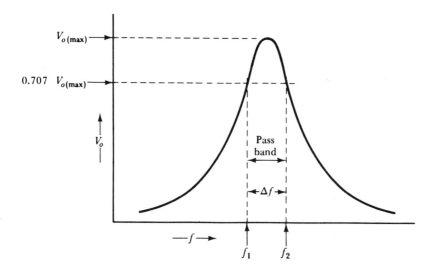

(c) Frequency response of band-pass filter

FIGURE 23-16. Band-pass filter using a series LCR resonance circuit. The output voltage developed across R is at a maximum for input frequencies within the bandwidth of the circuit.

The graph of V_o plotted versus frequency is shown in Figure 23-16(c). As discussed in Section 23-4, there is a band of frequencies over which the circuit is close to resonance. Thus, for the band-pass filter circuit of Figure 23-16(a), the output voltage remains high for the band of frequencies between f_1 and f_2 on Figure 23-16(c). This is said to be the *pass band* of the filter circuit. Recall that f_1 and f_2 are those frequencies at which V_o is 0.707 $V_{o\,(max)}$. Also recall that the width of the pass band is inversely proportional to the Q factor of the circuit (see Section 23-4). To calculate the circuit Q factor, the total resistance of the circuit must be taken as $R_1 + R_L$.

EXAMPLE 23-8 The band-pass filter circuit shown in Figure 23-16(a) has $L = 1$ mH, $C = 100$ pF, $R_1 = 150$ Ω, and $R_L = 15$ Ω. Determine the resonance frequency, Q factor, and pass band of the circuit.

SOLUTION

From Equation (23-3):

$$f_r = \frac{1}{2\pi\sqrt{LC}} = \frac{1}{2\pi\sqrt{1 \text{ mH} \times 100 \text{ pF}}}$$
$$\cong \textbf{500 kHz}$$

From Equation (23-10):

$$Q = \frac{1}{R_L + R_1}\sqrt{\frac{L}{C}}$$
$$= \frac{1}{15 \text{ }\Omega + 150 \text{ }\Omega}\sqrt{\frac{1 \text{ mH}}{100 \text{ pF}}}$$
$$= \textbf{19.2}$$

From Equation (23-17):

$$\Delta f = \frac{f_r}{Q} = \frac{500 \text{ kHz}}{19.2}$$
$$= \textbf{26 kHz}$$

$$f_1 = f_r - \frac{\Delta f}{2} = 500 \text{ kHz} - \frac{26 \text{ kHz}}{2}$$

$$= \textbf{487 kHz}$$

and $\quad f_2 = f_r + \dfrac{\Delta f}{2}$

$$= \textbf{513 kHz}$$

Another band-pass filter circuit, this time using a parallel resonance circuit, is shown in Figure 23-17(a). In this case the output voltage is developed across the resonance circuit. As illustrated in Figure 23-17(b), the input voltage is potentially divided across R_1 and Z, so that

$$V_o = \frac{V_i \times Z}{R_1 + Z}$$

At frequencies above and below resonance, Z is much smaller than R_1. At resonance Z is a maximum for a parallel LC circuit. If Z becomes much larger than R_1 at resonance, then V_o is just a little smaller than V_i. The frequency response for this filter circuit is exactly as shown for the previous circuit in Figure 23-16(c).

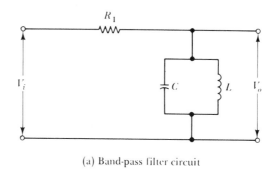

(a) Band-pass filter circuit

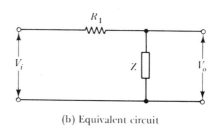

(b) Equivalent circuit

FIGURE 23-17. Band-pass filter using a parallel *LCR* resonance circuit. The output voltage developed across C and L is at a maximum for input frequencies within the bandwidth of the circuit.

BAND-STOP FILTER. Two *band-stop filter* circuits and their frequency response graph are shown in Figure 23-18. Consider the circuit in Figure 23-18(a). The output voltage V_o is developed across the series *LC* circuit. At frequencies away from resonance, the impedance of this series *LC* circuit will be much larger than R_1, while at resonance the impedance will be much smaller than R_1. Thus, at resonance, V_o becomes much smaller than V_i, and off resonance, V_o is approximately equal to V_i.

For the circuit of Figure 23-18(b), the parallel *LC* circuit has an impedance that is smaller than R_1 at frequencies off resonance, and much larger than R_1 at the resonance frequency. Again V_o is much smaller than V_i over a band of frequencies close to resonance. As shown in Figure 23-18(c), all frequencies except those in the *stop band* are passed by a band-stop filter. The frequency response graph shows $\Delta V_{o\,(max)}$ as the *voltage reduction* of the output at resonance. Frequenies f_1 and f_2 are those frequencies at which the *output reduction* is 0.707 of $\Delta V_{o\,(max)}$.

PRACTICE PROBLEM

23-10.1 A band-pass filter using a series resonance circuit, as in Figure 23-16 is to be designed to have a center frequency of 100 kHz and a bandwidth of 15 kHz. A 33 mH inductor with a 25 Ω

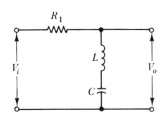

(a) Band-stop filter using series LC

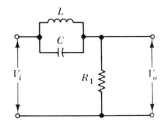

(b) Band-stop filter using parallel LC

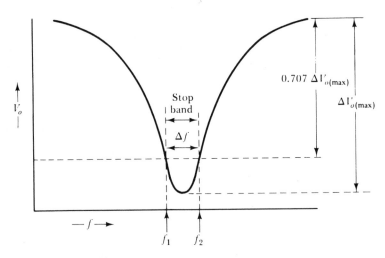

(c) Frequency response of band-stop filter

FIGURE 23-18. A series or parallel LCR resonance circuit may be employed as a band-stop filter. In each case, the output voltage is at a minimum for input frequencies within the bandwidth of the circuit.

coil resistance is to be used. Determine a suitable value for capacitor C and resistor R_1.

SUMMARY OF FORMULAS

Impedance of series LCR circuit:

$$Z = R + j(X_L - X_C)$$

At resonance:

$$Z = R$$

Resonance frequency for series resonance:

$$f_r = \frac{1}{2\pi\sqrt{LC}}$$

Q factor for series resonance:

$$Q = \frac{\omega L}{R}$$

$$Q = \frac{1}{\omega CR}$$

$$Q = \frac{1}{R}\sqrt{\frac{L}{C}}$$

Bandwidth:

$$\Delta f = f_2 - f_1$$

$$\Delta f = \frac{R}{2\pi L}$$

$$\Delta f = \frac{f_r}{Q}$$

For parallel LC circuit:

$$Y = \frac{1}{R_L + jX_L} + j\frac{1}{X_C}$$

At resonance:

$$Z = \frac{L}{CR_L}$$

Q factor for parallel resonance:

$$Q = \frac{\omega L}{R_L}$$

$$Q = \frac{1}{\omega CR_L}$$

Resonance frequency for parallel resonance:
When $Q > 10$:

$$f_r = \frac{1}{2\pi\sqrt{LC}}$$

When $Q < 10$:

$$f_r = \frac{1}{2\pi\sqrt{LC}}\sqrt{1 - \frac{CR_L^2}{L}}$$

For parallel LCR circuit:

$$Q = \frac{R_P}{\omega L}$$

For coupled coils:
Secondary current:

$$I_2 = \frac{\omega M I_1}{R_2 + jX_2}$$

Coupled impedance:

$$Z_2' = \frac{(\omega M)^2}{R_2 + jX_2}$$

Coupled resistance:

$$R_2' = \frac{(\omega M)^2 R_2}{R_2^2 + X_2^2}$$

At resonance:

$$R_2' = \frac{(\omega M)^2}{R_2}$$

Coupled reactance:

$$X_2' = -j\frac{(\omega M)^2 X_2}{R_2^2 + X_2^2}$$

At resonance:

$$Z_1 = R_1 + R_2'$$

For critical coupling:

$$R_1 = R_2'$$

$$M^2 = \frac{R_1 R_2}{\omega^2}$$

$$k = \frac{1}{\sqrt{Q_1 Q_2}}$$

REVIEW QUESTIONS

23-1 Sketch a series *LCR* circuit with an ac supply. Write the equation for the impedance of the circuit and explain what occurs at the resonance frequency.

23-2 Draw a sketch to show the graphs of inductive reactance and capacitive reactance plotted versus frequency. Also, show the graph of total impedance for a series *LCR* circuit. Briefly explain.

23-3 Sketch a phasor diagram for a series *LCR* circuit at resonance. Also, draw typical graphs showing the variation of impedance, current, and the current phase angle, above and below resonance.

23-4 Derive an equation for the resonance frequency of a series *LCR* circuit.

23-5 Define *Q* factor with respect to a series resonance circuit. Derive an expression for the *Q* factor of the inductance coil in a series resonance circuit. Also, derive an equation for the circuit *Q* factor in terms of *R*, *L*, and *C*.

23-6 Using illustrations, define half-power points, bandwidth, and selectivity, as applied to a resonance circuit. Derive an equation for the bandwidth of a resonance circuit in terms of the resonance frequency and the circuit Q factor.

23-7 Sketch a parallel LC circuit with some coil resistance and an ac supply. Write the equation for the circuit admittance and explain what occurs at resonance. Derive an expression for the impedance of the circuit at resonance.

23-8 Sketch the phasor diagram for the component currents in a parallel LC circuit at resonance. Briefly explain.

23-9 For a parallel LC circuit, sketch the typical graphs of circuit impedance, inductor current, and supply current, plotted versus frequency. Explain the shape of the graphs.

23-10 Derive equations for the Q factor and resonance frequency of a parallel LC circuit. Discuss any approximations made.

23-11 Explain resistance damping of a parallel resonant circuit, and write the equation for the Q factor of a parallel resonant circuit with resistance damping.

23-12 A coil supplied with an alternating voltage V is inductively coupled to another coil with a load resistance R_2. Sketch the circuit diagram for the coils, and sketch the equivalent circuit showing the voltage drops around the primary and secondary circuits. Derive equations for the resistance and reactance coupled from the secondary into the primary.

23-13 Two coupled coils each have a capacitor connected in series with them and are tuned to resonate at the same frequency. Explain the effect of resonance on the coupled reactance and resistance, and write the equation for the total primary circuit resistance at resonance.

23-14 Define the term *critical coupling* as applied to two coupled coils that are tuned to resonate at the same frequency. Derive an equation for the critical coupling factor, and sketch the graphs of primary current and secondary current versus frequency for various degrees of coupling between the coils. Explain the shapes of the graphs.

23-15 Sketch the circuits of a band-pass filter and a band-stop filter using series resonance. Sketch the frequency response graphs and explain briefly the operation of each circuit.

23-16 Sketch the circuits of a band-pass filter and a band-stop filter using parallel resonance. Sketch the frequency response graphs and explain briefly the operation of each circuit.

PROBLEMS

23-1   A series LCR circuit has $L = 506$ µH, $C = 200$ pF, $R = 32$ Ω, and a supply voltage of $V_s = 5$ V. Calculate the resonance frequency and determine the supply current at resonance.

23-2   For the resonance circuit in Problem 23-1, calculate the new capacitance value that will produce resonance at a frequency of 450 kHz.

23-3   For the resonance circuit in Problem 23-1, calculate the levels of supply current at frequencies of $0.25 f_r$, $0.5 f_r$, $0.8 f_r$, f_r, $1.25 f_r$, $2 f_r$, and $4 f_r$. Plot a graph of current to a logarithmic frequency base.

23-4   A series LCR circuit is to resonate at a frequency of 198 kHz. If $C = 320$ pF and $R = 390$ Ω, determine the required inductance. Also, calculate the circuit current at resonance when the signal voltage is 500 mV.

23-5   For the resonance circuit in Problem 23-1, calculate the component terminal voltages at frequencies of $0.25 f_r$, $0.5 f_r$, $0.8 f_r$, f_r, $1.25 f_r$, $2 f_r$, and $4 f_r$. Plot a graph of these voltages to a logarithmic frequency base and explain the resonance rise in voltage.

23-6   For the resonance circuit in Problem 23-4, calculate the component terminal voltages at frequencies of $0.25 f_r$, $0.5 f_r$, $0.8 f_r$, f_r, and $1.25 f_r$.

23-7   A series LCR circuit with an input of 7 V has $L = 600$ µH, $C = 150$ pF, $R = 20$ Ω, and a supply voltage of $V_s = 5$ V. Plot the graph of circuit current at frequencies of $0.25 f_r$, $0.5 f_r$, $0.8 f_r$, f_r, $1.25 f_r$, $2 f_r$, and $4 f_r$.

23-8   For the resonance circuit in Problem 23-7, calculate the component terminal voltages at frequencies of $0.25 f_r$, $0.5 f_r$, $0.8 f_r$, f_r, $1.25 f_r$, $2 f_r$, and $4 f_r$. Plot graph of these voltages to a logarithmic frequency base.

SECTION 23-2

23-9   A 750 µH inductance connected in series with a 47 Ω resistor and a variable capacitor is to be tuned to resonate at frequencies ranging from 100 kHz to 900 kHz. Determine the required range of adjustment of the capacitor.

23-10   Determine the range of adjustment of f_r for the circuit in Problem 23-9 if the capacitor is variable from 100 pF to 2000 pF, and a 10 pF stray capacitance is present in parallel with the capacitor.

23-11   The voltage across the capacitor in Problem 23-7 is monitored on an oscilloscope. If the oscilloscope has an input capacitance of 40 pF, determine its effect on the resonance frequency of the circuit.

23-12 If the 200 pF capacitor in Problem 23-1 is coupled into the resonance circuit by means of a series-connected 0.01 μF capacitor, calculate the circuit resonance frequency.

SECTION 23-3

23-13 Calculate the Q factors for the circuits in Problems 23-1 and 23-7. Also, determine the new Q factor for the circuit in Problem 23-7 when an oscilloscope is connected to the circuit as explained in Problem 23-11.

23-14 Calculate the Q factor of the circuit in Problem 23-4. Also, determine the Q factors of the circuit in Problem 23-9 at the maximum and minimum resonance frequencies.

SECTION 23-4

23-15 Determine the half-power frequencies and the bandwidth for the circuit described in Problem 23-1.

23-16 Determine the half-power frequencies and the bandwidth for the circuit described in Problem 23-4.

23-17 Calculate the half-power frequencies and the bandwidth for the circuit in Problem 23-7. Also, determine the new bandwidth when an oscilloscope is connected to the circuit as described in Problem 23-11.

23-18 The bandwidth in the circuit in Problem 23-1 is to be altered to 13 kHz. Determine how this can be done without affecting the resonance frequency of the circuit.

SECTION 23-5

23-19 An inductance with $L = 300$ μH and $R = 5$ Ω is connected in parallel with a capacitor having $C = 300$ pF. Determine the circuit impedance at resonance.

23-20 The capacitor in Problem 23-1 is reconnected in parallel with the series-connected inductor and resistor. Determine the circuit impedance at resonance. Also, calculate the effect of a 40 pF oscilloscope input capacitance in parallel with the capacitor.

SECTION 23-6

22-21 Calculate the resonance frequency Q factor for the circuit in Problem 23-19.

23-22 Calculate the Q factor for the circuit in Problem 23-20(a) without the 40 pF capacitance, (b) with the 40 pF capacitance.

SECTION 23-7

23-23 Determine the bandwidth of the circuit in Problem 23-19.

23-24 Calculate the upper and lower bandwidth frequency for the circuit in Problem 23-20, (a) without the 40 pF capacitance, (b) with the 40 pF capacitance.

23-25 An instrument with an unknown input capacitance is connected across the capacitor in Problem 23-19. The resonance frequency is found to be 505.8 kHz. Calculate the instrument input capacitance.

23-26 A parallel *LC* circuit which is to resonate at 270 kHz has a 600 μH inductor with a 37 Ω coil resistance. Calculate the required capacitance and determine the circuit *Q* factor and bandwidth.

SECTION 23-8

23-27 Calculate the damping resistance required to give a *Q* of 15 to the circuit described in Problem 23-19.

23-28 The parallel resonance circuit in Problem 23-26 is to be resistance damped to obtain a *Q* factor of 2. Calculate the required damping resistance.

SECTION 23-9

23-29 A 300 μH coil with a resistance of $R_1 = 15$ Ω is connected in series with a supply of 3.3 V and a 300 pF capacitor. A second coil with $L = 100$ μH and $R_2 = 10$ Ω is inductively coupled to the first coil and tuned to resonate at the same frequency as the first coil. Determine the critical value of the coefficient of coupling for the coils and calculate the current in each coil when critically coupled.

23-30 The resistance in the primary coil circuit in Problem 23-29 is to be adjusted to set the secondary current at 100 mA when the coils are critically coupled. Determine the required resistance for R_1 and the new coefficient of coupling.

SECTION 23-10

23-31 A band-stop filter using a parallel resonance circuit has $L = 5$ mH, $C = 250$ pF, $R_s = 12$ Ω, and $R_1 = 200$ Ω. Calculate the stop band of the circuit.

23-32 For the band-pass filter in Example 23-8, calculate the filter attenuation at frequencies of 300 kHz and 500 kHz.

23-33 The band-pass filter in Example 23-8 is to have its pass band expanded to 35 kHz by adjusting resistor R_1 [in Figure 23-16(a)]. Calculate the required resistance.

COMPUTER PROBLEM

23-34 Write a computer program to analyze a series resonance circuit to determine f_r. The program should also calculate the circuit current and component terminal voltages at frequencies of $0.25\,f_r$, $0.5\,f_r$, $0.8\,f_r$, f_r, $1.25\,f_r$, $2\,f_r$, and $4\,f_r$.

ANSWERS TO PRACTICE PROBLEMS

23-1.1	0.939 MHz
23-1.2	13.2 mA, 3.3 V, 13.2 V, 40.5 mA, 16.2 V, 25.4 V
23-2.1	544 kHz, 1.17 MHz
23-3.1	35, 75
23-4.1	15.6 kHz
23-5.1	53.8 kΩ, 10.1 kΩ
23-7.1	1.27 MHz, 550 kHz, 12.9 kHz, 99, 43
23-7.2	1.4 μH, 5 pF
23-8.1	1.41 kΩ
23-9.1	5.2 Ω
23-10.1	77 pF, 3.11 kΩ

24 TRANSFORMERS

Objectives

You will be able to:

Sketch the basic construction of a transformer and explain how it works.

Sketch the circuit diagram for a transformer and derive an equation relating the number of turns on each winding to the primary and secondary voltages.

Given input voltage and secondary load, calculate the secondary voltages and the primary and secondary currents of a transformer.

Derive an equation relating transformer primary or secondary voltages to core flux, supply frequency, and number of winding turns.

Calculate the level of flux in a transformer core.

Sketch and explain the phasor diagrams for a transformer under no-load conditions.

Sketch and explain the no-load equivalent circuit for a transformer.

Sketch and explain the complete equivalent circuit for a loaded transformer.

Sketch the phasor diagrams for the primary and secondary of a loaded transformer.

Given the parameters of a transformer and the load at the output terminals, determine the amplitudes and phase angles of the supply voltage and current.

Define referred resistance and referred reactance and show how the transformer equivalent circuit may be simplified by referring the secondary circuit components to the primary.

Given the parameters of a transformer and its load, refer all components to the primary and calculate the amplitude and phase angle of the input current.

Solve problems involving transformer efficiency and voltage regulation.

Analyze the results of open-circuit and short-circuit tests on transformers to determine the component values of the transformer equivalent circuit.

Explain autotransformers, current transformers, and audio transformers.

Introduction

A transformer basically consists of two coils wound on a single iron core. When an alternating voltage is applied to one of the coils, the mutual inductance causes an alternating voltage to be induced in the other coil. The ratio of the voltage amplitudes depends upon the number of turns on each coil. A transformer may be used either to increase or decrease an applied voltage, or to increase or decrease a current.

The emf induced in each winding of a transformer can be calculated from the *transformer emf equation*. The behavior of the transformer under *no-load* and *full-load* conditions is best understood by drawing the appropriate phasor diagrams and by studying the transformer equivalent circuit. Simplification of the equivalent circuit is possible by the process of *referring* the secondary resistance and reactance to the primary winding.

The performance of a transformer is described in terms of its *voltage regulation* and its *efficiency*. The transformer performance can be predicted from the results of two tests known as the *open-circuit test* and the *short-circuit test*.

24-1
PRINCIPLE OF TRANSFORMER OPERATION

TRANSFORMER CONSTRUCTION. The transformer is an application of mutual inductance. An alternating voltage applied to a *primary* winding generates an alternating magnetic flux which links with a *secondary* winding and induces an alternating emf in the secondary.

Figure 24-1(a) illustrates the basic construction of the transformer. The iron core forms a closed magnetic circuit, and the windings are simply coils of insulated wire wrapped around the core. The presence of the iron core causes virtually 100% of the magnetic flux generated by the primary to be linked with the secondary. The transformer is therefore similar to two *coupled coils* (see Section 14-4) in which the coefficient of coupling is 1. In fact, air-cored coupled coils are sometimes referred to as an *air-cored transformer*. The circuit symbol for the transformer is shown in Figure 24-1(b). The two lines between the coils indicate the presence of an iron core.

PRIMARY AND SECONDARY VOLTAGES. The alternating magnetic flux in the iron core of the transformer links with both the primary winding and the secondary winding. The flux linking with the primary winding generates a counter-emf in the primary. Since the counter-emf is essentially equal in magnitude to the supply voltage (see Sections 19-1 and 19-2), the counter-emf equation can be used to calculate the transformer primary voltage. From Equation (14-2),

$$E_p = \frac{\Delta \Phi}{\Delta t} N_p$$

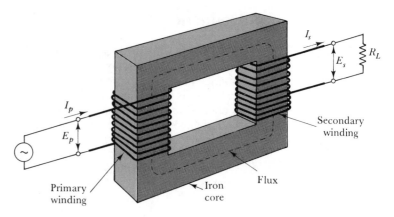

(a) Principle of transformer operation

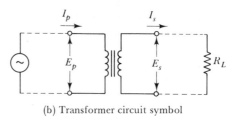

(b) Transformer circuit symbol

FIGURE 24-1. A transformer consists basically of two coils wound on an iron core. When an ac voltage is applied to the primary winding, the resultant core flux links with the secondary winding and induces a voltage in the secondary.

where E_p is the primary induced voltage and N_p is the number of turns on the primary winding. Also,

$$E_s = \frac{\Delta\Phi}{\Delta t} N_s$$

where E_s and N_s are the secondary induced voltage and the secondary turns, respectively. Combining the two equations,

$$\frac{E_p}{E_s} = \frac{\Delta\Phi/\Delta t}{\Delta\Phi/\Delta t} \times \frac{N_p}{N_s}$$

or

$$\frac{E_p}{E_s} = \frac{N_p}{N_s}$$

It is seen that the ratio of primary voltage to secondary voltage is the same as the ratio of primary coil turns to secondary turns. Where the primary (input) voltage and *turns ratio* are known, the secondary (output) voltage can be calculated from

$$E_s = E_p \times \frac{N_s}{N_p} \qquad (24\text{-}1)$$

When the primary and secondary windings have an equal number of turns, the output voltage is equal to the input voltage, and the transformer is referred to as a $1:1$ *transformer*. If the secondary has more turns than the primary, E_s is greater than E_p, and the device is termed a *step-up* transformer. This means that the transformer *steps up* the input voltage to produce a higher output level. It is also possible to have a secondary winding with fewer turns than the primary. In this case the output voltage is less than the input, and the device is a *step-down* transformer.

MULTI-OUTPUT TRANSFORMER. Figure 24-2 shows the circuit diagram of a transformer with three separate secondary windings. The output voltage from each winding depends upon the number of turns on that winding as well as upon the primary voltage and number of primary turns.

EXAMPLE 24-1 The transformer in Figure 24-2 has an alternating input of 100 V. Determine the output voltage from each secondary winding. Also, determine the total output voltage if all three secondary windings were connected series-aiding. The numbers of turns on each winding are $N_p=375$, $N_{s1} = 750$, $N_{s2} = 500$, and $N_{s3} = 75$.

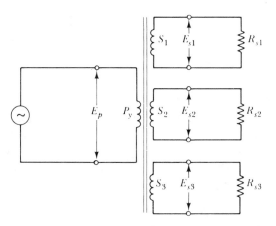

FIGURE 24-2. A transformer may have multiple secondary windings. The output voltage from each secondary depends upon the primary voltage and the ratio of primary to secondary turns.

SOLUTION

Equation (24-1):

$$E_s = E_p \times \frac{N_s}{N_p}$$

$$E_{s1} = E_p \times \frac{N_{s1}}{N_p} = 100 \text{ V} \times \frac{750}{375}$$

$$= \mathbf{200 \text{ V}}$$

$$E_{s2} = E_p \times \frac{N_{s2}}{N_p} = 100 \text{ V} \times \frac{500}{375}$$

$$= \mathbf{133 \text{ V}}$$

$$E_{s3} = E_p \times \frac{N_{s3}}{N_p} = 100 \text{ V} \times \frac{75}{375}$$

$$= \mathbf{20 \text{ V}}$$

$$\textit{total secondary turns} = N_{s1} + N_{s2} + N_{s3}$$

$$N_s = 750 + 500 + 75$$

$$= 1325$$

$$\textit{total secondary voltage, } E_s = E_p \times \frac{N_s}{N_p}$$

$$= 100 \text{ V} \times \frac{1325}{375}$$

$$= \mathbf{353 \text{ V}}$$

or

$$E_s = E_{s1} + E_{s2} + E_{s3}$$

$$= 200 \text{ V} + 133 \text{ V} + 20 \text{ V}$$

$$= \mathbf{353 \text{ V}}$$

It is important to note that the total output voltage calculation in Example 24-1 assumes that the secondary windings are connected *series-aiding*, that is, such that all the winding voltages are in phase. If one of the windings is connected so that its voltage is in antiphase to the others, this voltage will subtract from the others, and the connection is termed *series-opposing*. Figures 24-3(a) and (b) show a transformer that has two secondary windings. One of the secondaries gives a waveform with a peak level of 20 V, and the other has a peak level of 15 V. When the two are connected series-aiding, as illustrated in Figure 24-3(a), the resultant waveform has a peak output of (20 V + 15 V), i.e., 35 V. When connected series-opposing, as shown in Figure 24-3(b), the peak level of the

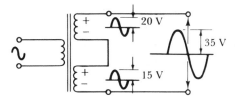

(a) Transformer with two secondary windings connected series-aiding

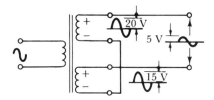

(b) Secondary windings connected series-opposing

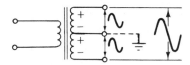

(c) Transformer with center-tapped secondary winding

FIGURE 24-3. A transformer with two secondary windings should have the secondaries connected series-aiding when the largest output voltage is required. A transformer with a center-tapped secondary produces two equal outputs.

lower winding is −15 V when the upper winding has a peak value of +20 V. Thus, the resultant output peak level is (20 V − 15 V) = 5 V.

The transformer shown diagrammatically in Figure 24-3(c) has a secondary that is said to be *center-tapped*. This means that the secondary winding consists of two equal-turn windings connected series-aiding. If the center tap is grounded, as shown dashed, the output from one terminal is positive-going with respect to ground, while the output from the other terminal is negative-going, and vice versa. The total output voltage from the two windings is the sum of the outputs from each.

PRIMARY AND SECONDARY CURRENTS. From Equation (21-7), the input power to a transformer is:

$$P_p = E_p I_p \cos \phi_p \quad \text{watts}$$

and the output power is

$$P_s = E_s I_s \cos \phi_s \quad \text{watts}$$

The primary and secondary circuit phase angles are closely equal, and as an approximation the transformer efficiency can be taken as 100%. Therefore,

$$P_p \cong P_s$$

or

$$E_p I_p = E_s I_s$$

which gives

$$\frac{I_p}{I_s} = \frac{E_s}{E_p} = \frac{N_s}{N_p}$$

or

$$\boxed{I_s = I_p \times \frac{N_p}{N_s}}$$ (24-2)

Comparing Equations (24-2) and (24-1), it is seen that the ratio of primary and secondary currents in terms of the turns ratio is the inverse of the ratio of primary and secondary voltages. This shows that *if a transformer steps up the voltage from the primary, it may be said to step down the current, and vice versa.*

EXAMPLE 24-2 The three secondary windings on the transformer described in Example 24-1 have the following loads: $R_{s1} = 1$ kΩ, $R_{s2} = 500$ Ω, and $R_{s3} = 100$ Ω. Assuming that the transformer is 100% efficient, determine the total primary current.

SOLUTION

For secondary S_1:

$$I_{s1} = \frac{E_{s1}}{R_{s1}} = \frac{200 \text{ V}}{1 \text{ k}\Omega}$$

$$= 200 \text{ mA}$$

From Equation (24-2),

$$I_p = I_s \times \frac{N_s}{N_p}$$

$$I_{p1} = I_{s1} \times \frac{N_{s1}}{N_p}$$

$$= 200 \text{ mA} \times \frac{750}{375}$$

$$= 400 \text{ mA}$$

For secondary S_2:

$$I_{s2} = \frac{E_{s2}}{R_{s2}} = \frac{133 \text{ V}}{500 \text{ }\Omega}$$

$$= 266 \text{ mA}$$

$$I_{p2} = I_{s2} \times \frac{N_{s2}}{N_p}$$

$$= 266 \text{ mA} \times \frac{500}{375}$$

$$= 355 \text{ mA}$$

For secondary S_3:

$$I_{s3} = \frac{E_{s3}}{R_{s3}} = \frac{20 \text{ V}}{100 \text{ }\Omega}$$

$$= 200 \text{ mA}$$

$$I_{p3} = I_{s3} \times \frac{N_{s3}}{N_p}$$

$$= 200 \text{ mA} \times \frac{75}{375}$$

$$= 40 \text{ mA}$$

Total primary current:

$$I_p = I_{p1} + I_{p2} + I_{p3}$$

$$= 400 \text{ mA} + 355 \text{ mA} + 40 \text{ mA}$$

$$\mathbf{I_p = 795 \text{ mA}}$$

PRACTICE PROBLEMS

24-1.1 A transformer is to produce peak outputs of 15 V and 22 V. The primary winding has 700 turns, and the input voltage is 115 V rms. Calculate the required number of turns for each of the secondary windings.

24-1.2 The transformer in Problem 24-1.1 has 470 Ω loads on each of the secondary windings. Determine the supply current.

24-2
EMF EQUATION

As discussed in the preceding section, the emfs induced in the primary and secondary windings are

$$E_p = \frac{\Delta\Phi}{\Delta t} N_p \quad \text{and} \quad E_s = \frac{\Delta\Phi}{\Delta t} N_s$$

The flux in the transformer core has a sinusoidal waveform, because the (primary) current producing it is sinusoidal. Therefore, as illustrated in Figure 24-4, the flux increases from zero to its maximum value Φ_m in a time period of

$$\Delta t = \tfrac{1}{4} T$$

or

$$\Delta t = \frac{1}{4f}$$

where f is the frequency of the flux waveform (i.e., the frequency of the primary voltage and current). Consequently, the average induced voltage is

$$E_{av} = \frac{\Phi_m}{1/(4f)} N$$

or

$$E_{av} = 4\Phi_m f N$$

For a sine wave:

$$E_{rms} = \frac{0.707}{0.637} E_{av} \qquad \text{(see Section 17-5)}$$

$$= 1.11 \times E_{av}$$

so

$$E_{rms} = 1.11 \times 4\Phi_m f N$$

$$= 4.44\Phi_m f N$$

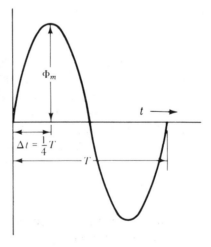

FIGURE 24-4. Because the flux in a transformer core has a sinusoidal waveform, it changes from zero to maximum in a time of $T/4$ or $1/(4f)$. The voltage induced in each winding is $4.44\Phi_m f N$.

This gives the primary rms voltage as:

$$E_p = 4.44\Phi_m f N_p \qquad \text{(24-3)}$$

and the secondary as:

$$E_s = 4.44\Phi_m f N_s \qquad \text{(24-4)}$$

where Φ_m is in webers, f is in hertz, and E_s and E_p are in volts.

Equations (24-3) and (24-4) enable the maximum value of flux in a transformer core to be calculated from a knowledge of the coil turns, coil voltage, and supply frequency. Once the maximum flux is known, the flux density in the core can be determined using the core dimensions.

EXAMPLE 24-3 The input voltage to the transformer described in Example 24-1 has a frequency of 400 Hz. Determine the peak value of the flux.

SOLUTION

From Equation (24-3):

$$\Phi_m = \frac{E_p}{4.44\, f\, N_p}$$

$$= \frac{100\,\text{V}}{4.44 \times 400\,\text{Hz} \times 375}$$

$$\Phi_m = 150\ \mu\text{Wb}$$

PRACTICE PROBLEMS

24-2.1 Determine the peak flux in the core of the transformer described in Problem 24-1.1 if the supply frequency is 60 Hz.

24-2.2 A transformer, designed to operate from a 50 V, 400 Hz supply, has 390 primary turns. If the transformer is used on a 60 Hz supply, determine the required level of primary voltage for the core flux to remain constant.

24-3

TRANSFORMER ON NO-LOAD

CORE LOSSES. A transformer is said to be on *no-load* when the output (secondary) terminals are open-circuited. In this condition there is no current flowing in the secondary windings. However, with an alternating

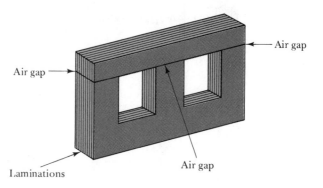

FIGURE 24-5. Energy losses occur in a transformer core due to eddy currents and air gaps. The primary no-load current must supply these core losses as well as creating a magnetic flux in the core.

voltage applied to the input (primary) terminals, a small primary current flows in order to create the magnetic flux in the core. This current is termed the *magnetizing current.*

In Section 12-8 it was explained that when the magnetic flux in a core is continuously increasing to a peak level in one direction and then reversing to a peak in the opposite direction, energy is aborbed by the core, due to *hysteresis* loss. The alternating magnetic flux also induces *eddy currents* in the transformer core, as discussed in Section 12-9. The eddy currents cause additional energy to be dissipated in the core. Hysteresis loss is kept to a minimum by the use of a magnetic material with a narrow hysteresis loop (see Section 12-8), and eddy current loss is minimized by constructing the core of *laminations* (see Section 12-9). Because of the laminated construction, there are air gaps in the core, as illustrated in Figure 24-5. The magnetizing current must create a flux in the air gaps as well as in the core. A further energy loss that must be supplied is the power dissipated in the transformer primary windings (the I^2R loss). The total primary *no-load current* is composed of the magnetizing current and the current required to supply the core losses.

NO-LOAD PHASOR DIAGRAM. An approximate phasor diagram for a transformer under no-load conditions is shown in Figure 24-6. Since the transformer windings are inductive, the input voltage V_p leads the magnetizing current I_{mag} by 90°. The flux Φ increases and decreases as the magnetizing current rises and falls; consequently, the flux phasor is shown in phase with I_{mag}. The current I_c is the component of the primary current which supplies the core losses and the small power loss in the primary winding. The total power losses are equal to $(I_c \times V_p)$, and so I_c is in phase with V_p. The no-load primary current I_o is the phasor sum of I_c and I_{mag}. Since I_c is normally much smaller than I_{mag}, the no-load power factor ($\cos \phi_1$) is very small. The voltages induced in the secondary and primary windings, E_s and E_p, respectively, lag the flux by 90°, and thus

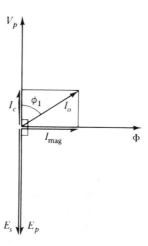

FIGURE 24-6. Approximate phasor diagram for a transformer on no-load. Core flux Φ is in-phase with the magnetizing current I_{mag}. The windings are inductive, so I_{mag} lags the primary voltage V_p by 90°.

the E_s and E_p phasors are drawn opposite to the V_p phasor. In Figure 24-6, E_s and E_p are shown as equal voltages (i.e., assuming a $1:1$ transformer). An approximation in the phasor diagram of Figure 24-6 occurs because E_p has been taken as exactly equal and opposite to V_p. In fact, E_p is equal and opposite to the phasor sum of V_p and the winding voltage drops due to I_o. This becomes more evident when a transformer *on-load* is considered.

NO-LOAD EQUIVALENT CIRCUIT. The no-load equivalent circuit for the transformer is shown in Figure 24-7. The transformer is replaced by

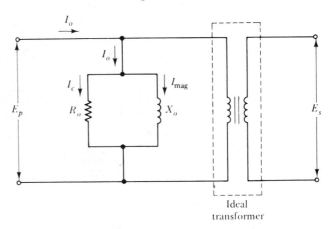

FIGURE 24-7. The no-load equivalent circuit of a transformer is made up of an ideal (no loss) transformer with its primary in parallel with a resistance R_o representing core losses, and an inductive reactance X_o to represent the winding inductance.

an ideal (no-loss) transformer, with a resistance R_o and an inductive reactance X_o in parallel with its primary. R_o represents the core losses, and so the current I_c which supplies the core losses is shown passing through R_o. The inductive reactance X_o represents a loss-free coil which passes the magnetizing current I_{mag}. Thus, the combination of R_o, X_o, and the ideal transformer simulates the actual transformer under no-load conditions.

24-4
TRANSFORMER ON LOAD

LEAKAGE INDUCTANCE. When a load is connected to a transformer secondary terminal, a secondary winding (load) current flows. As illustrated in Figure 24-8, the secondary current tends to generate a flux Φ_2 in the transformer core. To supply the secondary current, current must flow in the primary winding. The primary current generates a flux Φ_1 which is exactly equal and opposite to Φ_2. Thus, Φ_1 and Φ_2 cancel each other out, and the core flux remains at the level set up by the magnetizing current.

Figure 24-8 also shows that when primary and secondary currents flow, not all the flux set up by the currents passes through the iron core. Instead, there is some *leakage flux* passing through the air surrounding each coil. Since the magnetic path through the iron core has a very much smaller reluctance than the air path around each coil, the leakage flux is normally quite small. However, the leakage flux links with the winding turns in each coil, and sets up emfs that oppose the flow of current through each coil. Thus, the leakage flux produces the same effect as an unwanted inductance connected in series with each winding. The effect is termed the *leakage inductance*.

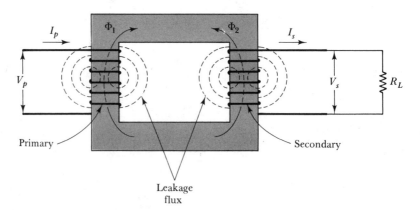

FIGURE 24-8. The currents flowing in the primary and secondary windings of a transformer set up core fluxes Φ_1 and Φ_2. Leakage flux also occurs around each winding.

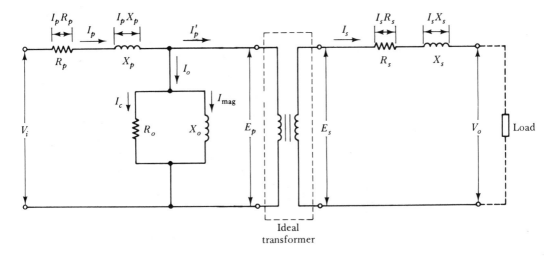

FIGURE 24-9. The complete equivalent circuit for a transformer is simply the no-load equivalent circuit (Figure 24-7) with components R_p, X_p, R_s, and X_s included to represent the winding resistances and leakage inductances.

COMPLETE EQUIVALENT CIRCUIT. The complete equivalent circuit for a transformer is shown in Figure 24-9. Inductive reactances X_p and X_s represent the leakage inductances of the primary and secondary windings, respectively, and R_p and R_s represent the windings resistances. An ideal (no-loss) transformer is shown with R_s and X_s connected in series with the secondary, so an output (load) current causes a voltage drop across R_s and X_s. Similarly, R_p and X_p are connected in series with the primary windings, and voltage drops are produced across them when a primary current flows. As before, R_o and X_o are shown in parallel with the primary, to simulate the no-load losses and the magnetizing current.

PHASOR DIAGRAM FOR SECONDARY. The phasor diagram for the secondary circuit of a transformer under load is shown in Figure 24-10. Referring to Figure 24-9 and Figure 24-10(a), V_o is the voltage at the transformer output terminals, and I_s is the secondary (load) current. For a load with a lagging phase angle ϕ_o, the I_s phasor is shown lagging the V_o phasor by angle ϕ_o. The secondary current flows through R_s and X_s and produces voltage drops across them. $I_s R_s$ is the resistive voltage drop; consequently, its phasor is in phase with I_s. $I_s X_s$ is an inductive voltage drop, so it leads the current by 90°, as shown in Figure 24-10(a).

Turning to Figure 24-10(b), the phasors of V_o, I_s, $I_s X_s$, and $I_s R_s$ are reproduced from Figure 24-10(a). The secondary induced voltage E_s (also see Figure 24-9) is the phasor sum of V_o, $I_s R_s$, and $I_s X_s$. From the discussion of phasor addition in Chapter 18:

$$E_s = \sqrt{(V_o \cos \phi_o + I_s R_s)^2 + (V_o \sin \phi_o + I_s X_s)^2} \qquad (24\text{-}5)$$

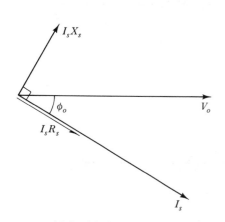

(a) Partial phasor diagram
for secondary circuit
with an inductive
(lagging φ) load.

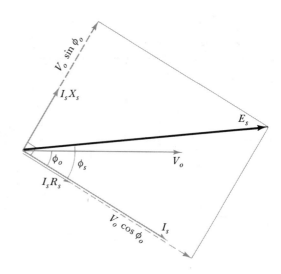

(b) E_s is the phasor sum
of V_o, $I_s X_s$ and $I_s R_s$

FIGURE 24-10. Phasor diagram for the secondary of a transformer under load. Secondary current I_s lags the output voltage V_o by the load phase angle ϕ_o. $I_s X_s$ and $I_s R_s$ are the voltage drops across X_s and R_s. The secondary induced voltage E_s is the phasor sum of V_o, $I_s X_s$, and $I_s R_s$.

and the phase angle between E_s and I_s is

$$\phi_s = \arctan\left(\frac{V_o \sin \phi_o + I_s X_s}{V_o \cos \phi_o + I_s R_s}\right) \qquad (24\text{-}6)$$

The secondary induced voltage always lags 90° behind the core flux Φ, so the flux phasor may be drawn 90° ahead of E_s.

EXAMPLE 24-4 A transformer with an output of 50 V and a load of $Z_L = 25 \ \Omega/30°$ has $R_s = 0.25 \ \Omega$, $X_s = 1 \ \Omega$, and $N_p/N_s = 2$. Determine the secondary current, the secondary induced voltage, and the angle between the two.

SOLUTION

Secondary circuit calculations (refer to Figures 24-9 and 24-10):

$$I_s = V_o/Z_L = 50 \text{ V}/(25 \ \Omega/30°)$$

$$= 2 \text{ A}/{-30°}$$

R_s volts drop,
$$I_s R_s = 2 \text{ A} \times 0.25 \text{ } \Omega$$
$$= 0.5 \text{ V}$$

X_s volts drop,
$$I_s X_s = 2 \text{ A} \times 1 \text{ } \Omega$$
$$= 2 \text{ V}$$

Phasor sum of V_o, $I_s R_s$, and $I_s X_s$, Equation 24-5:

$$E_s = \sqrt{(V_o \cos \phi_o + I_s R_s)^2 + (V_o \sin \phi_o + I_s X_s)^2}$$
$$= \sqrt{(50 \text{ V} \cos 30° + 0.5 \text{ V})^2 + (50 \text{ V} \sin 30° + 2 \text{ V})^2}$$
$$= \mathbf{51.5 \text{ V}}$$

Equation (24-6):

$$\phi_s = \arctan \left(\frac{V_o \sin \phi_o + I_s X_s}{V_o \cos \phi_o + I_s R_s} \right)$$

$$= \arctan \left(\frac{50 \text{ V} \sin 30° + 2 \text{ V}}{50 \text{ V} \cos 30° + 0.5 \text{ V}} \right)$$

$$= \mathbf{31.7°}$$

PHASOR DIAGRAM FOR PRIMARY. The phasor diagram for the transformer primary may be constructed in a similar way to that just discussed for the secondary. This time it is necessary to commence with the voltage E_p and current I_p' right at the primary winding (see Figure 24-9). E_p and I_p' may be calculated from a knowledge of the turns ratio and the secondary current and voltage. The phase angle between them is ϕ_s, the phase angle of the secondary circuit.

Referring to Figure 24-11(a), the $-E_p$ phasor is first drawn horizontally, and the I_p' current phasor is drawn at an angle of ϕ_s lagging $-E_p$. Note that because E_p is the voltage induced in the primary winding by the changing core flux, it is equal and opposite to the component of the applied voltage at the (ideal) transformer primary winding. Consequently, the applied voltage component is $-E_p$.

As well as I_p', the no-load current I_o (composed of I_c and I_{mag}) must be supplied. The actual current drawn from the supply is I_p, which is the phasor sum of I_p', I_c, and I_{mag} [see Figure 24-9 and 24-11(b)].

$$\boxed{I_p = \sqrt{(I_p' \cos \phi_s + I_c)^2 + (I_p' \sin \phi_s + I_{mag})^2}} \qquad \text{(24-7)}$$

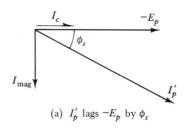

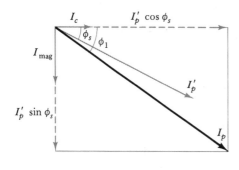

(a) I'_p lags $-E_p$ by ϕ_s

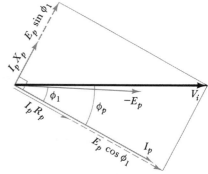

(b) I_p is the phasor sum of
I'_p, I_{mag}, and I_c

(c) V_i is the phasor sum of
$-E_p$, $I_p X_p$, and $I_p R_p$

FIGURE 24-11. Phasor diagram for the primary of a transformer under load. Primary current I'_p lags the primary voltage $-E_p$ by the secondary phase angle ϕ_s. No-load currents I_c and I_{mag} are also included, and the total primary current I_p is the phasor sum of I'_p, I_c, and I_{mag}. The supply voltage V_i is the phasor sum of $-E_p$, $I_p X_p$, and $I_p R_p$.

and

$$\phi_1 = \arctan\left(\frac{I'_p \sin \phi_s + I_{mag}}{I'_p \cos \phi_s + I_c}\right) \qquad (24\text{-}8)$$

I_p causes a voltage drop along R_p and X_p (Figure 24-9). The $I_p R_p$ phasor is in phase with I_p, and the $I_p X_p$ phasor leads I_p by 90°, as shown in Figure 24-11(c). The phasor sum of $I_p R_p$, $I_p X_p$, and $-E_p$ gives the supply voltage V_i. The primary input phase angle is then ϕ_p, which is the angle between V_i and I_p, as illustrated.

Given the secondary load and the parameters of the transformer equivalent circuit, the primary input voltage and current can be calculated by means of the phasor diagram.

EXAMPLE 24-5 For the transformer in Example 24-4, $R_p = 2.5\ \Omega$, $X_p = 6\ \Omega$, $R_o = 5\ k\Omega$, and $X_o = 2\ k\Omega$. Determine the supply voltage and current.

SOLUTION

Refer to Figures 24-9 and 24-11:

$$I_p' = \frac{N_s}{N_p} \times I_s = \frac{1}{2} \times 2 \text{ A}$$

$$= 1 \text{ A}$$

$$E_p = \frac{N_p}{N_s} \times E_s = \frac{2}{1} \times 51.5 \text{ V} = 103 \text{ V}$$

$$I_c = \frac{E_p}{R_o} = \frac{103 \text{ V}}{5 \text{ k}\Omega} = 20.6 \text{ mA}$$

$$I_{mag} = \frac{E_p}{X_o} = \frac{103 \text{ V}}{2 \text{ k}\Omega} = 51.5 \text{ mA}$$

Equation (24-7):

$$I_p = \sqrt{(I_p' \cos \phi_s + I_c)^2 + (I_p' \sin \phi_s + I_{mag})^2}$$

$$= \sqrt{(1 \text{ A} \cos 31.7° + 20.6 \text{ mA})^2 + (1 \text{ A} \sin 31.7° + 51.5 \text{ mA})^2}$$

$$= \mathbf{1.05 \ A}$$

Equation (24-8):

$$\phi_1 = \arctan \left(\frac{1 \text{ A} \sin 31.7° + 51.5 \text{ mA}}{1 \text{ A} \cos 31.7° + 20.6 \text{ mA}} \right)$$

$$= 33.5°$$

$$I_p R_p = 1.05 \text{ A} \times 2.5 \ \Omega = 2.63 \text{ V}$$

$$I_p X_p = 1.05 \text{ A} \times 6 \ \Omega = 6.3 \text{ V}$$

$$V_i = \sqrt{(E_p \cos \phi_1 + I_p R_p)^2 + (E_p \sin \phi_1 + I_p X_p)^2}$$

[*see* Figure 24-11(c)]

$$= \sqrt{(103 \text{ V} \cos 33.5° + 2.63 \text{ V})^2 + (103 \text{ V} \sin 33.5° + 6.3 \text{ V})^2}$$

$$= \mathbf{108.7 \ V}$$

$$\phi_p = \arctan \left(\frac{103 \text{ V} \sin 33.5° + 6.3 \text{ V}}{103 \text{ V} \cos 33.5° + 2.63 \text{ V}} \right)$$

$$= \mathbf{35.5°}$$

PRACTICE PROBLEMS

24-4.1 A transformer has an output of 35 V, $R_s = 0.3 \ \Omega$, $X_s = 0.8 \ \Omega$, and a load of $Z_L = 144 \ \Omega \underline{/45°}$. Calculate the secondary current and the secondary induced voltage.

24-4.2 The transformer in Problem 24-4.1 has $N_p/N_s = 4/3$, $R_p = 3.3 \ \Omega$, $X_p = 4.2 \ \Omega$, $R_o = 14 \text{ k}\Omega$, and $X_o = 3.6 \text{ k}\Omega$. Determine the input current and voltage.

24-5
REFERRED RESISTANCE AND REACTANCE

REFERRED QUANTITIES. The equivalent circuit of a transformer can be considerably simplified by replacing the secondary circuit resistive and reactive components by primary circuit components that have the same effect. Consider Figure 24-12(a). The secondary load resistance R_L can obviously be written as

$$R_L = \frac{E_s}{I_s}$$

Also, the primary circuit *sees* a resistance of

$$R = \frac{E_p}{I_p'}$$

But
$$E_p = E_s \left(\frac{N_p}{N_s}\right) \quad \text{and} \quad I_p' = I_s\left(\frac{N_s}{N_p}\right)$$

Therefore,
$$R = \frac{E_s\left(\dfrac{N_p}{N_s}\right)}{I_s\left(\dfrac{N_s}{N_p}\right)}$$

$$= R_L \left(\frac{N_p}{N_s}\right)^2$$

Writing,
$$\frac{N_p}{N_s} = a$$

$$\boxed{R = a^2 R_L} \tag{24-9}$$

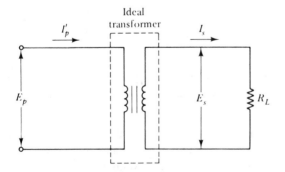

(a) Ideal transformer with load resistance R_L

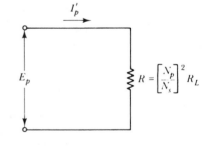

(b) Load resistance R_L referred to the primary

FIGURE 24-12. A load connected to the secondary of a transformer may be referred (or reflected) to the primary to simplify circuitry and calculations.

Thus, a primary resistance of $a^2 R_L$ would have the same effect on the transformer primary circuit as a secondary resistance of R_L [Figure 24-12(b)]. The primary equivalent of the secondary resistance calculated in this way is termed *the referred resistance* or the *reflected resistance* (i.e., the resistance of the secondary is said to be *referred* or *reflected* to the primary).

The secondary resistance, reactance, and load impedance may all be referred to the primary:

$$\boxed{\textit{referred reactance} = a^2 X_s} \qquad \textbf{(24-10)}$$

$$\boxed{\textit{referred load} = a^2 Z_L} \qquad \textbf{(24-11)}$$

SIMPLIFICATION OF EQUIVALENT CIRCUIT. Figure 24-13 illustrates the simplification of the transformer equivalent circuit by the technique of referring everything to the primary. R_s, X_s, and Z_L become $a^2 R_s$, $a^2 X_s$, and $a^2 Z_L$, respectively, when referred to the primary circuit [see Figures 24-13(a) and (b)]. Note that these referred components are *seen looking into* the primary winding of the ideal transformer. Consequently, the three of them (in series) must be shown in parallel with R_o and X_o in the primary equivalent circuit.

As an approximation to further simplify the equivalent circuit of a transformer on load, R_o and X_o may be omitted. This is because I_c and I_{mag} are each a very small percentage of the total load current I_p. The total primary resistive and reactive components may now be added together [Figure 24-13(c)] to give a primary equivalent resistance and equivalent reactance:

$$\boxed{R_e = R_p + a^2 R_s} \qquad \textbf{(24-12)}$$

and

$$\boxed{X_e = X_p + a^2 X_s} \qquad \textbf{(24-13)}$$

The referred load impedance remains $a^2 Z_L$; however, it may also be taken care of by adding its resistive and reactive components to R_e and X_e.

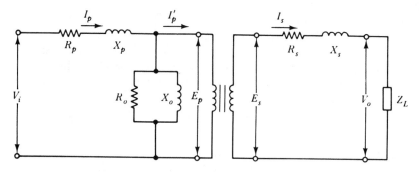

(a) Complete equivalent circuit

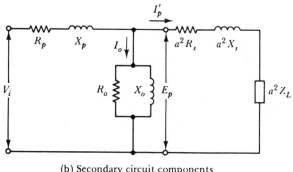

(b) Secondary circuit components
referred to the primary

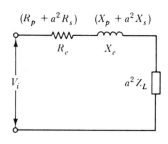

(c) Equivalent circuit simplified by
neglecting R_o and X_o, and by
combining R_p with $a^2 R_s$ and
X_p with $a^2 X_s$

FIGURE 24-13. The complete equivalent circuit of a transformer may be greatly
simplified by referring all of the secondary circuit components to the primary.

EXAMPLE 24-6 For the transformer circuit described in Examples 24-4 and 24-5, refer
all secondary components to the primary. Neglecting R_o and X_o,
calculate I_p and ϕ_p when V_i is 108.7 V.

SOLUTION

$$Z_L = 25 \ \Omega \underline{/30°}$$

$$= R_L + jX_L$$

$$= 25 \cos 30° + j25 \sin 30°$$

$$Z_L = 21.65 \ \Omega + j12.5 \ \Omega$$

total secondary resistance,

$$R_s' = R_s + R_L$$

$$= 0.25 \ \Omega + 21.65 \ \Omega$$

$$R_s' = 21.9 \ \Omega$$

total secondary reactance,

$$X'_s = X_s + X_L$$
$$= 1\ \Omega + 12.5\ \Omega$$
$$X'_s = 13.5\ \Omega$$

$$R'_s \text{ referred to primary} = a^2 R'_s$$
$$= 2^2 \times 21.9$$
$$= 87.6\ \Omega$$

$$X'_s \text{ referred to primary} = a^2 X'_s = 2^2 \times 13.5\ \Omega$$
$$= 54\ \Omega$$

total equivalent primary resistance,

$$R_e = R_p + a^2 R'_s = 2.5\ \Omega + 87.6\ \Omega$$
$$= 90.1\ \Omega$$

total equivalent primary reactance,

$$X_e = X_p + a^2 X'_s = 6\ \Omega + 54\ \Omega$$
$$= 60\ \Omega$$
$$R_e + jX_e = 90.1\ \Omega + j60\ \Omega$$
$$|I_p| = \frac{V_i}{\sqrt{R_e^2 + X_e^2}}$$
$$= \frac{108.7\ \text{V}}{\sqrt{(90.1\ \Omega)^2 + (60\ \Omega)^2}}$$
$$|I_p| = \mathbf{1.004\ A}$$

$$\phi_p = \arctan\left(\frac{X_e}{R_e}\right) = \arctan\left(\frac{60\ \Omega}{90.1\ \Omega}\right)$$
$$\boldsymbol{\phi_p = 33.66°}$$

Note that I_p and ϕ_p are slightly different from the values calculated in Example 24-5. This is because R_o and X_o have been neglected.

PRACTICE PROBLEMS

24-5.1   Refer all components in the secondary of the transformer described in Problems 24-4.1 and 24-4.2 to the primary. Neglecting the no-load circuit quantities, determine the primary equivalent circuit components.

24-5.2 A transformer is to supply 500 mW to a 12 Ω load. The input is 25 V peak-to-peak. Neglecting all core losses, and winding resistance and reactance, calculate the required turns ratio.

24-6

TRANSFORMER VOLTAGE REGULATION

It is clear from the transformer equivalent circuit in Figure 24-13(a) that the secondary current I_s produces voltage drops $I_s R_s$ and $I_s X_s$ across the resistive and reactive components. Also, the primary current I_p causes primary circuit voltage drops $I_p R_p$ and $I_p X_p$. Consequently, the effective primary voltage E_p is less than the input V_i, and the output voltage V_o is less than the calculated value of E_s.

When there is no load connected to the output terminals of the transformer, no secondary current flows, and therefore, no voltage drops occur across R_s and X_s. With zero secondary current, the primary current drops to the no-load current I_o, and the voltage drops across R_p and X_p become very small. Thus, in the no-load situation, E_p is almost equal to V_i, and V_o equals E_s.

It appears that the transformer output voltage is greatest on no-load, and that under loaded conditions the voltage drops across the resistive and reactive components of the equivalent circuit cause V_o to drop below its no-load level. (Note that, depending upon the power factor of the load, the output full-load voltage may actually be larger than the no-load voltage.) The percentage change in output voltage from no-load to full load is termed the *voltage regulation* of the transformer. Ideally, there should be no change in V_o from no-load to full-load (i.e., regulation = 0%). For best possible performance, the transformer should have the lowest possible regulation.

$$\text{voltage regulation} = \frac{V_{o(NL)} - V_{o(FL)}}{V_{o(FL)}} \times 100\% \qquad (24\text{-}14)$$

where $V_{o(NL)}$ is the transformer no-load output voltage, and $V_{o(FL)}$ is the full-load output voltage.

EXAMPLE 24-7 Neglecting the no-load current, calculate the voltage regulation for the transformer described in Examples 24-4 and 24-5.

SOLUTION

full-load output voltage, $V_{o(FL)} = 50$ V

input voltage, $V_i = 108.7$ V

At no-load:

$$E_p \cong V_i$$

and

$$E_s = \frac{N_s}{N_p} \times E_p$$

$$= \frac{1}{2} \times 108.7 \text{ V}$$

$$E_s = 54.35 \text{ V}$$

no-load output voltage, $V_{o(NL)} = E_s = 54.35 \text{ V}$

From Equation (24-14):

$$voltage\ regulation = \frac{54.35 \text{ V} - 50 \text{ V}}{50 \text{ V}} \times 100\%$$

voltage regulation = 8.7%

PRACTICE PROBLEM

24-6.1   A transformer with a 115 V input, $N_p = 600$, and $N_s = 329$, has a specified voltage regulation of 7%. Calculate the full-load output voltage.

24-7
TRANSFORMER EFFICIENCY

The efficiency of a transformer, like that of any other piece of equipment, is the output power expressed as a percentage of the input power.

$$\boxed{efficiency,\ \eta = \frac{P_o}{P_i} \times 100\%} \qquad (24\text{-}15)$$

or

$$\eta = \frac{V_o I_s \cos \phi_s}{V_i I_p \cos \phi_p} \times 100\%$$

Since

$$P_i = P_o + \text{losses}$$

$$\boxed{\eta = \frac{P_o}{P_o + \text{losses}} \times 100\%} \qquad (24\text{-}16)$$

The power losses in a transformer consist of core losses due to hysteresis and eddy currents, and copper losses due to the currents flowing in the primary and secondary windings (see Sections 24-3 and

24-4). As long as the supply frequency remains constant, the core losses tend to be a constant quantity. The copper losses have two components:

$$\text{primary winding copper loss} = I_p^2 R_p$$

$$\text{secondary winding copper loss} = I_s^2 R_s$$

The two copper losses can be lumped into one power loss proportional to the load current, if R_p is referred to the *secondary* winding, and added to R_s. Thus, where R_{es} is the equivalent of (R_p referred $+R_s$):

$$\text{total copper losses in both windings} = I_s^2 R_{es}$$

Now, rewriting Equation (24-16):

$$\eta = \frac{V_o I_s \cos \phi_s}{V_o I_s \cos \phi_s + I_s^2 R_{es} + P_c} \times 100\%$$

where $V_o I_s \cos \phi_s$ is the output power, $I_s^2 R_{es}$ is the total copper losses, and P_c is the core loss. Dividing the numerator and denominator by I_s,

$$\eta = \frac{V_o \cos \phi_s}{V_o \cos \phi_s + I_s R_{es} + (P_c/I_s)} \times 100\%$$

Since the output voltage V_o remains substantially constant (within the limits of the regulation), the maximum value of η is obtained from the equation above when the denominator has its minimum value. If I_s were zero, $I_s R_{es}$ becomes zero, but P_c/I_s becomes infinity. Similarly, if I_s were made very large, P_c/I_s might be made very small, but $I_s R_{es}$ would become very large. So, neither $I_s=0$, nor $I_s=$(a very large current) gives maximum efficiency. It can be shown by differential calculus, or by substituting practical values into $I_s R_{es}$ and P_c/I_s, that maximum transformer efficiency is obtained when

$$I_s R_{es} = \frac{P_c}{I_s}$$

or

$$\boxed{I_s^2 R_{es} = P_c} \qquad\qquad \text{(24-17)}$$

That is, *for maximum efficiency*,

$$(\textit{winding copper losses}) = (\textit{core loss})$$

EXAMPLE 24-8 Determine the efficiency of the transformer described in Examples 24-4 and 24-5. Also, calculate the level of output current at which maximum efficiency occurs.

SOLUTION

$$P_o = V_s I_s \cos \phi_s$$

$$= 50 \text{ V} \times 2 \text{ A} \times \cos 31.7°$$

$$= 85.3 \text{ W}$$

$$P_i = V_i I_p \cos \phi_p$$

$$= 108.7 \text{ V} \times 1.05 \text{ A} \times \cos 35.5°$$

$$= 92.9 \text{ W}$$

Equation (24-15):

$$\eta = \frac{P_o}{P_i} \times 100\%$$

$$= \frac{85.3 \text{ W}}{92.9 \text{ W}} \times 100\%$$

$$\cong \mathbf{92\%}$$

$$\textit{Core losses, } P_c = \frac{E_p^2}{R_o}$$

$$= \frac{103^2}{5 \text{ k}\Omega}$$

$$P_c = 2.12 \text{ W}$$

Referring R_p to the secondary, the equivalent secondary resistance is

$$R_{es} = R_s + R_p \left(\frac{N_s}{N_p}\right)^2$$

$$= 0.25 \text{ }\Omega + 2.5 \text{ }\Omega \left(\frac{1}{2}\right)^2$$

$$= 0.875 \text{ }\Omega$$

For maximum efficiency:
Equation (24-17):

$$I_s^2 R_{es} = P_c$$

Therefore,

$$I_s = \sqrt{\frac{P_c}{R_{es}}}$$

$$= \sqrt{\frac{2.12\,\text{W}}{0.875\,\Omega}}$$

$$I_s \cong 1.56\,\text{A}$$

PRACTICE PROBLEMS

24-7.1 A transformer has an output voltage of 30 V when supplying a load of $Z_L = 68\,\Omega\underline{/29°}$. The input voltage and current are measured as 115 V and 135 mA with a phase angle of 33°. Calculate the transformer efficiency.

24-7.2 For the transformer described in Problems 24-4.1 and 24-4.2, determine the level of output current at which maximum efficiency occurs.

24-8
OPEN-CIRCUIT AND SHORT-CIRCUIT TESTS

OPEN-CIRCUIT TEST. A transformer could be tested under no-load and full-load conditions to determine its turns ratio, regulation, and efficiency. However, without fully loading the transformer it is possible to perform two tests from which all important data can be derived.

Figure 24-14 shows the circuit for the transformer *open-circuit test.* The alternating input voltage is set to the normal primary level for the transformer, and the voltage at the open-circuited output terminals is monitored on a voltmeter, as shown. The input power is measured by the wattmeter (see Section 25-5), and the ammeter measures the primary current.

Since the secondary is effectively open-circuited, the primary current is very small, and the voltage drops across the ammeter and wattmeter can be assumed to be negligible. In this case the input voltage can be taken as the transformer primary voltage, and thus the ratio of the two voltmeter readings gives the turns ratio:

$$\frac{E_p}{E_s} = \frac{N_p}{N_s}$$

With a very small primary current, and near-zero secondary current (i.e., the voltmeter current), the copper loss in the windings can be assumed negligible. The input power measured on the wattmeter is then the total transformer core losses, and the ammeter indicates the no-load primary current I_o.

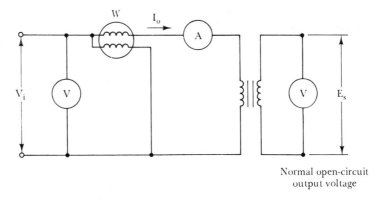

Normal open-circuit
output voltage

FIGURE 24-14. A transformer open-circuit test is performed by measuring the (no-load) secondary voltage, the primary current, and the input voltage and power.

From the measured values of input voltage, current, and power, the components of the no-load equivalent circuit can be determined:

$$true\ power,\ P = \frac{E_p^2}{R_o}$$

$$\boxed{R_o = \frac{E_p^2}{P}} \tag{24-18}$$

apparent power, $S = E_p I_o$

and $\quad S = \sqrt{(true\ power)^2 + (reactive\ power)^2} \quad$ [see Figure 21-7(b)]

$\qquad\quad = \sqrt{P^2 + Q^2}$

or $\quad Q = \sqrt{S^2 - P^2}$

or $\qquad\boxed{reactive\ power,\ Q = \sqrt{(E_p I_o)^2 - P^2}} \tag{24-19}$

and from Equation (21-6),

$$Q = \frac{E_p^2}{X_o}$$

Therefore,

$$\boxed{X_o = \frac{E_p^2}{Q}} \tag{24-20}$$

EXAMPLE 24-9   An open-circuit test on a certain transformer produced the following measurements: $E_p = 115$ V, $E_s = 57.5$ V, $P = 9.5$ W, and $I_o = 180$ mA. Determine the transformer turns ratio and the values of R_o and X_o.

SOLUTION

$$\frac{N_s}{N_p} = \frac{E_s}{E_p} = \frac{57.5\text{ V}}{115\text{ V}}$$

$$= \frac{1}{2}$$

Equation (24-18):

$$R_o = \frac{E_p^2}{P} = \frac{(115\text{ V})^2}{9.5\text{ W}}$$

$$= 1.39\text{ k}\Omega$$

Equation (24-19):

$$Q = \sqrt{(E_p I_o)^2 - P^2}$$

$$= \sqrt{(115\text{ V} \times 180\text{ mA})^2 - (9.5\text{ W})^2}$$

$$= 18.39\text{ var}$$

Equation (24-20):

$$X_o = \frac{E_p^2}{Q} = \frac{(115\text{ V})^2}{18.39\text{ vars}}$$

$$= 719\ \Omega$$

SHORT-CIRCUIT TEST. The transformer *short-circuit test* is performed with the secondary terminals short-circuited, as illustrated in Figure 24-15. Note that the primary voltage E_p is measured right at the transformer primary terminals, to avoid the error due to the voltage drops across the ammeter and wattmeter. The input voltage is increased from zero until the ammeter in the primary circuit indicates normal full-load primary current. When this occurs, the normal full-load secondary current is circulating in the secondary winding. Because the secondary terminals are short-circuited, the input voltage required to produce full-load primary and secondary currents is around 3% of the normal input voltage level. With such a low level of input voltage, the core flux is a minimum, and consequently the core losses are so small that they can be

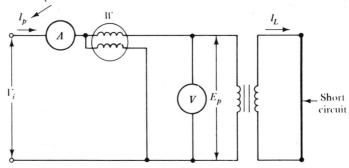

Normal full load
input current

FIGURE 24-15. To perform a transformer short-circuit test, the secondary is shorted, and the primary current is adjusted to the normal full-load level. The primary voltage and input power are measured.

neglected. [Referring to Figure 24-13(b), it is seen that the low input voltage causes the core loss current I_o to be small.] However, the windings are carrying normal full-load current, and thus the input is supplying the normal full-load copper losses.

The output power (to the short-circuit) is zero, so the wattmeter measuring true input power indicates the full-load copper losses. The product of the ammeter and voltmeter readings gives the apparent input power. From these quantities, calculation may be made of the resistive and reactive components of the full-load equivalent circuit referred to the primary [see Figure 24-13(c)]:

true power, $P = I_p^2 R_e$

$$R_e = \frac{P}{I_p^2} \qquad (24\text{-}21)$$

apparent power, $S = E_p I_p$

reactive power, $Q = \sqrt{(E_p I_p)^2 - P^2} \qquad (24\text{-}22)$

and $Q = I_p^2 X_e$

Therefore, $X_e = \frac{Q}{I_p^2} \qquad (24\text{-}23)$

EXAMPLE 24-10   Determine R_e and X_e for the transformer in Example 24-9 when the following measurements were made on a short-circuit test: $E_p = 5.5$ V, $I_p = 1$ A, and $P = 5.25$ W.

SOLUTION

Equation (24-21):

$$R_e = \frac{P}{I_p^2} = \frac{5.25\,\text{W}}{(1\,\text{A})^2}$$

$$= \mathbf{5.25\ \Omega}$$

Equation (24-22):

$$Q = \sqrt{(E_p I_p)^2 - P^2}$$

$$= \sqrt{(5.5\,\text{V} \times 1\,\text{A})^2 - (5.25)^2}$$

$$= 1.64\ \text{var}$$

Equation (24-23):

$$X_e = \frac{Q}{I_p^2} = \frac{1.64\,\text{var}}{(1\,\text{A})^2}$$

$$= \mathbf{1.64\ \Omega}$$

EXAMPLE 24-11   From the open-circuit and short-circuit test results given in Examples 24-9 and 24-10, determine the transformer regulation for a full load with a lagging phase angle of $\phi_s = 10°$.

SOLUTION

Equation (24-14):

$$\text{voltage regulation} = \frac{V_{o(\text{NL})} - V_{o(\text{FL})}}{V_{o(\text{FL})}} \times 100\%$$

$$V_{o(\text{NL})} = 57.5\ \text{V} \qquad (\textit{from open-circuit test})$$

$$V_{o(\text{FL})} = E_{p(\text{FL})} \times \frac{N_s}{N_p}$$

and

$$E_{p(\text{FL})} = E_{p(\text{NL})} - [\text{Primary volts drop due to } I_L]$$

$$= E_{p(\text{NL})} - [V_{i(\text{sc})} \cos \phi]$$

where ϕ is the phase difference between V_i and $-E_p$ [see Figure 24-11(c)]

$$\phi_p = \textit{primary phase angle on short-circuit}$$

From Equation (21-10):

$$\cos \phi_p = \frac{p}{E_p I_p}$$

$$= \frac{5.25 \text{ W}}{5.5 \text{ V} \times 1 \text{ A}}$$

$$\phi_p \cong 17.3°$$

Again referring to Figure 24-11(c), *the phase angle between V_i and $-E_p$ is*

$$\phi = (\phi_p - \phi_1)$$

$$\cong (\phi_p - \phi_s) \quad \text{[neglecting core losses]}$$

Therefore, full-load primary voltage

$$E_{p(FL)} = E_{p(NL)} - [V_{i(sc)}\cos(\phi_p - \phi_s)]$$

$$= 115 \text{ V} - [5.5 \text{ V} \cos(17.3° - 10°)]$$

$$= 109.5 \text{ V}$$

and full-load output voltage

$$V_{o(FL)} = E_{p(FL)} \times \frac{N_s}{N_p}$$

$$= 109.5 \text{ V} \times \frac{1}{2}$$

$$= 54.75 \text{ V}$$

$$\text{voltage regulation} = \frac{57.5 \text{ V} - 54.75 \text{ V}}{54.75 \text{ V}} \times 100\%$$

$$\cong \mathbf{5\%}$$

EXAMPLE 24-12   From the open-circuit and short-circuit test results given in Examples 24-9, 24-10, and 24-11, calculate the transformer efficiency on full-load with a load phase angle of $\phi_s = 10°$ lagging.

SOLUTION

full-load output current, $I_{s(FL)} = I_{p(FL)} \times \frac{N_p}{N_s}$

$$= 1 \text{ A} \times \frac{2}{1}$$

$$= 2 \text{ A}$$

$$P_o = V_{o(FL)} \times I_{s(FL)} \times \cos \phi_s$$

$$= 54.75 \text{ V} \times 2 \text{ A} \times \cos 10°$$

$$= 107.8 \text{ W}$$

$$core\ losses = 9.5 \text{ W} \quad (from\ open\text{-}circuit\ test)$$

$$full\text{-}load\ copper\ losses = 5.25 \text{ W} \quad (from\ short\text{-}circuit\ test)$$

Equation (24-15):

$$efficiency,\ \eta = \frac{P_o}{P_i} \times 100\%$$

$$= \frac{P_o}{P_o + \text{losses}} \times 100\%$$

$$= \frac{107.8 \text{ W}}{107.8 \text{ W} + 9.5 \text{ W} + 5.25 \text{ W}} \times 100\%$$

$$= \mathbf{88\%}$$

PRACTICE PROBLEMS

24-8.1 A transformer with its secondary open-circuited has $E_p = 40$ V and $E_s = 9$ V. The input current is measured as 5 mA, and the input power is 120 mW. Determine the transformer turns ratio and the no-load equivalent circuit components.

24-8.2 When the transformer in Problem 24-8.1 is short-circuited, the input voltage required to maintain full load primary current is found to be 3 V. The primary current and input power are 125 mA and 250 mW. Determine the primary equivalent resistance and reactance.

24-8.3 Using the results of the open-circuit and short-circuit tests given in Problems 24-8.1 and 24-8.2, determine the voltage regulation and efficiency of the transformer when supplying full load to a pure resistance.

24-9

AUTO-TRANSFORMER

The *autotransformer* has a single winding on an iron core. One of the coil terminals is common to both input and output, and the other output terminal is movable so that it can make contact with any turn on the winding. The principle is illustrated in Figure 24-16(a). Several fixed output terminals (or *taps*) are sometimes used instead of a continuously variable output.

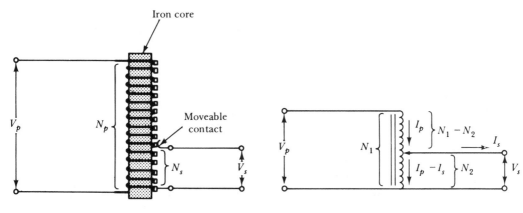

(a) Basic construction of an autotransformer

(b) Currents in an autotransformer

FIGURE 24-16. An autotransformer produces a secondary voltage V_s which is variable to a maximum level of V_P.

The induced primary voltage, as in the case of a transformer with two windings, is approximately

$$E_p = \frac{V_p}{N_p} \quad \text{volts/turn [see Figure 24-16(a)]}$$

and the total induced secondary voltage is

$$E_s = N_s\left(\frac{E_p}{N_p}\right) \quad \text{volts}$$

Therefore,
$$\frac{E_s}{E_p} = \frac{N_s}{N_p}$$

which is the same as in the case of a double-wound transformer.

The obvious advantage of the autotransformer is the facility for adjusting the output voltage to any desired level. The obvious disadvantage is that

the output is no longer dc isolated from the input.

Another advantage of the autotransformer is illustrated in Figure 24-16(b). The input current I_p is seen to flow through turns $(N_1 - N_2)$. Because the secondary current flows out at the output terminal, the current flowing through N_2 turns is $(I_p - I_s)$. When I_p and I_s are almost equal, $(I_p - I_s)$ can be a very small quantity, and the N_2 portion of the winding can be constructed of relatively thin copper wire. This cannot apply in the case of a continuously variable output transformer designed to supply a wide range of loads. However, in the case of a fixed-output autotransformer, the reduced thickness of the copper windings can result in a significant saving of copper compared to a double-wound transformer designed to do the same job.

24-10
CURRENT TRANSFORMER

A high-level alternating current can be most easily measured by accurately transforming the current to a much lower level. Also, because conductors carrying large currents are frequently at high voltage levels, a measuring instrument directly connected to the conductor would have to be very well insulated. Another problem that arises when rectifier ammeters are to be used (see Section 25-4) is the relatively large voltage drop across each rectifier. The dc isolation afforded by a transformer solves the insulation and rectifier problems.

Figure 24-17 illustrates the principle of the current transformer. A conductor carrying a large current passes through a circular laminated iron core. The conductor constitutes a one-turn primary winding. The secondary winding consists of a number of turns of much finer wire wrapped around the core as shown. The secondary current is given by Equation (24-2):

$$I_s = I_p \times \frac{N_p}{N_s}$$

For example, suppose that $I_p = 100$ A in Figure 24-17, and let the ammeter be capable of measuring a maximum of 1 A. Then

$$N_s = N_p \times \frac{I_p}{I_s}$$

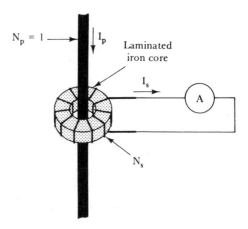

FIGURE 24-17. Use of a current transformer allows very high current levels to be measured on low-current instruments.

$$= 1 \times \frac{100 \text{ A}}{1 \text{ A}}$$

$$N_s = 100$$

It is very important to note that a current transformer must never be operated with its secondary winding open-circuited. This is because, when there is no secondary current to oppose the core flux generated by the primary, serious overheating of the core can occur. **It is also possible that the secondary open-circuit voltage could reach a** *dangerously* **high level.**

24-11
AUDIO TRANSFORMER

An *audio transformer* is designed to operate over the range of audible frequencies. These devices are usually employed in coupling an *audio amplifier* to the *speaker* of a sound system. The audio range is approximately from 50 Hz to 12 kHz; however, because other factors reduce the frequency range of the system, most audio amplifiers are designed to amplify frequencies ranging from about 20 Hz to 20 kHz, or higher.

As discussed in Section 24-5, a load Z_L connected to the secondary appears as an impedance of $(a^2 Z_L)$ in series with the primary input terminals. Most speakers used in sound-reproduction systems have impedance values on the order of 8 Ω to 16 Ω. By careful selection of the transformer turns ratio, the low-value speaker impedance can be made to appear in the transformer primary as a more convenient larger-value impedance. The process is referred to as *impedance matching*.

Over most of the audio range of frequencies, the load impedance referred to the primary dominates the transformer equivalent circuit. Thus the input impedance remains approximately $a^2 Z_L$, and a constant level of ac input voltage to the transformer primary produces a constant level of output voltage. Over this range, the transformer is said to have a *flat frequency response*. A typical transformer frequency response is shown in Figure 24-18, and it is seen that the output voltage level is constant from about 100 Hz to 10 kHz.

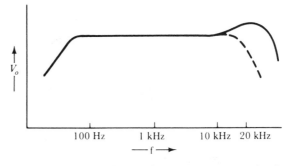

FIGURE 24-18. An audio transformer has an approximately flat frequency response over much of the audio range of frequencies.

Remembering that the transformer is an ac device which does not give an output change for direct input voltages, it is obvious that neither will it respond to very low input frequencies. Thus, as the frequency of the input voltage is decreased below a lower limit, the transformer output voltage falls off, as illustrated in Figure 24-18. Similarly, when the input frequency becomes high enough, energy losses in the iron core increase to the point at which the output voltage falls off as the input frequency increases. Resonance of the leakage reactance and the capacitance between the turns of the winding may cause the frequency response to rise initially at the high frequency end before falling off.

SUMMARY OF FORMULAS

Primary/secondary voltage ratio:

$$\frac{E_s}{E_p} = \frac{N_s}{N_p}$$

Primary/secondary current ratio:

$$\frac{I_s}{I_p} = \frac{N_p}{N_s}$$

emf equations:

$$E_p = 4.44\Phi_m f N_p$$

$$E_s = 4.44\Phi_m f N_s$$

Secondary induced voltage:

$$E_s = \sqrt{(V_o \cos \phi_o + I_s R_s)^2 + (V_o \sin \phi_o + I_s X_s)^2}$$

Phase angle between E_s and I_s:

$$\phi_s = \arctan \left(\frac{V_o \sin \phi_o + I_s X_s}{V_o \cos \phi_o + I_s R_s} \right)$$

Primary current:

$$I_p = \sqrt{(I_p' \cos \phi_s + I_c)^2 + (I_p' \sin \phi_s + I_{mag})^2}$$

Phase angle between I_p and E_p:

$$\phi_1 = \arctan \left(\frac{I_p' \sin \phi_s + I_{mag}}{I_p' \cos \phi_s + I_c} \right)$$

Referred resistance:

$$R = \left(\frac{N_p}{N_s} \right)^2 R_L = a^2 R_L$$

Referred reactance:

$$X = a^2 X_s$$

Referred impedance:

$$Z = a^2 Z_L$$

Primary equivalent resistance:

$$R_e = R_p + a^2 R_s$$

Primary equivalent reactance:

$$X_e = X_p + a^2 X_s$$

Voltage regulation:

$$\frac{V_{o(NL)} - V_{o(FL)}}{V_{o(FL)}} \times 100\%$$

Efficiency:

$$\eta = \frac{P_o}{P_i} \times 100\%$$

For maximum efficiency:

$$I_s^2 R_{es} = P_c$$

From open-circuit test:

$$R_o = \frac{E_p^2}{P}$$

$$Q = \sqrt{(E_p I_o)^2 - P^2}$$

$$X_o = \frac{E_p^2}{Q}$$

From short-circuit test:

$$R_e = \frac{P}{I_p^2}$$

$$Q = \sqrt{(E_p I_p)^2 - P^2}$$

$$X_e = \frac{Q}{I_p^2}$$

REVIEW QUESTIONS

24-1 Draw a sketch to show the basic construction of a transformer and explain how it functions. Sketch the circuit diagram for a simple two-winding transformer and derive the approximate relationship between the numbers of coil turns and the primary and secondary voltages.

24-2 Define primary winding, secondary winding, turns ratio, 1:1 transformer, step-up transformer, and step-down transformer.

24-3 Derive an expression relating the number of primary and secondary turns on a transformer to the primary and secondary currents.

24-4 Derive the transformer emf equation relating primary or secondary voltage to the core flux, the supply frequency, and the number of turns on the winding.

24-5 Discuss the power losses that occur in a transformer, and explain measures that may be taken to minimize the losses.

24-6 Sketch the approximate phasor diagram of transformer primary voltage and currents under no-load conditions. Also, sketch the no-load equivalent circuit for the transformer. Explain briefly the phasor diagram and equivalent circuit.

24-7 Explain the terms *leakage flux* and *leakage inductance*. Sketch the complete equivalent circuit for a transformer under load. Identify all components of the circuit and all voltages and currents. Explain the origin of each quantity.

24-8 Sketch a complete phasor diagram for the secondary circuit of a transformer under load. Briefly explain the diagram.

24-9 Sketch a complete phasor diagram for the primary circuit of a transformer under load. Briefly explain the diagram.

24-10 Explain the principle of *referred resistance* and *referred reactance*. Show how the transformer equivalent circuit may be simplified by referring the secondary circuit components to the primary.

24-11 Explain what is meant by the *voltage regulation* of a transformer, and write the equation for calculating the voltage regulation.

24-12 Discuss the efficiency of transformers, and derive an equation for transformer efficiency in terms of the output voltage and current, and the core losses.

24-13 Sketch the circuit diagrams for performing open and short-circuit tests on a transformer. Explain the testing procedure and develop the necessary equation from which R_o, X_o, R_e, and X_e may be calculated from the test results.

24-14 Sketch the circuit diagram for an *autotransformer* and explain its operation. Also, discuss the advantages and disadvantages of an autotransformer.

24-15 Describe a *current transformer*. Explain its application and discuss any precautions necessary when using a current transformer.

24-16 Explain the purpose of an audio transformer. Sketch the typical frequency response graph of an audio transformer, and explain its shape.

PROBLEMS

SECTION 24-1

24-1 A transformer with a primary winding of 250 turns has an input of 115 V rms. There are two secondary windings, one with 65 turns and the other with 80 turns. Determine the output voltage from each winding. Also, calculate the combined output voltage:
a. When the secondary windings are connected series-aiding.
b. When the secondary windings are connected series-opposing.

24-2 A transformer with a 115 V, 60 Hz primary voltage is to have 35 V, 15 V, and 5 V secondaries which have loads of 100 Ω, 500 Ω, and 25 Ω respectively. If the primary winding has 1000 turns, determine the required number of turns for each secondary. Also, calculate the primary current assuming 100% efficiency.

24-3 A transformer with a primary winding consisting of 250 turns has four secondary windings with the following numbers of turns: $N_{s1} = 13$, $N_{s2} = 52$, $N_{s3} = 125$, and $N_{s4} = 1042$. The secondary windings have the following loads connected to them: $R_{s1} = 31.5$ Ω, $R_{s2} = 50$ Ω, $R_{s3} = 1.2$ kΩ, and $R_{s4} = 500$ kΩ. If the primary voltage is 120 V, determine the output voltage from each secondary winding. Also, calculate the total primary current. Assume that the transformer is 100% efficient.

24-4 A transformer with a 50 turn secondary and a 270 turn primary is to deliver 5 W to a 16 Ω load. Assuming 100% efficiency, calculate the required primary voltage and current.

24-5 A secondary current of 10 mA is measured on a transformer which has 5000 secondary turns and one turn on its primary. Calculate the primary current.

24-6 The center-tapped secondary winding of a transformer has a total of 200 turns. The primary has 1000 turns and an input of 5 V (rms). Calculate the secondary peak outputs when the center-tap is grounded. If the whole of the secondary is accidentally connected to the input voltage supply, calculate the output voltage (from the primary).

SECTION 24-2

24-7 The transformer described in Problem 24-3 has a supply frequency of 60 Hz. Calculate the peak core flux.

24-8 Determine the peak core flux for the transformer in Problem 24-2.

24-9 The input frequency for the transformer in Problem 24-4 ranges from 75 Hz to 10 kHz. Calculate the maximum and minimum levels of peak core flux.

24-10 Determine the peak core flux for the transformer in Problem 24-5 if the supply frequency is 60 Hz, and the secondary output is 18 V.

24-11 Calculate the peak core flux for the transformer in Problem 24-6 when correctly connected, and when the secondary is connected as a primary. The minimum frequency of the input voltage is 50 Hz.

SECTION 24-4

24-12 A transformer with an output of 75 V supplies a load consisting of $R_L = 33.5 \, \Omega$ and $X_L = 22 \, \Omega$. The transformer parameters are $R_s = 0.25 \, \Omega$, $X_s = 1.2 \, \Omega$, and $N_p/N_s = 3/2$. Calculate the secondary current, the secondary induced voltage, and the angle between the two.

24-13 The transformer in Problem 24-12 has $R_p = 2 \, \Omega$, $X_p = 5 \, \Omega$, $R_o = 7.5 \, k\Omega$, and $X_o = 3 \, k\Omega$. Determine the supply voltage and current, and the phase angle between the two.

24-14 The load connected to the secondary of the transformer in Example 24-4 is changed to $Z_L = 40 \, \Omega \underline{/20°}$. If the output remains at 50 V, calculate the new secondary current, induced voltage, and phase angle.

24-15 For the load change described in Problem 24-14, determine the new primary current, voltage, and phase angle for the transformer in Examples 24-4 and 24-5.

24-16 The transformer described in Problems 24-12 and 24-13 has the inductive part of its load doubled to 44 Ω. If the output remains at 75 V, calculate the new secondary current, induced voltage, and phase angle.

24-17 Determine the new primary current, voltage, and phase angle for the transformer in Problems 24-12 and 24-13, when the load is changed as described in Problem 24-16.

SECTION 24-5

24-18 For the transformer in Examples 24-4 and 24-5 with its load changed as described in Problem 24-14, refer all secondary components to the primary. Then, neglecting R_o and X_o, calculate I_p and ϕ_p when $V_i = 108.7$ V.

24-19 Refer all the secondary components of the transformer described in Problems 24-12 and 24-13 to the primary winding. Neglecting the components representing the core loss, calculate the primary currents and phase angles with respect to the input voltage when the input voltage is 120 V.

24-20 For the transformer in Problems 24-12 and 24-13 with its load changed as described in Problem 24-16, refer all secondary components to the primary. Then, neglecting R_o and X_o, calculate I_p and ϕ_p when $V_i = 120.85$ V.

SECTION 24-6

24-21 Neglecting the core losses, determine the voltage regulation for the transformer specified in Problems 24-12 and 24-13.

24-22 The transformer in Problem 24-4 has an output of 9.5 V when the load is disconnected. Calculate its regulation.

24-23 A transformer has a primary input of 115 V, a regulation of 4%, and a turns ratio of 4.6:1. Calculate its no-load and full-load output voltages.

SECTION 24-7

24-24 The input power to a transformer is measured as 22 W when the output voltage is 31.6 V and the secondary load is 50 Ω. Calculate the transformer efficiency.

24-25 Calculate the efficiency of the transformer specified in Problems 24-12 and 24-13, and determine the output current for maximum efficiency.

24-26 Determine the efficiency of the transformer in Examples 24-4 and 24-5 when it is operated with the new load described in Problem 24-14.

24-27 A transformer has a core loss of 0.95 W, a secondary winding resistance of 0.18 Ω, and a primary winding resistance of 1.3 Ω. The primary and secondary turns are $N_p = 720$ and $N_s = 288$, and the secondary output voltage is 50 V. Calculate the load resistance that will allow the transformer to operate at maximum efficiency.

SECTION 24-8

24-28 An open-circuit test on a transformer gave the following results: $E_p = 120$ V, $E_s = 35$ V, $P = 5$ W, and $I_o = 125$ mA. A short-circuit test on the same transformer produced: $E_p = 4$ V, $I_p = 0.8$ A, and $P = 3$ W. Calculate R_o, X_o, N_p/N_s, R_e, and X_e.

24-29 From the open-circuit and short-circuit test results given in Problem 24-28 calculate the transformer full-load regulation when the load has a phase angle of 15° lagging.

24-30 Using the open-circuit and short-circuit test results given in Problem 24-28 calculate the efficiency of the transformer under full-load with a load phase angle of 15° lagging.

24-31 Write a computer program to analyze the results of transformer open-circuit and short-circuit tests to determine R_o, X_o, N_p/N_s, R_e and X_e.

24-32 Write a computer program to analyze the results of transformer open-circuit and short-circuit tests to determine the full-load regulation and efficiency when the load has a given phase angle.

ANSWERS TO PRACTICE PROBLEMS

24-1.1	65, 95
24-1.2	6.6 mA
24-2.1	617 μWb
24-2.2	7.5 V
24-4.1	243 mA, 35.2 V
24-4.2	194 mA, 47.9 V
24-5.1	184.8 Ω, 186.6 Ω
24-5.2	3.6/1
24-6.1	58.6 V
24-7.1	89%
24-7.2	274 mA
24-8.1	4.4:1, 13.3 kΩ, 10 kΩ
24-8.2	16 Ω, 18 Ω
24-8.3	5%, 93%

25 AC MEASURING INSTRUMENTS

Objectives You will be able to:

Sketch half-wave and bridge rectifier circuits. Sketch the input and output waveforms and explain the operation of each circuit.

Sketch the circuits of ac voltmeters using half-wave and bridge recitifier circuits. Explain the operation of each voltmeter.

Given the parameters of a permanent magnet moving-coil (PMMC) instrument, calculate the values of multiplier resistors required to construct an ac voltmeter using half-wave or full-wave rectification.

Sketch the circuit of a rectifier ammeter using a current transformer. Explain the operation of the circuit.

Solve problems involving the design of rectifier ammeters.

Explain why a dynamometer instrument gives a positive deflection for current flow in either direction.

Show how a dynamometer wattmeter is connected to measure the power delivered to a load. Explain why the instrument measures true power in an ac circuit.

Solve problems involving measured power, load voltage, load current, and phase angle between voltage and current.

Sketch the circuit diagrams of a series-resistance capacitance bridge, a Maxwell bridge, and a Maxwell-Wein bridge. Analyze each circuit to determine the balance equations and explain the operation of each circuit.

Solve problems involving measurements using ac bridges.

Introduction

The permanent magnet moving-coil (PMMC) instrument operates only on dc. For ac measurements, the alternating quantity must be converted to dc by rectification before being applied to a PMMC instrument. The dynamometer instrument can be used directly to measure *dc* quantities or *rms ac* quantities. Its most important application is as a wattmeter, and on *ac* the dynamometer wattmeter measures *true power*. Precise measurements of inductance and capacitance can be made by the use of an *ac bridge*.

25-1
PMMC INSTRUMENT ON AC

In Chapter 13 it is pointed out that the *permanent magnet moving-coil* (PMMC) instrument is *polarized*. This means that the terminals of the instrument are identified as + and − , and that for correct deflection the meter must be connected with that polarity. When incorrectly connected, the pointer attempts to deflect to the left of zero (i.e., *off-scale*).

Now consider what occurs when a PMMC instrument is connected directly to an alternating current source, as shown in Figure 25-1(a). For the illustration showing instantaneous pointer position [Figure 25-1(b)] it is assumed that the frequency of the alternating current is very low, on

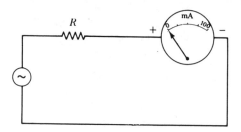

(a) PMMC ammeter connected to alternating current source

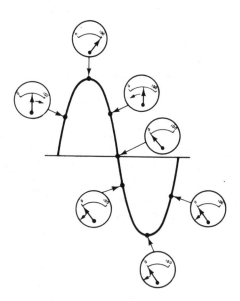

(b) Behavior of PMMC ammeter when measuring a very low frequency alternating current

FIGURE 25-1. A permanent magnet moving-coil (PMMC) instrument used directly on sinusoidal alternating current normally indicates zero. On very low frequency ac, the pointer would move up and down the scale.

the order of 0.1 Hz. As the current level increases positively from zero, the meter pointer moves to the right until it indicates peak current. Then the pointer falls back toward zero again as the current level drops. When the current direction reverses during the second half-cycle of the waveform, the meter attempts to deflect to the left, which, of course, is not possible. Consequently, the pointer remains at (or just below) the zero mark during the negative half-cycle of the current.

Most ac frequencies are much greater than 0.1 Hz, 60 Hz being the normal ac supply frequency in North America. At this frequency, the meter pointer would have to rise and fall 60 times in every second. The damping mechanism of the instrument, as well as the inertia of the moving system, prevents the moving coil and pointer from moving this fast. Consequently, the instrument pointer settles at the average level of the alternating current passing through the windings. For normal sinusoidal alternating current, the average level is zero. Therefore, a PMMC instrument used directly to measure 60 Hz ac indicates zero.

25-2
RECTIFICATION

A rectifier is usually a *semiconductor diode,* although it may also be a vacuum tube. Without going into the theory of operation of the semiconductor diode, it is sufficient to note that it has two terminals, *anode* and *cathode,* as illustrated in Figure 25-2(a), and that current can flow through the device in only one direction. When the anode terminal is positive with respect to the cathode, current flows (in the conventional direction) from anode to cathode [see Figure 25-2(b)]. Current will not flow through the rectifier when the cathode is positive with respect to the anode [Figure 25-2(c)]. When *forward-biased* (i.e., conducting), there is a small volt drop from anode to cathode. This is normally on the order of 0.7 V or less, depending upon the semiconductor material employed in manufacturing the rectifier. The conventional direction of

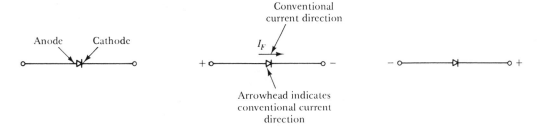

(a) Circuit symbol for a rectifier

(b) Current flows when anode is + and cathode is −

(c) Effectively zero current flows when the cathode is + and the anode is −

FIGURE 25-2. The semiconductor rectifier used in ac instruments is a one-way device; it passes current in one direction and blocks current flow in the other direction.

current flow through the device is indicated by the arrowhead in the circuit symbol, pointing from anode to cathode.

Figure 25-3 shows how a rectifier is used to convert alternating current into a series of pulses which all have the same polarity. In Figure 25-3(a) a single rectifier is connected in series with a resistor. Only the positive half-cycles of current are passed through the rectifier, so that the negative portion of the ac waveform is cut off. The waveform of the

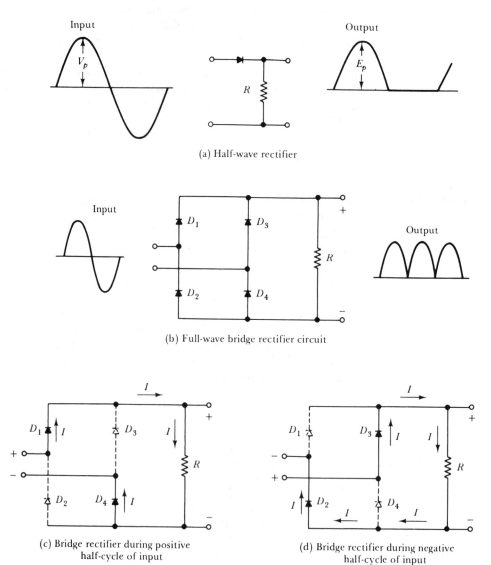

(a) Half-wave rectifier

(b) Full-wave bridge rectifier circuit

(c) Bridge rectifier during positive half-cycle of input

(d) Bridge rectifier during negative half-cycle of input

FIGURE 25-3. Half-wave rectifiers circuits pass only one half of each cycle of sine wave input. Full-wave rectifier circuits convert all half-cycles to the same polarity and pass them to the output.

current that passes through the rectifier and the resistor is a continuous series of positive pulses with intervening spaces (see the illustration). This waveform is termed *half-wave-rectified.*

The circuit shown in Figure 25-3(b) employs four rectifiers and is termed a *bridge rectifier circuit.* In this case, when the input waveform is positive [see Figure 25-3(c)], rectifiers D_1 and D_4 are forward-biased, while D_2 and D_3 are reverse-biased. Thus, current flows through D_1, resistor R, and D_4, as illustrated. Conversely, when the input waveform is negative [Figure 25-3(d)], D_2 and D_3 are forward-biased and D_1 and D_4 are reverse-biased. Current now flows through D_3, resistor R, and D_2, and it is seen that the current direction through the resistor is the same as before. The result of this is that the output waveform is a continuous series of unidirectional pulses, and the process is termed *full-wave rectification.*

25-3
RECTIFIER VOLTMETER

The circuit shown in Figure 25-4(a) is that of an ac voltmeter using a PMMC instrument and a full-wave rectifier circuit. As in the case of a dc voltmeter, a *multiplier* resistance must be included in the circuit to limit the current through the instrument. The actual current that flows through the deflection instrument has the full-wave rectified waveform shown in Figure 25-4(b). Since the deflection of a PMMC instrument is proportional to the average level of the current through its coil, the pointer tends to indicate 0.637 of V_p. However, the effective value, or rms value of the alternating waveform, is the quantity that is normally required in any measurement. Because there is a direct relationship between rms and average values, the instrument can be designed to have its scale marked to indicate rms volts. The following design example shows how this is accomplished.

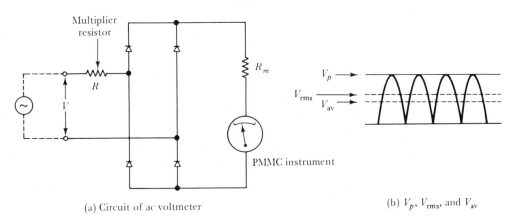

(a) Circuit of ac voltmeter

(b) V_p, V_{rms}, and V_{av}

FIGURE 25-4. AC voltmeter using a full-wave rectifier circuit and a PMMC meter. The meter deflection is proportional to the average of the rectified waveform, but the scale is calibrated to indicate rms values.

EXAMPLE 25-1   An ac voltmeter to indicate 100 V rms at full scale is to be constructed using a deflection instrument which has full-scale deflection (FSD) at 500 μA. The coil resistance of the instrument is 1 kΩ, and the rectifiers used each have a forward voltage drop of 0.7 V. Employing the full-wave rectifier circuit shown in Figure 25-4(a), determine the required multiplier resistance.

SOLUTION

At FSD, the average current through the deflection instrument is

$$I_{av} = 500 \ \mu A$$

For a half cycle of sinusoidal waveform:
From Equation (17-13):

$$peak \ current, \quad I_p = \frac{\pi}{2} I_{av}$$

$$= \frac{\pi}{2} \times 500 \ \mu A$$

$$\cong 785 \ \mu A$$

and $\quad I_p = \dfrac{\text{(applied peak voltage)} - \text{(rectifier voltage drops)}}{\text{total circuit resistance}}$

$$= \frac{(1.414 V_{rms}) - 2V_r}{R + R_m} , \quad (2V_r \ for \ two \ rectifiers \ in \ series)$$

Therefore, $\quad R = \dfrac{(1.414 V_{rms}) - 2V_r}{I_p} - R_m$

$$= \frac{(1.414 \times 100 \ V) - (2 \times 0.7 \ V)}{785 \ \mu A} - 1 \ k\Omega$$

$$\cong \mathbf{177 \ k\Omega}$$

A half-wave rectifier circuit can also be employed with a deflection instrument, to produce an ac voltmeter (see the circuit in Figure 25-5). The procedure for calculating the multiplier resistor for the half-wave instrument is not exactly the same as in the full-wave case. This is because, as shown in the illustration, the average current through the meter is now

$$I_{av} = \tfrac{1}{2} \times 0.637 \times I_p$$

It is important to note that rectifier voltmeters designed for sine-wave operation *can be used only where pure sine waves are involved.*

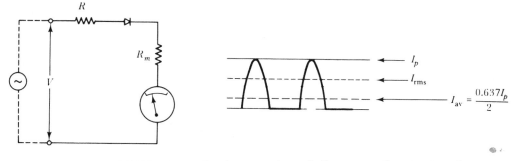

FIGURE 25-5. AC voltmeter using a half-wave rectifier circuit and a PMMC meter. Here again, the meter deflection is proportional to the average of the rectified waveform, and the scale is calibrated in rms.

Where the voltage to be measured has any other waveform, the voltmeter will *not* correctly indicate rms voltage. This is because the 1.11 relationship between rms and average current levels applies only to pure sine waves.

For multirange ac voltmeters, a rotary switch is used to select one of several values of multiplier resistor, exactly as in the case of dc voltmeters.

PRACTICE PROBLEMS

25-3.1 Determine the multiplier resistance for the half-wave rectifier voltmeter circuit in Figure 25-5 to meet the specification in Example 25-1.

25-3.2 A full-wave rectifier voltmeter circuit uses a 75 μA meter with a coil resistance of 2.98 kΩ and diodes which have a forward voltage drop of 0.3 V. If the multiplier resistance is 352 kΩ, determine the full-scale rms voltage.

25-4
RECTIFIER
AMMETER

Recall from Chapter 13 that a dc ammeter must have a very low resistance, because it is always connected in series with the circuit in which current is to be measured. For the same reasons, an ac ammeter must also have a very low resistance. The low resistance requirement implies that when measuring current, the voltage drop across the ammeter must be very small. In most circumstances this voltage drop should be not greater than about 100 mV. But the voltage drop across one rectifier is typically 0.7 V, and for a bridge rectifier circuit in which there are two rectifiers in series, the total rectifier voltage drop becomes typically 1.4 V. Clearly, the ordinary rectifier instrument cannot be directly applied as an ac ammeter.

The use of a *current transformer* (Section 24-10) gives the ammeter a low terminal resistance and low voltage drop while providing sufficient voltage to operate the rectifier instrument. Figure 25-6 shows a full-wave rectifier ammeter using a current transformer. The transformer currents might typically be 0.1 A in the primary and 500 μA in the secondary, as illustrated. In this case, from Equation (24-2), the ratio of secondary to primary turns is

$$\frac{N_s}{N_p} = \frac{I_{py}}{I_{sy}}$$

$$= \frac{0.1\,\text{A}}{500\,\mu\text{A}} = 200$$

This would normally mean a one-turn primary, with 200 turns on the secondary. The meter scale would, of course, be marked as 0.1 A for full-scale deflection.

When a load resistance R_1 (shown broken in Figure 25-6) is connected across the secondary winding, a larger secondary current is drawn from the transformer. This allows the use of a larger number of primary turns, thus giving a more accurate current transformation. For example, if the inclusion of R_1 increased the total secondary current to 5 mA, then with 200 turns on the secondary the number of primary turns would be

$$N_p = \frac{I_{sy}}{I_{py}} \times N_s$$

$$= \frac{5\,\text{mA}}{0.1\,\text{A}} \times 200$$

$$N_p = 10\,\text{turns}$$

If additional terminals are provided on the primary winding so that its number of turns can be altered, as illustrated in Figure 25-6, the range

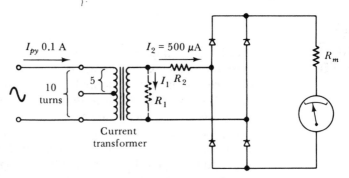

FIGURE 25-6. Rectifier ammeter circuit using a current transformer. The primary winding of the transformer offers the necessary low resistance, while the secondary produces sufficient output voltage to operate the rectifier circuit and the meter.

of the ammeter can be changed. When N_p is 5 turns and I_{sy} is to be 5 mA to give full-scale deflection on the instrument, then

$$I_{py} = \frac{N_s}{N_p} \times I_{sy}$$

$$= \frac{200}{5} \times 5 \text{ mA}$$

$$= 0.2 \text{ A}$$

It is seen that halving the number of primary turns doubled the range of current that the ammeter can measure. Thus, the provision of several terminals on the primary winding allows the selection of several ac current ranges. Note that the range of the instrument may also be changed by switching to different values of secondary load resistance (R_1). For range switching, make-before-break rotary switches should be employed, as already discussed for dc ammeters.

Like the rectifier voltmeter, the rectifier ammeter is suitable only for use with pure sinusoidal waveforms.

EXAMPLE 25-2 A rectifier ammeter has the circuit shown in Figure 25-6. The deflection instrument has full-scale deflection for a coil current of 1 mA and has a coil resistance of 100 Ω. The series resistance R_2 is 10 kΩ and R_1 is 1.37 kΩ. If the current transformer has $N_s = 250$ turns and $N_p = 5$ turns, determine the primary current required to give full-scale deflection on the instrument.

SOLUTION

For FSD:
$$I_{av} = 1 \text{ mA}$$

From Equation (17-13),

peak current, $\qquad I_p = \frac{\pi}{2} \times I_{av}$

$$= \frac{\pi}{2} \times 1 \text{ mA} = 1.57 \text{ mA}$$

and from Equation (17-15), $\qquad I_2 = I_{rms} = 0.707 I_p$

$$= 0.707 \times 1.57 \text{ mA} = 1.11 \text{ mA}$$

Transformer secondary peak voltage:

$$V_p = I_p(R_2 + R_m) + 2V_r$$
$$= 1.57 \text{ mA}(10 \text{ k}\Omega + 100 \text{ }\Omega) + (2 \times 0.7 \text{ V})$$
$$= 17.26 \text{ V}$$

Transformer secondary rms voltage:

$$V_s = 0.707 \times V_p$$
$$= 0.707 \times 17.26 \text{ V}$$
$$= 12.2 \text{ V}$$

rms current through R_1:

$$I_1 = \frac{V_s}{R_1} = \frac{12.2 \text{ V}}{1.37 \text{ k}\Omega}$$
$$= 8.9 \text{ mA}$$

Total secondary rms current:

$$I_{sy} = I_2 + I_1$$
$$= 1.11 \text{ mA} + 8.9 \text{ mA}$$
$$\cong 10 \text{ mA}$$

Primary rms current:

$$I_{py} = \frac{N_s}{N_p} \times I_{sy}$$
$$= \frac{250}{5} \times 10 \text{ mA}$$
$$\mathbf{= 500 \text{ mA}}$$

PRACTICE PROBLEMS

25-4.1 The ammeter circuit described in Example 25-2 is to have its range changed to 300 mA. Determine a new secondary load resistance R_1 to effect the change.

25-4.2 A half-wave rectifier circuit is substituted in place of the full-wave rectifier in Example 25-2. Calculate the transformer primary current for full-scale deflection.

25-5
DYNAMOMETER INSTRUMENT ON AC

As mentioned in Section 13-7, the dynamometer instrument can be employed as a voltmeter or ammeter or in its major application as a wattmeter. Consider the illustrations in Figure 25-7, which shows the field coils and moving coil of a dynamometer instrument connected in

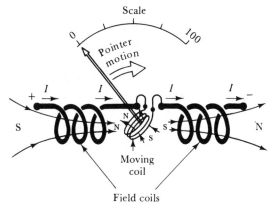

(a) Current flowing from left to right
produces positive deflection

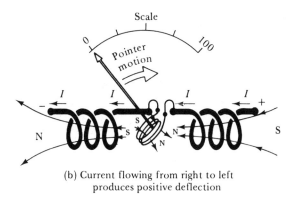

(b) Current flowing from right to left
produces positive deflection

FIGURE 25-7. When the current flow through a dynamometer instrument reverses, both the field flux and the moving coil flux reverse directions. The pointer deflection is unaffected. This makes the instrument suitable for direct application to ac measurements.

series. With current flowing in the direction shown in Figure 25-7(a), the field coils set up fluxes which have their N poles on the right-hand side and their S poles on the left-hand side. Also, the moving coil flux has the polarity illustrated, S at the bottom and N at the top. The N pole of the moving coil flux is repelled from the adjacent N pole of the field coil flux, and the two adjacent S poles also repel each other. The result of this is that the moving coil is deflected clockwise, causing the pointer to move over the scale from left to right.

Now consider the effect of reversing the direction of the current through the coils. As illustrated in Figure 25-7(b), the field coil fluxes are now S at the right-hand side, and N at the left-hand side. The moving coil flux is also reversed, having S at the top and N at the bottom. Once again there are like poles adjacent to each other, and the pointer moves from left to right over the scale.

It is seen that the dynamometer instrument has a positive deflection, regardless of the direction of the current through the meter. This means that the meter terminals are not marked + and − ; that is, the dynamometer instrument is *unpolarized*. It also means that the instrument gives a positive deflection when either direct or alternating current flows through the coils. As an ammeter, the scale of the instrument can be read as direct current when measuring dc, and as the rms value of alternating current when measuring ac. Similarly, as a voltmeter, the dynamometer instrument indicates dc volts or rms ac volts. The scale of the instrument can be conveniently calibrated on dc and then used to measure ac.

For ac as well as dc, the major application of the dynamometer instrument is as a wattmeter. As illustrated in Figure 25-8(a), the connection for measuring ac power is similar to that for measuring power in a dc circuit. In Chapter 13 it is explained that the *input* terminals of the voltage and current coils are identified with a ±, ↓, or * sign. The marked terminal of the current coil should be connected to the supply, and the voltage coil marked terminal connected to the load side of the current coil.

In alternating-current applications, the load current could lead or lag the load voltage by a phase angle ϕ. The deflection of the instrument is proportional to the in-phase components of current and voltage. Thus, as shown in Figure 25-8(b), the instrument indication is proportional to *EI* cos ϕ.

As explained in Section 21-5, the *true power* dissipated in a load with an ac supply is

$$P = EI \cos \phi$$

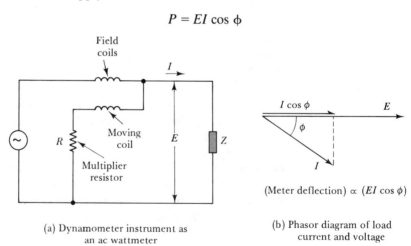

(a) Dynamometer instrument as an ac wattmeter

(b) Phasor diagram of load current and voltage

FIGURE 25-8. The deflection of the pointer on a dynamometer wattmeter is proportional to *EI* cos ϕ; it measures true ac power.

Therefore, when used as an ac wattmeter, the dynamometer instrument measures the true power supplied to the load.

EXAMPLE 25-3 A wattmeter measures the ac power delivered to a certain load as 100 W, and an ammeter and voltmeter monitor the load current as 1.5 A and 100 V, respectively. Calculate the phase angle between the load current and voltage.

SOLUTION

$$true\ power = P = EI \cos \phi$$

Therefore,
$$\cos \phi = \frac{P}{EI}$$

$$P = 100\ \text{W}$$

$$EI = 100\ \text{V} \times 1.5\ \text{A}$$

Therefore,
$$\cos \phi = \frac{100\ \text{W}}{100\ \text{V} \times 1.5\ \text{A}}$$

$$= 0.667$$

and
$$\phi = 48.2°$$

25-6
AC BRIDGES

SIMPLE CAPACITANCE BRIDGE. AC bridges are used for measurement of inductance and capacitance. All ac bridge circuits are based on the Wheatstone bridge (see Section 13-8).

Figure 25-9(a) shows the circuit of a *simple capacitance bridge*. C_s

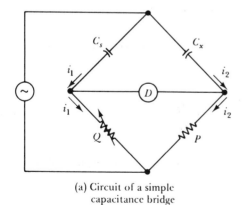

(a) Circuit of a simple capacitance bridge

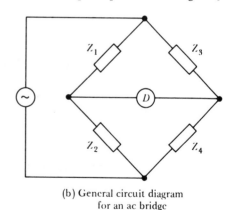

(b) General circuit diagram for an ac bridge

FIGURE 25-9. A simple capacitance bridge is similar to a Wheatstone bridge, except that two capacitors are used instead of two of the resistors. The bridge must be supplied from an ac source, and the null detector must be an ac instrument.

is a precise standard capacitor and C_x is an unknown capacitance, while Q and P are standard resistors, one or both of which is adjustable. An ac supply is used, and the *null detector* (D) must be an ac instrument. A low current rectifier ammeter is frequently employed as null detector. Q is adjusted until the null detector indicates zero, and when this is obtained, the bridge is said to be *balanced*.

When the detector indicates null, the voltage drop across C_s must equal that across C_x, and similarly the voltage across Q must be equal to the voltage across P. Therefore,

$$V_{cs} = V_{cx}$$

or

$$i_1 X_{cs} = i_2 X_{cx} \tag{1}$$

and

$$V_Q = V_p$$

or

$$i_1 Q = i_2 P \tag{2}$$

Dividing Equation (1) by Equation (2):

$$\boxed{\frac{X_{cs}}{Q} = \frac{X_{cx}}{P}} \tag{25-1}$$

Referring to Equation (25-1) and to Figure 25-9(b), the general *balance equation* for all ac bridges can be written

$$\boxed{\frac{Z_1}{Z_2} = \frac{Z_3}{Z_4}} \tag{25-2}$$

Substituting $1/\omega C_s$ for X_{cs}, and $1/\omega C_x$ for X_{cx} in Equation (25-1),

$$\frac{1}{\omega C_s Q} = \frac{1}{\omega C_x P}$$

or

$$C_x = \frac{Q \omega C_s}{P \omega}$$

giving,

$$C_x = \frac{QC_s}{P}$$ (25-3)

It is seen that the unknown capacitance C_x can now be calculated from the known values of Q, C_s, and P.

SERIES-RESISTANCE CAPACITANCE BRIDGE. One disadvantage of the simple capacitance bridge is that perfect balance of the bridge is obtained only when C_s and C_x are both pure capacitances (i.e., they have virtually no resistive component). In general, this occurs only with capacitors that have air or mica dielectrics. Capacitors with other types of dielectric have a leakage current, and consequently the equivalent circuits for the capacitors have resistive components that must be included in the bridge circuit.

The circuit of the *series resistance capacitance bridge* shown in Figure 25-10 eliminates the balance problems that can occur with the simple capacitance bridge. Resistance r_x in series with the unknown capacitance represents the resistive component of the capacitor equivalent circuit. The standard capacitor C_s normally has mica dielectric, and thus has a very small resistive component. Consequently, the adjustable resistance S must be included in the circuit to balance the effect of r_x.

The *balance equations* for the series resistance capacitance bridge are derived as follows:

Equation (25-2): $$\frac{Z_1}{Z_2} = \frac{Z_3}{Z_4}$$

Therefore, $$\frac{S - j\dfrac{1}{\omega C_s}}{Q} = \frac{r_x - j\dfrac{1}{\omega C_x}}{P}$$

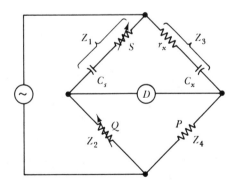

FIGURE 25-10. The use of series resistors with a capacitance bridge makes balance easy to obtain and allows the resistive component of the capacitors to be measured.

$$\frac{S}{Q} - j\frac{1}{\omega C_s Q} = \frac{r_x}{P} - j\frac{1}{\omega C_x P}$$ (25-4)

For the above equation to be correct, the real parts on each side must be equal, *and* the imaginary parts on each side must be equal.
Equating the real parts,

$$\frac{S}{Q} = \frac{r_x}{P}$$

and

$$r_x = \frac{PS}{Q}$$ (25-5)

Equating the imaginary parts,

$$\frac{1}{\omega C_s Q} = \frac{1}{\omega C_x P}$$

and

$$C_x = \frac{QC_s}{P}$$ (25-6)

The resistive and capacitive components of the unknown capacitor can now be calculated by means of Equations (25-5) and (25-6). Note that neither the supply voltage nor the frequency of the ac supply is involved in the balance equations for the bridge.

Because of the need to balance the real and imaginary components of the bridge impedances, the process of obtaining balance in an ac bridge is a little more complicated than with the Wheatstone bridge. One of the adjustable components (Q or S in Figure 25-10) is first altered to obtain the lowest possible indication of the null meter. Then the other adjustable component is varied to obtain a lower reading. The process is repeated until further adjustment of either component cannot produce a lower reading on the null meter. At this point the bridge is balanced.

EXAMPLE 25-4   The capacitance bridge shown in Figure 25-10 has a 0.1 μF standard capacitor (C_s) and a standard resistor of $P = 1$ kΩ. Zero deflection is obtained on the null detector when $Q = 10.25$ kΩ and $S = 2.25$ kΩ. Calculate the value of the unknown capacitance and its resistive component.

SOLUTION

Equation (25-6):

$$C_x = \frac{QC_s}{P}$$

$$= \frac{10.25\ \mathrm{k\Omega} \times 0.1\ \mu\mathrm{F}}{1\ \mathrm{k\Omega}}$$

$$= \mathbf{1.025\ \mu F}$$

Equation (25-5):

$$r_x = \frac{PS}{Q}$$

$$= \frac{1\ \mathrm{k\Omega} \times 2.25\ \mathrm{k\Omega}}{10.25\ \mathrm{k\Omega}}$$

$$= \mathbf{219.5\ \Omega}$$

INDUCTANCE BRIDGES. For measurement of inductance, the *Maxwell bridge* shown in Figure 25-11(a) can be employed. It is seen that the circuit of the Maxwell bridge is simply a repeat of the series resistance capacitance bridge, with the capacitors replaced by inductors. A disadvantage of this bridge is that standard inductors are larger and more difficult to manufacture than standard capacitors. Consequently, a variation of this circuit, known as the *Maxwell-Wein bridge*, is most often employed for inductance measurement.

The circuit of the Maxwell-Wein bridge is shown in Figure 25-11(b). L_X is the unknown inductance to be measured and r_x is the resistance of its windings. C_s is again a precise standard capacitor, and P is a standard resistor. Q and S are accurate adjustable resistors. At balance: Equation (25-2):

$$\frac{Z_1}{Z_2} = \frac{Z_3}{Z_4}$$

Therefore

$$\boxed{\frac{S}{1/\left(\dfrac{1}{Q} + j\omega C_s\right)} = \frac{r_x + j\omega L_x}{P}} \tag{25-7}$$

or

$$S\left(\frac{1}{Q} + j\omega C_s\right) = \frac{r_x + j\omega L_x}{P}$$

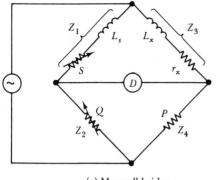

(a) Maxwell bridge

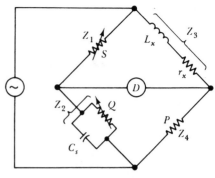

(b) Maxwell-Wein bridge

FIGURE 25-11. A Maxwell bridge is an inductance bridge arranged similarly to a capacitance bridge. A Maxwell-Wein bridge uses resistors and a standard capacitor to measure inductance.

$$\frac{S}{Q} + j\omega C_s S = \frac{r_x}{P} + j\frac{\omega L_x}{P} \tag{25-8}$$

Equating the real terms and the imaginary terms:

$$\frac{S}{Q} = \frac{r_x}{P}$$

$$r_x = \frac{SP}{Q} \tag{25-9}$$

and

$$\omega C_s S = \frac{\omega L_x}{P}$$

$$L_x = PC_s S \tag{25-10}$$

Once again it is seen that the supply voltage and frequency are not involved in the balance equations for the bridge. This is *not* always the case with ac bridges; indeed, one particular bridge can be used to measure the frequency of the supply in terms of the bridge component values at balance.

EXAMPLE 25-5   The Maxwell-Wein bridge shown in Figure 25-11(b) uses a 0.1 μF standard capacitor (C_s) and a standard resistor of $Q = 1$ kΩ. Zero deflection of the null detector is obtained when $P = 1.33$ kΩ and $S = 870$ Ω. Calculate the inductance and resistance of Z_3.

SOLUTION

Equation (25-9):

$$r_x = \frac{SP}{Q}$$

$$= \frac{870\,\Omega \times 1.33\ \text{k}\Omega}{1\,\text{k}\Omega}$$

$$= \mathbf{1.16\ k\Omega}$$

Equation (25-10):

$$L_x = PC_s S$$

$$= 1.33\ \text{k}\Omega \times 0.1\ \mu\text{F} \times 870\ \Omega$$

$$= \mathbf{115.7\ mH}$$

Comparing the circuits shown in Figure 25-10 and 25-11(b), it is seen that a series capacitance bridge and a Maxwell-Wein bridge could each be constructed from the same set of components. In fact, many commercial ac bridges use one set of components, which are connected in the form of a series resistance capacitance bridge for capacitance measurement, and are switched into the Maxwell-Wein bridge configuration for inductance measurement.

PRACTICE PROBLEMS

25-6.1 In the capacitance bridge in Example 25-4, resistor Q is variable from 500 Ω to 10 kΩ, and S ranges from 1 kΩ to 3 kΩ. Calculate the range of measurement of C_x and r_x.

25-6.2 An ac bridge has the following components: $C_s = 0.2\ \mu\text{F}$, $S = 500\ \Omega$ to 1.6 kΩ, $Q = 500\ \Omega$ to 15 kΩ, and $P = 100\ \Omega$ to 1 MΩ. If the bridge can be either a series-resistance capacitance circuit or a Maxwell-Wein circuit, determine the range of unknown C and L that can be measured.

REVIEW QUESTIONS

25-1 Explain what a rectifier is used for. Sketch the circuit symbol for a rectifier and show the current direction and voltage polarity when the rectifier is conducting.

25-2 Sketch half-wave and bridge rectifier circuits. Show the input and output waveform and explain briefly the operation of each circuit.

25-3 Sketch the circuit of a rectifier voltmeter that uses a bridge rectifier circuit. Exaplin how the instrument operates and discuss its limitations.

25-4 Sketch the circuit for an *ac* voltmeter using a PMMC instrument and half-wave rectification.

25-5 Sketch the circuit of a rectifier ammeter. Explain how the instrument operates and discuss its limitations.

25-6 Using illustrations, explain why a dynamometer instrument gives a positive deflection for current flow through the meter in either direction. What changes are necessary to produce a negative deflection on a dynamometer instrument?

25-7 Show how a dynamometer wattmeter is connected to measure the power delivered to a load and explain why the instrument measures the true power in an ac circuit.

25-8 Sketch the circuit of a simple capacitance bridge. Explain how it operates and derive an expression for the unknown capacitance.

25-9 Sketch the circuit of a Maxwell bridge. Explain the steps involved in balancing the bridge and derive the balance equations.

25-10 Sketch the circuits of two ac bridges that may be constructed from a standard capacitor and three variable standard resistors. Identify each bridge by name and derive the balance equations for the bridge.

PROBLEMS

SECTION 25-3

25-1 A PMMC voltmeter having a full-scale deflection of 100 V is used to measure an alternating voltage with an rms value of 70.7 V. What will the meter reading be:
a. When the ac frequency is 60 Hz?
b. When the frequency is 0.05 Hz? Explain.

25-2 A 200 μA PMMC instrument with a coil resistance of 75 Ω is to be connected to a bridge rectifier circuit to construct an ac voltmeter with a range of 100 V. If the rectifiers each have a forward volts drop of 0.3 V, calculate the required multiplier resistance. Also, determine the sensitivity of the voltmeter.

25-3 The voltmeter in Problem 25-2 is to have additional ranges of 250 V and 50 V. Determine the multiplier resistance for each range.

25-4 Recalculate the multiplier resistance values for Problems 25-2 and 25-3 when a half-wave rectifier circuit is employed. Also, determine the new voltmeter sensitivity.

25-5 A PMMC instrument that gives full-scale deflection for 50 μA and has a coil resistance of 850 Ω is to be used as an ac voltmeter. A bridge rectifier circuit is to be used, with the rectifiers having a forward voltage drop of 0.7 V each. If the voltmeter is to have full-scale deflection at 150 V, calculate the value of the required multiplier resistance.

25-6 An ac voltmeter consists of a 150 μA PMMC meter with a 100 Ω coil resistance, three series-connected resistors, and a bridge rectifier with rectifier voltage drops of 0.7 V. The resistor values are $R_1 = 174$ kΩ, $R_2 = 719$ kΩ, and $R_3 = 1.8$ MΩ. Determine the voltmeter range when the multiplier resistance is (a) R_1 alone, (b) $R_1 + R_2$, (c) $R_1 + R_2 + R_3$.

25-7 For the voltmeter in Problem 25-6, determine the voltmeter ranges if two rectifiers become open-circuited so that bridge rectifier functions as a half-wave rectifier circuit with two rectifiers connected in series.

25-8 Calculate the multiplier resistance for the voltmeter in Problem 25-5 is a half-wave rectifier circuit is used.

25-9 An ac voltmeter consists of a half-wave rectifier with a forward voltage drop of 1 V, a 37.5 μA PMMC meter with a 862 Ω coil resistance, and a 591 kΩ multiplier resistance. Calculate the voltmeter range.

25-10 The ac voltmeter in Problem 25-2 is employed to measure the voltage drop across each of two resistors in a potential divider circuit with a 115 V, 60 Hz supply. The resistors are $R_1 = 220$ kΩ and $R_2 = 390$ kΩ. Calculate the measured voltages and determine V_{R1} and V_{R2} when the voltmeter is not connected.

25-11 The resistive potential divider in Problem 25-10 is replaced with a capacitive potential divider with $C_1 = 0.012$ μF and $C_2 = 0.008$ μF. Calculate the new voltage levels with and without the voltmeter in the circuit.

SECTION 25-4

25-12 A 200 μA PMMC instrument with a coil resistance of 900 Ω is used in a rectifier ammeter which is to indicate 1 A rms at full scale. The ammeter circuit has a bridge rectifier and a current transformer with 20 primary turns and 180 secondary turns. If a 120 kΩ resistance is connected in series with the PMMC instrument, calculate the required value of the resistance that must be connected in parallel with the secondary winding.

25-13 The ammeter in Problem 25-12 is to have its range changed to 100 mA by adding turns to the transformer primary. Determine the required number of additional turns.

25-14 The ammeter in Problem 25-12 is to have its range changed to 100 mA by changing the parallel-connected secondary load resistance. Determine the new load resistance.

25-15 A half-wave rectifier is substituted in place of the full-wave rectifier in Problem 25-12. Calculate the new level of transformer primary current for full-scale deflection.

SECTION 25-5

25-16 A wattmeter measures the ac power delivered to a load as 250 W. The supply voltage is 115 V and the load is known to have a phase angle of $\phi = 33°$. Calcualte the load current.

25-17 A wattmeter with a field coil resistance of 0.9 Ω, and a moving coil resistance of 100 kΩ, is designed to give full-scale deflection when the field coil current is 1 A and the moving coil current is 1 mA. (a) Calculate the maximum power that can be measured by the instrument. (b) Determine the voltage applied to the moving coil circuit if the instrument indicates full scale when measuring the power dissipated in a 1 A load with a phase angle of 45°.

25-18 The wattmeter in Problem 25-17 has its moving coil contact moved to the supply side of the field coils (see Figure 25-8). Determine the measurement error that results when the load current and moving coil current are 1 A and 1 mA respectively.

SECTION 25-6

25-19 A series-connected capacitance bridge (Figure 25-10) uses a 0.15 μF standard capacitor, a series resistance $S = 1$ kΩ to 1.8 kΩ, and a standard resistor $Q = 12$ kΩ. If the capacitance to be measured is known to be between 2 μF and 2.5 μF, calculate the required range of adjustment for resistor P. Also, determine the range of measurement for the resistive component of the unknown capacitor.

25-20 A simple capacitance bridge, as in Figure 25-9, has $C_s = 0.2$ μF, $P = 10$ kΩ, and Q adjustable from 3 kΩ to 15 kΩ. Calculate the range of unknown capacitance C_x that might be measured with the bridge.

25-21 A 1 μF capacitor and three decade resistors which may be adjusted from 100 Ω to 5 kΩ are available for construction of the two bridges referred to in Review Question 25-10. Calculate the measurable ranges of capacitance and inductance, and their resistive components.

25-22 The capacitance bridge in Problem 25-19 is to be used to measure a small capacitance outside the range of the bridge. A capacitor C_1 with an approximate capacitance of 2 μF is first measured. Then, the small capacitance C_2 is connected in parallel with C_1, and a new measurement is made. With only C_1 in the circuit, $P = 875\ \Omega$ at balance. With C_1 and C_2 in parallel, $P = 869$ Ω. Calculate C_2.

25-23 The capacitance bridge in Problem 25-19 is to be used to measure a large capacitance outside the range of the bridge. A capacitor C_1 with an approximate capacitance of 2 μF is first measured. Then, the large capacitance C_2 is connected in series with C_1, and a new measurement is made. With only C_1 in the circuit, $P = 875\ \Omega$ at balance. If C_2 has a capacitance of 56.5 μF, determine the expected setting for P at balance.

25-24 An inductance L_1 measured on the bridge in Problem 25-21 gives balance with $P = 1$ kΩ, $Q = 1.98$ kΩ, and $S = 122\ \Omega$. When another inductance L_2 is connected in series with L_1, the new balance is found with $P = 852\ \Omega$, $Q = 1.31$ kΩ, and $S = 149\ \Omega$. Determine the inductance of L_1 and L_2 and their coil resistances.

ANSWERS TO PRACTICE PROBLEMS

25-3.1 88.6 kΩ

25-3.2 30 V

25-4.1 2.49 kΩ

25-4.2 915 mA

25-6.1 (0.05 μF to 1 μF), (100 Ω to 6 kΩ)

25-6.2 (100 pF to 30 μF), (10 mH to 320 H)

26

THREE-PHASE AC SYSTEMS

Objectives   You will be able to:

Explain the generation of a three-phase alternating voltage, sketch the output voltage waveforms, and draw a phasor diagram of the voltages.

Sketch the circuits of Y-connected and Δ-connected generators, sketch phasor diagrams for generator line and phase voltages, and sketch phasor diagrams for line and phase currents.

Analyze Y-connected and Δ-connected four-wire and three-wire loads to determine load and line voltages, load and line currents, neutral current, and the phase relationships between the various quantities.

Explain the effects of phase sequence reversal and analyze circuits with different phase sequences.

Explain the process of power factor correction in three-phase loads and calculate the power factor correction components.

Describe various methods of measuring the power dissipated in a three-phase load and calculate wattmeter indications.

Introduction

A three-phase generator produces three separate sinusoidal alternating voltage waveforms from three output terminals. The terminals are identified as *A, B,* and *C;* sometimes a neutral terminal (*N*) is also available. The three waveforms have 120° phase differences, with a particular phase sequence depending on the generator terminal connections. The generator output voltages can be measured as: *phase voltage*—the output at the terminals of a single generator, and *line voltage*—the voltage measured at the terminals of two interconnected generators. The three generator coils may be connected in one of two possible configurations, Y-connection and Δ-connection. Similarly, the load on a three-phase generator may be connected in Y or Δ fashion.

26-1
GENERATION OF THREE-PHASE VOLTAGES

The simple rotating-loop ac generator described in Section 17-1 produces an output that is one cycle of sine wave continuously repeating. In Figures 26-1(a) and (b) three similar conducting loops are shown situated between two magnetic poles. Each loop generates a sinusoidal alternating voltage as it rotates in the magnetic field, exactly as described in Section 17-1. Three separate sinusoidal voltages are produced at the three pairs of output terminals. Because the loops are set at an angle to each other, the individual output peak voltages occur at differ-

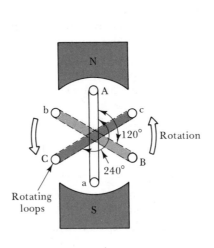

(a) Three-phase generation

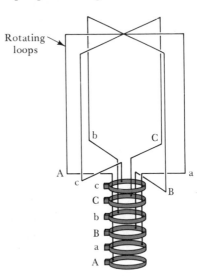

(b) Output connections of generator

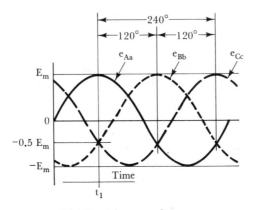

(c) Three-phase waveform

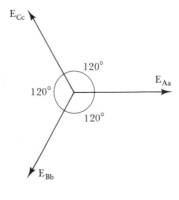

(d) Phasor diagram of voltages

FIGURE 26-1. A three-phase generator circuit consists basically of three conducting loops rotated in a magnetic field. Because the loops are 120° apart, the three output voltage waveforms have 120° phase angle differences.

ent times. This means that there are phase differences between the three output sine waves.

Referring to Figures 26-1(a) and (b), note that the two sides (or rotating conductors) of each loop are identified as A and a, B and b, and C and c. The loops are rotated in a counterclockwise direction. With conductor A moving past the N pole and conductor a moving past the S pole, as illustrated, the instantaneous emf (e_{Aa}) generated in a and A is a maximum. Also, terminal A is positive with respect to terminal a. (This is explained in Section 17-1, where conductors 1 and 2 correspond to conductors A and a respectively.)

At the instant when the output from A and a is a positive maximum, conductor B is moving away from the S pole and b is leaving the N pole. Consequently, the voltage output from terminals B and b (e_{Bb}) is past its negative peak value and is moving toward zero (terminal B being negative with respect to b).

Loop Bb in Figure 26-1 is 120° behind loop Aa. The output voltage e_{Bb} can be written with respect to voltage e_{Aa}. From Equation (17-4),

$$e_{Aa} = E_m \sin \alpha$$

and
$$e_{Bb} = E_m \sin(\alpha - 120°)$$

Now consider loop Cc in Figure 26-1(a). At the instant illustrated, conductors C and c are seen to be approaching the S and N poles respectively. Thus, the output from terminals C and c is a negative voltage (C being negative with respect to c), growing toward its negative peak level.

Loop Cc is 120° behind loop Bb, and 240° behind loop Aa. This is the same as 120° ahead of loop Aa. Therefore,

$$e_{Cc} = E_m \sin(\alpha - 240°)$$
$$= E_m \sin(\alpha + 120°)$$

The three output voltage waveforms are shown in Figure 26-1(c) with the correct phase relationships. Waveform e_{Bb} reaches its positive peak 120°, or one-third of a cycle, after e_{Aa} peaks. The positive peak level of e_{Cc} occurs 240°, or two-thirds of a cycle, after the peak of e_{Aa}. The output voltages at the instant illustrated in Figure 26-1(a) are shown at t_1 in Figure 26-1(c). As already discussed, at this point e_{Aa} is a positive maximum, e_{Bb} is negative moving toward zero, and e_{Cc} is negative moving in the direction of its negative peak level. Looking at Figure 26-1(c), it is seen that the waveforms peak with the sequence e_{Aa}, e_{Bb}, and e_{Cc}. The voltages are said to have a phase sequence of ABC.

An ac generator that produces a single sine wave output voltage is referred to as a *single-phase* generator. The type of generator discussed above, which produces three separate sine wave output voltages with

120° phase differences, is a *three-phase* generator. (*Three-phase* is sometimes written 3φ.)

As in the case of single-phase ac circuits, rms values of voltage and current are normally used in three-phase calculations rather than instantaneous or peak quantities. The phasor diagram of the three-phase rms output voltages is drawn in Figure 26-1(d). E_{Aa} is the rms output voltage from loop Aa. E_{Bb} is 120° behind E_{Aa}, and E_{Cc} is 240° behind E_{Aa}. It is seen that the phasors are arranged symmetrically with 120° phase differences between them.

26-2
WYE-CONNECTED GENERATOR

PHASE VOLTAGES AND CURRENTS. The output voltage levels produced by a three-phase generator depend on the connection arrangement of the three individual generator loops (or coils). The two possible connection methods are known as *wye* (Y), and *delta* (Δ), because of circuit resemblances to those symbols. Figure 26-2 represents a Y-connected three-phase generator, in which the coil terminals a, b, and c are connected to a single *neutral terminal* (N). There are four output terminals: A, B, C, and N.

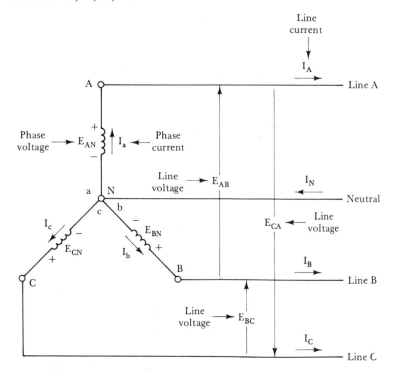

FIGURE 26-2. For a Y-connected three-phase generator, the *phase voltage* and *phase current* are the output voltage and current from each coil. The *line voltage* is the voltage measured between any two output lines (conductors). The *line current* is the current flowing in a line.

The system illustrated in Figure 26-2 is termed a *three-phase, four-wire supply.* In some cases the neutral conductor may not be connected, and the system becomes a *three-phase, three-wire supply.*

All three coils in a three-phase generator are identical, and all three produce the same level of rms output voltage. This is termed the *phase voltage* (V_p). Figure 26-2 shows E_{Aa}, E_{Bb}, and E_{Cc} now identified as E_{AN}, E_{BN}, and E_{CN} respectively. All three are equal to V_p.

The *phase current* (I_p) is the current level supplied by each of the three generator coils. The directions of the phase currents $(I_a, I_b,$ and $I_c)$ shown in Figure 26-2 are those that occur when each coil has a positive output voltage. Thus, the phase current for coil Aa is shown flowing out of terminal A and into a.

DOUBLE SUBSCRIPT NOTATION. Double subscript notation is used with the voltage symbols to indicate the direction of current flow when each coil (or loop) output is positive. Subscript Aa used with voltage E_{Aa} identifies terminal A as positive with respect to terminal a during the time that this coil output is positive. E_{Aa} is rewritten as E_{AN} in Figure 26-2, and this indicates that terminal A is positive with respect to the neutral terminal while the coil output is positive. Similarly, when the output of coil Bb is a positive quantity (see Figure 26-1), terminal B is positive with respect to b, and the internal current direction is from b to B. This voltage is identified as E_{Bb}, or E_{BN}, in Figure 26-2. The voltage labeled E_{Cc} or E_{CN} shows that terminal C is positive with respect to c when coil Cc has a positive output. In a different subscipt system, the order of the double subscripts indicates the current direction, instead of the voltage polarity.

LINE CURRENTS AND VOLTAGES. The conductors that connect a three-phase generator to a load are referred to as *lines,* and the current carried by each of these conductors is known as the *line current* (I_L). Figure 26-2 shows that in a Y-connected generator the phase currents and line currents are the same quantities.

The neutral conductor is the return conductor for all three individual coils in a Y-connected system. Therefore, the neutral current (I_N) is the phasor sum of all three line currents. This would seem to make I_N larger than I_L and therefore require a neutral conductor that is thicker than the line conductors. However, in Example 26-1 it is demonstrated that when all three coils have identical loads (*a balanced load*), I_N is zero. When a balanced load condition does not exist, current flows in the neutral conductor.

In a Y-connected generator, the *phase voltage* is the voltage measured between any line and neutral. The *line voltage* is the potential difference measured between any two lines. Thus, the line voltage is the phasor difference of two phase voltages. Figure 26-3 shows how the line voltages are calculated.

In Figure 26-3(a) the individual coil voltage phasors are reproduced

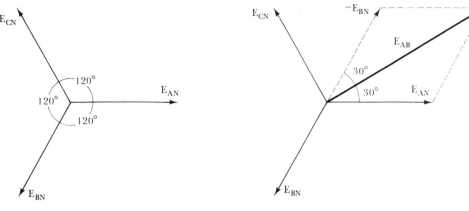

(a) Relationship of phase voltages

(b) Line voltage E_{AB} is the phasor difference $E_{AN} - E_{BN}$

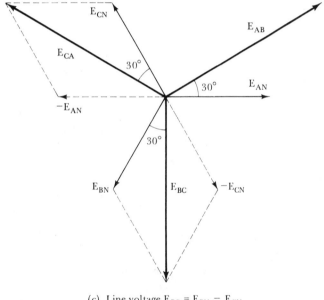

(c) Line voltage $E_{BC} = E_{BN} - E_{CN}$
and $E_{CA} = E_{CN} - E_{AN}$

FIGURE 26-3. For a Y-connected three-phase generator, the line voltages are the phasor sums of pairs of phase voltages. This makes each line voltage equal to $\sqrt{3} \times$ (phase voltage).

from Figure 26-1(d). Line voltage E_{AB} in Figure 26-2 is the phasor difference of E_{AN} and E_{BN}. To determine E_{AB}, $-E_{BN}$ is first drawn equal and opposite to E_{BN} [see Figure 26-3(b)]. Because of the 120° phase difference between E_{AN} and E_{BN}, there is a 60° angle between E_{AN} and $-E_{BN}$.

Voltages E_{AN} and E_{BN} are equal in magnitude; therefore, phasors

E_{AN} and $-E_{BN}$ are equal in length. Because of this equality, phasor E_{AB} (representing the sum of E_{AN} and $-E_{BN}$) is situated 30° from E_{AN} and 30° from $-E_{BN}$, as illustrated. This gives

$$E_{AB} = E_{AN}\cos 30° + (-E_{BN}\cos 30°)$$

or,

$$E_{AB} = V_p\cos 30° + V_p\cos 30°$$
$$= 2(V_p \times 0.866)$$
$$= 1.732V_p$$

or,

$$E_{AB} = \sqrt{3}\, V_p$$

For a Y-connected generator:

$$line\ voltage = \sqrt{3}\ (phase\ voltage)$$

that is,

$$\boxed{V_L = \sqrt{3}\ V_p} \tag{26-1}$$

Also,

$$line\ current = phase\ current$$

that is,

$$\boxed{I_L = I_p} \tag{26-2}$$

Figure 26-3(b) also shows that line voltage E_{AB} leads phase voltage E_{AN} by 30°.

Figure 26-3(c) shows the determination of E_{BC} and E_{CA} as the phasor differences $E_{BN} - E_{CN}$ and $E_{CN} - E_{AN}$ respectively. Note that for line voltage E_{AB}, the subscript indicates that the generator terminal A is positive with respect to terminal B when E_{AB} is a positive quantity (see Figure 26-2).

WYE-CONNECTED LOADS. The circuit of a Y-connected generator with Y-connected load resistors is shown in Figure 26-4(a). Note that the generator phase voltages are once again identified as E_{AN}, E_{BN}, and E_{CN}, and the load voltages are similarly identified. The reasoning that developed Equations (26-1) and (26-2) for a Y-connected generator can also be applied to determine the voltage and current relationships for the

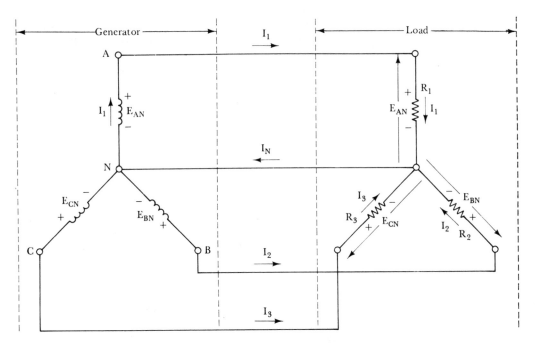

(a) Y-connected three-phase generator with y-connected resistive loads

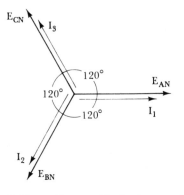

(b) Phasor diagram of load voltages and currents for purely resistive loads

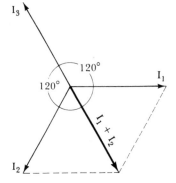

(c) When $I_1 = I_2 = I_3$,
$I_1 + I_2 = -I_3$
giving $I_1 + I_2 + I_3 = 0$

FIGURE 26-4. Circuit and phasor diagrams for a Y-connected three-phase, four-wire generator with a Y-connected balanced load. In this case, the load voltages and currents are equal to the generator phase voltage and phase current.

Y-connected *load*. Referring to Figure 26-4(a), the voltages across the individual loads are obviously equal to the generator phase voltages. Therefore, Equation (26-1) can be rewritten:

$$line\ voltage = \sqrt{3}\ (load\ voltage)$$

Also, the individual load currents are clearly the same currents that flow in the lines:

$$line\ current = load\ current$$

EXAMPLE 26-1 Load resistors, R_1, R_2, and R_3 in Figure 26-4(a) each have a value of 100 Ω, and the phase voltage is $V_p = 100$ V. Determine (*a*) the line current, (*b*) the neutral current, and (*c*) the line voltage.

SOLUTION

(*a*)
$$E_{AN} = E_{BN} = E_{CN} = V_p = 100\ \text{V}$$
$$I_L = I_1 = I_2 = I_3$$
$$I_L = \frac{V_p}{R_1} = \frac{100\ \text{V}}{100\ \Omega} = 1\ A$$

(*b*) *The phasor diagram for the currents is drawn in Figure 26-4(b). Because the loads are resistive, each load current is in phase with the phase voltages.*
     Figure 26-4(c) shows that the phasor sum of I_1 and I_2 equals $-I_3$.
     Therefore,

$$I_N = I_1 + I_2 + I_3$$
$$= 0$$

Alternatively:

$$I_1 = \frac{E_{AN}}{R_1} = \frac{100\ \text{V}\underline{/0°}}{100\ \Omega} = 1\ A\ \underline{/0°}$$
$$= 1\ A + j0\ A$$
$$I_2 = \frac{E_{BN}}{R_2} = \frac{100\ \text{V}\underline{/-120°}}{100\ \Omega} = 1\ A\ \underline{/-120°}$$
$$= 1\ A[\cos(-120°) + j\ \sin(-120°)]$$
$$= -0.5\ A - j0.866\ A$$
$$I_3 = \frac{E_{CN}}{R_3} = \frac{100\ \text{V}\underline{/-240°}}{100\ \Omega} = 1\ A\ \underline{/-240°}$$

$$= 1\ A\left[\cos(-240°) + j\sin(-240°)\right]$$

$$= -0.5\ A + j0.866\ A$$

$$I_N = I_1 + I_2 + I_3$$

$$= 1\ A + (-0.5\ A - j0.866\ A) + (-0.5\ A + j0.866\ A)$$

$$= 0$$

(c) *From* Equation (26-1),

$$V_L = \sqrt{3}\ V_p = \sqrt{3} \times 100\ V$$

$$= 173.2\ V$$

EXAMPLE 26-2 Recalculate the line and neutral currents if R_2 in Figure 26-4 is changed to 200 Ω, and R_3 becomes 50 Ω.

SOLUTION

$$I_1 = \frac{E_{AN}}{R_1} = \frac{100\ V}{100\ \Omega}$$

$$= 1\ A\ \underline{/0°}$$

$$I_2 = \frac{E_{BN}}{R_2} = \frac{100\ V\ \underline{/-120°}}{200\ \Omega}$$

$$= 0.5\ A\ \underline{/-120°}$$

$$= 0.5\ A\left[\cos(-120°) + j\sin(-120°)\right]$$

$$= -0.25\ A - j0.433\ A$$

$$I_3 = \frac{E_{CN}}{R_3} = \frac{100\ V\ \underline{/-240°}}{50\ \Omega}$$

$$= 2\ A\ \underline{/-240°}$$

$$= 2\ A\left[\cos(-240°) + j\sin(-240°)\right]$$

$$= -1\ A + j1.73\ A$$

$$I_N = I_1 + I_2 + I_3$$

$$= 1\ A + (-0.25\ A - j0.433\ A) + (-1\ A + j1.73\ A)$$

$$= -0.25\ A + j1.3\ A$$

$$= 1.32\ A\ \underline{/100.9°}$$

EXAMPLE 26-3 In the circuit in Figure 26-5, the generator phase voltage is $V_p = 100$ V, and its frequency is 60 Hz. The load impedance components are $R_1 = 100$ Ω, $R_2 = 100$ Ω, $C_2 = 66.3$ μF, $R_3 = 100$ Ω, and $L_3 = 159.2$ mH.

Calculate the three line currents and the neutral current.

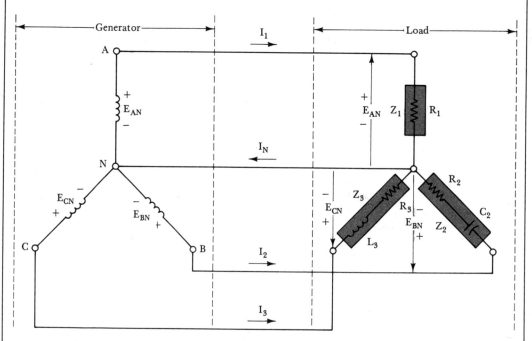

FIGURE 26-5. Circuit of a Y-connected three-phase, four-wire generator with a Y-connected unbalanced load. The load voltages and currents, once again, equal the generator phase voltages and individual phase currents, but the neutral current is no longer zero.

SOLUTION

$$I_1 = \frac{E_{AN}}{Z_1} = \frac{100 \text{ V}}{100 \text{ Ω}} = 1 \text{ A} \underline{/0°}$$

$$= 1 \text{ A}$$

$$X_{C2} = \frac{1}{2\pi fC} = \frac{1}{2\pi \times 60 \text{ Hz} \times 66.3 \text{ μF}}$$

$$= 40 \text{ Ω}$$

$$Z_2 = R_2 - jX_{C2} = 100 \text{ Ω} - j40 \text{ Ω}$$

$$= 107.7 \text{ Ω} \underline{/-21.8°}$$

$$I_2 = \frac{E_{BN}}{Z_2} = \frac{100 \text{ V}\underline{/-120°}}{107.7 \text{ }\Omega\underline{/-21.8°}}$$

$$= 0.929 \text{ A }\underline{/-98.2°}$$

$$= 0.929 \text{ A}[\cos(-98.2°) + j\sin(-98.2°)]$$

$$= -0.133 \text{ A} - j0.92 \text{ A}$$

$$X_{L3} = 2\pi f L_3 = 2\pi \times 60 \text{ Hz} \times 159.2 \text{ mH}$$

$$= 60 \text{ }\Omega$$

$$Z_3 = R_3 + jX_{L3} = 100 \text{ }\Omega + j60 \text{ }\Omega$$

$$= 116.6 \text{ }\Omega\underline{/31°}$$

$$I_3 = \frac{E_{CN}}{Z_3} = \frac{100 \text{ V}\underline{/-240°}}{116.6 \text{ }\Omega\underline{/31°}}$$

$$= 0.858 \text{ A }\underline{/-271°}$$

$$= 0.858 \text{ A}[\cos(-271°) + j\sin(-271°)]$$

$$= 0.015 \text{ A} + j0.858 \text{ A}$$

$$I_N = I_1 + I_2 + I_3$$

$$= 1 \text{ A} + (-0.133 \text{ A} - j0.92 \text{ A}) + (0.015 \text{ A} + j0.858 \text{ A})$$

$$= 0.882 \text{ A} - j0.062 \text{ A}$$

$$= 0.884 \text{ A }\underline{/-4°}$$

Example 26-1 shows that for a balanced load, the neutral current is zero. Examples 26-2 and 26-3 show that when the load is unbalanced, the neutral current can be substantial.

PRACTICE PROBLEMS

26-2.1 A balanced Y-connected four-wire load, as in Figure 26-4, has a line voltage of 200 V and a line current of 3.5 A. Calculate the load resistances.

26-2.2 Determine the neutral current in the circuit of Problem 26-2.1 when one of the load resistors is open-circuited.

26-2.3 A Y-connected four-wire load, as in Figure 26-5, has a line voltage of 300 V and the following loads: $Z_1 = 86.6 \text{ }\Omega$, $Z_2 = 86.6 \text{ }\Omega\underline{/-30°}$, $Z_3 = 86.6 \text{ }\Omega\underline{/60°}$. Determine the line and neutral currents.

26-3

DELTA-CONNECTED GENERATOR

VOLTAGES AND CURRENTS. As an alternative to Y-connection, the individual coils of a three-phase generator may be connected in Δ (delta) configuration. In the Δ-connected generator shown in Figure 26-6, the coil terminals are connected *A* to *c*, *B* to *a*, and *C* to *b*. The three output lines are identified as *A*, *B*, and *C*. No neutral conductor is involved. The line voltage (V_L) is the voltage measured as E_{AB}, E_{BC}, or E_{CA}. Once again, the order of the subscripts indicates the terminal polarity when each line voltage is positive.

Recall that the phase voltage (V_p) is the rms voltage measured across the output terminal of each of the three coils (or loops). It is obvious from Figure 26-6 that the line voltage and phase voltage are the same quantities in a Δ-connected generator.

For a Δ-connected generator:

$$line\ voltage = phase\ voltage$$

that is,

$$\boxed{V_L = V_p}$$

(26-3)

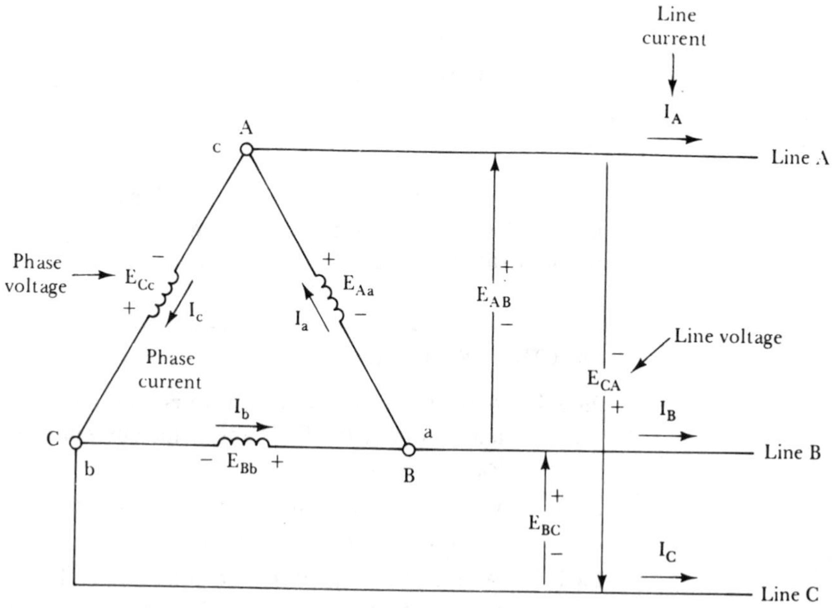

FIGURE 26-6. For a Δ-connected three-phase generator, the *phase voltage* and *phase current* are the output voltage and current from each coil. The *line voltage* is the voltage measured between any two output lines. The *line current* is the current flowing in a line.

The line current (I_L) is the current level measured in each of the line conductors. The direction indicated for I_L in each line is the direction in which current flows when the line voltage is positive. When E_{AB} is positive, line A is positive with respect to line B. The direction of line current I_A is then out of terminal A, as indicated in Figure 26-6. Similarly, when B is positive with respect to C, I_B flows out of terminal B, and when C is more positive than A, the direction of I_C is away from terminal C.

The phase current (I_p) is the current level produced in each of the three generator coils. Once again, the current directions indicated in Figure 26-6 are those that occur when each individual coil output is positive. When E_{Aa} is positive, I_a flows out of the terminal A and into terminal a. When E_{Bb} is positive, the I_b direction within the coil is from b to B. Similarly, for E_{Cc}, I_c flows from c to C so that the output current is out of terminal C and into c.

With the phase currents having the directions indicated in Figure 26-6, it appears that a circulating current might flow around the coils of a Δ-connected generator. This would not be acceptable, because it would produce unnecessary and wasteful power dissipation in the generator coils. Consider the waveforms in Figure 26-1(c) once again, and note that at t_1:

$$e_{Aa} = E_m$$
$$e_{Bb} = -0.5 E_m$$

and

$$e_{Cc} = -0.5 E_m$$

These instantaneous voltage levels are easily confirmed by calculation:

$$e_{Aa} = E_m \sin \alpha$$
$$= E_m \sin 90°$$
$$= E_m$$
$$e_{Bb} = E_m \sin(\alpha - 120°)$$
$$= E_m \sin(90° - 120°)$$
$$= -0.5 E_m$$
$$e_{Cc} = E_m \sin(90° - 240°)$$
$$= -0.5 E_m$$

The instantaneous level of voltage producing a circulating current in Figure 26-6 is:

$$e = e_{Aa} + e_{Bb} + e_{Cc}$$

At time t_1 in Figure 26-1(c),

$$e = E_m - 0.5\,E_m - 0.5\,E_m$$

$$= 0$$

At time t_1 there is no voltage acting around the closed loop formed by the coils, consequently there is no current circulating in the coils. By taking any other instant in the three-phase waveforms illustrated in Figure 26-1(c), it can be demonstrated that the circulating current is always zero in a Δ-connected generator.

Referring again to Figure 26-6, note that I_a flows toward terminal A, while both I_c and I_A flow away from A. This gives

$$I_a = I_A + I_c$$

or, the line current is

$$I_A = I_a - I_c$$

Since the currents are phasor quantities, this is a phasor difference. Figure 26-7(a) shows the phasor diagram for the phase currents that

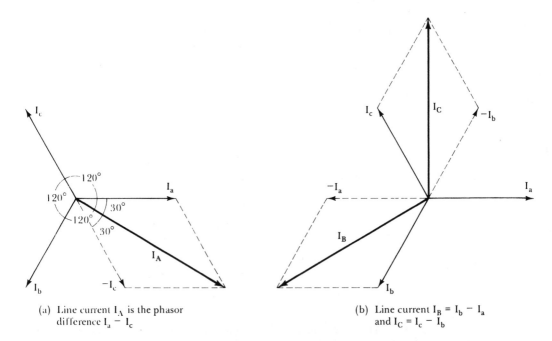

(a) Line current I_A is the phasor difference $I_a - I_c$

(b) Line current $I_B = I_b - I_a$ and $I_C = I_c - I_b$

FIGURE 26-7. For a Δ-connected three-phase generator, the line currents are the phasor sums of pairs of phase currents. With a balanced load, each line current is equal to $\sqrt{3} \times$ (phase current).

Chap. 26   Three-Phase AC Systems

might be produced by a three-phase generator with a *balanced load.* As already explained, when a balanced load occurs, all three line currents are equal and have the same phase relationship to their respective line voltages. If the load consists of three equal Δ-connected resistors, then a balanced load condition exists.

In Figure 26-7(a), the phasors for the three equal phase currents (I_a, I_b, and I_c) are drawn with 120° phase angle separation. Line current I_A is shown in Figure 26-7(a) as $I_a - I_c$ (as determined above). It is seen that I_a and $-I_c$ are separated by 60°, and because they are equal in magnitude:

$$I_A = I_a \cos 30° + I_c \cos 30°$$

$$= 2(I_p \cos 3°)$$

$$= 1.732 \, I_p$$

or,

$$I_L = \sqrt{3} \, I_p$$

Therefore, with a balanced load,

$$\text{line current} = \sqrt{3} \ (\text{phase current})$$

that is,

$$\boxed{I_L = \sqrt{3} \, I_p} \tag{26-4}$$

Figure 26-7(a) also shows that (for a balanced load) the phase current I_a leads the line current I_A by an angle 30°.

In Figure 26-7(b) the determination of I_B and I_C are shown as the phasor differences $I_b - I_a$ and $I_c - I_b$ respectively.

DELTA-CONNECTED LOADS. The reasoning used in deriving Equations (26-3) and (26-4) for a Δ-connected generator can also be applied to a Δ-connected load. In Figure 26-8(a), the individual Δ-connected load voltages are obviously equal to the line voltages:

load voltage = line voltage

Also, where all three load impedances are identical, giving a balanced load, the individual load currents combine to produce:

line current $= \sqrt{3}$ *(load current)*

EXAMPLE 26-4 Load resistors, R_1, R_2, and R_3 in Figure 26-8(a) are each 100 Ω, and the generator phase voltage is $V_p = 100$ V. Calculate (*a*), the current in each load resistor, and (*b*), the line currents.

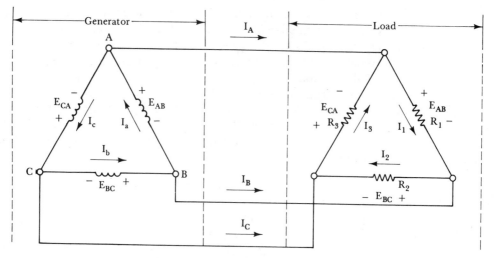

(a) Δ-connected three-phase generator with
Δ-connected resistive loads

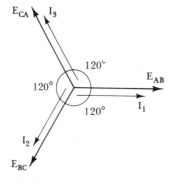

(b) Phasor diagram of load voltages and
currents for purely resistive loads

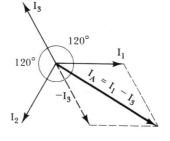

(c) Line current I_A is the phasor
difference $I_1 - I_3$

FIGURE 26-8. Circuit and phasor diagrams for a Δ-connected three-phase generator with a Δ-connected balanced load. The load voltage equals the generator phase voltage, and the line current is $\sqrt{3} \times$ (load current).

SOLUTION

(a) Load currents
From Equation (26-3),

$$V_L = V_p = 100 \text{ V}$$

Taking voltage E_{AB} as the reference voltage for all phase angles:

Chap. 26   Three-Phase AC Systems

$$I_1 = \frac{E_{AB}}{R_1} = \frac{100 \text{ V}}{100 \text{ }\Omega}$$

$$= 1 \text{ A} \underline{/0°}$$

$$I_2 = \frac{E_{BC}}{R_2} = \frac{100 \text{ V}\underline{/-120°}}{100 \text{ }\Omega}$$

$$= 1 \text{ A} \underline{/-120°}$$

$$= 1 \text{ A}[\cos(-120°) + j\sin(-120°)]$$

$$= -0.5 \text{ A} - j0.866 \text{ A}$$

$$I_3 = \frac{E_{CA}}{R_3} = \frac{100 \text{ V}\underline{/-240°}}{100 \text{ }\Omega}$$

$$= 1 \text{ A} \underline{/-240°}$$

$$= 1 \text{ A}[\cos(-240°) + j\sin(-240°)]$$

$$= -0.5 \text{ A} + j0.866 \text{ A}$$

The phasor diagram for load voltages and currents is drawn in Figure 26-8(b). Since the loads are purely resistive, the load currents are in phase with the load voltages.

(b) *Line currents*

$$I_A = I_1 - I_3$$

$$= 1 \text{ A} - (-0.5 \text{ A} + j0.866 \text{ A})$$

$$= 1.5 \text{ A} - j0.866 \text{ A}$$

$$= \sqrt{(1.5 \text{ A})^2 + (0.866 \text{ A})^2} \underline{/\arctan(-0.866/1.5)}$$

$$= \mathbf{1.732 \text{ A} \underline{/-30°}}$$

Figure 26-8(c) shows the phasor derivation of I_A as $I_1 - I_3$.

$$I_B = I_2 - I_1$$

$$= (-0.5 \text{ A} - j0.866 \text{ A}) - 1 \text{ A}$$

$$= -1.5 \text{ A} - j0.866 \text{ A}$$

$$= \mathbf{1.732 \text{ A} \underline{/-150°}}$$

$$I_C = I_3 - I_2$$

$$= (-0.5 \text{ A} + j0.866 \text{ A}) - (-0.5 \text{ A} - j0.866 \text{ A})$$

$$= 0 + j1.732 \text{ A}$$

$$= \mathbf{1.732 \text{ A} \underline{/-270°}}$$

EXAMPLE 26-5 Recalculate the line currents in the circuit of Figure 26-8(a) if R_2 is changed to 200 Ω, and R_3 becomes 50 Ω.

SOLUTION

$$I_1 = \frac{E_{AB}}{R_1} = \frac{100\text{ V}\angle 0°}{100\ \Omega}$$

$$= 1\text{ A}\angle 0°$$

$$I_2 = \frac{E_{BC}}{R_2} = \frac{100\text{ V}\angle -120°}{200\ \Omega}$$

$$= 0.5\text{ A}\angle -120°$$

$$= 0.5\text{ A}[\cos(-120°) + j\sin(-120°)]$$

$$= -0.25\text{ A} - j0.433\text{ A}$$

$$I_3 = \frac{E_{CA}}{R_3} = \frac{100\text{ V}\angle -240°}{50\ \Omega}$$

$$= 2\text{ A}\angle -240°$$

$$= 2\text{ A}[\cos(-240°) + j\sin(-240°)]$$

$$= -1\text{ A} + j1.732\text{ A}$$

$$I_A = I_1 - I_3$$

$$= 1\text{ A} - (-1\text{ A} + j1.732\text{ A})$$

$$= 2\text{ A} - j1.732\text{ A}$$

$$= \sqrt{(2\text{ A})^2 + (1.732\text{ A})^2}\ \angle\arctan(-1.732/2)$$

$$\mathbf{= 2.65\text{ A}\angle -40.9°}$$

$$I_B = I_2 - I_1$$

$$= (-0.25\text{ A} - j0.433\text{ A}) - 1\text{ A}$$

$$= -1.25\text{ A} - j0.433\text{ A}$$

$$= \sqrt{(1.25\text{ A})^2 + (0.433\text{ A})^2}\ \angle\arctan(-0.433/-1.25)$$

$$\mathbf{= 1.32\text{ A}\angle -161°}$$

$$I_C = I_3 - I_2$$

$$= (-1\text{ A} + j1.732\text{ A}) - (-0.25\text{ A} - j0.433\text{ A})$$

$$= -0.75\text{ A} + j2.165\text{ A}$$

$$= \sqrt{(0.75\text{ A})^2 + (2.165\text{ A})^2}\ \angle\arctan(2.165/-0.75)$$

$$\mathbf{= 2.29\text{ A}\angle -251°}$$

EXAMPLE 26-6 The circuit in Figure 26-9 has: $R_1 = R_2 = R_3 = 200\ \Omega$, $C_2 = 10\ \mu F$, $L_3 = 400\ mH$. The line voltage is $V_L = 250\ V$, and the supply frequency is 60 Hz. Calculate the load currents and the line currents.

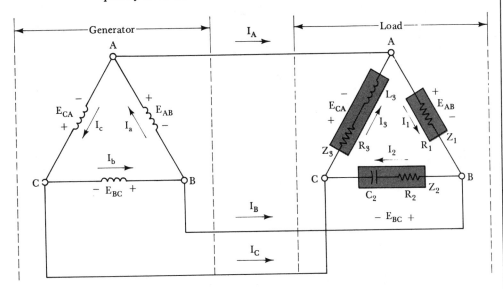

FIGURE 26-9. Circuit of a Δ-connected three-phase generator with a Δ-connected unbalanced load. The load voltage, once again, equals the generator phase voltage, but the individual line currents depend upon the load.

SOLUTION

Taking E_{AB} as reference for all phase angles:

$$I_1 = \frac{E_{AB}}{Z_1} = \frac{250\ V}{200\ \Omega}$$

$$= 1.25\ A\ \underline{/0°}$$

$$X_{C2} = \frac{1}{2\pi fC} = \frac{1}{2\pi \times 60\ Hz \times 10\ \mu F}$$

$$= 265.3\ \Omega$$

$$Z_2 = R_2 - jX_{C2} = 200\ \Omega - j265.3\ \Omega$$

$$= \sqrt{(200\ \Omega)^2 + (265.3\ \Omega)^2}\ \underline{/\arctan(-265.3/200)}$$

$$= 332.2\ \Omega\underline{/-53°}$$

$$I_2 = \frac{E_{BC}}{Z_2} = \frac{250\ V\underline{/-120°}}{332.2\ \Omega\underline{/-53°}}$$

$$= 0.753\ A\ \underline{/-67°}$$

$$= 0.753\ A[\cos(-67°) + j\sin(-67°)]$$

$$= 0.294\ A - j0.693\ A$$

$$X_{L3} = 2\pi f L = 2\pi \times 60 \text{ Hz} \times 400 \text{ mH}$$

$$= 151 \; \Omega$$

$$Z_3 = R_3 + jX_{L3} = 200 \; \Omega + j151 \; \Omega$$

$$= 251 \; \Omega \underline{/37.1°}$$

$$I_3 = \frac{E_{CA}}{Z_3} = \frac{250 \text{ V} \underline{/-240°}}{251 \; \Omega \underline{/37.1°}}$$

$$\cong \mathbf{0.996 \; A \underline{/-277°}}$$

$$= 0.996 \; A[\cos(-277°) + j \sin(277°)]$$

$$= 0.123 \; A + j0.988 \; A$$

$$I_A = I_1 - I_3$$

$$= 1.25 \; A - (0.123 \; A + j0.988 \; A)$$

$$= 1.127 \; A - j0.988 \; A$$

$$= \sqrt{(1.127 \; A)^2 + (0.988 \; A)^2} \; \underline{/\arctan(-0.988/1.127)}$$

$$= \mathbf{1.5 \; A \underline{/-41.2°}}$$

$$I_B = I_2 - I_1$$

$$= (0.294 \; A - j0.693 \; A) - 1.25 \; A$$

$$= -0.956 \; A - j0.693 \; A$$

$$= \sqrt{(0.956 \; A)^2 + (0.693 \; A)^2} \; \underline{/\arctan(-0.693/-0.956)}$$

$$I_B = \mathbf{1.18 \; A \underline{/-144°}}$$

$$I_C = I_3 - I_2$$

$$= (0.123 \; A + j0.988 \; A) - (0.294 \; A - j0.693 \; A)$$

$$= -0.171 \; A + j1.68 \; A$$

$$= \sqrt{(0.171 \; A)^2 + (1.68 \; A)^2} \; \underline{/\arctan(1.68/-0.171)}$$

$$= \mathbf{1.69 \; A \underline{/95.8°}}$$

PRACTICE PROBLEMS

26-3.1 A balanced Δ-connected load, as in Figure 26-8(a), has a line voltage of 200 V and a line current of 3.5 A. Calculate the load resistances.

26-3.2 Determine the line currents in the circuit of Problem 26-3.1 when the load resistors are $R_1 = 80 \; \Omega$, $R_2 = 60 \; \Omega$, and $R_3 = 120 \; \Omega$.

26-3.3 A Δ-connected load, as in Figure 26-9, has a line voltage of 150 V and the following loads: $Z_1 = 86.6 \; \Omega$, $Z_2 = 86.6 \; \Omega \underline{/-30°}$, $Z_3 = 86.6 \; \Omega \underline{/60°}$. Determine the line currents.

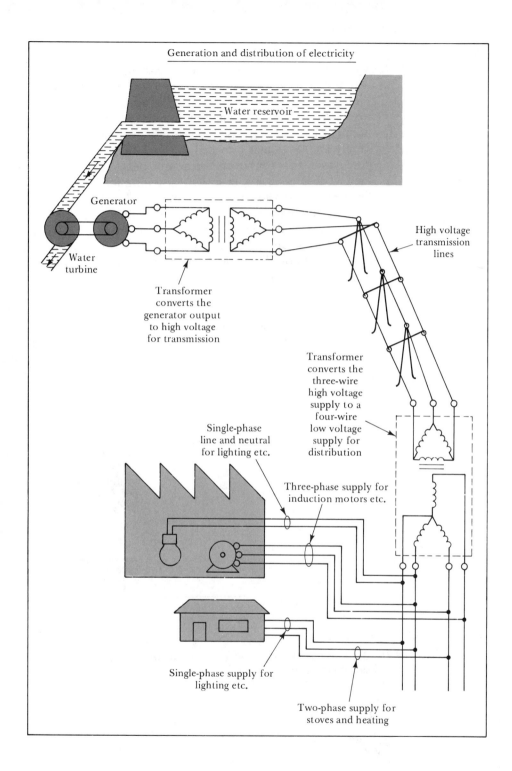

Generation and distribution of electricity

Water reservoir

Generator

Water turbine

Transformer converts the generator output to high voltage for transmission

High voltage transmission lines

Transformer converts the three-wire high voltage supply to a four-wire low voltage supply for distribution

Single-phase line and neutral for lighting etc.

Three-phase supply for induction motors etc.

Single-phase supply for lighting etc.

Two-phase supply for stoves and heating

26-4
Y–Δ AND Y–Y SYSTEMS

Y–Δ SYSTEM. Electricity is usually distributed in a three-phase, four-wire system from a Y-connected generator (or transformer). Each of the three lines may be used along with the neutral as a single-phase supply for relatively low power applications such as lighting. Individual homes normally have single-phase supplies, although two lines of a three-phase supply are sometimes provided for domestic cooking and heating purposes. Industrial consumers use a three-phase, four-wire supply. This is then distributed locally in single-phase form for low power loads and in three-phase form for induction motors and high power loads.

A three-phase generator may have several loads, some of which are Y-connected and some Δ-connected. In this case each Y- or Δ-connected load may be analyzed separately to determine the required line currents at the generator. The superposition theorem can then be employed to calculate the total generator currents.

EXAMPLE 26-7 A balanced Δ-connected load is supplied from a Y-connected generator, as illustrated in Figure 26-10. The load consists of $R_1 = R_2 = R_3 = 33.3 \; \Omega$, and $L_1 = L_2 = L_3 = 523$ mH. The supply has a phase voltage of 115.5 V and a frequency of 60 Hz. Calculate the line current.

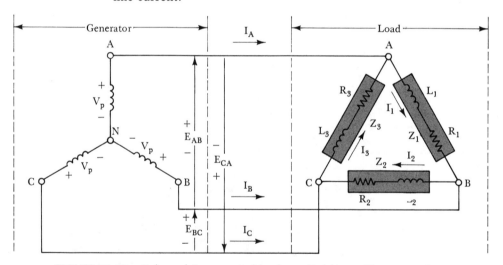

FIGURE 26-10. Balanced Δ-connected load supplied from a Y-connected generator. The load voltage is the generator line voltage, and the line current (for a balanced load) is $\sqrt{3} \times$ (load current).

SOLUTION

From Equation (26-1),

line voltage,

$$V_L = \sqrt{3}\, V_p = \sqrt{3} \times 115.5 \text{ V}$$
$$= 200 \text{ V}$$
$$X_L = 2\pi f L = 2\pi \times 60 \text{ Hz} \times 523 \text{ mH}$$
$$= 197.2 \ \Omega$$
$$Z_1 = Z_2 = Z_3 = R + jX_L$$
$$= 33.3 \ \Omega + j197.2 \ \Omega$$

or

$$|Z| = \sqrt{33.3^2 + 197.2^2}$$
$$= 200 \ \Omega$$

load current,

$$|I_p| = I_1 = I_2 = I_3$$
$$= \frac{V_L}{Z} = \frac{200 \text{ V}}{200 \ \Omega}$$
$$= 1 \text{ A}$$

From Equation (26-4),

line current,

$$I_L = \sqrt{3}\, I_p = \sqrt{3} \times 1 \text{ A}$$
$$\mathbf{= 1.732 \text{ A}}$$

EXAMPLE 26-8 Recalculate the line currents in the circuit of Figure 26-10 to determine the phase angle of each current with respect to phase voltage E_{AN}.

SOLUTION

From Figure 26-3(b),

$$E_{AB} \text{ leads } E_{AN} \text{ by } 30°,$$

and
$$E_{AB} = \sqrt{3}\, E_{AN} = \sqrt{3} \times 115.5 \text{ V}\underline{/30°}$$
$$= 200 \text{ V}\underline{/30°}$$

and from Figure 26-3(c),

$$E_{BC} = 200 \text{ V}\underline{/-90°} \quad \text{with respect to } E_{AN}$$

and

$$E_{CA} = 200 \text{ V} \underline{/-210°}$$

$$Z_1 = Z_2 = Z_3 = 33.3 \ \Omega + j197.2 \ \Omega$$

$$= 200 \ \Omega \underline{/80.4°}$$

$$I_1 = \frac{E_{AB}}{Z_1} = \frac{200 \text{ V} \underline{/30°}}{200 \ \Omega \underline{/80.4°}}$$

$$= 1 \text{ A} \underline{/-50.4°} = 0.637 \text{ A} - j0.771 \text{ A}$$

$$I_2 = \frac{E_{BC}}{Z_2} = \frac{200 \text{ V} \underline{/-90°}}{200 \ \Omega \underline{/80.4°}}$$

$$= 1 \text{ A} \underline{/-170.4°} = -0.986 \text{ A} - j0.167 \text{ A}$$

$$I_3 = \frac{E_{CA}}{Z_3} = \frac{200 \text{ V} \underline{/-210°}}{200 \ \Omega \underline{/80.4°}}$$

$$= 1 \text{ A} \underline{/-290.4°} = 0.349 \text{ A} + j0.937 \text{ A}$$

$$I_A = I_1 - I_3$$

$$= (0.637 \text{ A} - j0.771 \text{ A}) - (0.349 \text{ A} + j0.973 \text{ A})$$

$$= 0.288 \text{ A} - j1.71 \text{ A}$$

$$\mathbf{= 1.73 \text{ A} \underline{/-80.4°}}$$

$$I_B = I_2 - I_1$$

$$= (-0.986 \text{ A} - j0.167 \text{ A}) - (0.637 \text{ A} - j0.771 \text{ A})$$

$$= -1.62 \text{ A} + j0.604 \text{ A}$$

$$\mathbf{= 1.73 \text{ A} \underline{/-200.4°}}$$

$$I_C = I_3 - I_2$$

$$= (0.349 \text{ A} + j0.937 \text{ A}) - (-0.986 \text{ A} - j0.167 \text{ A})$$

$$= 1.34 \text{ A} + j1.1 \text{ A}$$

$$\mathbf{= 1.73 \text{ A} \underline{/-320.4°}}$$

The phasor diagrams of voltages and currents for this circuit are drawn in Figure 26-11. Note that each line current is 80.4° behind each phase voltage.

Y-Y SYSTEM. Figure 26-12(a) shows a Y-connected load with a three-wire supply (no neutral conductor). To calculate the line currents for such a load, the Y-connected circuit should first be converted to its

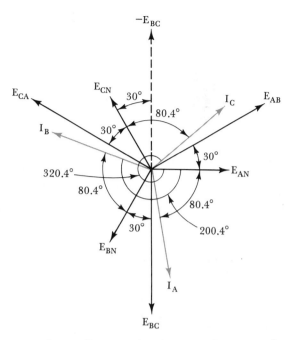

FIGURE 26-11. Phasor diagram of voltages and currents for a balanced Δ-connected load supplied from a Y-connected generator.

Δ-connected equivalent circuit, Figure 26-12(b). Equations (22-6), (22-7), and (22-8) for Y to Δ conversion are applied to determine the equivalent Δ-connected components:

Equation (22-6):

$$Z_{ab} = \frac{Z_a Z_b + Z_a Z_c + Z_b Z_c}{Z_c}$$

Equation (22-7):

$$Z_{ac} = \frac{Z_a Z_b + Z_a Z_c + Z_b Z_c}{Z_b}$$

Equation (22-8):

$$Z_{bc} = \frac{Z_a Z_b + Z_a Z_c + Z_b Z_c}{Z_a}$$

The load phase currents I_1, I_2, and I_3 in Figure 26-12(b) are calculated next. Then the line currents are determined as the phasor difference of the phase currents: $I_A = I_1 - I_3$, $I_B = I_2 - I_1$, $I_C = I_3 - I_2$.

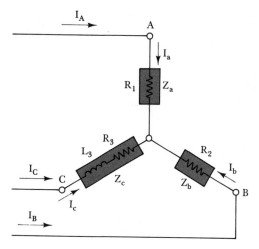

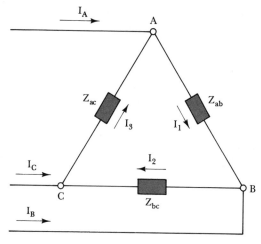

(a) Unbalanced y-connected load with three-wire supply

(b) Δ-connected equivalent circuit of y-connected load

FIGURE 26-12. For analysis of a Y-connected load with a three-wire supply, the Y-Δ transformation equations should be used to convert it into an equivalent Δ-connected load.

EXAMPLE 26-9 An unbalanced Y-connected load has a three-wire supply, as in Figure 26-12(a). The load components are $R_1 = R_2 = R_3 = 50\ \Omega$, $L_3 = 398$ mH. The supply frequency is 60 Hz, and the line voltage is 200 V. Calculate the line currents I_A, I_B, and I_C and the load currents I_a, I_b, and I_c.

SOLUTION

$$Z_a = R_1 = 50\ \Omega\underline{/0°}$$
$$= 50\ \Omega + j0\ A$$
$$Z_b = R_2 = 50\ \Omega\underline{/0°}$$
$$= 50\ \Omega + j0\ A$$
$$X_{L3} = 2\pi f L_3 = 2\pi \times 60\ \text{Hz} \times 398\ \text{mH}$$
$$= 150\ \Omega$$
$$Z_c = R_3 + jX_{L3}$$
$$= 50\ \Omega + j150\ \Omega = 158\ \Omega\underline{/71.6°}$$
$$Z_a Z_b = 50\ \Omega\underline{/0°} \times 50\ \Omega\underline{/0°}$$
$$= 2500\underline{/0°} = 2500 + j0\ A$$
$$Z_a Z_c = 50\ \Omega\underline{/0°} \times 158\ \Omega\underline{/71.6°}$$
$$= 7900\underline{/71.6°} = 2494 + j7496\ \Omega$$

$$Z_b Z_c = 50 \ \Omega \underline{/0°} \times 158 \ \Omega \underline{/71.6°}$$

$$= 7900 \underline{/71.6°} = 2494 + j7496 \ \Omega$$

$$Z_a Z_b + Z_a Z_c + Z_b Z_c = 7500 + j15 \ 000$$

$$\cong 16 \ 800 \underline{/63.4°}$$

Equation (22-6):

$$Z_{ab} = \frac{Z_a Z_b + Z_a Z_c + Z_b Z_c}{Z_c} = \frac{16 \ 800 \underline{/63.4°}}{158 \ \Omega \underline{/71.6°}}$$

$$= 106 \ \Omega \underline{/-8.2°}$$

Equation (22-7):

$$Z_{ac} = \frac{Z_a Z_b + Z_a Z_c + Z_b Z_c}{Z_b} = \frac{16 \ 800 \underline{/63.4°}}{50 \ \Omega \underline{/0°}}$$

$$= 336 \ \Omega \underline{/63.4°}$$

Equation (22-8):

$$Z_{bc} = \frac{Z_a Z_b + Z_a Z_c + Z_b Z_c}{Z_a} = \frac{16 \ 800 \underline{/63.4°}}{50 \ \Omega \underline{/0°}}$$

$$= 336 \ \Omega \underline{/63.4°}$$

$$I_1 = \frac{E_{AB}}{Z_{ab}} = \frac{200 \ V}{106 \ \Omega \underline{/-8.2°}}$$

$$= 1.88 \ A \underline{/8.2°} = 1.86 \ A + j0.268 \ A$$

$$I_2 = \frac{E_{BC}}{Z_{bc}} = \frac{200 \ V \underline{/-120°}}{336 \ \Omega \underline{/63.4°}}$$

$$= 0.595 \ A \underline{/-183.4°} = -594 \ mA + j35.3 \ mA$$

$$I_3 = \frac{E_{CA}}{Z_{ac}} = \frac{200 \ V \underline{/-240°}}{336 \ \Omega \underline{/63.4°}}$$

$$= 0.595 \ A \underline{/-303.4°} = 328 \ mA + j497 \ mA$$

$$I_A = I_1 - I_3 = 1.53 \ A - j229 \ mA$$

$$= 1.55 \ A \underline{/-8.5°}$$

$$\boldsymbol{I_a = I_A = 1.55 \ A \underline{/-8.5°}}$$

$$I_B = I_2 - I_1 = -2.45 \ A - j233 \ mA$$

$$= 2.46 \ A \underline{/-174.6°}$$

$$\boldsymbol{I_b = I_B = 2.46 \ A \underline{/-174.6°}}$$

$$I_C = I_3 - I_2 = 922 \ mA + j461 \ mA$$

$$= 1.03 \ A \underline{/26.6°}$$

$$\boldsymbol{I_c = I_C = 1.03 \ A \underline{/26.6°}}$$

26-4.1 A balanced Δ-connected load is supplied from a Y-connected generator, as in Figure 26-10. The generator phase voltage and phase current are 115.5 V and 3.46 A. Calculate the load impedances.

26-4.2 A Y-connected load, as in Figure 26-12(a) has a three-wire supply with a line voltage of 150 V. The loads are $R_a = 86.6 \ \Omega$, $R_b = 173.2 \ \Omega$, and $R_c = 43.3 \ \Omega$. Determine the load curents.

26-5
PHASE SEQUENCE

CORRECT PHASE SEQUENCE. In Section 26-1 it was explained that the three output waveforms are generated in a particular sequence. Refer to the waveforms in Figure 26-1(c) again and recall that voltage e_{Bb} reaches its peak 120° or one-third of a cycle after the peak of e_{Aa} occurs. Also, e_{Cc} peaks 240° after e_{Aa} and 120° after e_{Bb}. The phase sequence of the voltages can be listed as $e_{Aa} \ e_{Bb} \ e_{Cc}$, or ABC. The phasor diagram is drawn in Figure 26-1(d) for a phase sequence ABC. Since phasors are thought of as rotating in a counterclockwise direction, those illustrated would pass a stationary point with the sequence ABC. Reading the phasors in a clockwise direction is the same as thinking of them rotating in a counterclockwise direction. Either way, the phase sequence is ABC, and it can also be correctly stated as BCA or CAB, since the actual repeating sequence is $ABCABCABC$, etc.

Figure 26-13(a) shows a Y-connected generator together with its voltage waveforms and phasor diagram. The sequence of the phase voltages is ABC, and the line voltages E_{AB}, E_{BC}, and E_{CA} also have the phase sequence ABC. This is easily verified from the phasor diagram by reading either the first or second of the double subscripts clockwise around the diagram. The first subscripts read ABC, and the second read BCA, which means that both give the phase sequence as ABC. The supply lines from the generator in Figure 26-13(a) are identified as *line A*, *line B*, and *line C*, and they are connected respectively to generator terminals A, B, and C.

REVERSED PHASE SEQUENCE. Now refer to Figure 26-13(b), which shows a generator in which the terminals have been incorrectly connected. The supply line identified as line A is connected to terminal C, and line C is connected to terminal A. This change in terminal connections has a significant effect on the waveforms and on the phasor diagram. The voltage measured between line A and neutral is identified as e_{AN}, but is actually the output of terminal C, which reaches its peak 120° behind E_{BN}. Similarly, the voltage measured between line C and neutral is now 120° ahead of E_{BN}, because it is actual E_{AN} (the output of terminal

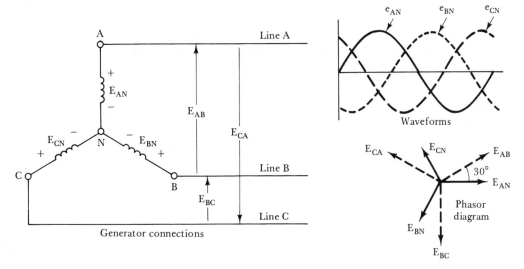

(a) Generator connections, waveforms and phasor diagrams for phase sequence ABC

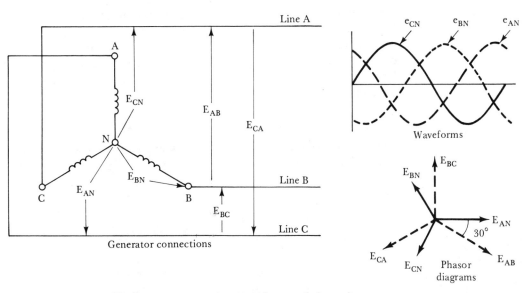

(b) Generator connections, waveforms and phasor diagrams for phase sequence CBA

FIGURE 26-13. When two line connections are interchanged on a three-phase generator, the phase sequence is reversed. This can reverse the direction of three-phase induction motors and can have *serious* consequences for unbalanced loads.

A). Thus, the waveforms in Figure 26-13(b) show e_{CN} leading e_{BN} by $120°$, and e_{AN} lagging e_{BN} by $120°$. The phase sequence is CBA. This is, in fact, a reversal of the phase sequence ABC.

The phase sequence reversal also shows in the phasor diagram in Figure 26-13(b). Comparing the phase voltages to those in Figure 26-13(a), it is seen that E_{CN} (the output from terminals N and C) is now $120°$ behind E_{NA}, whereas previously it was $240°$ behind. Also, E_{BN} is now $240°$ behind E_{AN}, instead of $120°$, as before.

The line voltage phasor E_{AB} in Figure 26-13(b), which is the phasor difference of E_{AN} and E_{BN}, is seen to be $30°$ *lagging* E_{AN}. The phasor diagram in Figure 26-13(a) shows that before the terminal connection change, E_{AB} *led* E_{AN} by $30°$. Also, in Figure 26-13(b) E_{CA} and E_{BC} are now reversed in their phase relationships with respect to E_{AB}.

PHASE SEQUENCE EFFECTS. Where balanced resistive loads are involved, the phase sequence is normally not important. However, with unbalanced loads, and especially with three-phase induction motors, any change in the phase sequence of a supply can have *serious* consequences. An induction motor is usually required to run in a particular direction. When initially connected, its direction of rotation can be checked. If it is not correct, the direction can be reversed by interchanging any two line connections at the motor. If the phase sequence of the supply is subsequently altered, the motor direction will be reversed again. Example 26-10 demonstrates the effect of phase sequence reversal on an unbalanced load.

EXAMPLE 26-10 The ABC phase sequence used in Example 26-9 is reversed to ACB. Recalculate the load currents I_a, I_b, and I_c.

SOLUTION

$$I_1 = \frac{E_{AB}}{Z_{ab}} = \frac{200 \text{ V}}{106 \ \Omega \underline{/-8.2°}} = 1.88 \ A \underline{/8.2°}$$

$$= 1.86 \ A + j0.268 \ A$$

$$I_2 = \frac{E_{BC}}{Z_{bc}} = \frac{200 \text{ V} \underline{/-240°}}{336 \ \Omega \underline{/63.4°}} = 0.595 \ A \underline{/-303.4°}$$

$$= 328 \text{ mA} + j497 \text{ mA}$$

$$I_3 = \frac{E_{CA}}{Z_{ac}} = \frac{200 \text{ V} \underline{/-120°}}{336 \ \Omega \underline{/63.4°}} = 0.595 \ A \underline{/-183.4°}$$

$$= -594 \text{ mA} + j35.3 \text{ mA}$$

$$I_a = I_A = I_1 - I_3$$

$$= 2.45 \ A + j233 \text{ mA}$$

$$= 2.46 \, A \, \underline{/5.4°}$$

$$I_b = I_B = I_2 - I_1$$

$$= -1.53 \, A + j229 \, mA$$

$$= 1.55 \, A \, \underline{/-188.5°}$$

$$I_c = I_C = I_3 - I_2$$

$$= -922 \, mA - j462 \, mA$$

$$= 1.03 \, A \, \underline{/-153.4°}$$

PHASE SEQUENCE TESTER. Comparing the results of Example 26-10 to those obtained in Example 26-9, it is seen that the phase sequence reversal caused I_a to change from 1.55 A to 2.46 A, I_b to change from 2.46 A to 1.55 A, and I_c to remain 1.03 A. The results can be employed to demonstrate the operation of a *phase sequence tester*.

The circuit of Figure 26-12(a) is redrawn in Figure 26-14 with resistors R_1 and R_2 replaced by lamps. Each lamp has a filament resistance of 50 Ω, as for R_1 and R_2. Therefore, the circuit is essentially the same as in Figure 26-12(a), and the analysis performed in Examples 26-9 and 26-10 applies. In Example 26-9, with phase sequence ABC it was found that $I_a = 1.55 \, A$ and $I_b = 2.46 \, A$. The larger current through lamp b would cause it to glow more brightly than lamp a. When the phase sequence is changed to CBA (Example 26-10), I_a becomes 2.46 A while I_b is reduced to 1.55 A. Now lamp a is brighter than lamp b. Thus, the circuit in Figure 26-14 is a phase sequence tester. With phase sequence ABC, lamp b is brighter than lamp a. With sequence CBA, lamp a is brightest.

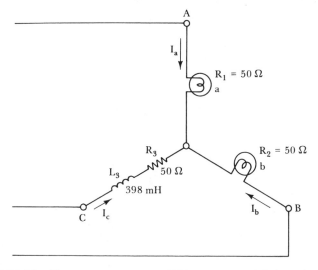

FIGURE 26-14. Phase sequence tester. With sequence ABC, lamp b is brighter than lamp a. When the sequence is CBA, lamp a is brightest.

26-5.1   Recalculate the load currents in the circuit in Problem 26-4.2 when the phase sequence is reversed.

26-6

POWER IN THREE-PHASE SYSTEMS

TRUE POWER. The power dissipated in each of the three loads or branches in a three-phase system is:

$$\boxed{P_p = V_p I_p \cos \phi \text{ (watts)}} \tag{26-5}$$

where V_p and I_p are the phase voltage and current, and ϕ is the phase angle bettween V_p and I_p.

In the case of an unbalanced load, whether Y- or Δ-connected, the individual phase (load) power dissipations, P_1, P_2, and P_3 are calculated and added to determine the total power dissipated in the load.

$$\boxed{P = P_1 + P_2 + P_3} \tag{26-6}$$

For a balanced load, the power dissipation is the same in all three branches. Thus, from Equation (26-5),

$$\boxed{P = 3 V_p I_p \cos \phi} \tag{26-7}$$

In the case of a Y-connected load, Equations (26-1) and (26-2) give:

$$V_p = \frac{V_L}{\sqrt{3}} \quad \text{and} \quad I_p = I_L$$

For a Δ-connected load, Equations (26-3) and (26-4) give:

$$V_p = V_L \quad \text{and} \quad I_p = \frac{I_L}{\sqrt{3}}$$

Thus, for both Y- and Δ-connected balanced loads, the total power is:

$$P = 3 \frac{V_L I_L}{\sqrt{3}} \cos \phi$$

or,

$$\boxed{P = \sqrt{3} \, V_L I_L \cos \phi} \tag{26-8}$$

Equation (26-8) allows the total power (in watts) to be calculated from the measured line current and line voltage. Note that ϕ is still the phase angle between V_p and I_p, which is the phase angle of each individual load. It is *not* the phase angle between the line quantities V_L and I_L.

REACTIVE POWER. The reactive power per phase [from Equation (21-8)] is:

$$Q_p = V_p I_p \sin \phi$$

or, total reactive power with a balanced load,

$$Q = 3 V_p I_p \sin \phi$$

Substituting the line voltage and current into the equation, as before:

$$\boxed{Q = \sqrt{3} \, V_L I_L \sin \phi} \tag{26-9}$$

Equation (26-9) gives the total reactive power (in var) for a balanced load. Once again, ϕ is the phase angle of each of the three individual loads.

APPARENT POWER. The apparent power per phase [Equation (21-9)] is calculated as

$$S = V_p I_p \qquad \text{(volt amp)},$$

and the total apparent power with a balanced load is

$$S = 3 V_p I_p$$

Once again the line voltage and current may be substituted into the equation:

$$\boxed{S = \sqrt{3} \, V_L I_L} \tag{26-10}$$

Dividing Equation (26-8) by Equation (26-10),

$$\frac{P}{S} = \frac{\sqrt{3} \, V_L I_L \cos \phi}{\sqrt{3} \, V_L I_L}$$

giving

$$\boxed{\cos \phi = \frac{P}{S}} \tag{26-11}$$

EXAMPLE 26-11   Calculate the power dissipated in the load shown in Figure 26-5. As
for Example 26-3, the generator phase voltage is $V_p = 100$ V, and its
frequency is 60 Hz. The load components are $R_1 = R_2 = R_3 = 100\ \Omega$,
$C_2 = 66.3\ \mu F$, and $L_3 = 159.2$ mH.

SOLUTION

From Example 26-3:

$$Z_1 = 100\ \Omega\underline{/0°}, \qquad Z_2 = 107.7\ \Omega\underline{/-21.8°},$$

$$Z_3 = 166.6\ \Omega\underline{/31°}, \qquad I_1 = 1\ A,$$

$$I_2 = 0.929\ A \quad \text{and} \quad I_3 = 0.858\ A.$$

Equation (26-5):

$$P_1 = V_{P1}\,I_{P1}\cos\,\phi_1$$

$$= 100\ V \times 1\ A \times \cos 0$$

$$= 100\ W$$

$$P_2 = V_{P2}\,I_{P2}\cos\,\phi_2$$

$$= 100\ V \times 0.929\ A \times \cos(21.8°)$$

$$\cong 86.3\ W$$

$$P_3 = V_{P3}\,I_{P3}\cos\,\phi_3$$

$$= 100\ V \times 0.858\ A \times \cos 31°$$

$$= 73.5\ W$$

Equation (26-6):

$$P_T = P_1 + P_2 + P_3$$

$$= \mathbf{259.8\ W}$$

EXAMPLE 26-12   For the load shown in Figure 26-10, calculate the total power dissipated, the reactive power, and the apparent power. The circuit quantities are as listed in Examples 26-7 and 26-8.

SOLUTION

From Examples 26-7 *and* 26-8:

$$V_L = 200\ V, \qquad I_L = 1.732\ A, \quad \text{and} \quad \phi = 80.4°$$

Equation (26-8):

$$P = \sqrt{3}\, V_L I_L \cos \phi$$
$$= \sqrt{3} \times 200 \text{ V} \times 1.732 \text{ A} \times \cos 80.4°$$
$$= \mathbf{100 \text{ W}}$$

Equation (26-9):

$$Q = \sqrt{3}\, V_L I_L \sin \phi$$
$$= \sqrt{3} \times 200 \text{ V} \times 1.732 \text{ A} \times \sin 80.4°$$
$$= \mathbf{591.6 \text{ var}}$$

Equation (26-10),

$$S = \sqrt{3}\, V_L I_L$$
$$= \sqrt{3} \times 200 \text{ V} \times 1.732 \text{ A}$$
$$= \mathbf{600 \text{ VA}}$$

PRACTICE PROBLEMS

26-6.1 Calculate the power dissipated in the circuit described in Problem 26-3.3.

26-6.2 Determine the apparent power, reactive power, and true power for a Y-connected load consisting of $Z_a = Z_b = Z_c = 47\ \Omega\underline{/45°}$. The line voltage is 122 V.

26-7

POWER FACTOR CORRECTION

In Section 21-6, it is explained that if the power factor ($\cos \phi$) of a circuit can be improved, or *corrected,* the current carried by the conductors can be reduced without any reduction in the power dissipated in the load. The discussion in Section 21-6 referred to a single-phase circuit, but it is also applicable to three-phase circuits.

Figure 26-15(a) shows a Y-connected circuit consisting of resistors and inductors, in parallel with a Δ-connected capacitor circuit. The *RL* circuit could represent an induction motor or other inductive circuit. Operating without the capacitors, the *RL* circuit would tend to have a large lagging power factor. Figure 26-15(b) shows the power triangle for the *RL* circuit. When the capacitors are included in the circuit, the capacitor reactive volt-amp (Q_C) subtracts from the inductive reactive volt-amp (Q_L) to give a reduced phase angle, Figure 26-15(c). As in the case of single-phase circuits, this results in an increased power factor and

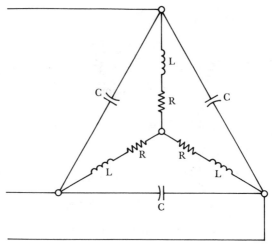

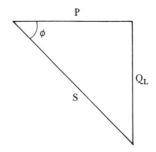

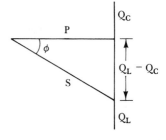

(a) Power factor correction capacitors connected in parallel with inductive-resistive load

(b) Power triangle for RL circuit

(c) Power triangle for RLC circuit

FIGURE 26-15. Power factor correction capacitors for three-phase loads are usually Δ-connected. The effect of the capacitors is to increase the power factor and reduce the line current.

reduced line current. Example 26-13 demonstrates the effect of the capacitors.

Instead of Δ-connecting the capacitors, a Y-connected arrangement could be used. However, it is found that larger capacitance values are required for the Y-connected circuit.

EXAMPLE 26-13 The circuit shown in Figure 26-15(a) has $L = 530$ mH, $R = 200$ Ω, $C = 1$ μF, $V_L = 400$ V and $f = 60$ Hz. Calculate P, I_L, and cos ϕ with (a) only the inductors and resistors in the circuit, (b) only the capacitors in the circuit, and (c) all components present.

SOLUTION

(*a*) *Inductor and resistor circuit*

$$X_L = 2\pi fL = 2 \times \pi \times 60 \text{ Hz} \times 530 \text{ mH}$$
$$= 200 \ \Omega$$
$$Z = \sqrt{R^2 + X_L^2} = \sqrt{200^2 + 200^2}$$
$$\cong 283 \ \Omega$$
$$\cos \phi = \frac{R}{Z} = \frac{200 \ \Omega}{283 \ \Omega}$$
$$= \mathbf{0.707}$$

Equation (26-1):

$$V_p = \frac{V_L}{\sqrt{3}} = \frac{400 \text{ V}}{\sqrt{3}}$$
$$\cong 231 \text{ V}$$
$$I_L = I_p = \frac{V_p}{Z} = \frac{231 \text{ V}}{283 \ \Omega}$$
$$= \mathbf{816 \ mA}$$

Equation (26-8):

$$P = \sqrt{3} \ V_L I_L \cos \phi$$
$$= \sqrt{3} \times 400 \text{ V} \times 816 \text{ mA} \times 0.707$$
$$P_L = \mathbf{400 \ W}$$

Equation (26-9):

$$\sin \phi = \frac{X_L}{Z} = \frac{200 \ \Omega}{238 \ \Omega}$$
$$= 0.707$$
$$Q = \sqrt{3} \ V_L I_L \sin \phi$$
$$= \sqrt{3} \times 400 \text{ V} \times 816 \text{ mA} \times 0.707$$
$$Q_L = 400 \text{ var (inductive)}$$

(*b*) *Capacitor circuit*

$$X_C = \frac{1}{2\pi fC} = \frac{1}{2\pi \times 60 \text{ Hz} \times 1 \ \mu\text{F}}$$
$$= 2.65 \text{ k}\Omega$$

$$I_p = \frac{V_L}{X_C} = \frac{400 \text{ V}}{2.65 \text{ k}\Omega}$$

$$= 151 \text{ mA}$$

$$I_L = \sqrt{3} \; I_p = \sqrt{3} \times 151 \text{ mA}$$

$$\mathbf{= 261 \text{ mA}}$$

$$\cos \phi = \frac{R}{Z} = \frac{0}{2.65 \text{ k}\Omega}$$

$$= 0$$

$$P = \sqrt{3} \; V_L I_L \cos \phi$$

$$= \sqrt{3} \; V_L I_L \times 0$$

$$\mathbf{P_C = 0}$$

$$\sin \phi = \frac{X_L}{Z} = \frac{2.65 \text{ k}\Omega}{2.65 \text{ k}\Omega}$$

$$= 1$$

$$Q = \sqrt{3} \; V_L I_L \sin \phi$$

$$= \sqrt{3} \times 400 \text{ V} \times 261 \text{ mA} \times 1$$

$$Q_C = 181 \text{ var (capacitive)}$$

(c) *All components in circuit*

$$P = P_L + P_C = 400 \text{ W} + 0$$

$$\mathbf{= 400 \text{ W}}$$

$$Q = Q_L - Q_C = 400 \text{ var} - 181 \text{ var}$$

$$= 219 \text{ var (inductive)}$$

$$\phi = \arctan \frac{Q}{P} = \arctan \frac{219}{400}$$

$$= 28.7°$$

$$\mathbf{\cos \phi = 0.877}$$

$$P = \sqrt{3} \; V_L I_L \cos \phi$$

or

$$I_L = \frac{P}{\sqrt{3} \; V_L \cos \phi}$$

$$= \frac{400 \text{ W}}{\sqrt{3} \times 400 \text{ V} \times 0.877}$$

$$\mathbf{= 658 \text{ mA}}$$

It is seen that the capacitors increase the power factor from 0.707 to 0.877, and this produced a reduction in line current from 816 mA to 658 mA without any reduction in load power dissipation.

Chap. 26   Three-Phase AC Systems

26-7.1   For the circuit in Problem 26-6.2, determine the size of power
factor correction capacitors [connected as in Figure 26-15(a)]
which will reduce the line current by 10%. The supply frequency is 60 Hz.

26-8

THREE-PHASE POWER MEASUREMENT

Several methods are available for measuring the power delivered to
a three-phase load. These depend on whether the load is Y- or
Δ-connected, the accessibility of the load terminals, and whether the
load is balanced or unbalanced.

As with all instruments, some care must be taken to connect the
wattmeter correctly. The current coil terminal identified as ± should be
connected to the voltage source. (Sometimes a ↓ or * is used to identify
this terminal.) The ± voltage coil terminal is connected to the load
terminal of the current coil. If the instrument is not connected correctly,
a negative deflection may occur.

THREE-WATTMETER METHODS. Figure 26-16 shows three wattmeters
employed to measure the power in a Y-connected four-wire system with
an unbalanced load. Each wattmeter measures the power dissipated in
one phase of the load ($V_p I_p \cos \phi$), and the total power is then the sum of

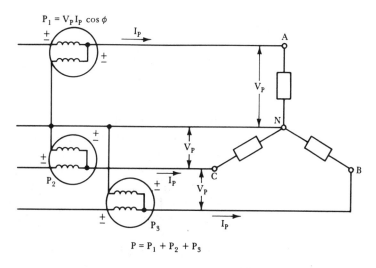

FIGURE 26-16. Three wattmeters connected to measure the power delivered to
each section of a three-phase, four-wire load. The total load power is the sum of
the three readings. For a balanced load, only one meter is required, and the load
power is three times the meter reading.

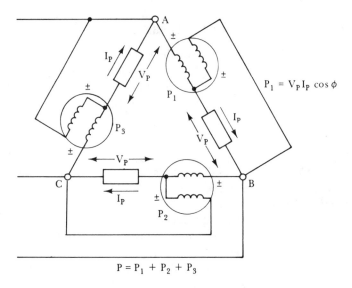

$$P = P_1 + P_2 + P_3$$

FIGURE 26-17. Three wattmeters connected to measure the power delivered to each section of a three-phase Δ-connected load. The total load power is the sum of the three readings. For a balanced load, only one meter is required.

the three meter readings. This circuit could also be used in the case of a three-wire load, if the neutral point of the load is accessible. When the load is balanced, all three meters in Figure 26-16 would read the same. Thus, only one wattmeter is needed, and the total power is three times the meter reading.

In Figure 26-17, a three-wattmeter arrangement is shown for measuring the power in a Δ-connected load. Here again, each meter indicates $(V_p I_p \cos \phi)$ for each of the three load sections. The total power dissipated is the sum of the three meter readings. When the system is balanced, only one wattmeter is required, and the meter indication is multiplied by 3 to determine the total power.

SINGLE-WATTMETER METHOD. In a Y-connected three-wire system and in a Δ-connected system, no neutral conductor is available for connecting the common point of the wattmeter voltage coils. Figure 26-18 shows a circuit in which three equal resistors are Y-connected to provide a neutral point. The single wattmeter now measures the power (P_1) dissipated in one branch of the load, and the total power is $P = 3P_1$. This circuit is suitable only for a balanced load.

TWO-WATTMETER METHOD. The most commonly used method of measuring power in a three-phase load is the *two-wattmeter method*. The circuit for this method is shown in Figure 26-19. The wattmeter current coils are each connected in series with one line, and the voltage coils are commoned at the remaining line, as illustrated.

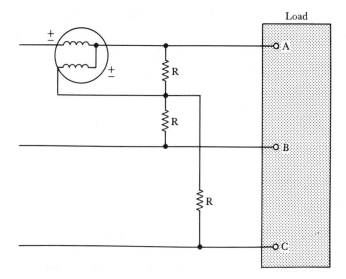

FIGURE 26-18. Single-wattmeter method of measuring the power delivered to a balanced three-wire load. The load power is three times the meter reading.

If the load is Y-connected, the instantaneous power dissipations in each phase can be written in terms of the instantaneous phase current and instantaneous phase voltage:

$$P_A = i_A v_{AN} = i_A v_{Aa}$$

$$P_B = i_B v_{BN} = i_B v_{Bb}$$

$$P_C = i_C v_{CN} = i_C v_{Cc}$$

The total instantaneous power is:

$$P = i_A v_{Aa} + i_B v_{Bb} + i_C v_{Cc} \qquad (1)$$

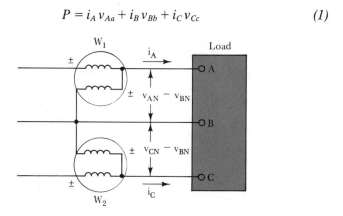

FIGURE 26-19. The two-wattmeter method of three-phase power measurement is most often used. It is suitable for Y-connected or Δ-connected, balanced or unbalanced loads. The load power is the sum of the meter readings.

The instantaneous current in meter W_1 is i_A, and the instantaneous voltage across the voltage coil of W_1 is $V_{AN} - V_{BN}$. Similarly, the instantaneous quantities for W_2 are i_C and $V_{CN} - V_{BN}$. Therefore, the instantaneous powers measured by each of these instruments are:

$$P_1 = i_A (v_{Aa} - v_{Bb})$$

and

$$P_2 = i_C (v_{Cc} - v_{Bb})$$

The sum of P_1 and P_2 is:

$$P = i_A (v_{Aa} - v_{Bb}) + i_C (v_{Cc} - v_{Bb})$$
$$= i_A v_{Aa} - i_A v_{Bb} + i_C v_{Cc} - i_C v_{Bb}$$
$$= i_A v_{Aa} + i_C v_{Cc} - v_{Bb}(i_A + i_C) \qquad (2)$$

With balanced or unbalanced loads, the algebraic sum of the currents at the neutral point of a Y-connected load must be zero (for a three-wire supply). Thus,

$$i_A + i_B + i_C = 0$$

or

$$i_A + i_C = -i_B \qquad (3)$$

Substituting from Equation (3) into Equation (2):

$$P = i_A v_{Aa} + i_C v_{Cc} + i_B v_{Bb} \qquad (4)$$

Equation (4) is identical to Equation (1), showing that the sum of the instantaneous powers measured by the two wattmeters is equal to the total instantaneous power dissipated in the load. A wattmeter indicates the average of the instantaneous power dissipations in a load. Consequently, the sum of the two wattmeter readings is the average power dissipated in the three-phase load.

It can also be shown that for a Δ-connected load the two wattmeters indicate the total power dissipated. In this case the wattmeter current coils carry instantaneous currents of $i_A - i_C$ and $i_C - i_B$. The voltage coil instantaneous potential differences are $-v_{Aa}$ and $-v_{Bb}$. The sum of the meter readings is:

$$P = v_{Aa}(i_A - i_C) - v_{Bb}(i_C - i_B)$$
$$= v_{Aa} i_A + v_{Bb} i_B - i_C(v_{Aa} + v_{Bb})$$

For a Δ-connected system, there is no circulating current, and:

$$v_{Aa} + v_{Bb} + v_{Cc} = 0$$

Chap. 26   Three-Phase AC Systems

or

$$v_{Aa} + v_{Bb} = -v_{Cc}$$

giving

$$P = v_{Aa}i_A + v_{Bb}i_B + v_{Cc}i_C \qquad (5)$$

Equation (5) is, once again, the same as Equation (1), showing that the sum of the meter readings gives the total load power.

It is seen that the two-wattmeter method measures the power delivered to a three-phase load, whether the load is Y- or Δ-connected, balanced or unbalanced.

It should be noted that the phase angle between the voltage and current applied to the wattmeter may be greater than 90°. In this case cos φ is a negative quantity, and the wattmeter tends to read in reverse. In order to obtain a positive (on scale) deflection the terminal connections of the wattmeter voltage (or current) coil are reversed. The meter reading must then be recorded as a negative quantity. Example 26-14 demonstrates this situation.

EXAMPLE 26-14   The load described in Example 26-7 has its power dissipation measured by the two-wattmeter method illustrated in Figure 26-19. Calculate the load power and the power indicated by each wattmeter:

SOLUTION

From Examples 26-7 and 26-8:

$$V_L = 200 \text{ V}, \qquad I_P = 1 \text{ A}, \qquad I_L = 1.73 \text{ A}, \qquad \text{and} \qquad \phi = 80.4°$$

load phase voltage,

$$V_P = V_L = 200 \text{ V}$$

$$P = 3 \, V_P I_P \cos \phi$$

$$= 3 \times 200 \text{ V} \times 1 \text{ A} \times \cos 80.4°$$

$$= \textbf{100 W}$$

or,

$$P = \sqrt{3} \, V_L I_L \cos \phi$$

$$= \sqrt{3} \times 200 \text{ V} \times 1.73 \text{ A} \times \cos 80.4°$$

$$= 100 \text{ W}$$

and, $$P_1 = E_{AB} I_A \cos \phi_A$$

Where ϕ_A is the phase angle between E_{AB} and I_A (see Figure 26-19):
From Figure 26-11,

$$\phi_A = (80.4° + 30°)$$

$$P_1 = 200 \text{ V} \times 1.73 \text{ A} \times \cos(80.4° + 30°)$$

$$= \textbf{-120.6 W}$$

Sec. 26-8   Three-Phase Power Measurement

As noted above, the meter terminal connections must be reversed in this case to obtain a positive reading.

$$P_2 = E_{CB} I_C \cos \phi_C$$

Where ϕ_C is the phase angle between E_{CB} and I_C (see Figure 26-19):

$$E_{CB} = -E_{BC}$$

So from Figure 26-11,

$$\phi_C = (80.4° - 30°)$$
$$P_2 = 200 \text{ V} \times 1.73 \text{ } A \times \cos(80.4° - 30°)$$
$$= \textbf{220.5 W}$$
$$P_1 + P_2 = -120.6 \text{ W} + 220.5 \text{ W}$$
$$\cong 100 \text{ W}$$

POWER FACTOR DETERMINATION. When the two-wattmeter method is employed to measure the power dissipated in a balanced load, the power factor of the circuit can be calculated from the meter readings.

Consider the phasor diagram for a balanced Y-connected load, as shown in Figure 26-20. The phase voltages E_{AN}, E_{BN}, and E_{CN} have 120° phase differences, and line voltages E_{AB}, E_{BC}, and E_{CA} are 30° ahead of E_{AN}, E_{BN}, and E_{CN} respectively. (These phasors are explained in Section 26-2). The line currents I_A, I_B, and I_C (which are also the phase currents in a Y-connected load) each lag the related phase voltage by phase angle ϕ.

Now look at the circuit diagram in Figure 26-19 again. The current coil of wattmeter W_1 carries line current I_A, and the potential difference across its voltage coil is line voltage E_{AB}. The phase difference between E_{AB} and I_A is $30° + \phi$ (see Figure 26-20). Consequently, the power indicated by W_1 is:

$$P_1 = E_{AB} I_A \cos(30° + \phi)$$

Since E_{AB} is line voltage V_L, and I_A is line current I_L,

$$\boxed{P_1 = V_L I_L \cos(30° + \phi)} \tag{26-12}$$

In Figure 26-19, the current flowing in the current coil of wattmeter W_2 is line current I_C. The potential difference across its voltage coil is $-E_{BC}$. If the voltage coil of W_2 had been connected with its $\pm$ terminal

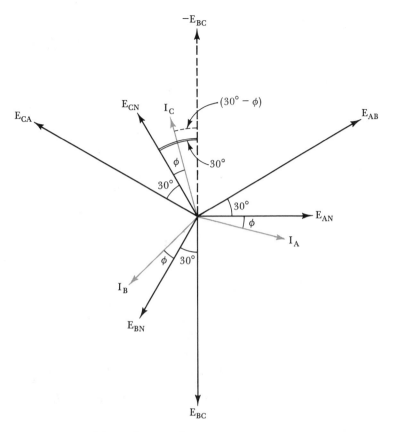

FIGURE 26-20. Phasor diagram for a two-wattmeter power measurement method with a balanced Y-connected load. The load power factor can also be determined from the meter readings.

to line B, the voltage would be $+E_{BC}$. But, when connected as illustrated, the voltage coil potential difference is $-E_{BC}$. As shown in the phasor diagram in Figure 26-20, there is a 30° difference between $-E_{BC}$ and E_{CN}. Since I_C lags E_{CN} by angle ϕ, the phase difference between $-E_{BC}$ and I_C is $30° - \phi$. The power indicated by W_2 can now be written as:

$$P_2 = E_{BC} I_C \cos(30° - \phi)$$

or,

$$\boxed{P_2 = V_L I_L \cos(30° - \phi)} \tag{26-13}$$

$$P_2 + P_1 = V_L I_L [\cos(30° - \phi) + \cos(30° + \phi)]$$
$$= V_L I_L [2 \cos 30° \cos \phi]$$
$$= \sqrt{3} \, V_L I_L \cos \phi$$

and

$$P_2 - P_1 = V_L I_L [\cos(30° - \phi) - \cos(30° + \phi)]$$

$$= V_L I_L \sin \phi$$

$$\frac{P_2 - P_1}{P_2 + P_1} = \frac{V_L I_L \sin \phi}{\sqrt{3} \, V_L I_L \cos \phi}$$

$$= \frac{1}{\sqrt{3}} \tan \phi$$

$$\boxed{\phi = \arctan \ \sqrt{3} \left[\frac{P_2 - P_1}{P_2 + P_1}\right]} \qquad \qquad \textbf{(26-14)}$$

In deriving Equation (26-14), it was assumed that ϕ is less than 60°. When ϕ is greater than 60°, $(30° + \phi)$ exceeds 90°, and $\cos(30° + \phi)$ becomes a negative quantity. Consequently, the reading on wattmeter W_1 is negative. As already noted, to get an on-scale indication, the connections to either the current coil or the voltage coil must be reversed, and the power measured by this wattmeter must be recorded as a negative quantity.

The two-wattmeter method does not indicate whether the calculated power factor is leading or lagging. This must be determined by considering the load: lagging power factor for a largely inductive load, or leading power factor with a predominantly capacitive load.

EXAMPLE 26-15 From the wattmeter indications in Example 26-14, calculate the load power factor.

SOLUTION

Equation (26-14): $\phi = \arctan \left[\sqrt{3} \left(\frac{P_2 - P_1}{P_2 + P_1}\right)\right]$

$$= \arctan \left[\sqrt{3} \left(\frac{220.5 \ W - (-120.6 \ W)}{220.5 \ W + (-120.6 \ W)}\right)\right]$$

$$= \arctan 5.91$$

$$= 80.4°$$

power factor $= \cos \phi$

$$= \cos 80.4°$$

$$= \textbf{0.167}$$

26-8.1 The power supplied to the circuit described in Problem 26-6.2 is measured by the two-wattmeter method, as illustrated in Figure 26-19. Determine the power indicated by each meter.

26-8.2 From the wattmeter readings for Problem 26-8.1, calculate the load power factor.

SUMMARY OF FORMULAS

For a Y-connected generator or load:

$$V_L = \sqrt{3}\, V_P$$

$$I_L = I_P$$

For a Δ-connected generator or load:

$$V_L = V_P$$

$$I_L = \sqrt{3}\, I_P$$

Power dissipated:

$$P = \sqrt{3}\, V_L I_L \cos \phi$$

Reactive power:

$$Q = \sqrt{3}\, V_L I_L \sin \phi$$

Apparent power:

$$S = \sqrt{3}\, V_L I_L$$

Power factor:

$$\cos \phi = \frac{P}{S}$$

REVIEW QUESTIONS

26-1 Sketch illustrations to show how three-phase ac is generated. Also, sketch the output voltage waveforms and phasor diagrams. Briefly explain.

26-2 Sketch the circuit of a Y-connected three-phase generator. Identify phase currents, phase voltages, line currents, and line voltages. Explain.

26-3 Sketch the phasor diagram for the phase voltage in a Y-connected generator. Show how the line voltage phasors are derived from the phase voltage phasors.

26-4 Sketch the circuit of a Y-connected generator providing a four-wire supply to a Y-connected load. Identify phase and line currents and voltages at the generator and at the load.

26-5 Draw the phasor diagram for the load voltages and currents for a purely resistive balanced Y-connected four-wire load. Show that the phasor sum of the load currents is zero.

26-6 Sketch the phasor diagram of load currents and voltages for a balanced Y-connected load with $\phi = 20°$ lagging. Show that the phasor sum of the load currents is zero.

26-7 Sketch the circuit of a Δ-connected three-phase generator. Identify phase currents, phase voltages, line currents, and line voltages. Explain.

26-8 Sketch the phasor diagram for the phase currents in a Δ-connected generator. Show how the line current phasors are derived from the phase current phasors.

26-9 Sketch the circuit of a Δ-connected generator supplying a Δ-connected load. Identify phase and line currents and voltages at the generator and at the load.

26-10 For a Δ-connected purely resistive balanced load, sketch the phasor diagram for the load voltages and currents. Show that the phasor sum of the load currents is zero.

26-11 Sketch the phasor diagram of line voltages, phase currents, and line currents for a balanced Δ-connected load with $\phi = 40°$ lagging.

26-12 Sketch the circuit of a Y-connected generator supplying a Δ-connected load. Identify phase and line currents and voltages, at the generator and at the load.

26-13 Briefly discuss the procedure for analyzing Y-connected three-wire unbalanced loads and combinations of Y- and Δ-connected loads.

26-14 Sketch the circuit waveforms and the line voltage phasor diagram for a Y-connected generator with phase sequence ABC. Repeat for a generator with sequence CBA. Explain why the phase sequence is important.

26-15 Sketch the circuit of a phase sequence tester and briefly explain its operation.

26-16 For a three-phase load, write equations for (a) power dissipated per phase, (b) total load power in an unbalanced load, (c) total balanced load power in terms of V_P and I_P, and (d) total balanced load power in terms of V_L and I_L. Briefly explain each equation.

26-17 For a three-phase balanced load, write equations for (a) reactive power per phase, (b) total load reactive power in terms of V_L and I_L, (c) apparent power per phase, (d) total apparent power in terms of V_P and I_P, (e) total apparent power in terms of V_L and I_L, and (f) cos ϕ. Briefly explain each equation.

Chap. 26   Three-Phase AC Systems

26-18 Sketch the circuit of a balanced Y-connected resistive-inductive load. Show how capacitors may be connected to increase the load power factor. Explain.

26-19 Sketch the following wattmeter connection diagrams for three-phase power measurement in a Y-connected load: (a) three-wattmeter method for an unbalanced four-wire load, (b) single-wattmeter method for a balanced load. Briefly explain each diagram.

26-20 Sketch a three-wattmeter circuit for measuring power in an unbalanced Δ-connected load, and a one-wattmeter circuit for measuring power in a balanced Δ-connected load. Explain.

26-21 Draw the circuit diagram of the two-wattmeter method for measuring power in a three-phase load. Show that the sum of the meter readings equals the total power dissipated in the load, whether the load is balanced or unbalanced, Y-connected or Δ-connected.

26-22 Show that for a balanced load the two-wattmeter method can be used to determine the load power factor.

PROBLEMS

SECTION 26-2

26-1 Three Y-connected 155 Ω resistors have a three-phase, four-wire supply with a line voltage of $V_L = 400$ V. Determine the phase voltage, the line current, the phase current, and the neutral current.

26-2 For Problem 26-1, draw the phasor diagram showing line and load voltages and line and load currents.

26-3 For Problem 26-1, recalculate the line and neutral currents when the resistors are changed to 300 Ω, 75 Ω, and 400 Ω.

26-4 A Y-connected load with a three-phase, four-wire supply consists of $Z_A = 180$ Ω, $Z_B = (180$ Ω in series with $C = 33$ μF), and $Z_C = (180$ Ω in series wth $L = 300$ mH). The supply voltage has $V_L = 200$ V and $f = 60$ Hz. Calculate the line currents and the neutral current.

26-5 Draw a phasor diagram for the circuit in Problem 26-4.

26-6 Three impedances consist of: $Z_A = R_1 + jX_{L1}$; $Z_B = R_2 - jX_{C2}$; and $Z_C = R_3$. $R_1 = 330$ Ω; $R_2 = 250$ Ω; $R_3 = 190$ Ω; $L_1 = 600$ mH; and $C_2 = 10$ μF. The impedances are Y-connected and have a three-phase, four-wire supply. The supply voltage has $V_L = 150$ V and $f = 60$ Hz. Determine the line currents and the neutral current.

26-7 Draw a phasor diagram for the circuit in Problem 26-6.

26-8 Three Δ-connected 155 Ω resistors are supplied from a Δ-connected generator with a line voltage of $V_L = 400$ V. Calculate current in each resistor and the line currents.

26-9 Draw the phasor diagram for the circuit of Problem 26-8, showing individual load voltages, load currents, and line currents.

26-10 For Problem 26-8, recalculate the load and line currents when the resistors are changed to 300 Ω, 75 Ω, and 400 Ω.

26-11 The load described in Problem 26-4 is reconnected in Δ form. The supply remains $V_L = 200$ V and $f = 60$ Hz. Calculate the load and line currents.

26-12 Draw the phasor diagram of line voltages, line currents, and load currents for Problem 26-11.

26-13 The load described in Problem 26-6 is reconnected in Δ form. The supply remains $V_L = 150$ V and $f = 60$ Hz. Calculate the load and line currents.

SECTION 26-4

26-14 A balanced Δ-connected load consists of three impedances, each of which have $R = 75$ Ω and $C = 2$ μF connected in series. The supply is Y-connected, and has a phase voltage of 50 V and frequency of 400 Hz. Calculate the line currents, and the phase angle of each line current with respect to the line voltages.

26-15 A Y-connected load with a 220 V, 60 Hz three-wire supply has branch impedances of $Z_a = 66$ Ω$\underline{/15°}$, $Z_b = 85$ Ω$\underline{/-30°}$, and $Z_c = 90$ Ω$\underline{/20°}$. Calculate the line currents and their phase angle with respect to line voltage E_{AB}.

SECTION 26-5

26-16 A Y-connected load has three impedances: $Z_a =$ a 100 Ω resistor in series with a 400 mH inductor, $Z_b =$ a 120 Ω resistor in series with a 10 μF capacitor, and $Z_c =$ a 250 Ω resistor. The three-wire supply has $V_L = 75$ V, $f = 60$ Hz, and phase sequence ABC. Calculate the line currents.

26-17 Recalculate the line currents for Problem 26-16 when the phase sequence is changed to CBA.

26-18 Recalculate the line currents in Problem 26-13 when the phase sequence is reversed.

26-19 A Y-connected load consists of $Z_a = Z_b =$ a 120 Ω resistor, and $Z_c =$ a 60 Ω resistor in series with a 200 mH inductor. The supply has $V_L = 220$ V and $f = 100$ Hz. Calculate the line currents: (a) when the phase sequence is ABC, and (b) when the sequence is CBA.

26-20 Repeat Problem 26-19 with the 200 mH inductor replaced by a 15 μF capacitor.

SECTION 26-6

26-21 Calculate the total power dissipated in the load described in Problem 26-1.

26-22 Determine the total power dissipated in the circuit referred to in Problem 26-3.

26-23 Calculate the total power dissipated in the load described in Problem 26-4. Also, calculate the reactive power and the apparent power.

26-24 Calculate the total power dissipated in the load described in Problem 26-6. Also, calculate the reactive power and the apparent power.

26-25 For Problem 26-11, calculate the true power, apparent power, and reactive power.

26-26 For Problem 26-13, determine true power, apparent power, and reactive power.

26-27 Calculate the apparent power, true power, and reactive power for the load described in Problem 26-14.

26-28 Calculate the apparent power, true power, and reactive power for the load described in Problem 26-19.

SECTION 26-7

26-29 A balanced Y-connected load consists of three impedances, each of which has $R = 75 \ \Omega$ and $L = 40$ mH. The supply has $V_L = 80$ V and $f = 400$ Hz. Calculate the line currents, the power factor, and the load power dissipation.

26-30 Three Δ-connected 0.5 μF capacitors are connected in parallel with the Y-connected components of Problem 26-29. The circuit is as illustrated in Figure 26-15(a). Recalculate the line currents, power factor, and load power.

26-31 Three Δ-connected impedances each consist of a 180 Ω resistor in series with a 330 mH inductor. The supply voltage has $V_L = 300$ V and $f = 50$ Hz. Three Δ-connected 5 μF capacitors are connected in parallel with the load. Calculate the line current, load power, and power factor: (a) without the capacitors in the circuit, and (b) with the capacitors connected.

SECTION 26-8

26-32 Calculate each wattmeter indication when the two-wattmeter method is used to measure the power in the load described in

Problem 26-29. Also, use the meter indications to determine the load power factor.

26-33 Calculate each wattmeter indication when the two-wattmeter method is used to measure the power delivered to the load described in Problem 26-16.

COMPUTER PROBLEMS

26-34 Write a computer program to analyze a Y-connected three-phase, four-wire load to determine line and neutral currents.

26-35 Write a computer program to analyze a Δ-connected three-phase load to determine the line currents.

ANSWERS TO PRACTICE PROBLEMS

26-2.1 33 Ω

26-2.2 3.5 A

26-2.3 2 A, 3.01 A

26-3.1 99 Ω

26-3.2 3.64 A$/-23.5°$, 5.07 A$/-145°$, 4.41 A$/79°$

26-3.3 1.732 A$/-60°$, 2.45 A$/-135°$, 3.35 A$/75°$

26-4.1 100 Ω

26-4.2 1.13 A$/-49°$, 657 mA$/-139°$, 1.31 A$/101°$

26-5.1 1.13 A$/49°$, 657 mA$/139°$, 1.31 A$/-101°$

26-6.1 615 W

26-6.2 317 VA, 224 var, 224 W

26-7.1 2.8 μF

26-8.1 47 W, 177 W

26-8.2 0.707

27

NONSINUSOIDAL WAVEFORMS

Objectives You will be able to:

Sketch common nonsinusoidal waveforms and discuss the characteristics of each waveform.

Sketch waveforms to show that a square wave is made up of many sinusoidal harmonic components, and explain how waveform distortion can result when harmonics are not present in the correct proportion.

Use Fourier equations to analyze various nonsinusoidal waveforms to determine the amplitudes of harmonic components.

Calculate rms values of nonsinusoidal waveforms.

Calculate the effect of a nonsinusoidal waveform applied as an input to *CR* and *LCR* circuits.

Introduction

Although preceding ac-related chapters refer exclusively to voltages and currents with sinusoidal waveforms, much electrical and electronic equipment involves waveforms that are nonsinusoidal. Some of these are deliberately created; others are the results of distortion added by equipment. All repetitive waveforms can be shown to be made up of combinations of many sinusoidal waves. Any waveform can be analyzed to determine the component quantities. These components may be used when investigating the response of an impedance network to a nonsinusoidal input.

27-1

MISCELLA-NEOUS WAVEFORMS

Figure 27-1 shows several common nonsinusoidal waveforms. All of the waveforms illustrated are *repetitive* waves, also termed *periodic*. This means that they are made up of identical cycles repeating over and over again. A waveform that does not have identical repeating cycles is termed *nonrepetitive* or *aperiodic*.

Figures 27-1(a) and (b) show half-wave and full-wave rectified sine waves. These are discussed in Section 25-2. Although they are sinusoidal during half-cycles, rectified waves must be classified as nonsinusoidal. Many voltmeters are calibrated to indicate the rms value of voltages having a pure sine wave. When a rectified waveform is applied to such an instrument, the indicated voltage will *not* be the rms value.

A *square wave* and a *pulse wave* are illustrated in Figures 27-1(c) and (d) respectively. In both cases, the voltage rises rapidly from a low level to a high level, where it remains constant for a brief time. Then it returns rapidly to its low level, where it again remains constant for some time. The square wave has a high level for a time $T/2$, and a low level for an equal time $(T/2)$. In the case of the pulse wave, the high level and low level times are unequal. Another way of stating this is: the *pulse width and space width* are different.

A square wave with severe distortion of its high and low levels is illustrated in Figure 27-1(e). This type of distortion occurs most often when the wave is *capacitor coupled* from one point to another in an electronic circuit. The distortion results when the capacitor value is too small. When an extremely small capacitor is used in this situation, the spike waveform shown in Figure 27-1(f) might result.

The exponential waveform in Figure 27-1(g) occurs when a capacitor is charged and discharged via a resistance. It could also be produced at the output of a low-frequency amplifier when a square wave input is applied.

Figure 27-1(h) shows a *triangular wave*. This is produced for certain applications by some electronic signal generators. It usually results from a capacitor being charged and discharged by means of a constant current. The *sawtooth waveform* in Figure 27-1(i) partially resembles the triangular wave. It can be generated by charging a capacitor with a constant current, then rapidly discharging. This is the type of waveform used to deflect the electron beam in an oscilloscope from one side of the screen to the other.

The distorted sine wave shown in Figure 27-1(j) is the result of adding a low-amplitude sine wave to a larger-amplitude lower-frequency sine wave. This subject is further discussed in Section 27-2.

Figures 27-1(k) and (l) show the kind of output waveforms produced by radio transmitter. A high-frequency *carrier* wave is generated for both *amplitude modulated* (AM) and *frequency modulated* (FM) transmissions. In the AM case, the low-frequency (audio) information is transmitted by increasing and decreasing the amplitude of the carrier.

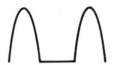

(a) Half-wave rectifier sine wave
(b) Full-wave rectifier sine wave
T
T/2
T/2
(c) Square wave
Pulse width
Space width
(d) Pulse wave
(e) Square wave with severe distortion
(f) Spike wave
(g) Exponential wave
(h) Triangular wave
(i) Sawtooth wave
(j) Distorted sine wave
(k) Amplitude modulated sine wave
(l) Frequency modulated sine wave

FIGURE 27-1. A wide variety of nonsinusoidal waveforms are to be found in electronics equipment. Many of these are deliberately created for useful purposes; some are the result of unwanted distortion. All repetitive waveforms can be shown to be made up of combinations of sine waves.

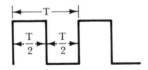

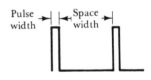

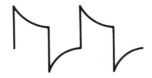

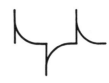

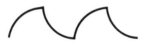

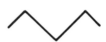

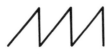

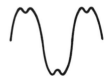

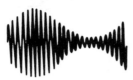

For FM radio, the frequency of the carrier is modulated by the audio signal, thus producing a sinusoidal wave with a changing time period.

The waveforms discussed above are representative of many types of waves that occur in electronic equipment. Although many of those illustrated in Figure 27-1 do not resemble sine waves, it can be shown that *they are all combinations of many sine waves,* sometimes with a dc component.

27-2
HARMONICS IN WAVEFORMS

FREQUENCY SYNTHESIS AND HARMONIC ANALYSIS. In Figure 27-2 two sine wave generators are shown connected in series. The output from generator B has a frequency three times the frequency of generator A. The amplitude of the output from B is also much less than that from A. The waveform produced by the two generators in series is the larger-amplitude (and lower-frequency) signal with the smaller-amplitude signal superimposed. It is seen that the combination approximately resembles a square wave with its peaks dented.

Figure 27-3 shows a third generator connected in series with A and B. The output from generator C is smaller in amplitude than that from B, and the frequency of this third generator is five times the frequency of the output from generator A. The waveform produced by the three generators in series now more closely resembles a square wave. If appropriate additional higher-frequency waveforms are included, the final wave becomes increasingly more square.

The process of constructing a waveform by combining several sine waves of different frequencies and amplitudes is referred to as *frequency synthesis.* The converse of frequency synthesis is *harmonic analysis:*

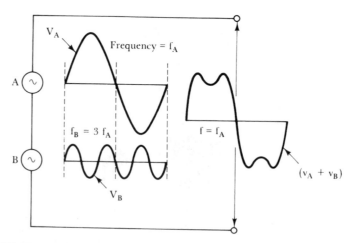

FIGURE 27-2. A sine wave with (fundamental) frequency f_A, when combined with a smaller amplitude sine wave with (third harmonic) frequency $f_B = 3 f_A$, produces an approximately square waveform.

Chap. 27   Nonsinusoidal Waveforms

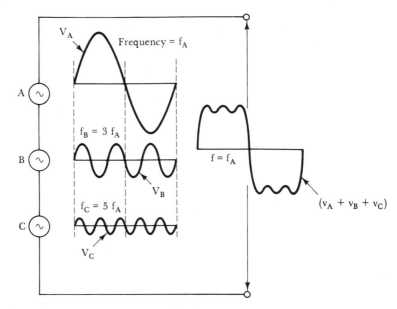

FIGURE 27-3. When a (fifth harmonic) sinewave with frequency $f_C = 5\ f_A$ is added to the two waveforms in Figure 27-2, the resultant output more closely resembles a square wave.

analyzing a waveform to discover its component sine waves. By harmonic analysis, periodic nonsinusoidal waveforms can be shown to consist of combinations of pure sine waves, sometimes with a dc component. One major component, a large-amplitude sine wave having the same frequency as the periodic wave being analyzed, is termed the *fundamental*. The other components are sine waves with frequencies that are exact multiples of the fundamental frequency. These waves, referred to as *harmonics,* are numbered according to the ratio of their frequency to that of the fundamental. For example, a harmonic with a frequency double that of the fundamental is called the *second harmonic*. The frequency of the *third harmonic* is obviously three times the fundamental frequency.

GENERATION OF NONSINUSOIDAL WAVEFORMS. Although it has been explained that nonsinusoidal waveforms are made up of sinusoidal components, generation of these waves does not normally involve the addition of several sine waves. As discussed in Section 27-1, for example, a triangular wave is produced by charging a capacitor with a constant current. Square waves are generated by causing a transistor to switch its output from a low voltage level to a high level. Many nonsinusoidal waveforms are the result of distortion produced when an input signal (sinusoidal or nonsinusoidal) is processed through electronic circuitry.

When a sinusoidal input signal is applied to an amplifier, the output may be expected to be an amplified version of the input waveform.

(a) Sine wave with peaks
    clipped by excessive
    amplification

(b) Sine wave with rounded
    negative peak due to
    second harmonic

FIGURE 27-4. A sine wave with clipped peaks has approximately the same harmonic content as a square wave. Second harmonic distortion produces a sine wave with one half-cycle slightly peaked and the other half-cycle flattened.

For this to occur, the signal amplitude must be within an acceptable range for the amplifier. Also, the signal frequency must be within the frequency range (or *bandwidth*) of the amplifier.

If the input to an amplifier is much larger than it should be, the output is likely to look like the *clipped sine wave* illustrated in Figure 27-4(a). This ouptut is fairly similar to a square wave, so its harmonic content is close to that of a square wave. When the sinusoidal input has an amplitude that is just a little greater than the acceptable maximum for the amplifier, the output might have the waveshape shown in Figure 27-4(b). In this case, the upper peak is almost sinusoidal while the lower peak is rounded. The waveform can be constructed by adding a second harmonic to the fundamental. Consequently, this particular type of distortion is termed *second harmonic distortion*.

When a square wave is applied as input to a circuit that does not pass all the necessary frequency components, the resultant output is a distorted square wave. The type of distortion depends on whether the circuit has poor low-frequency response or poor high-frequency response. In Figure 27-5(a) the long rise and fall times of the square wave show that the high-frequency harmonics are attenuated. Thus, the circuit

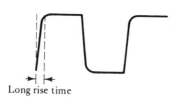

Long rise time

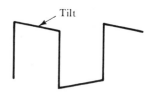

Tilt

(a) Square wave with high-
    frequency components
    attenuated

(b) Square wave with low-
    frequency components
    attenuated

FIGURE 27-5. A square wave with excessive rise time and rounded corners has its high-frequency harmonic components attenuated. Attenuation of low-frequency components produces tilt on a square wave.

Chap. 27   Nonsinusoidal Waveforms

has poor high-frequency response. Figure 27-5(b) shows the output of a circuit that has good high-frequency response but poor low-frequency response. The *tilt* on the top and bottom of the waveform is the result of low-frequency components being attenuated by the circuit.

27-3
HARMONIC ANALYSIS

By the mathematical operation known as *Fourier analysis,* waveforms can be analyzed to determine their harmonic content. The amplitude of each harmonic and its phase relationship to the fundamental can be found. Also, the level of any dc component can be calculated.

SQUARE WAVE. A perfect square wave that is symmetrical above and below ground level [Figure 27-6(a)] can be shown by Fourier analysis to be representable as:

$$e = \frac{4E_m}{\pi} \left[\sin \omega t + \frac{\sin 3\omega t}{3} + \frac{\sin 5\omega t}{5} + \frac{\sin 7\omega t}{7} + \cdots \right] \quad (27\text{-}1)$$

where e is the instantaneous voltage at time t, E_m is the peak value of the wave, and ω is $2\pi \times$ (fundamental frequency). The $\sin \omega t$ component is the fundamental, the $\sin 3\omega t$ quantity is the third harmonic, $\sin 5\omega t$ represents the fifth harmonic, etc. The symmetrical square wave can be said to be made up of a fundamental, odd-numbered harmonics, no even harmonics, and no dc component.

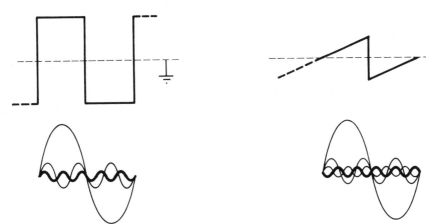

(a) Harmonic analysis of a symmetrical square wave shows that it contains a fundamental and odd-numbered harmonics

(b) Harmonic analysis of a symmetrical sawtooth wave reveals that it is composed of a fundamental and all harmonics

FIGURE 27-6. Fourier analysis can be applied to all repetitive waveforms to determine their harmonic content.

SAWTOOTH WAVE. The Fourier equation for the sawtooth waveform in Figure 27-6(b) is:

$$e = \frac{2E_m}{\pi}\left[\sin \omega t - \frac{\sin 2\omega t}{2} + \frac{\sin 3\omega t}{3} - \frac{\sin 4\omega t}{4} + \frac{\sin 5\omega t}{5}\cdots\right]$$

(27-2)

In this case, all the harmonics are present, and once again there is no dc component. In general, a waveform has no dc component when it is symmetrical above and below ground level.

RECTIFIED WAVE. The full-wave rectified sine wave in Figure 27-1(b) can be represented by the equation:

$$e = \frac{4E_m}{\pi}\left[\frac{1}{2} + \frac{\cos 2\omega t}{3} - \frac{\cos 4\omega t}{15} + \frac{\cos 6\omega t}{35}\cdots\right] \qquad (27\text{-}3)$$

Equation (27-3) shows that the waveform has a dc component $4E_m/(2\pi)$ and even-numbered harmonics $2\omega t$, $4\omega t$, $6\omega t$, etc. It would appear that there is no fundamental frequency component. However, in this case the fundamental frequency is taken as the input frequency (f) of the waveform prior to rectification. It could be argued that the fundamental frequency of the rectified waveform is actually $2f$. For example, a 60 Hz sine wave when full-wave rectified, produces a succession of sinusoidal half-cycles with a frequency of 120 Hz.

Examination of the above equations shows that in all cases the amplitudes of the harmonic components decrease as the frequency increases. Thus, the higher-order harmonics appear to have decreasing importance. This is certainly true in terms of the contribution of these components to the rms value of the waveform and to power dissipated in a load. However, for good reproduction of the waveform, many of the higher-order harmonics must be present. For example, in the case of a square wave, all components up to the eleventh harmonic (or higher) may be required. For a pulse waveform, harmonics up to the one hundredth may have to be present to create a good output waveshape.

EXAMPLE 27-1 A square wave with a peak-to-peak amplitude of 2 V is symmetrical above and below ground level. Calculate the amplitudes of each harmonic component up to the seventh.

SOLUTION

Equation (27-1):

Chap. 27   Nonsinusoidal Waveforms

$$e = \frac{4\,E_m}{\pi}\left[\sin \omega t + \frac{\sin 3\omega t}{3} + \frac{\sin 5\omega t}{5} + \frac{\sin 7\omega t}{7}\right]$$

$$E_m = 1 \text{ V (peak)}$$

$$\frac{4\,E_m}{\pi} = \frac{4 \times 1 \text{ V}}{\pi} = 1.27 \text{ V}$$

Fundamental,

$$E_{m1} = \frac{4\,E_m}{\pi}$$

$$= \mathbf{1.27 \text{ V}}$$

Third harmonic,

$$E_{m3} = \frac{4\,E_m}{\pi} \times \frac{1}{3}$$

$$= \mathbf{0.42 \text{ V}}$$

Fifth harmonic,

$$E_{m5} = \frac{4\,E_m}{\pi} \times \frac{1}{5}$$

$$= \mathbf{0.25 \text{ V}}$$

Seventh harmonic,

$$E_{m7} = \frac{4\,E_m}{\pi} \times \frac{1}{7}$$

$$= \mathbf{0.18 \text{ V}}$$

Note that the harmonic voltage components calculated above are all peak values. Each must be multiplied by 0.707 to determine the rms values.

EXAMPLE 27-2 A full-wave-rectified sine wave has an amplitude of 30 V and a (pre-rectified) frequency of 60 Hz. Calculate the dc component and the rms values of the harmonic components up to the sixth. Also, determine the harmonic frequencies.

SOLUTION

Equation (27-3):

$$e = \frac{4\,E_m}{\pi}\left[\frac{1}{2} + \frac{\cos 2\omega t}{3} - \frac{\cos 4\omega t}{15} + \frac{\cos 6\omega t}{35}\right]$$

$$\frac{4\,E_m}{\pi} = \frac{4 \times 30 \text{ V}}{\pi}$$

$$= 38.2 \text{ V}$$

dc component,

$$E_{dc} = \frac{4\,E_m}{\pi} \times \frac{1}{2}$$

$$= 38.2 \text{ V} \times \frac{1}{2}$$

$$= \mathbf{19.1\ V}$$

Fundamental,

$$E_{m1} = \mathbf{0\ V}$$

Second harmonic,

$$E_{m2} = \frac{4\,E_m}{\pi} \times \frac{1}{3}$$

$$= 12.7 \text{ V}$$

$$E_2 = 0.707 \times 12.7 \text{ V}$$

$$= \mathbf{9\ V}$$

$$f_2 = 2f_1 = 2 \times 60 \text{ Hz}$$

$$= \mathbf{120\ Hz}$$

Fourth harmonic,

$$E_{m4} = \frac{4\,E_m}{\pi} \times \frac{1}{15}$$

$$= 2.5 \text{ V}$$

$$E_4 = 0.707 \times 2.5 \text{ V}$$

$$= \mathbf{1.8\ V}$$

$$f_4 = 4f_1 = 4 \times 60 \text{ Hz}$$

$$= \mathbf{240\ Hz}$$

Sixth harmonic,

$$E_{m6} = \frac{4\,E_m}{\pi} \times \frac{1}{35}$$

$$= 1.1 \text{ V}$$

$$E_6 = 0.707 \times 1.1 \text{ V}$$

$$= \mathbf{0.78 \text{ V}}$$

$$f_6 = 6f_1 = 6 \times 60 \text{ Hz}$$

$$= \mathbf{360 \text{ Hz}}$$

PRACTICE PROBLEMS

27-3.1 A sawtooth wave, as in Figure 27-6(b), has a peak-to-peak amplitude of 3 V. Determine the amplitude of all component waves up to the fifth harmonic.

27-3.2 The Fourier equation for a half-wave rectified sine wave is,

$$e = \frac{2E_m}{\pi}\left[\frac{1}{2} + \frac{\pi \sin \omega t}{4} - \frac{\cos 2\omega t}{3} - \frac{\cos 4\omega t}{15} \cdots\right]$$

Determine the amplitudes of all components up to the fourth harmonic for a wave with $E_m = 18$ V.

27-4

RMS VALUE OF NON-SINUSOIDAL WAVEFORMS

Three ac voltage sources operating in series to produce a flow of current I through a resistor R are shown in Figure 27-7. The total current flowing can be determined by means of the superposition theorem. However, since these are ac quantities, a phasor addition is involved, and the relative phase angles of the currents must be known. A calculation of total power dissipated in the resistors can be made without any knowledge of the current phase angles:

Equation (21-2):

$$P = \frac{E^2}{R}, \tag{1}$$

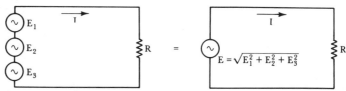

FIGURE 27-7. The rms value of nonsinusoidal waveforms can be determined from a knowledge of the harmonic components. The rms value of any waveform is the square root of the sum of the individual harmonic rms values squared.

$$E = \sqrt{E_1^2 + E_2^2 + E_3^2 + \cdots}$$

where E is the rms value of the E_1, E_2, and E_3 combined.

The power dissipation due to each individual source can be determined as:

$$P_1 = \frac{E_1^2}{R}, \qquad P_2 = \frac{E_2^2}{R}, \qquad \text{and} \qquad P_3 = \frac{E_3^2}{R}$$

or total output power,

$$P = \frac{E_1^2}{R} + \frac{E_2^2}{R} + \frac{E_3^2}{R} \qquad (2)$$

$$P = \frac{1}{R}[E_1^2 + E_2^2 + E_3^2] \qquad (3)$$

From Equations (1) and (3):

$$\frac{E^2}{R} = \frac{1}{R}[E_1^2 + E_2^2 + E_3^2]$$

$$E^2 = E_1^2 + E_2^2 + E_3^2$$

giving,

$$E = \sqrt{E_1^2 + E_2^2 + E_3^2}$$

A waveform composed of a fundamental and a number of harmonics can be treated as several series-connected voltage generators, as for Figure 27-7. The generators have voltages E_1, E_2, E_3, etc., which are the rms values of the fundamental and harmonics. Consequently, the rms value of a nonsinusoidal voltage wave is:

$$\boxed{E = \sqrt{E_1^2 + E_2^2 + E_3^2 + \cdots}} \qquad \textbf{(27-4)}$$

The above reasoning can also be applied to the determination of the rms value of a nonsinusoidal current wave:

Equation (21-3):

$$P = I^2 R$$
$$= I_1^2 R + I_2^2 R + I_3^2 R$$
$$= R[I_1^2 + I_2^2 + I_3^2]$$
$$I^2 = I_1^2 + I_2^2 + I_3^2$$

or

$$I = \sqrt{I_1^2 + I_2^2 + I_3^2 + \cdots}$$ (27-5)

EXAMPLE 27-3 Determine the rms value of the square wave analyzed in Example 27-1.

SOLUTION

Equation (27-4):

$$E = \sqrt{E_1^2 + E_2^2 + E_3^2}$$

The harmonic voltage components determined in Example 27-1 are peak values. They must be converted to rms values before substituting into Equation (27-4).

$$E_{1(rms)} = 0.707 \times 1.27 \text{ V} \cong 0.9 \text{ V}$$

$$E_{3(rms)} = 0.707 \times 0.42 \text{ V} \cong 0.3 \text{ V}$$

$$E_{5(rms)} = 0.707 \times 0.25 \text{ V} \cong 0.18 \text{ V}$$

$$E_{7(rms)} = 0.707 \times 0.18 \text{ V} \cong 0.13 \text{ V}$$

$$E_{(rms)} = \sqrt{(0.9 \text{ V})^2 + (0.3 \text{ V})^2 + (0.18 \text{ V})^2 + (0.13 \text{ V})^2}$$

$$\cong \textbf{0.97 V}$$

PRACTICE PROBLEMS

27-4.1 Calculate the rms value of the sawtooth wave analyzed for Problem 27-3.1

27-4.2 Calculate the rms value of the half-wave rectified sine wave analyzed for Problem 27-3.2.

27-5
NON-SINUSOIDAL VOLTAGE AS A CIRCUIT INPUT

When a nonsinusoidal voltage is applied to a purely resistive circuit, the waveform of the current that flows is identical to the input voltage wave [see Figure 27-8(a)]. The peak level of the current can be calculated as $I_m = E_m/R$, and the instantaneous current at any time is $i = e/R$, where e is the instantaneous input voltage.

When the circuit is not purely resistive, the current waveform is a distorted version of the input voltage wave [Figures 27-8(b) and (c)]. This is because the circuit offers different impedances to the (different frequency) harmonic components of the input. Consequently, the vari-

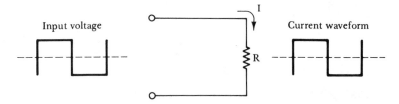

(a) Square wave input to a purely resistive circuit produces a square current waveform

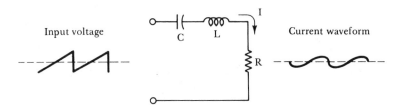

(b) The current waveform is distorted when a square wave input voltage is applied to an impedance circuit

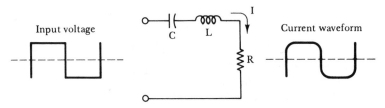

(c) A distorted current wave results from a sawtooth input voltage to an impedance circuit

FIGURE 27-8. A purely resistive circuit has no effect on the harmonic components of a nonsinusoidal waveform. An impedance circuit attenuates different input frequencies by different amounts, thus producing distortion on the current waveform.

ous current components do not have the same proportional relationship as the input voltage components.

To determine the effect of a nonsinusoidal input voltage on an impedance network, the voltage must first be analyzed to find its harmonic components. The individual voltage components are then treated as separate inputs, and the (different) circuit impedances must be calculated for each harmonic frequency. When the current components have been determined, the superposition theorem may be applied to find the resultant current. Once again, note that because phasor quantities are involved, the component phase relationships must be taken into account.

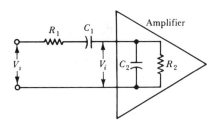

FIGURE 27-9. A *CR* network exists at the input of electronic amplifiers. R_1 is the signal source resistance, C_1 is a coupling capacitor, R_2 is the amplifier input resistance, and C_2 is the input capacitance.

The rms value of the resultant current can be calculated using Equation (27-5), without reference to the phase angle of each component.

The circuit in Figure 27-9 illustrates the conditions at the input of many electronic amplifiers. The input resistance of the amplifier is R_2, but this is shunted by an input capacitance C_2. Capacitor C_1 is a *coupling capacitor* for coupling the signal source to the amplifier input while maintaining dc isolation. R_1 is the signal source resistance. At low frequencies, the impedance of C_2 is normally so much larger than R_2 that it has virtually no effect on the signal. At high frequencies, where X_{C2} is not very much larger than R_2, there can be significant attenuation of the signal. Conversely capacitor C_1 causes insignificant signal attenuation at high frequencies but can produce attenuation at low frequencies. For a nonsinusoidal signal, the coupling capacitor may affect the fundamental frequency and perhaps the lower harmonics, while the amplifier input capacitance is most likely to attenuate the high frequency harmonics.

PRACTICE PROBLEMS

27-5.1 The circuit in Figure 27-8(b) has $C = 1.6\ \mu F$, $L = 48$ mH, and $R = 2.2\ k\Omega$. A square wave with a frequency of 1 kHz and an amplitude of 2 V (as analyzed in Example 27-1) is applied as input. Determine the rms value of the current.

27-5.2 The circuit in Figure 27-9 has $C_1 = 1\ \mu F$, $C_2 = 300$ pF, $R_1 = 10\ k\Omega$, and $R_2 = 100\ k\Omega$. A 1 kHz square wave is applied as input signal. Determine the attenuation (from signal source to amplifier input) of the fundamental and of the 15th harmonic of the signal.

SUMMARY OF FORMULAS

Rms values of nonsinusoidal waveforms

$$E = \sqrt{E_1^2 + E_2^2 + E_3^2 + \cdots}$$

$$I = \sqrt{I_1^2 + I_2^2 + I_3^2 \cdots}$$

27-1 Sketch several repetitive nonsinusoidal waveforms. Briefly discuss the shape of each wave and how it is generated. Define the terms *periodic* and *aperiodic*.

27-2 Sketch the circuit of two sine-wave generators connected in series. Show how the waveforms combine to produce an approximation of a square wave. Explain the significance of this effect.

27-3 Draw sketches to show two kinds of distortion on a sine wave. Briefly explain the origin of the distortion in each case.

27-4 Repeat Question 27-3 for a square wave.

27-5 Refer to Fourier analysis Equations (27-1), (27-2), and (27-3) in the text. Briefly discuss each component of each equation in turn, and explain how the component quantities relate to the waveform represented.

27-6 Derive the equations for the rms values of nonsinusoidal voltages and currents.

27-7 Explain how an impedance network may be analyzed to determine its response to a nonsinusoidal input voltage.

PROBLEMS

SECTION 27-3

27-1 A sawtooth waveform which is symmetrical above and below ground has a peak-to-peak amplitude of 20 V and a frequency of 1 kHz. Calculate the amplitude of each ac component up to the fifth harmonic.

27-2 The amplitude and frequency of a full-wave rectified sinusoidal voltage is adjusted to give a fourth harmonic with a frequency of 18 kHz and an rms value of 150 mV. Determine the frequency and rms level of the second harmonic.

27-3 The Fourier equation for a half-wave rectified sinusoidal wave is,

$$e = \frac{E_m}{\pi} + \frac{E_m \sin \omega t}{2} - \frac{2E_m}{\pi}\left[\frac{\cos 2\omega t}{3} + \frac{\cos 4\omega t}{15} + \frac{\cos 6\omega t}{35} + \cdots\right]$$

Determine the amplitude of each component of a 15 V, 400 Hz half-wave rectified sine wave up to the sixth harmonic.

27-4 A square wave which is symmetrical above and below ground has a peak amplitude of 6 V. Determine the amplitude of each component up to the seventh harmonic.

27-5 A 400 Hz sinusoidal voltage wave with an rms value of 10 V is half-wave rectified. Calculate the amplitude of each component of the wave up to the sixth harmonic.

27-6 The Fourier equation for a triangular wave which is symmetrical above and below ground is,

$$e = \frac{8 E_m}{\pi^2} \left[\cos \omega t + \frac{\cos 3\omega t}{3^2} + \frac{\cos 5\omega t}{5^2} + \frac{\cos 7\omega t}{7^2} + \cdots \right]$$

Calculate the peak values of all components up to the seventh harmonic of such a triangular waveform, if its peak-to-peak amplitude is 25 V, and its frequency is 1 kHz.

27-7 A 50 V, 400 Hz sine wave is full-wave rectified. Determine the dc component and the peak values of the harmonics up to the sixth.

27-8 The fifth harmonic of a symmetrical triangular wave has an rms value of 162 mV. Calculate the peak value of the waveform.

SECTION 27-4

27-9 The sawtooth waveform in Problem 27-1 is applied to a 3.3 kΩ resistor. Determine the rms value of the current that flows.

27-10 Calculate the rms value of the half-wave rectified sine wave in Problem 27-3.

27-11 Calculate the rms value of the symmetrical square wave in Problem 27-4.

27-12 Calculate the rms value of the full-wave rectified waveform in Problem 27-7.

27-13 Determine the rms value of the symmetrical triangular wave in Problem 27-6.

SECTION 27-5

27-14 An impedance network consists of a resistor R_1 in series with parallel-connected components R_2 and C_1. $R_1 = 1.5$ kΩ, $R_2 = 15$ kΩ, and $C_1 = 10$ μF. Calculate the dc and rms values of the capacitor voltage when the waveform in Problem 27-3 is applied as input.

27-15 A symmetrical square wave with a peak-to-peak amplitude of 15 V and a frequency of 700 Hz is applied to the network in Problem 27-14. Calculate the rms current level.

27-16 The waveform in Problem 27-6 is capacitor-coupled via a 0.01 μF capacitor to an amplifier with a 10 kΩ input resistance. Calculate the rms voltage level at the amplifier input terminals.

27-17 The sawtooth waveform in Problem 27-1 is investigated by means of an instrument which measures the dc level and the rms values of the harmonics. The signal source resistance is 10 kΩ, and the measuring instrument has an input resistance of 1 MΩ in parallel with a 40 pF input capacitance. Calculate the rms values of the fundamental and the fifth harmonic and the measured value of each.

27-18 For Problem 27-17, recalculate the measured component values for a sawtooth waveform frequency of 100 kHz.

27-19 A full-wave rectified sine wave with an amplitude of 50 V and a (pre-rectified) frequency of 400 Hz is applied to an impedance network. The network consists of a resistor R_1 in series with parallel-connected components R_2 and C_1. If $R_1 = 1.5$ kΩ, $R_2 = 15$ kΩ, and $C_1 = 0.02$ μF, determine the dc and rms values of the voltage developed across R_2.

27-20 In the circuit in Problem 27-19, an inductor $L = 800$ mH with a coil resistance of $R_L = 18$ Ω is connected in series with R_1, and the capacitor is removed. Recalculate the dc and rms output voltages.

27-21 A half-wave rectified sine wave with an amplitude of 50 V and a frequency of 400 Hz is applied to an impedance network. The network consists of a resistor $R_1 = 1.2$ kΩ in series with parallel-connected components $R_2 = 10$ kΩ and $C_1 = 1000$ pF. Determine the dc and rms levels of the voltage developed across R_2.

27-22 An inductor $L = 1$H with a coil resistance of $R_L = 10$ Ω is connected in series with R_1 in the circuit of Problem 27-21. The capacitor is disconnected. Recalculate the dc and rms output voltages.

COMPUTER PROBLEMS

27-23 Write a computer program to analyze a symmetrical square wave to determine its dc and harmonic components and its rms value.

27-24 Write a computer program to calculate the rms voltage developed across R in Figure 27-8(c) when a symmetrical sawtooth wave is applied, as illustrated.

ANSWERS TO PRACTICE PROBLEMS

27-3.1 0.955 V, 0.477 V, 0.318 V, 0.239 V, 0.191 V

27-3.2 5.73 V, 9 V, 3.82 V, 0.764 V

27-4.1 1.16 V

27-4.2 11.36 V

27-5.1 430 μA

27-5.2 0.91, 0.88

APPENDIX 1
CIRCUIT SYMBOLS

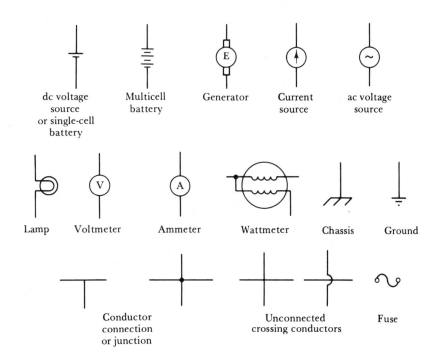

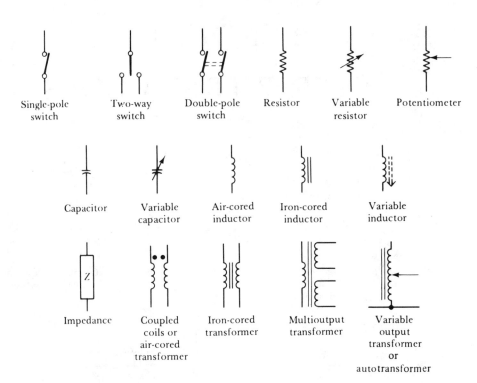

Single-pole switch   Two-way switch   Double-pole switch   Resistor   Variable resistor   Potentiometer

Capacitor   Variable capacitor   Air-cored inductor   Iron-cored inductor   Variable inductor

Impedance   Coupled coils or air-cored transformer   Iron-cored transformer   Multioutput transformer   Variable output transformer or autotransformer

APPENDIX 2

THE INTERNATIONAL SYSTEM OF UNITS (SI UNITS)

Before standard systems of measurement were invented, many approximate units were used. A long distance was often measured by the number of *days* it would take to ride a horse over the distance; a horse's height was measured in *hands;* liquid was measured by the *bucket* or *barrel.*

With the development of science and engineering, more accurate units had to be devised. The English-speaking peoples adopted the *foot* and the *mile* for measuring distances, the *pound* for mass, and the *gallon* for liquid. Other nations followed the lead of the French in adopting a *metric system,* in which large and small units are very conveniently related by a factor of 10.

With the increase of world trade and the exchange of scientific information between nations, it became necessary to establish a single system of units of measurement that would be acceptable internationally. Ater several world conferences on the matter, a metric system which uses the *meter, kilogram,* and *second* as fundamental units has now been generally adopted around the world. This is known, from the French term "système international," as the *SI* or *international system.*

FUNDAMENTAL UNITS

In the SI system the three basic units are:

unit of *length:*   the **meter** (m)*

unit of *mass:*   the **kilogram** (kg)

unit of *time:*   the **second** (s)

These are known as *fundamental units.* Other units derived from the fundamental units are termed *derived units.* Because of the use of the meter, kilogram, and second, the system is sometimes referred to as an *MKS system.*

The *meter* was originally defined as 1 ten-millionth of a meridian passing through Paris from the North Pole to the equator. The *kilogram* was defined as 1000 times the mass of 1 cubic centimeter of distilled water. The *liter*[†] is 1000 times the volume of 1 cubic centimeter of liquid. Consequently, 1 liter of water has a mass of 1 kilogram. Because of the possibility of error in the original definitions, the meter was redefined in terms of atomic radiation. Also, the kilogram is now defined as the mass of a certain platinum-iridium standard bar kept at the International Bureau of Weights and Measures in France. The second is, of course, 1/(86 400) of a mean solar day, but it is also accurately defined by atomic radiation.

Other fundamental SI units include the unit of electrical current and the unit of temperature.

UNIT OF FORCE

*The SI unit of force is the **newton**[‡] (N), defined as that force which will give a mass of 1 kilogram an acceleration of 1 meter per second per second.*

When a body is to be accelerated or decelerated, a force must be applied proportional to the desired rate of change of velocity, that is, proportional to the acceleration (or deceleration).

force = mass × acceleration
$F = ma$

When the mass is in kg and the acceleration is in m/s², the above equation gives the force in newtons.

---

* Canadian spelling is metre.

[†] Canadian spelling is litre.

[‡] Named for the great English philosopher and mathematician Sir Isaac Newton (1642–1727).

If the body is to be accelerated vertically from the earth's surface, the *acceleration due to gravity* (*g*) must be overcome before any vertical motion is possible. In SI units:

$$g = 9.81 \text{ m/s}^2$$

Thus, a mass of 1 kg has a gravitational force of 9.81 N.

WORK, ENERGY, AND POWER

WORK. When a body is moved, a force is exerted to overcome the body's resistance to motion.

The **work** *done in moving a body is the product of the force and the distance through which the body is moved in the direction of the force.*

$$\text{work} = \text{force} \times \text{distance}$$
$$W = Fd$$

The SI unit of work is the **joule*** *(J), defined as the amount of work done when a force of 1 newton acts through a distance of 1 meter.*

Thus, the *joule* may also be termed a *newton-meter. For the equation W = Fd, work is expressed in joules when F is in newtons and d is in meters.*

ENERGY

Energy *is defined as the capacity for doing work.*

Energy is measured in the same units as work.

POWER

Power *is the time rate of doing work.*

If a certain amount of work *W* is to be done in a time *t*, the power required is

$$power = \frac{work}{time}$$
$$P = \frac{W}{t}$$

---

* Named after the English physicist James P. Joule (1818–1899).

*The SI unit of power is the **watt*** (W), defined as the power developed when 1 joule of work is done in 1 second.*

For P = W/t, P is in watts when W is in joules and t is in seconds.

TEMPERATURE AND HEAT

There are two SI temperature scales, the *Celsius scale*** and the *Kelvin scale.**** These are illustrated below, alongside the non-SI *Fahrenheit*† *scale.*

The Celsius scale has 100 equal divisions (or *degrees*) between the freezing temperature and the boiling temperature of water. At normal atmospheric pressure, water freezes at 0°C (*zero degrees Celsius*) and boils at 100°C.

The Kelvin temperature scale, also known as the *absolute scale,* commences at absolute zero of temperature, which corresponds to -273.15°C. Therefore, 0°C is equal to 273.15 K, and 100°C is the same temperature as 373.15 K. A temperature difference of 1 K is the same as a temperature difference of 1°C. With the Fahrenheit scale, 32°F is the freezing temperature of water and 212°F is the boiling temperature.

JOULES EQUIVALENT. *To raise 1 liter of water through 1°C requires an energy input of 4187 J*. This is known as **Joules equivalent,** or the **mechanical equivalent of heat.** Using this figure, the energy required to raise any quantity of water through a given temperature change can be easily calculated.

When water is heated, the container must also be raised to the same temperature as the water. The container is usually defined as having a certain *water equivalent.* This is a quantity of water that would absorb the same amount of energy as the water container when heated through a given temperature change.

UNITS OF CURRENT AND CHARGE

As explained in Chapter 1, electric current (I) is a flow of charge carriers. Therefore, current could be defined in terms of the quantity of electricity (Q) that passes a given point in a conductor during a time of 1 s.

*The **coulomb**‡ (C) is the unit of electrical charge or quantity of electricity.*

---

* Named after the Scottish engineer and inventor James Watt (1736–1819).

** Invented by the Swedish astronomer and scientist Anders Celsius (1701–1744).

*** Named for the Irish-born scientist and mathematician William Thomson, who became Lord Kelvin (1824–1907).

† Devised by German physicist Gabriel Daniel Fahrenheit (1686–1736).

‡ Named after the French physicist Charles Augustin de Coulomb (1736–1806).

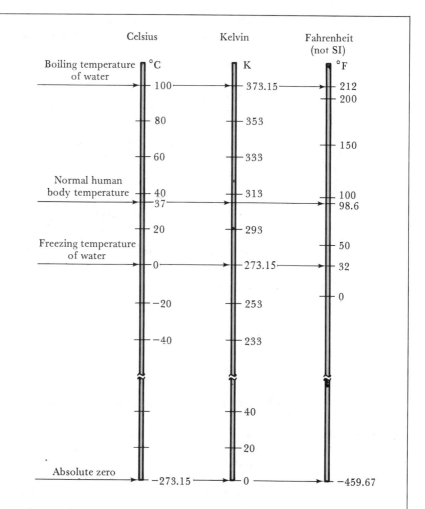

The Celsius and Kelvin temperature scales are SI scales. The Fahrenheit scale is not an SI scale. The freezing point of water is 0°C, 273.15 K, and 32°F. Boiling point of water occurs at 100°C; 373.15 K, and 212°F.

The coulomb was originally selected as the fundamental electrical unit from which all other units were derived. However, since it is much easier to measure current accurately than it is to measure charge, the unit of *current* is now the *fundamental electrical unit* in the SI system. Thus, the coulomb is a *derived unit,* defined in terms of the unit of electric current.

*The **ampere*** *(A) is the unit of electric current.*

---

* Named after the French physicist and mathematician André Marie Ampère (1775–1836).

*The **ampere** is defined as that constant current which, when flowing in each of two infinitely long parallel conductors 1 meter apart, exerts a force of 2×10^{-7} newtons per meter of length on each conductor.*

*The **coulomb** is defined as that charge which passes a given point in a conductor each second, when a current of 1 ampere flows.*

These definitions show that the coulomb could be termed an *ampere-second.* Conversely, the ampere can be described as a *coulomb per second.*

$$\text{amperes} = \frac{\text{coulombs}}{\text{seconds}}$$

It has been established experimentally that *1 coulomb is equal to the total charge carried by 6.24×10^{18} electrons.* Therefore, the charge carried by one electron is

$$Q = \frac{1}{6.24 \times 10^{18}}$$

$$= 1.602 \times 10^{-19} \text{ C}$$

EMF, POTENTIAL, DIFFERENCE, AND VOLTAGE

*The **volt*** (V) is the unit of emf and potential difference.*

*The **volt** (V) is defined as the potential difference between two points on a conductor carrying a constant current of 1 ampere when the power dissipated between these points is 1 watt.*

As already noted, the coulomb is the charge carried by 6.24×10^{18} electrons. One joule of work is done when 6.24×10^{18} electrons are moved through a potential difference of 1 V. One electron carries a charge of $1/(6.24 \times 10^{18})$ coulombs. If only one electron is moved through 1 V, the energy involved is an *electron volt* (eV).

$$1 \text{ eV} = \frac{1}{6.24 \times 10^{18}} \text{ J}$$

The electron-volt is frequently used in the case of the very small energy levels associated with electrons in orbit around the nucleus of an atom.

---

* Named in honor of the Italian physisict Count Alessandro Volta (1745–1827), inventor of the voltaic pile.

RESISTANCE AND CONDUCTANCE

The **ohm*** is the unit of resistance, and the symbol used for ohms is Ω, the Greek capital letter omega.

The **ohm** is defined as that resistance which permits a current flow of 1 ampere when a potential difference of 1 volt is applied to the resistance.

The term *conductance (G)* is applied to the reciprocal of resistance. The **siemens** **(S)** *is the unit of conductance.*

$$\text{conductance} = \frac{1}{\text{resistance}}$$

MAGNETIC FLUX AND FLUX DENSITY

The **weber*** **(Wb)** *is the SI unit of magnetic flux.*

The **weber** *is defined as the magnetic flux which, linking a single-turn coil, produces an emf of 1 V when the flux is reduced to zero at a constant rate in 1 s.*

The **tesla**† **(T)** *is the SI unit of magnetic flux density.*

The **tesla** *is the flux density in a magnetic field when 1 Wb of flux occurs in a plane of 1 m^2; that is, the* **tesla** *can be described as 1 Wb/m^2.*

INDUCTANCE

The SI unit of inductance is the **henry**‡ **(H)**.

The inductance of a circuit is **one henry** *(1 H), when an emf of 1 V is induced by the current changing at the rate of 1 A/s.*

CAPACITANCE

The **farad**φ **(F)** *is the SI unit of capacitance.*

The **farad** *is the capacitance of a capacitor that contains a charge of 1 coulomb when the potential difference between its terminals is 1 volt.*

---

*Named after the German physicist George Simon Ohm (1787–1854), whose investigations led to his statement of "Ohm's law of resistance."

** Named after Sir William Siemens (1823–1883), a British engineer who was born Karl William von Siemens in Germany. The unit of conductance was previously the mho (ohm spellled backwards).

*** Named after the German physicist Wilhelm Weber (1804–1890).

†Named for the Croatian-American researcher and inventor Nikola Tesla (1856–1943).

‡Named for the American physicist Joseph Henry (1797–1878).

φNamed for the English chemist and physicist Michael Farraday (1791–1867).

APPENDIX 3

UNIT CONVERSION FACTORS

The following factors may be used to convert non-SI units to SI units.

TO CONVERT	TO	MULTIPLY BY
acres	square meters	4047
acres	hectares	0.4047
amperes/inch	amperes/meter	39.37
angstroms	meters	1×10^{-10}
atmospheres	kg/m^2	10 332
bars	kg/m^2	1.020×10^4
Btu	joules	1054.8
Btu	kW.h	2.928×10^{-4}
bushels	m^3	0.035 24
circular mils	m^2	5.067×10^{-2}
cubic feet	m^3	0.028 32
cubic inches	liters	0.016 39
dynes	grams	1.020×10^{-3}
dynes	newtons	10^{-5}

TO CONVERT	TO	MULTIPLY BY
ergs	joules	10^{-7}
ergs	kW.h	0.2778×10^{-13}
Fahrenheit (°F)	°C	$(°F - 32) \times \frac{5}{9}$
feet	meters	0.3048
foot-pounds	joules	1.356
foot-pounds	kgm	0.1383
gallons (U.S.)	m^3	3.785×10^{-3}
gallons (imperial)	m^3	4.546×10^{-3}
gallons (U.S.)	liters	3.785
gallons (imperial)	liters	4.546
gausses	Teslas	10^{-4}
gilberts	ampere (turns)	0.7958
gills	liters	0.1183
grams	newtons	9.807×10^{-3}
horsepower	watts	745.7
inches	cm	2.54
knots	km/h	1.853
lines/sq. in.	Teslas	1.55×10^{-5}
Maxwells	webers	10^{-8}
mhos	Siemens	1
microns	meters	10^{-6}
miles (nautical)	km	1.853
miles (statute)	km	1.609
miles/hour	km/h	1.609
mils	cm	2.54×10^{-3}
ounces	grams	28.35
pints (U.S.)	liters	0.4732
pints (imperial)	liters	0.5683
poundals	newtons	0.1383
pounds	grams	453.59
pounds-force	newtons	4.448
pounds/sq. ft.	kg/m^2	4.882
pounds/sq. in.	kg/m^2	703
quarts (U.S.)	liters	0.9463
quarts (imperial)	liters	1.137
slug	kg	14.59
sq. ft.	m^2	0.0929
sq. in.	cm^2	6.452
sq. miles	km^2	2.59
tons (long)	kg	1016
tons (short)	kg	907.18

APPENDIX 4

AMERICAN WIRE GAUGE SIZES AND METRIC EQUIVALENTS

GAUGE	DIA. (mm)	COPPER WIRE RESISTANCE (Ω/km)	DIA. (mil)	COPPER WIRE RESISTANCE (Ω/1000 ft)
0000	11.68	0.160	460	0.049
000	10.40	0.203	409.6	0.062
00	9.266	0.255	364.8	0.078
0	8.252	0.316	324.9	0.098
1	7.348	0.406	289.3	0.124
2	6.543	0.511	257.6	0.156
3	5.827	0.645	229.4	0.197
4	5.189	0.813	204.3	0.248
5	4.620	1.026	181.9	0.313
6	4.115	1.29	162	0.395
7	3.665	1.63	144.3	0.498
8	3.264	2.06	128.5	0.628
9	2.906	2.59	114.4	0.792
10	2.588	3.27	101.9	0.999
11	2.30	4.10	90.7	1.26
12	2.05	5.20	80.8	1.59
13	1.83	6.55	72	2

GAUGE	DIA. (mm)	RESISTANCE (Ω/km)	DIA. (mil)	RESISTANCE (Ω/1000 ft)
14	1.63	8.26	64.1	2.52
15	1.45	10.4	57.1	3.18
16	1.29	13.1	50.8	4.02
17	1.15	16.6	45.3	5.06
18	1.02	21.0	40.3	6.39
19	0.912	26.3	35.9	8.05
20	0.813	33.2	32	10.1
21	0.723	41.9	28.5	12.8
22	0.644	52.8	25.3	16.1
23	0.573	66.7	22.6	20.3
24	0.511	83.9	20.1	25.7
25	0.455	106	17.9	32.4
26	0.405	134	15.9	41
27	0.361	168	14.2	51.4
28	0.321	213	12.6	64.9
29	0.286	267	11.3	81.4
30	0.255	337	10	103
31	0.227	425	8.9	130
32	0.202	537	8	164
33	0.180	676	7.1	206
34	0.160	855	6.3	261
35	0.143	1071	5.6	329
36	0.127	1360	5	415
37	0.113	1715	4.5	523
38	0.101	,2147	4	655
39	0.090	2704	3.5	832
40	0.080	3422	3.1	1044

APPENDIX 5
COLOR CODES

RESISTOR COLOR CODE

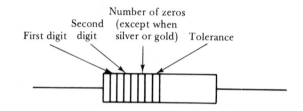

First digit · Second digit · Number of zeros (except when silver or gold) · Tolerance

	First Three Bands			Fourth Band	
Black	– 0	Blue	– 6	Gold	± 5%
Brown	– 1	Violet	– 7	Silver	± 10%
Red	– 2	Grey	– 8	none	± 20%
Orange	– 3	White	– 9		
Yellow	– 4	Silver	0.01		
Green	– 5	Gold	0.1		

CAPACITOR COLOR CODE.

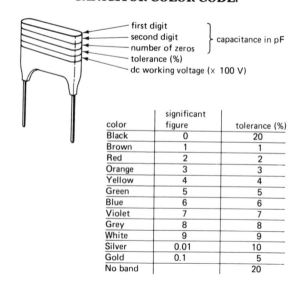

first digit
second digit } capacitance in pF
number of zeros
tolerance (%)
dc working voltage (× 100 V)

color	significant figure	tolerance (%)
Black	0	20
Brown	1	1
Red	2	2
Orange	3	3
Yellow	4	4
Green	5	5
Blue	6	6
Violet	7	7
Grey	8	8
White	9	9
Silver	0.01	10
Gold	0.1	5
No band		20

INDUCTOR COLOR CODE.

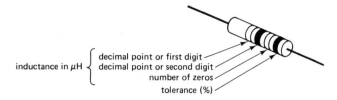

inductance in μH $\left\{\begin{array}{l} \text{decimal point or first digit} \\ \text{decimal point or second digit} \\ \text{number of zeros} \\ \text{tolerance (\%)} \end{array}\right.$

color	significant figure	tolerance (%)
Black	0	
Brown	1	
Red	2	
Orange	3	
Yellow	4	
Green	5	
Blue	6	
Violet	7	
Grey	8	
White	9	
Silver		10
Gold	decimal point	5
No Band		20

COLOR CODE MEMORY AID

Black bruins relish ornery young greenhorns.
Blue violets growing wild smell good.

Memory aid	Color	Number	
Black	black	0	
bruins	brown	1	
relish	red	2	
ornery	orange	3	
young	yellow	4	
greenhorns	green	5	
Blue	blue	6	
violets	violet	7	
growing	grey	8	
wild	white	9	
smell	silver	0.01	10%
good	gold	0.1	5%

APPENDIX 6
RESISTOR AND
CAPACITOR VALUES

TYPICAL STANDARD RESISTOR VALUES

Ω	Ω	Ω	$k\Omega$	$k\Omega$	$k\Omega$	$M\Omega$	$M\Omega$
—	10	100	1	10	100	1	10
—	12	120	1.2	12	120	1.2	—
—	15	150	1.5	15	150	1.5	15
—	18	180	1.8	18	180	1.8	—
—	22	220	2.2	22	220	2.2	22
2.7	27	270	2.7	27	270	2.7	—
3.3	33	330	3.3	33	330	3.3	—
3.9	39	390	3.9	39	390	3.9	—
4.7	47	470	4.7	47	470	4.7	—
5.6	56	560	5.6	56	560	5.6	—
6.8	68	680	6.8	68	680	6.8	—
—	82	820	8.2	82	820	—	—

TYPICAL STANDARD CAPACITOR VALUES

pF	pF	pF	pF	μF	μF	μF	μF	μF	μF	μF
5	50	500	5000		0.05	0.5	5	50	500	5000
—	51	510	5100		—	—	—	—	—	—
—	56	560	5600		0.056	0.56	5.6	56	—	5600
—	—	—	6000		0.06	—	6	—	—	6000
—	62	620	6200		—	—	—	—	—	—
—	68	680	6800		0.068	0.68	6.8	—	—	—
—	75	750	7500		—	—	—	75	—	—
—	—	—	8000		—	—	8	80	—	—
—	82	820	8200		0.082	0.82	8.2	82	—	—
—	91	910	9100		—	—	—	—	—	—
10	100	1000		0.01	0.1	1	10	100	1000	10 000
—	110	1100		—	—	—	—	—	—	
12	120	1200		0.012	0.12	1.2	—	—	—	
—	130	1300		—	—	—	—	—	—	
15	150	1500		0.015	0.15	1.5	15	150	1500	
—	160	1600		—	—	—	—	—	—	
18	180	1800		0.018	0.18	1.8	18	180	—	
20	200	2000		0.02	0.2	2	20	200	2000	
22	220	2200		—	0.22	2.2	22	—	—	
24	240	2400		—	—	—	—	240	—	
—	250	2500		—	0.25	—	25	250	2500	
27	270	2700		0.027	0.27	2.7	27	270	—	
30	300	3000		0.03	0.3	3	30	300	3000	
33	330	3300		0.033	0.33	3.3	33	330	3300	
36	360	3600		—	—	—	—	—	—	
39	390	3900		0.039	0.39	3.9	39	—	—	
—	—	4000		0.04	—	4	—	400	—	
43	430	4300		—	—	—	—	—	—	
47	470	4700		0.047	0.47	4.7	47	—	—	

ANSWERS TO ODD-NUMBERED PROBLEMS

CHAPTER 1

1-1 1.5×10^{-2}, 1.6×10^4,
6.26×10^3, 7×10^{-4},
9.89×10^5

1-3   7.8 km, 60 μm, 19 mg,
91.5 kg, 50 mm

CHAPTER 2

2-1   320 mA, 570 mA, 820 mA

2-3   80 mA, 142 mA, 205 mA

2-5   16 V, 28.5 V, 41 V

2-7   (a) 20 V, 59 V, 82.5 V,
(b) 64 mV, 185 mV,
260 mV

2-9   *(a)* $R \times 10$, (b) $R \times 1$ k,
(c) $R \times 10$ k, (d) $R \times 1$

2-11   680 Ω, 250 Ω, 130 Ω,

2-13   0.43 mA, 1.67 mA,
2.38 mA

2-15   9 mA, 33.5 mA, 47.5 mA

2-17   120 Ω, 11.7 Ω, 1 Ω

2-19   1.2 MΩ, 117 kΩ, 10 kΩ

2-21   43 V, 167 V, 238 V

2-23   0.38 mA to 0.48 mA,
1.62 mA to 1.72 mA,
2.33 mA to 2.43 mA

2-25   (a) 220 Ω to 280 Ω,
(b) 220 kΩ to 280 kΩ

2-27   (a) 9.398 V to 9.438 V,
(b) 1.286 V to 1.294 V,
(c) 3.06 V to 3.074 V

2-29   (a) 4.634 kΩ to 4.692 kΩ,
(b) 556.6 Ω to 563.6 Ω,
(c) 27.59 Ω to 27.95 Ω

CHAPTER 3

3-1 1 μS, 10 S, 667 μS,
3.7 mS, 30.3 μS

3-3 15 μA, 150 A, 10 mA,
55.6 mA, 455 μA

3-5 33 V

3-7 0.8 Ω

3-9 100 V

3-11 1.8 mS, 90 mA

3-13 0.11 Ω

3-15 180 W

3-17 10 W

3-19 870 mA, 132 Ω

3-21 0.2 A, 250 Ω

3-23 200 V, 2 kΩ

3-25 10.9 A, 6.25 kWh

3-27 8.7 A, 26.5 Ω, 750 Wh

CHAPTER 4

4-1 100 kV/m

4-3 1.15×10^{-4} cm

4-5 45 kV

4-7 54 kV

4-9 4.16 μm

4-11 115 V, 25 W

4-13 #12

4-15 2.19×10^{-2} Ω, 5.47×10^{-3} Ω, 2.43×10^{-3} Ω, 1.37×10^{-2} Ω, 8.76×10^{-4} Ω

4-17 0.06 Ω, 0.1 Ω

4-19 0.47 mm

4-21 0.516 mW

4-23 1.84 mm

4-25 13.12 Ω

4-27 440 Ω, 470 Ω, 530 Ω, 560 Ω, 590 Ω, 620 Ω, 650 Ω, 680 Ω, 710 Ω, 740 Ω

4-29 #10

4-31 56 kΩ ± 20%,
820 Ω ± 10%,
3.9 MΩ ± 5%

4-33   17.46 mA, 14.32 V;
            0.707 mA, 707 V

4-35   294 mW, 5.1 W, 893 mW,
            6.1 W, 24 mW

4-37   2.64 Ω

4-39   56.5 kΩ, 827 Ω, 3.93 MΩ

CHAPTER 5

5-1   20 mA

5-3   1 A

5-5   3 V, 5 V, 2.5 V, 1.5 V

5-7   8.96 V

5-9   0.4 V, −12 V, −3.8 V

5-11   22.2 V, 52.8 V

5-13   25 kΩ

5-15   ±1.27 V

5-17   40.6 V

5-19   11.5 V, 5.16 V, 82 mW,
            148 mW, 120 mW

5-21   0.55 Ω, 45.9 W

5-23   46 mW, 0.34 W

5-25   (a) 1.08 W, 1.98 W,
            1.44 W, 4.5 W
            (b) 43.2 mW, 79.2 mW,
            57.6 mW, 180 mW

5-27   200 Ω, 45 mW

5-29   6.75 mA, 12.13 V, 4.58 V,
            10.1 V, 3.17 V

5-31   2.31 A, 30.9 W

5-33   (a) −0.46 V, −0.46 V,
            −3.47 V, (b) 6 V, −6 V,
            −6 V

CHAPTER 6

6-1   115 mA, 250 mA, 365 mA

6-3   555 mA, 221 mA, 455 mA,
            1.23 A

6-5   123.2 Ω, 365 mA

6-7   4.3 V

6-9   115 mA, 250 mA, 365 mA

6-11   252 Ω

6-13   95 Ω

6-15	115 mA, 250 mA
6-17	25.4 mA, 7.13 mA, 12.43 mA
6-19	5.19 W, 11.25 W, 16.43 W
6-21	870 mA, 348 mA, 522 mA, 217 mA, 225 W
6-23	44 mA, 31 mA, 27.28 V, 1.2 W, 0.85 W
6-25	271 mW, 76.3 mW, 133 mW, 480.3 mW
6-27	407 Ω
6-29	30.8 V
6-31	169.4 mA, 88 mA, 323.4 mA, 0

CHAPTER 7

7-1	958 Ω, 83.5 mA
7-3	6.8 kΩ, 1.32 mA
7-5	47.4 kΩ, 738 μA
7-7	83.5 mA, 52.5 mA, 31 mA
7-9	1.32 mA, 0.38 mA, 0.68 mA, 0.26 mA
7-11	427 μA, 311 μA, 169 μA, 142 μA, 311 μA, 185 μA, 126 μA
7-13	62.6 V, 17.4 V, 17.4 V
7-15	4.4 V, 4.6 V
7-17	35 V, 3.1 V, 7.95 V, 7.95 V, 21.15 V, 2.77 V, 2.77 V
7-19	427 μA, 403 μA, 4 V, 0, 27.4 V, 3.6 V
7-21	3.73 mA
7-23	875 Ω
7-25	-770 μA, 146 μA, 58 μA, 88 μA, 145 μA, -10 V, 1.4 V, 0.6 V, 0.6 V, 0.8 V, 1.2 V
7-27	-12 V
7-29	3.5 V, -1.04 V, -3.9 V
7-31	726 μA, 406 μA, 329 μA, 1.45 mA, 6.9 V, 6.9 V, 26.1 V
7-33	27.1 kΩ

CHAPTER 8

8-1 75 V, 5 Ω, 62.5 V, 2.5 A

8-3 30.6 μA, 98 kΩ

8-5 4.8 V, 600 Ω

8-7 18.06 V, 42 kΩ

8-9 13.59 μA

8-11 775 μA

8-13 821 mV

8-15 1.55 mA

8-17 69 μA

8-19 13.59 μA

8-21 775 μA

8-23 821 mV

8-25 1.55 mA

8-27 69 μA

8-29 13.59 μA

8-31 775 μA

8-33 0.528 V

8-35 1.55 mA

8-37 68 μA

8-39 309 kΩ, 334 kΩ, 409 kΩ

CHAPTER 9

9-1 1.29 mA

9-3 374 μA

9-5 1.56 mA

9-7 1.41 mA (right to left)

9-9 543 μA

9-11 489 μA

9-13 902 μA

9-15 5.08 V

9-17 3.04 V

9-19 396 μA

9-21 1.29 mA

9-23 1.04 V

9-25 748 μA

9-27 939 μA, 664 μA

9-29 1.29 mA

9-31 6.3 kΩ

9-33 5.8 kΩ

CHAPTER 10

10-1	7.5 Ah, 30 h
10-3	5.3 h
10-5	1.4 V, 33 A
10-7	77 Ω
10-9	8 V, 7.5 V
10-11	8 cells, 0.4 Ω, 11.54 V
10-13	12 cells, 6 V
10-15	10 cells
10-17	1.7 V, 6 A, 72 Ah, 8.75×10^{-3} Ω
10-19	60 km
10-21	6664 Ah
10-23	16 cells
10-25	174
10-27	77 Ah
10-29	(a) 33.3 h, (b) 37.5 min, (c) 22.5 min
10-31	(a) 50 A, (b) 11.75 V

CHAPTER 11

11-1	3.3 mT, 0.2 mT
11-3	6.16 mWb
11-5	45 μWb
11-7	1.7 cm
11-9	1.8 mWb
11-11	562 μT
11-13	2100 A, 4200 A/m
11-15	858
11-17	100 mA
11-19	66.7 mA
11-21	0.9 N
11-23	12.5 N, 4.9 N
11-25	9.9 μN.m
11-27	1.04 mA
11-29	18.2 mN
11-31	1.96 mA

CHAPTER 12

12-1	7.95×10^4 A/m, 0.1 T, 22.5 μWb

12-3   44.5 mA

12-5   28.4 A

12-7   45.07 mA

12-9   746 mA

12-11 191, 796

12-13 387 µWb

12-15 1.5×10^3 A/m, 0.4 T, 360 µWb

12-17 202 mA

12-19 2.26 A

12-21 222 mA

12-23 411×10^3

12-25 8.8 kg

12-27 73 kg

12-29 681 mA

12-31 92.1 mA

CHAPTER 13

13-1   0.27 Ω

13-3   (a) 0.027 Ω, (b) 0.0027 Ω

13-5   13.3 A

13-7   2.5 A, 3.3 A, 5 A, 10 A

13-9   2.5 MΩ, 1 MΩ, 500 kΩ

13-11 25 V, 75 V, 150 V

13-13 95 V, 142.5 V

13-15 8.18 V, 10.98 V

13-17 1.22 MΩ, 2.4 MΩ

13-19 3.33 MΩ, 20 kΩ/V

13-21 40 kΩ, 20 kΩ, 10 kΩ

13-23 70 kΩ, 30 kΩ, 16.7 kΩ

13-25 30 kΩ, 7.5 kΩ

13-27 79.5 Ω

13-29 100 Ω, 100 kΩ

CHAPTER 14

14-1   40 V

14-3   6 V

14-5   1.13 A

14-7   24

14-9   10 V

14-11   712

14-13 1.76×10^{-8} Wb, 0.25 mV

14-15   4.74 mH

14-17   2.7 mH, 1.46 mH

14-19   17.35 mH, 17.35 mH

14-21   0.047

14-23   251 mH, 98.2 mH, 157 mH

14-25   14.2 J

14-27   3.8 H

14-29   563 mJ

14-31   (a) 1.15 H, (b) 733 μH

14-33   (a) 1.2 mH, (b) 800 μH

14-35   150 μH, 275 μH

CHAPTER 15

15-1   40 000 V/m, 2 μC/m^2

15-3   50 V

15-5   (a) 88.5 pF, (b) 6.64 nF

15-7   37.7 m^2

15-9   0.27 μm

15-11   0.884 μF

15-13   0.796 μF, 9.55 μC

15-15   44.25 pF

15-17   67.9

15-19   0.62 mm

15-21   5.7

15-23   5.88 μF, 1.471 V, 2.94 V, 5.88 V, 14.7 V, 147 μC

15-25   21.4 μF

15-27   637 μJ, 319 μJ, 5.74 mJ

15-29   108 μJ, 216 μJ, 432 μJ, 1.08 mJ

15-31   1.59 nJ, 99 pJ

CHAPTER 16

16-1   63.2 mA, 86.5 mA, 95 mA, 98.2 mA, 99.3 mA

16-3   14.3 V, 4.1 V, 1.2 V, 0.34 V

16-5   92.9 H

16-7   1.6 ms, 0.36 ms

16-9	1.39 H
16-11	1.23 kΩ
16-13	62.5 kV, 24 kΩ
16-15	24.7 V, 37.2 V, 43.5 V, 46.7 V, 48.3 V, 49.2 V, 49.6 V
16-17	22.7 mA, 9.16 mA, 3.69 mA, 1.49 mA, 0.6 mA, 0.24 mA
16-19	492 Ω
16-21	128 ms
16-23	111 V
16-25	0.46 μF
16-27	34.6 s
16-29	5 kΩ, 1.175 s

CHAPTER 17

17-1	2.33 Hz, 88 mV, 880 mV
17-3	0.62 V, 0.62 V, −0.62V
17-5	0.82 V
17-7	64.4 mV, 128.6 mV, 315.4 mV, 374.7 mV
17-9	7.7 V, −9 V, 7.28 V
17-11	1 kHz, 2.5 km, 25 m
17-13	277 W, 277 W, 277 W, 370 W
17-15	0.44 W, 0.6 W, 0.3 W
17-17	414 Ω, 1.62 mW, 813 μW
17-19	2.62 A, 2.35 A, 185 W
17-21	107.7 mA, 49.8 mA, 76.1 mA, 22.6 W
17-23	580 mV, 0, 336 Ω
17-25	3.15 V, 0, 1.11 V, 33.3 Hz, 1.5 V
17-27	2.7, 20
17-29	5 div, 0.5 div
17-31	45°, 272°
17-33	25 Hz, 12 000 km, 46.5 Hz, 6450 km

CHAPTER 18

18-1 $77.8 \sin(\phi + 8.2°)$

18-3 $30 \sin(\phi - 15.3°)$

18-5 $180 \underline{/-39°}$

18-7 $5.5 \underline{/-68°}$

18-9 $44 \underline{/16°}$

18-11 $144 \underline{/34°}$

18-13 $90.1 \underline{/-33.7°}$, $25.5 \underline{/11.3°}$, $2298 \underline{/22.4°}$, $3.5 \underline{/-45°}$

18-15 $139.1 + j\, 56.2$, $0 + j\, 85$, $19.8 + j\, 60.9$

18-17 $39 + j\, 22.5$, $25.2 + j\, 9.7$, $22.3 + j\, 20.1$

18-19 $77.8 \sin(\phi + 8.2°)$

18-21 $40.5 \underline{/9°}$

18-23 $143.8 \underline{/33.9°}$

18-25 $81.07 \underline{/-22.9°}$

18-27 $19 \underline{/25.1°}$

18-29 $5.5 \underline{/-68°}$

18-31 $-34.2 - j\, 36.4$

18-33 $29.4 \underline{/122°}$

18-35 $249.7 \underline{/17.6°}$

CHAPTER 19

19-1 471.2 Ω, 212 mA

19-3 1 MHz

19-5 21.2 Ω, 4.7 A

19-7 99.5 Hz

19-9 356 mA, 24.2 V, 22.4 V, 42.7°

19-11 16.1 mA, 10.7 V

19-13 92.8 mA, 11.1 V, 4.47 V, 21.9°

19-15 0.71 μF

19-17 72.1 mA, 16°, 14.4 V, 5.44 V, 9.56 V,

19-19 139 Ω

19-21 88.42 Hz, 71.9 mA, $47.2 \text{ V} \underline{/19.3°}$

19-23 30.3 mA, 72.5 mA,
50.3 mA, 37.6 mA$\underline{/-36.2°}$

19-27 578 mA, 839 mA$\underline{/-76.7°}$

19-29 (a) 0.85, (b) 6.35×10^{-2}

19-31 0.39 V, 1.22 mV, 319

19-33 0.159 μF, 45°

CHAPTER 20

20-1 389 mA$\underline{/18.8°}$

20-3 103 V$\underline{/-71.2°}$,
58.7 V$\underline{/67.3°}$

20-5 47.3 V$\underline{/-29.9°}$,
27.4 V$\underline{/59.4°}$

20-7 2.2 kΩ$\underline{/17.1°}$,
8.63 mA$\underline{/-17.1°}$

20-9 18.9 mA$\underline{/-29.9°}$,
10.86 mA$\underline{/59.4°}$

20-11 2.93 mA$\underline{/30.4°}$,
6.65 mA$\underline{/-50.4°}$,
18.49 mA$\underline{/37.1°}$

20-13 534 Ω$\underline{/14.4°}$

20-15 8.14 V$\underline{/43.3°}$,
2.9 V$\underline{/-94.6°}$

20-17 3.9 kΩ$\underline{/-29.3°}$

20-19 9.94 mA$\underline{/14.4°}$

20-21 5.83 kΩ$\underline{/68.6°}$

20-23 3.9 kΩ$\underline{/-0.3°}$

20-25 0.574 V$\underline{/-75.2°}$,
0.71 V$\underline{/-74°}$

20-29 35 kΩ‖(14 kΩ + 703 mH)

20-31 79.6 mA$\underline{/-90°}$,
101 mA$\underline{/-90°}$,
128.9 mA$\underline{/0°}$,
221.9 mA$\underline{/-54.5°}$

20-33 1.17 kΩ$\underline{/-6.8°}$

CHAPTER 21

21-1 (a) 24.2 Ω, (b) 9.09 A,
(c) 4 kW

21-3 117 W

21-5 499 mW

21-7 (a) 457 var, (b) 576 W,
(c) 3.6 kvar

21-9 75 VA, 37.5 var, 65 W

21-11 0.161 VA, 69.1 mW,
0.145 var

21-13 150.7 VA, 42.8 W, 144 var

21-15 1.11 VA, 1.03 W, 0.414 var

21-17 880 VA, 607.2 W,
637.3 var, 15.9 μF

21-19 1.8 kVA, 1.67 kW,
663 var

21-21 256.4 A, 196 A, 1530 μF

21-23 1.8 W, 0.047 A

21-25 65.2 A, 79.5 μF

21-27 12 dB, 7 dB, 19 dBm

21-29 31 dB, 3500

CHAPTER 22

22-1 9.25 mA $\underline{/45.1°}$

22-3 29 μA $\underline{/-66.2°}$

22-5 3.2 V $\underline{/0.1°}$

22-7 1.3 mA $\underline{/-16.5°}$

22-9 $[E_1Z_2 + E_2(Z_1 + Z_3)]/$
$[Z_1Z_2 + Z_2Z_3 + Z_4$
$(Z_1 + Z_2 + Z_3)]$

22-11 59.1 μA $\underline{/72.8°}$

22-13 1.39 mA $\underline{/-14.3°}$

22-15 0.586 mA $\underline{/88.3°}$

22-17 1.48 V $\underline{/23°}$

22-19 9.21 mA $\underline{/45.4°}$

22-21 as for 22-9

22-23 1.3 mA $\underline{/-16.5°}$

22-25 4.2 A $\underline{/-10.5°}$

22-27 286.6 μA $\underline{/98.2°}$

22-29 125.5 μA $\underline{/-20°}$

22-31 $(E_2 - I_1Z_1)/(Z_1 + Z_2 + Z_3)$

22-33 $E_1Z_2/(Z_1Z_2 + Z_1Z_3 + Z_2Z_3)$

22-35 72.3 μA $\underline{/31.2°}$

22-37 500.5 Ω, 2.5 mH

22-39 $R = 17.5$ kΩ, $L = 60.5$ mH

22-41 12.8 mA $\underline{/91.05°}$

22-43 3.2 mA $\underline{/-124°}$

CHAPTER 23

23-1   500 kHz, 156 mA

23-3

	$0.25 f_r$	$0.5 f_r$	$0.8 f_r$	f_r	$1.25 f_r$	$2 f_r$	$4 f_r$
I (mA)	0.84	2.1	6.9	156	6.9	2.15	0.84

23-5

V_L (V)	0.33	1.67	8.76	248	13.7	6.71	5.34
V_C (V)	5.35	6.68	13.7	248	8.76	1.7	0.33
V_R (mV)	27	67	221	4990	221	69	27

23-7 I (mA)   0.93   2.3   7.8   350   7.8   2.3   0.93

23-9   40 pF to 3400 pF

23-11   471 kHz

23-13   49.7, 100, 88.9

23-15   505 kHz, 495 kHz, 10 kHz

23-17   527.3 kHz, 532.7 kHz,
468.4 kHz, 473.7 kHz

23-19   200 kΩ

23-21   530.5 kHz, 200

23-23   2.65 kHz

23-25   30 pF

23-27   15 kΩ

23-29   0.02, 110 mA, 130 mA

23-31   141.8 kHz to 142.2 kHz

23-33   206 Ω

CHAPTER 24

24-1   29.9 V, 36.8 V, 66.7 V,
6.9 V

24-3   6.24 V, 25 V, 60 V, 500 V,
143.5 mA

24-5   50 A

24-7   1.8 mWb

24-9   537 μWb, 4 μWb

24-11   22.5 μWb, 113 μWb

24-13   121 V, 1.28 A, 37.3°

24-15   663.5 mA, 104.5 V, 23°

24-17   946.7 mA, 119.7 V, 53.3°

24-19   1.24 A, 36.3°

24-21   7.6%

24-23   25 V, 24 V

24-25   93.8%, 1.24 A

24-27   32 Ω

24-29   3.6%

CHAPTER 25

25-1   (a) 0, (b) 0 to 100 V peak

25-3   1.12 MΩ, 223 kΩ

25-5   2.7 MΩ

25-7   59 V, 299 V, 895 V

25-9   50 V

25-11   44.1 V, 66.2 V, 46 V, 69 V

25-13   180

25-15   1.95 A

25-17   100 W, 141.4 V

25-19   720 Ω to 900 Ω, 60 Ω to 135 Ω

25-21   0.02 μF to 50 μF, 10 mH to 25 H, 2 Ω to 250 kΩ

25-23   919.5 Ω

CHAPTER 26

26-1   231 V, 1.49 mA, 1.49 mA, 0

26-3   0.77 A, 3.08 A, 0.578 A, 2.42 A

26-11   1.11 A, 1.95 A, 0.941 A, 1.43 A, 1.58 A, 1.95 A

26-13   375 mA, 412 mA, 790 mA, 1.14 A, 264 mA, 1.19 A

26-15   1.24 A$\underline{/-34.7°}$, 1.72 A$\underline{/-139°}$, 1.85 A$\underline{/81.5°}$

26-17   358 mA$\underline{/16°}$, 286 mA$\underline{/-162°}$, 73.2 mA$\underline{/-170°}$

26-19   (a) 644 mA, 1.37 A, 1.1 A, (b) 1.37 A, 644 mA, 1.1 A

26-21   1.03 kW

26-23   189 W, 5.7 var, 204 VA

26-25   567 W, 623 VA, 17.1 var

26-27   106 VA, −99 var, 37.4 W

26-29   368 mA, 0.598, 30.5 W

26-31   (a) 2.5 A, 1.125 kW, 0.866; (b) 2.21 A, 1.125 kW, 0.98

26-33   7.61 W, 16.32 W

CHAPTER 27

27-1 6.37 V, 3.18 V, 2.21 V, 1.59 V, 1.27 V

27-3 4.77 V, 7.5 V, 3.18 V, 0.637 V, 0.27 V

27-5 4.5 V, 7.07 V, 3 V, 0.6 V, 0.26 V

27-7 45 V, 30 V, 6 V, 2.6 V

27-9 5.44 V

27-11 5.85 V

27-13 7.34 V

27-15 103 mV

27-17 4.5 V, 0.88 V, 4.45 V, 0.87 V

27-19 28.9 V, 32 V

27-21 14.2 V, 22.3 V

INDEX